D1118966

# An Introduction to Optimization

## WILEY-INTERSCIENCE
## SERIES IN DISCRETE MATHEMATICS AND OPTIMIZATION

**ADVISORY EDITORS**

RONALD L. GRAHAM
*AT & T Bell Laboratories, Murray Hill, New Jersey, U.S.A.*

JAN KAREL LENSTRA
*Centre for Mathematics and Computer Science, Amsterdam, The Netherlands*
*Erasmus University, Rotterdam, The Netherlands*

ROBERT E. TARJAN
*Princeton University, New Jersey, and*
*NEC Research Institute, Princeton, New Jersey, U.S.A.*

AARTS AND KORST • Simulated Annealing and Boltzmann Machines: A Stochastic Approach to Combinatorial Optimization and Neural Computing
ALON, SPENCER, AND ERDŐS • The Probabilistic Method
ANDERSON AND NASH • Linear Programming in Infinite-Dimensional Spaces: Theory and Application
BARTHÉLEMY AND GUÉNOCHE • Trees and Proximity Representations
BAZARRA, JARVIS, AND SHERALI • Linear Programming and Network Flows
CHONG AND ZAK • An Introduction to Optimization
COFFMAN AND LUEKER • Probabilistic Analysis of Packing and Partitioning Algorithms
DINITZ AND STINSON • Contemporary Design Theory: A Collection of Surveys
GLOVER, KLINGHAM, AND PHILLIPS • Network Models in Optimization and Their Practical Problems
GOLSHTEIN AND TRETYAKOV • Modified Lagrangians and Monotone Maps in Optimization
GONDRAN AND MINOUX • Graphs and Algorithms *(Translated by S. Vajdā)*
GRAHAM, ROTHSCHILD, AND SPENCER • Ramsey Theory, Second Edition
GROSS AND TUCKER • Topological Graph Theory
HALL • Combinatorial Theory, Second Edition
JENSEN AND TOFT • Graph Coloring Problems
LAWLER, LENSTRA, RINNOOY KAN, AND SHMOYS, EDITORS • The Traveling Salesman Problem: A Guided Tour of Combinatorial Optimization
LEVITIN • Perturbation Theory in Mathematical Programming Applications
MAHMOUD • Evolution of Random Search Trees
MARTELLO AND TOTH • Knapsack Problems: Algorithms and Computer Implementations
MINC • Nonnegative Matrices
MINOUX • Mathematical Programming: Theory and Algorithms *(Translated by S. Vajdā)*
MIRCHANDANI AND FRANCIS, EDITORS • Discrete Location Theory
NEMHAUSER AND WOLSEY • Integer and Combinatorial Optimization
NEMIROVSKY AND YUDIN • Problem Complexity and Method Efficiency in Optimization *(Translated by E. R. Dawson)*
PACH AND AGARWAL • Combinatorial Geometry
PLESS • Introduction to the Theory of Error-Correcting Codes, Second Edition
SCHRIJVER • Theory of Linear and Integer Programming
TOMESCU • Problems in Combinatorics and Graph Theory *(Translated by R. A. Melter)*
TUCKER • Applied Combinatorics, Second Edition

# An Introduction to Optimization

EDWIN K. P. CHONG

and

STANISLAW H. ŻAK

A Wiley-Interscience Publication
JOHN WILEY & SONS, INC.
New York · Chichester · Brisbane · Toronto · Singapore

This text is printed on acid-free paper.

Copyright © **1996** by John Wiley & Sons, Inc.

All rights reserved. Published simultaneously in Canada.

Reproduction or translation of any part of this work beyond that permitted by Section 107 or 108 of the 1976 United States Copyright Act without the permission of the copyright owner is unlawful. Requests for permission or further information should be addressed to the Permissions Department, John Wiley & Sons, Inc., 605 Third Avenue, New York, NY 10158-0012.

*Library of Congress Cataloging in Publication Data:*

Chong, Edwin K. P.

An introduction to optimization / Edwin K. P. Chong and Stanislaw H. Żak.

p. cm. — (Wiley-Interscience series in discrete mathematics and optimization)

Includes bibliographical references (p. – – ) and index.

ISBN 0-471-08949-4 (cloth : acid-free)

1. Mathematical optimization. I. Żak, Stanislaw H.

II. Title. III. Series.

QA402.5.Z35 1995

519.3-dc20

95-6111

CIP

Printed in the United States of America

10 9 8 7 6 5 4

*E. K. P. Chong dedicates this book to his wife Yat-Yee and to his parents Paul and Julienne Chong.*

*S. H. Żak dedicates this book to his wife Mary Ann and to his parents Janina and Konstanty Żak.*

# Contents

# Preface

Optimization is central to any problem involving decision making, whether in engineering or in economics. The task of decision making entails choosing between various alternatives. This choice is governed by our desire to make the "best" decision. The measure of goodness of the alternatives is described by an objective function or performance index. Optimization theory and methods deal with selecting the best alternative in the sense of the given objective function.

The area of optimization has received enormous attention in recent years, primarily because of the rapid progress in computer technology, including the development and availability of user-friendly software, high-speed and parallel processors, and artificial neural networks. A clear example of this phenomenon is the wide accessibility of optimization software tools such as MATLAB[1] and the many other commercial software packages.

There are currently several excellent graduate textbooks on optimization theory and methods [e.g., 4, 19, 22, 28, 51, 52, 59, 76], as well as undergraduate textbooks on the subject with an emphasis on engineering design [e.g., 2, 62]. However, there is a need for an introductory textbook on optimization theory and methods at a senior undergraduate or beginning graduate level. The present text was written with this goal in mind. The material is an outgrowth of our lecture notes for a one-semester course in optimization methods for seniors and first-year graduate students at Purdue University, West Lafayette. In our presentation, we assume a working knowledge of basic linear algebra and multivariable calculus. For the reader's convenience, a part of this book (Part I) is devoted to a review of the required mathematical background material. The many figures throughout the text complement the written presentation of the material. We also include a variety of exercises at the end of each chapter. A solutions manual with complete solutions to the exercises is available from the publisher to instructors who adopt this text. Some of the exercises require using MATLAB. The student edition of MATLAB is sufficient for almost all the MATLAB exercises included in the text. The MATLAB source listings for the MATLAB exercises are also included in the solutions manual.

The purpose of the book is to give the reader a working knowledge of optimization theory and methods. To accomplish this goal, we include numerous

[1] MATLAB is a registered trademark of The MathWorks, Inc.

examples that illustrate the theory and algorithms discussed in the text. However, it is not our intention to provide a cookbook of the most recent numerical techniques for optimization; rather, our goal is to equip the reader with sufficient background for further study of advanced topics in optimization.

The field of optimization is a very active research area. In recent years, various new approaches to optimization have been proposed. In this text, we have tried to reflect at least some of the flavor of recent activity in the area. For example, in our treatment of linear programming, we discuss not only the classical simplex method, but also the more recent methods of Khachiyan and Karmarkar for solving linear programs. There has also been a recent surge of applications of optimization methods to a variety of new problems. A prime example of this is the use of descent algorithms for the training of feedforward neural networks. An entire chapter in the book is devoted to this topic. The area of neural networks is an active area of ongoing research, and many books have been devoted to this subject. The topic of neural network training fits perfectly into the framework of unconstrained optimization methods. Therefore, the chapter on feedforward neural networks provides not only an example of application of unconstrained optimization methods, it also gives the reader an accessible introduction to what is currently a topic of wide interest. We also include a discussion of genetic algorithms, a topic becoming of increasing importance in the study of complex adaptive systems.

The material in this book is organized into four independent parts. Part I contains a review of some basic definitions, notations, and relations from linear algebra, geometry, and calculus that we use frequently throughout the book. In Part II, we consider unconstrained optimization problems. We first discuss some theoretical foundations of unconstrained optimization, including necessary and sufficient conditions for minimizers and maximizers. This is followed by a treatment of various iterative optimization algorithms, together with their properties. A discussion of genetic algorithms is included in this part. We also analyze the least-squares optimization problem and the associated recursive least-squares algorithm. Parts III and IV are devoted to constrained optimization. Part III deals with linear programming problems, which form an important class of constrained optimization problems. We give examples and analyze properties of linear programs and then discuss the simplex method for solving linear programs. We also provide a brief treatment of dual linear programming problems. We wrap up Part III by presenting non-simplex algorithms for solving linear programs, namely, Khachiyan's and Karmarkar's methods. In Part IV, we treat nonlinear constrained optimization. Here, as in Part II, we first present some theoretical foundations of nonlinear constratined optimization problems. We then discuss different algorithms for solving constrained optimization problems with equality as well as inequality constraints.

We are grateful to several people for their help during the course of writing this book. In particular, we thank Dennis Goodman of Lawrence Livermore Laboratories for his comments on early versions of Part II, and for making available to us his lecture notes on nonlinear optimization. We thank Moshe Kam of Drexel

University for pointing out some useful references on non-simplex methods. We are grateful to Ed Silverman and Russell Quong of Purdue University for their valuable remarks on Part I. We also thank the students of EE 580 for their many helpful comments and suggestions. In particular, we are grateful to Christopher Taylor for his diligent proofreading of early versions of the book.

E. K. P. Chong
S. H. Żak

*West Lafayette, Indiana*

# Part I

# MATHEMATICAL REVIEW

# 1

# Methods of Proof and Some Notation

## 1.1. METHODS OF PROOF

Consider two statements, "A" and "B," which could be either true or false. For example, let "A" be the statement "John is an engineering student," and let "B" be the statement "John is taking a course on optimization." We can combine these statements to form other statements, like "A and B" or "A or B." In our example, "A and B" means "John is an engineering student, and he is taking a course on optimization." We can also form statements like "not A," "not B," "not (A and B)," and so on. For example, "not A" means "John is not an engineering student." The truth or falsity of the combined statements depends on the truth or falsity of the original statements, "A" and "B." This relationship is expressed by means of truth tables (e.g., see Tables 1.1 and 1.2). From Tables 1.1 and 1.2, it is easy to see that the statement "not (A and B)" is equivalent to "(not A) or (not B)" (see Exercise 1.3). This is called *DeMorgan's Law*.

In proving statements, it is convenient to express a combined statement by a *conditional*, such as "A implies B," which we denote "$A \Rightarrow B$." The conditional "$A \Rightarrow B$" is simply the combined statement "(not A) or B," and is often also read "A only if B," "if A then B," "A is sufficient for B," or "B is necessary for A."

We can combine two conditional statements to form a *biconditional* statement of the form "$A \Leftrightarrow B$," which simply means "$(A \Rightarrow B)$ and $(B \Rightarrow A)$." The statement "$A \Leftrightarrow B$" reads "A if, and only if, B," or "A is equivalent to B," or "A is necessary and sufficient for B." Truth tables for conditional and biconditional statements are given in Table 1.3.

It is easy to verify, using the truth table, that the statement "$A \Rightarrow B$" is equivalent to the statement "(not B) $\Rightarrow$ (not A)." The latter is called the *contrapositive* of the former.

If we take the contrapositive to DeMorgan's Law, we obtain the assertion that "not (A or B)" is equivalent to "(not A) and (not B)."

Most statements we deal with have the form "$A \Rightarrow B$." To prove such a statement, we may use one of the following three different techniques:

1. The direct method

**Table 1.1 Truth Table for "A and B" and "A or B"**

| A | B | A and B | A or B |
|---|---|---|---|
| F | F | F | F |
| F | T | F | T |
| T | F | F | T |
| T | T | T | T |

**Table 1.2 Truth Table for "not A"**

| A | Not A |
|---|---|
| F | T |
| T | F |

**Table 1.3 Truth Table for Conditionals and Biconditionals**

| A | B | $A \Rightarrow B$ | $A \Leftarrow B$ | $A \Leftrightarrow B$ |
|---|---|---|---|---|
| F | F | T | T | T |
| F | T | T | F | F |
| T | F | F | T | F |
| T | T | T | T | T |

2. Proof by contraposition
3. Proof by contradiction or *reductio ad absurdum.*

In the case of the *direct method*, we start with "A," then deduce a chain of various consequences to end with "B."

A useful method for proving statements is *proof by contraposition*, based on the equivalence of the statements "$A \Rightarrow B$" and "(not B)$\Rightarrow$(not A)." We start with "not B," then deduce various consequences to end with "not A" as a conclusion.

Another method of proof that we use is *proof by contradiction*, based on the equivalence of the statements "$A \Rightarrow B$" and "not (A and (not B))." Here we begin with "A and (not B)" and derive a contradiction.

Occasionally, we will use the *Principle of Induction* to prove statements. This principle may be stated as follows. Assume that a given property of positive integers satisfies the following conditions:

The number 1 possesses this property;

If the number $n$ possesses this property, then the number $n + 1$ possesses it too.

The Principle of Induction states that under these assumptions any positive integer possesses the property.

The Principle of Induction is easily understood using the following intuitive argument. If the number 1 possesses the given property, then the second condition implies that the number 2 possesses the property. But, then again, the second condition implies that the number 3 possesses this property, and so on. The Principle of Induction is a formal statement of this intuitive reasoning.

## 1.2. NOTATION

Throughout, we use the following notation. If $X$ is a set, then we write $x \in X$ to mean that $x$ is an element of $X$. When an object $x$ is not an element of a set $X$, then we write $x \notin X$. We also use the "curly bracket notation" for sets, writing down the first few elements of a set followed by three dots. For example, $\{x_1, x_2, x_3, \ldots\}$ is the set containing the elements $x_1$, $x_2$, $x_3$, and so on. Alternatively, we can explicitly display the law of formation. For example, $\{x : x \in \mathbb{R}, x > 5\}$ reads "the set of all $x$ such that $x$ is real and $x$ is greater than 5." The colon following $x$ reads "such that." An alternative notation for the same set is $\{x \in \mathbb{R} : x > 5\}$.

If $X$ and $Y$ are sets, then we write $X \subset Y$ to mean that every element of $X$ is also an element of $Y$. In this case, we say that $X$ is a *subset* of $Y$. If $X$ and $Y$ are sets, then we denote by $X \backslash Y$ the set of all points in $X$ that are not in $Y$. Note that $X \backslash Y$ is a subset of $X$. The notation $f : X \rightarrow Y$ means "$f$ is a function from the set $X$ into the set $Y$." The symbol $:=$ denotes arithmetic assignment. Thus, a statement of the form $x := y$ means "$x$ becomes $y$." The symbol $\triangleq$ means "equals by definition."

Throughout the text, we mark the end of theorems, lemmas, propositions, and corollaries using the symbol □. We mark the end of proofs, definitions, and examples by ■.

## EXERCISES

**1.1** Construct the truth table for the statement "(not B)⇒(not A)," and use it to show that this statement is equivalent to the statement "A⇒B."

**1.2** Construct the truth table for the statement "not (A and (not B))," and use it to show that this statement is equivalent to the statement "A⇒B."

**1.3** Prove DeMorgan's Law by constructing the appropriate truth tables.

**1.4** Prove that for any statements A and B, we have "A ⇔ (A and B) or (A and (not B))." This is useful because it allows us to prove a statement A by proving the two separate cases "(A and B)," and "(A and (not B))." For example, to prove that $|x| \geq x$ for any $x \in \mathbb{R}$, we separately prove the cases "$|x| \geq x$ and $x \geq 0$," and "$|x| \geq x$ and $x < 0$." Proving the two cases turns out to be easier than directly proving the statement $|x| \geq x$ (see Section 2.4 and Exercise 2.4).

**1.5** Suppose you are shown four cards, laid out in a row. Each card has a letter on one side and a number on the other. On the visible side of the cards are printed the symbols:

$$S \quad 8 \quad 3 \quad A$$

Determine which cards you should turn over to decide if the following rule is true or false: "If there is a vowel on one side of the card, then there is an even number on the other side."

# 2

# Real Vector Spaces and Matrices

## 2.1. REAL VECTOR SPACES

We define a column $n$ vector to be an array of $n$ numbers, denoted by

$$\boldsymbol{a} = \begin{bmatrix} a_1 \\ a_2 \\ \vdots \\ a_n \end{bmatrix}.$$

The number $a_i$ is called the $i$th component of the vector $\boldsymbol{a}$. Denote by $\mathbb{R}$ the set of real numbers, and by $\mathbb{R}^n$ the set of column $n$-vectors with real components. We call $\mathbb{R}^n$ an $n$-dimensional *real vector space*. We commonly denote elements of $\mathbb{R}^n$ by lower case bold letters, for example, $\boldsymbol{x}$. The components of $\boldsymbol{x} \in \mathbb{R}^n$ are denoted by $x_1, \ldots, x_n$.

We define a *row* $n$-vector as

$$[a_1, a_2, \ldots, a_n].$$

The *transpose* of a given column vector $\boldsymbol{a}$ is a row vector with corresponding elements, denoted $\boldsymbol{a}^T$. For example, if

$$\boldsymbol{a} = \begin{bmatrix} a_1 \\ a_2 \\ \vdots \\ a_n \end{bmatrix},$$

then

$$\boldsymbol{a}^T = [a_1, a_2, \ldots, a_n].$$

Equivalently, we may write $\boldsymbol{a} = [a_1, a_2, \ldots, a_n]^T$. Throughout the text, we adopt the convention that the term "vector" (without the qualifier "row" or "column") refers to a column vector.

Two vectors $\boldsymbol{a} = [a_1, a_2, \ldots, a_n]^T$ and $\boldsymbol{b} = [b_1, b_2, \ldots, b_n]^T$ are equal if $a_i = b_i$, $i = 1, 2, \ldots, n$.

The sum of the vectors $\boldsymbol{a}$ and $\boldsymbol{b}$, denoted $\boldsymbol{a} + \boldsymbol{b}$, is the vector

$$\boldsymbol{a} + \boldsymbol{b} = [a_1 + b_1, a_2 + b_2, \ldots, a_n + b_n]^T.$$

The operation of addition of vectors has the following properties:

**1.** The operation is commutative:

$$\boldsymbol{a} + \boldsymbol{b} = \boldsymbol{b} + \boldsymbol{a},$$

**2.** The operation is associative:

$$(\boldsymbol{a} + \boldsymbol{b}) + \boldsymbol{c} = \boldsymbol{a} + (\boldsymbol{b} + \boldsymbol{c}),$$

**3.** There is a zero vector

$$\mathbf{0} = [0, 0, \ldots, 0]^T$$

such that

$$\boldsymbol{a} + \mathbf{0} = \mathbf{0} + \boldsymbol{a} = \boldsymbol{a}.$$

The vector

$$[a_1 - b_1, a_2 - b_2, \ldots, a_n - b_n]^T$$

is called the difference between $\boldsymbol{a}$ and $\boldsymbol{b}$, and is denoted $\boldsymbol{a} - \boldsymbol{b}$.

The vector $\mathbf{0} - \boldsymbol{b}$ is denoted $-\boldsymbol{b}$. Note that

$$\boldsymbol{b} + (\boldsymbol{a} - \boldsymbol{b}) = \boldsymbol{a},$$

$$-(-\boldsymbol{b}) = \boldsymbol{b},$$

$$-(\boldsymbol{a} - \boldsymbol{b}) = \boldsymbol{b} - \boldsymbol{a}.$$

The vector $\boldsymbol{b} - \boldsymbol{a}$ is the unique solution of the vector equation

$$\boldsymbol{a} + \boldsymbol{x} = \boldsymbol{b}.$$

Indeed, suppose $\boldsymbol{x} = [x_1, x_2, \ldots, x_n]^T$ is a solution to $\boldsymbol{a} + \boldsymbol{x} = \boldsymbol{b}$. Then

$$\begin{aligned} a_1 + x_1 &= b_1, \\ a_2 + x_2 &= b_2, \\ &\vdots \\ a_n + x_n &= b_n, \end{aligned}$$

and thus

$$x = b - a.$$

We define an operation of multiplication of a vector $\boldsymbol{a} \in \mathbb{R}^n$ by a real scalar $\alpha \in \mathbb{R}$ as

$$\alpha \boldsymbol{a} = [\alpha a_1, \alpha a_2, \ldots, \alpha a_n]^T.$$

This operation has the following properties:

**1.** The operation is distributive: For any real scalars $\alpha$ and $\beta$,

$$\alpha(\boldsymbol{a} + \boldsymbol{b}) = \alpha \boldsymbol{a} + \alpha \boldsymbol{b},$$

$$(\alpha + \beta)\boldsymbol{a} = \alpha \boldsymbol{a} + \beta \boldsymbol{a}.$$

**2.** The operation is associative:

$$\alpha(\beta \boldsymbol{a}) = (\alpha\beta)\boldsymbol{a}.$$

**3.** The scalar 1 satisfies:

$$1\boldsymbol{a} = \boldsymbol{a}.$$

**4.** Any scalar $\alpha$ satisfies:

$$\alpha \boldsymbol{0} = \boldsymbol{0}.$$

**5.** The scalar 0 satisfies:

$$0\boldsymbol{a} = \boldsymbol{0}.$$

**6.** The scalar $-1$ satisfies:

$$(-1)\boldsymbol{a} = -\boldsymbol{a}.$$

Note that $\alpha \boldsymbol{a} = \boldsymbol{0}$ if, and only if, $\alpha = 0$ or $\boldsymbol{a} = \boldsymbol{0}$. To see this, observe that $\alpha \boldsymbol{a} = \boldsymbol{0}$ is equivalent to $\alpha a_1 = \alpha a_2 = \cdots = \alpha a_n = 0$. If $\alpha = 0$ or $\boldsymbol{a} = \boldsymbol{0}$, then $\alpha \boldsymbol{a} = \boldsymbol{0}$. If $\boldsymbol{a} \neq \boldsymbol{0}$, then at least one of its components $a_k \neq 0$. For this component, $\alpha a_k = 0$, and hence we must have $\alpha = 0$. Similar arguments can be applied to the case when $\alpha \neq 0$.

A set of vectors $\{\boldsymbol{a}_1, \ldots, \boldsymbol{a}_k\}$ is said to be *linearly independent* if the equality

$$\alpha_1 \boldsymbol{a}_1 + \alpha_2 \boldsymbol{a}_2 + \cdots + \alpha_k \boldsymbol{a}_k = \boldsymbol{0}$$

implies that all coefficients $\alpha_i$, $i = 1, \ldots, k$, are equal to zero. A set of the vectors $\{\boldsymbol{a}_1, \ldots, \boldsymbol{a}_k\}$ is *linearly dependent* if it is not linearly independent.

Note that the set composed of the single vector $\mathbf{0}$ is linearly dependent, for if $\alpha \neq 0$ then $\alpha\mathbf{0} = \mathbf{0}$. In fact, any set of vectors containing the vector $\mathbf{0}$ is linearly dependent.

A set composed of a single nonzero vector $\boldsymbol{a} \neq \mathbf{0}$ is linearly independent since $\alpha\boldsymbol{a} = \mathbf{0}$ implies $\alpha = 0$.

A vector $\boldsymbol{a}$ is said to be a *linear combination* of vectors $\boldsymbol{a}_1, \boldsymbol{a}_2, \ldots, \boldsymbol{a}_k$ if there are scalars $\alpha_1, \ldots, \alpha_k$ such that

$$\boldsymbol{a} = \alpha_1\boldsymbol{a}_1 + \alpha_2\boldsymbol{a}_2 + \cdots + \alpha_k\boldsymbol{a}_k.$$

**Proposition 2.1.** *A set of vectors $\{\boldsymbol{a}_1, \boldsymbol{a}_2, \ldots, \boldsymbol{a}_k\}$ is linearly dependent if, and only if, one of the vectors from the set is a linear combination of the remaining vectors.* □

*Proof.* $\Rightarrow$: If $\{\boldsymbol{a}_1, \boldsymbol{a}_2, \ldots, \boldsymbol{a}_k\}$ is linearly dependent then

$$\alpha_1\boldsymbol{a}_1 + \alpha_2\boldsymbol{a}_2 + \cdots + \alpha_k\boldsymbol{a}_k = \mathbf{0},$$

where at least one of the scalars $\alpha_i \neq 0$, whence

$$\boldsymbol{a}_i = -\frac{\alpha_1}{\alpha_i}\boldsymbol{a}_1 - \frac{\alpha_2}{\alpha_i}\boldsymbol{a}_2 - \cdots - \frac{\alpha_k}{\alpha_i}\boldsymbol{a}_k.$$

$\Leftarrow$: Suppose

$$\boldsymbol{a}_1 = \alpha_2\boldsymbol{a}_2 + \alpha_3\boldsymbol{a}_3 + \cdots + \alpha_k\boldsymbol{a}_k,$$

then

$$(-1)\boldsymbol{a}_1 + \alpha_2\boldsymbol{a}_2 + \cdots + \alpha_k\boldsymbol{a}_k = \mathbf{0}.$$

Since the first scalar is nonzero, then the set of vectors $\{\boldsymbol{a}_1, \boldsymbol{a}_2, \ldots, \boldsymbol{a}_k\}$ is linearly dependent. The same argument holds if $\boldsymbol{a}_i$, $i = 2, \ldots, k$, is a linear combination of the remaining vectors. ■

A subset $\mathscr{V}$ of $\mathbb{R}^n$ is called a *subspace* of $\mathbb{R}^n$ if $\mathscr{V}$ is closed under the operations of vector addition and scalar multiplication. That is, if $\boldsymbol{a}$ and $\boldsymbol{b}$ are vectors in $\mathscr{V}$, then the vectors $\boldsymbol{a} + \boldsymbol{b}$ and $\alpha\boldsymbol{a}$ are also in $\mathscr{V}$ for every scalar $\alpha$.

Every subspace contains the zero vector $\mathbf{0}$; for if $\boldsymbol{a}$ is an element of the subspace, so is $(-1)\boldsymbol{a} = -\boldsymbol{a}$. Hence, $\boldsymbol{a} - \boldsymbol{a} = \mathbf{0}$ also belongs to the subspace.

Let $\boldsymbol{a}_1, \boldsymbol{a}_2, \ldots, \boldsymbol{a}_k$ be arbitrary vectors in $\mathbb{R}^n$. The set of all their linear combinations is called the *span* of $\boldsymbol{a}_1, \boldsymbol{a}_2, \ldots, \boldsymbol{a}_k$ and is denoted

$$\operatorname{span}[\boldsymbol{a}_1, \boldsymbol{a}_2, \ldots, \boldsymbol{a}_k] = \left\{ \sum_{i=1}^{k} \alpha_i\boldsymbol{a}_i : \alpha_1, \ldots, \alpha_k \in \mathbb{R} \right\}.$$

Given a vector $\boldsymbol{a}$, the subspace span $[\boldsymbol{a}]$ is composed of the vectors $\alpha\boldsymbol{a}$, where $\alpha$ is an arbitrary real number ($\alpha \in \mathbb{R}$). Also observe that if $\boldsymbol{a}$ is a linear combination of $\boldsymbol{a}_1, \boldsymbol{a}_2, \ldots, \boldsymbol{a}_k$ then

$$\operatorname{span}[\boldsymbol{a}_1, \boldsymbol{a}_2, \ldots, \boldsymbol{a}_k, \boldsymbol{a}] = \operatorname{span}[\boldsymbol{a}_1, \boldsymbol{a}_2, \ldots, \boldsymbol{a}_k].$$

The span of any set of vectors is a subspace.

Given a subspace $\mathscr{V}$, any set of linearly independent vectors $\{\boldsymbol{a}_1, \boldsymbol{a}_2, \ldots, \boldsymbol{a}_k\} \subset \mathscr{V}$ such that $\mathscr{V} = \operatorname{span}[\boldsymbol{a}_1, \boldsymbol{a}_2, \ldots, \boldsymbol{a}_k]$ is referred to as a *basis* of the subspace $\mathscr{V}$. All bases of a subspace $\mathscr{V}$ contain the same number of vectors. This number is called the *dimension* of $\mathscr{V}$, denoted $\dim \mathscr{V}$.

**Proposition 2.2.** *If $\{\boldsymbol{a}_1, \boldsymbol{a}_2, \ldots, \boldsymbol{a}_k\}$ is a basis of $\mathscr{V}$, then any vector $\boldsymbol{a}$ of $\mathscr{V}$ can be represented uniquely as*

$$\boldsymbol{a} = \alpha_1\boldsymbol{a}_1 + \alpha_2\boldsymbol{a}_2 + \cdots + \alpha_k\boldsymbol{a}_k,$$

*where $\alpha_i \in \mathbb{R}$, $i = 1, 2, \ldots, k$.* □

*Proof.* To prove the uniqueness of the representation of $\boldsymbol{a}$ in terms of the basis vectors, assume that

$$\boldsymbol{a} = \alpha_1\boldsymbol{a}_1 + \alpha_2\boldsymbol{a}_2 + \cdots + \alpha_k\boldsymbol{a}_k$$

and

$$\boldsymbol{a} = \beta_1\boldsymbol{a}_1 + \beta_2\boldsymbol{a}_2 + \cdots + \beta_k\boldsymbol{a}_k.$$

We will now show that $\alpha_i = \beta_i$, $i = 1, \ldots, k$. We have

$$\alpha_1\boldsymbol{a}_1 + \alpha_2\boldsymbol{a}_2 + \cdots + \alpha_k\boldsymbol{a}_k = \beta_1\boldsymbol{a}_1 + \beta_2\boldsymbol{a}_2 + \cdots + \beta_k\boldsymbol{a}_k,$$

or

$$(\alpha_1 - \beta_1)\boldsymbol{a}_1 + (\alpha_2 - \beta_2)\boldsymbol{a}_2 + \cdots + (\alpha_k - \beta_k)\boldsymbol{a}_k = \boldsymbol{0}.$$

Because the set $\{\boldsymbol{a}_i : i = 1, 2, \ldots, k\}$ is linearly independent, $\alpha_1 - \beta_1 = \alpha_2 - \beta_2 = \cdots = \alpha_k - \beta_k = 0$, that is, $\alpha_i = \beta_i$, $i = 1, \ldots, k$. ■

Suppose we are given a basis $\{\boldsymbol{a}_1, \boldsymbol{a}_2, \ldots, \boldsymbol{a}_k\}$ of $\mathscr{V}$ and a vector $\boldsymbol{a} \in \mathscr{V}$ such that

$$\boldsymbol{a} = \alpha_1\boldsymbol{a}_1 + \alpha_2\boldsymbol{a}_2 + \cdots + \alpha_k\boldsymbol{a}_k.$$

The coefficients $\alpha_i$, $i = 1, \ldots, k$, are called the *coordinates* of $\boldsymbol{a}$ with respect to the basis $\{\boldsymbol{a}_1, \boldsymbol{a}_2, \ldots, \boldsymbol{a}_k\}$.

The *natural basis* for $\mathbb{R}^n$ is the set of vectors

$$e_1 = \begin{bmatrix} 1 \\ 0 \\ 0 \\ \vdots \\ 0 \\ 0 \end{bmatrix}, \quad e_2 = \begin{bmatrix} 0 \\ 1 \\ 0 \\ \vdots \\ 0 \\ 0 \end{bmatrix}, \ldots, e_n = \begin{bmatrix} 0 \\ 0 \\ 0 \\ \vdots \\ 0 \\ 1 \end{bmatrix}.$$

The reason for calling these vectors the natural basis is that

$$x = \begin{bmatrix} x_1 \\ x_2 \\ \vdots \\ x_n \end{bmatrix} = x_1 e_1 + x_2 e_2 + \cdots + x_n e_n.$$

We can similarly define *complex vector spaces*. For this, let $\mathbb{C}$ denote the set of complex numbers, and $\mathbb{C}^n$ the set of column $n$ vectors with complex components. As the reader can easily verify, the set $\mathbb{C}^n$ has similar properties to $\mathbb{R}^n$, where scalars can take complex values.

## 2.2. RANK OF A MATRIX

A *matrix* is a rectangular array of numbers, commonly denoted by upper case bold letters, for example, $\boldsymbol{A}$. A matrix with $m$ rows and $n$ columns is called an $m \times n$ matrix, and we write

$$\boldsymbol{A} = \begin{bmatrix} a_{11} & a_{12} & \cdots & a_{1n} \\ a_{21} & a_{22} & \cdots & a_{2n} \\ \vdots & \vdots & \ddots & \vdots \\ a_{m1} & a_{m2} & \cdots & a_{mn} \end{bmatrix}.$$

Let us denote the $k$th column of $\boldsymbol{A}$ by $\boldsymbol{a}_k$, that is,

$$\boldsymbol{a}_k = \begin{bmatrix} a_{1k} \\ a_{2k} \\ \vdots \\ a_{mk} \end{bmatrix}.$$

The maximal number of linearly independent columns of $\boldsymbol{A}$ is called the *rank* of the matrix $\boldsymbol{A}$, denoted $\operatorname{rank} \boldsymbol{A}$. Note that $\operatorname{rank} \boldsymbol{A}$ is the dimension of span $[\boldsymbol{a}_1, \ldots, \boldsymbol{a}_n]$.

**Proposition 2.3.** *The rank of a matrix $\boldsymbol{A}$ is invariant under the following operations:*

1. *Multiplication of the columns of $\boldsymbol{A}$ by nonzero scalars*
2. *Interchange of the columns*
3. *Addition to a given column a linear combination of other columns* □

*Proof.*

1. Let $\boldsymbol{b}_k = \alpha_k \boldsymbol{a}_k$, where $\alpha_k \neq 0$, $k = 1, \ldots, n$, and let $\boldsymbol{B} = [\boldsymbol{b}_1, \boldsymbol{b}_2, \ldots, \boldsymbol{b}_n]$. Obviously

$$\operatorname{span}[\boldsymbol{a}_1, \boldsymbol{a}_2, \ldots, \boldsymbol{a}_n] = \operatorname{span}[\boldsymbol{b}_1, \boldsymbol{b}_2, \ldots, \boldsymbol{b}_n],$$

and thus

$$\operatorname{rank} \boldsymbol{A} = \operatorname{rank} \boldsymbol{B}.$$

2. The number of linearly independent vectors does not depend on their order.
3. Let

$$\begin{aligned} \boldsymbol{b}_1 &= \boldsymbol{a}_1 + c_2 \boldsymbol{a}_2 + \cdots + c_n \boldsymbol{a}_n, \\ \boldsymbol{b}_2 &= \boldsymbol{a}_2, \\ &\vdots \\ \boldsymbol{b}_n &= \boldsymbol{a}_n. \end{aligned}$$

So, for any $\alpha_1, \ldots, \alpha_n$,

$$\alpha_1 \boldsymbol{b}_1 + \alpha_2 \boldsymbol{b}_2 + \cdots + \alpha_n \boldsymbol{b}_n = \alpha_1 \boldsymbol{a}_1 + (\alpha_2 + \alpha_1 c_2) \boldsymbol{a}_2 + \cdots + (\alpha_n + \alpha_1 c_n) \boldsymbol{a}_n,$$

and hence

$$\operatorname{span}[\boldsymbol{b}_1, \boldsymbol{b}_2, \ldots, \boldsymbol{b}_n] \subset \operatorname{span}[\boldsymbol{a}_1, \boldsymbol{a}_2, \ldots, \boldsymbol{a}_n].$$

On the other hand

$$\begin{aligned} \boldsymbol{a}_1 &= \boldsymbol{b}_1 - c_2 \boldsymbol{b}_2 - \cdots - c_n \boldsymbol{b}_n, \\ \boldsymbol{a}_2 &= \boldsymbol{b}_2, \\ &\vdots \\ \boldsymbol{a}_n &= \boldsymbol{b}_n. \end{aligned}$$

Hence,

$$\operatorname{span}[\boldsymbol{a}_1, \boldsymbol{a}_2, \ldots, \boldsymbol{a}_n] \subset \operatorname{span}[\boldsymbol{b}_1, \boldsymbol{b}_2, \ldots, \boldsymbol{b}_n].$$

Therefore

$$\operatorname{rank} \boldsymbol{A} = \operatorname{rank} \boldsymbol{B}. \qquad \blacksquare$$

Associated with each square ($n \times n$) matrix $\boldsymbol{A}$ is a scalar called the *determinant* of the matrix $\boldsymbol{A}$, denoted $\det \boldsymbol{A}$ or $|\boldsymbol{A}|$. The determinant of a square matrix is a function of its columns and has the following properties:

**1.** The determinant of the matrix $\boldsymbol{A} = [\boldsymbol{a}_1, \boldsymbol{a}_2, \ldots, \boldsymbol{a}_n]$ is a linear function of each column; that is

$$\begin{aligned} &\det[\boldsymbol{a}_1, \ldots, \boldsymbol{a}_{k-1}, \alpha \boldsymbol{a}_k^{(1)} + \beta \boldsymbol{a}_k^{(2)}, \boldsymbol{a}_{k+1}, \ldots, \boldsymbol{a}_n] \\ &\quad = \alpha \det[\boldsymbol{a}_1, \ldots, \boldsymbol{a}_{k-1}, \boldsymbol{a}_k^{(1)}, \boldsymbol{a}_{k+1}, \ldots, \boldsymbol{a}_n] \\ &\qquad + \beta \det[\boldsymbol{a}_1, \ldots, \boldsymbol{a}_{k-1}, \boldsymbol{a}_k^{(2)}, \boldsymbol{a}_{k+1}, \ldots, \boldsymbol{a}_n], \end{aligned}$$

for each $\alpha, \beta \in \mathbb{R}$, $\boldsymbol{a}_k^{(1)}, \boldsymbol{a}_k^{(2)} \in \mathbb{R}^n$.

**2.** If for some $k$ we have $\boldsymbol{a}_k = \boldsymbol{a}_{k+1}$, then

$$\det \boldsymbol{A} = \det[\boldsymbol{a}_1, \ldots, \boldsymbol{a}_k, \boldsymbol{a}_{k+1}, \ldots, \boldsymbol{a}_n] = \det[\boldsymbol{a}_1, \ldots, \boldsymbol{a}_k, \boldsymbol{a}_k, \ldots, \boldsymbol{a}_n] = 0.$$

**3.** Let

$$\boldsymbol{I}_n = [\boldsymbol{e}_1, \boldsymbol{e}_2, \ldots, \boldsymbol{e}_n] = \begin{bmatrix} 1 & 0 & \cdots & 0 \\ 0 & 1 & \cdots & 0 \\ \vdots & \vdots & \ddots & \vdots \\ 0 & 0 & \cdots & 1 \end{bmatrix},$$

where $\{\boldsymbol{e}_1, \ldots, \boldsymbol{e}_n\}$ is the natural basis for $\mathbb{R}^n$. Then

$$\det \boldsymbol{I}_n = 1.$$

Note that if $\alpha = \beta = 0$ in property 1, then

$$\det[\boldsymbol{a}_1, \ldots, \boldsymbol{a}_{k-1}, \boldsymbol{0}, \boldsymbol{a}_{k+1}, \ldots, \boldsymbol{a}_n] = 0.$$

Thus, if one of the columns is $\boldsymbol{0}$, then the determinant is equal to zero.

The determinant does not change its value if we add to a column another column multiplied by a scalar. This follows from properties 1 and 2 as shown below:

$$\begin{aligned} &\det[\boldsymbol{a}_1, \ldots, \boldsymbol{a}_{k-1}, \boldsymbol{a}_k + \alpha \boldsymbol{a}_j, \boldsymbol{a}_{k+1}, \ldots, \boldsymbol{a}_j, \ldots, \boldsymbol{a}_n] \\ &\quad = \det[\boldsymbol{a}_1, \ldots, \boldsymbol{a}_{k-1}, \boldsymbol{a}_k, \boldsymbol{a}_{k+1}, \ldots, \boldsymbol{a}_j, \ldots, \boldsymbol{a}_n] \\ &\qquad + \alpha \det[\boldsymbol{a}_1, \ldots, \boldsymbol{a}_{k-1}, \boldsymbol{a}_j, \boldsymbol{a}_{k+1}, \ldots, \boldsymbol{a}_j, \ldots, \boldsymbol{a}_n] \\ &\quad = \det[\boldsymbol{a}_1, \ldots, \boldsymbol{a}_n]. \end{aligned}$$

However, the determinant changes its sign if we interchange columns. To show this property, note that

$$\begin{aligned}
&\det[\boldsymbol{a}_1,\ldots,\boldsymbol{a}_{k-1},\boldsymbol{a}_k,\boldsymbol{a}_{k+1},\ldots,\boldsymbol{a}_n]\\
&\quad=\det[\boldsymbol{a}_1,\ldots,\boldsymbol{a}_k+\boldsymbol{a}_{k+1},\boldsymbol{a}_{k+1},\ldots,\boldsymbol{a}_n]\\
&\quad=\det[\boldsymbol{a}_1,\ldots,\boldsymbol{a}_k+\boldsymbol{a}_{k+1},\boldsymbol{a}_{k+1}-(\boldsymbol{a}_k+\boldsymbol{a}_{k+1}),\ldots,\boldsymbol{a}_n]\\
&\quad=\det[\boldsymbol{a}_1,\ldots,\boldsymbol{a}_k+\boldsymbol{a}_{k+1},-\boldsymbol{a}_k,\ldots,\boldsymbol{a}_n]\\
&\quad=-\det[\boldsymbol{a}_1,\ldots,\boldsymbol{a}_k+\boldsymbol{a}_{k+1},\boldsymbol{a}_k,\ldots,\boldsymbol{a}_n]\\
&\quad=-(\det[\boldsymbol{a}_1,\ldots,\boldsymbol{a}_k,\boldsymbol{a}_k,\ldots,\boldsymbol{a}_n]+\det[\boldsymbol{a}_1,\ldots,\boldsymbol{a}_{k+1},\boldsymbol{a}_k,\ldots,\boldsymbol{a}_n])\\
&\quad=-\det[\boldsymbol{a}_1,\ldots,\boldsymbol{a}_{k+1},\boldsymbol{a}_k,\ldots,\boldsymbol{a}_n].
\end{aligned}$$

A $p$th-order *minor* of an $m \times n$ matrix $\boldsymbol{A}$, with $p \leqslant \min(m, n)$, is the determinant of a $p \times p$ matrix obtained from $\boldsymbol{A}$ by deleting $m-p$ rows and $n-p$ columns.

One can use minors to investigate the rank of a matrix. In particular, we have the following proposition.

**Proposition 2.4.** *If an $m \times n$ $(m \geqslant n)$ matrix $\boldsymbol{A}$ has a nonzero* $n$th-*order minor, then the columns of $\boldsymbol{A}$ are linearly independent, that is,* rank $\boldsymbol{A} = n$. □

*Proof.* Suppose $\boldsymbol{A}$ has a nonzero $n$th-order minor. Without loss of generality, we assume that the $n$th-order minor corresponding to the first $n$ rows of $\boldsymbol{A}$ is nonzero. Let $x_i$, $i = 1, \ldots, n$, be scalars such that

$$x_1\boldsymbol{a}_1 + x_2\boldsymbol{a}_2 + \cdots + x_n\boldsymbol{a}_n = \boldsymbol{0},$$

The above vector equality is equivalent to the following set of $m$ equations

$$\begin{aligned}
a_{11}x_1 + a_{12}x_2 + \cdots + a_{1n}x_n &= 0\\
a_{21}x_1 + a_{22}x_2 + \cdots + a_{2n}x_n &= 0\\
&\ \ \vdots\\
a_{n1}x_1 + a_{n2}x_2 + \cdots + a_{nn}x_n &= 0\\
&\ \ \vdots\\
a_{m1}x_1 + a_{m2}x_2 + \cdots + a_{mn}x_n &= 0.
\end{aligned}$$

For $i = 1, \ldots, n$, let

$$\tilde{\boldsymbol{a}}_i = \begin{bmatrix} a_{1i} \\ \vdots \\ a_{ni} \end{bmatrix}.$$

Then, $x_1\tilde{\boldsymbol{a}}_1 + \cdots + x_n\tilde{\boldsymbol{a}}_n = \boldsymbol{0}$.

The $n$th-order minor is $\det[\tilde{\boldsymbol{a}}_1, \tilde{\boldsymbol{a}}_2, \ldots, \tilde{\boldsymbol{a}}_n]$, assumed to be nonzero. From the properties of determinants it follows that the columns $\tilde{\boldsymbol{a}}_1, \tilde{\boldsymbol{a}}_2, \ldots, \tilde{\boldsymbol{a}}_n$ are linearly independent. Therefore, all $x_i = 0$, $i = 1, \ldots, n$. Hence, the columns $\boldsymbol{a}_1, \boldsymbol{a}_2, \ldots, \boldsymbol{a}_n$ are linearly independent. ■

From the above it follows that if there is a nonzero minor, then the columns associated with this nonzero minor are linearly independent.

If a matrix $\boldsymbol{A}$ has an $r$th-order minor $|\boldsymbol{M}|$ with the properties (i) $|\boldsymbol{M}| \neq 0$ and (ii) any minor of $\boldsymbol{A}$ that is formed by adding a row and a column of $\boldsymbol{A}$ to $\boldsymbol{M}$ is zero, then

$$\operatorname{rank} \boldsymbol{A} = r.$$

Thus, the rank of a matrix is equal to the highest order of its nonzero minor(s).

A *nonsingular* (or *invertible*) matrix is a square matrix whose determinant is nonzero. Suppose that $\boldsymbol{A}$ is an $n \times n$ square matrix. Then, $\boldsymbol{A}$ is nonsingular if, and only if, there is another $n \times n$ matrix $\boldsymbol{B}$ such that

$$\boldsymbol{A}\boldsymbol{B} = \boldsymbol{B}\boldsymbol{A} = \boldsymbol{I}_n,$$

where $\boldsymbol{I}_n$ denotes the $n \times n$ *identity matrix*:

$$\boldsymbol{I}_n = \begin{bmatrix} 1 & 0 & \cdots & 0 \\ 0 & 1 & \cdots & 0 \\ \vdots & \vdots & \ddots & \vdots \\ 0 & 0 & \cdots & 1 \end{bmatrix}.$$

We call the above matrix $\boldsymbol{B}$ the *inverse matrix* of $\boldsymbol{A}$, and write $\boldsymbol{B} = \boldsymbol{A}^{-1}$.

Consider the $m \times n$ matrix

$$\boldsymbol{A} = \begin{bmatrix} a_{11} & a_{12} & \cdots & a_{1n} \\ a_{21} & a_{22} & \cdots & a_{2n} \\ \vdots & \vdots & \ddots & \vdots \\ a_{m1} & a_{m2} & \cdots & a_{mn} \end{bmatrix}.$$

The *transpose* of $\boldsymbol{A}$, denoted $\boldsymbol{A}^T$, is the $n \times m$ matrix

$$\boldsymbol{A}^T = \begin{bmatrix} a_{11} & a_{21} & \cdots & a_{m1} \\ a_{12} & a_{22} & \cdots & a_{m2} \\ \vdots & \vdots & \ddots & \vdots \\ a_{1n} & a_{2n} & \cdots & a_{mn} \end{bmatrix};$$

that is, the columns of $\boldsymbol{A}$ are the rows of $\boldsymbol{A}^T$, and vice versa.

## 2.3. LINEAR EQUATIONS

Suppose we are given $m$ equations in $n$ unknowns of the form

$$\begin{aligned} a_{11}x_1 + a_{12}x_2 + \cdots + a_{1n}x_n &= b_1 \\ a_{21}x_1 + a_{22}x_2 + \cdots + a_{2n}x_n &= b_2 \\ &\vdots \\ a_{m1}x_1 + a_{m2}x_2 + \cdots + a_{mn}x_n &= b_m. \end{aligned}$$

We can represent the above set of equations as a vector equation

$$x_1\boldsymbol{a}_1 + x_2\boldsymbol{a}_2 + \cdots + x_n\boldsymbol{a}_n = \boldsymbol{b},$$

where

$$\boldsymbol{a}_j = \begin{bmatrix} a_{1j} \\ a_{2j} \\ \vdots \\ a_{mj} \end{bmatrix}, \quad \boldsymbol{b} = \begin{bmatrix} b_1 \\ b_2 \\ \vdots \\ b_m \end{bmatrix}.$$

Associated with the above system of equations are the following matrices

$$\boldsymbol{A} = [\boldsymbol{a}_1, \boldsymbol{a}_2, \ldots, \boldsymbol{a}_n],$$

and an augmented matrix

$$[\boldsymbol{A}, \boldsymbol{b}] = [\boldsymbol{a}_1, \boldsymbol{a}_2, \ldots, \boldsymbol{a}_n, \boldsymbol{b}].$$

We can also represent the above system of equations as

$$\boldsymbol{A}\boldsymbol{x} = \boldsymbol{b},$$

where

$$\boldsymbol{x} = \begin{bmatrix} x_1 \\ x_2 \\ \vdots \\ x_n \end{bmatrix}.$$

**Theorem 2.1.** *The system of equations $\boldsymbol{A}\boldsymbol{x} = \boldsymbol{b}$ has a solution if, and only if,*

$$\operatorname{rank} \boldsymbol{A} = \operatorname{rank} [\boldsymbol{A}, \boldsymbol{b}]. \qquad \square$$

*Proof.* $\Rightarrow$: Suppose the system $\boldsymbol{A}\boldsymbol{x} = \boldsymbol{b}$ has a solution. Therefore, $\boldsymbol{b}$ is a linear combination of the columns of $\boldsymbol{A}$, that is, there exist $x_1, \ldots, x_n$ such that

$x_1\boldsymbol{a}_1 + x_2\boldsymbol{a}_2 + \cdots + x_n\boldsymbol{a}_n = \boldsymbol{b}$. It follows that $\boldsymbol{b}$ belongs to span $[\boldsymbol{a}_1, \ldots, \boldsymbol{a}_n]$ and hence

$$\begin{aligned}\operatorname{rank} \boldsymbol{A} &= \dim \operatorname{span}[\boldsymbol{a}_1, \ldots, \boldsymbol{a}_n] \\ &= \dim \operatorname{span}[\boldsymbol{a}_1, \ldots, \boldsymbol{a}_n, \boldsymbol{b}] \\ &= \operatorname{rank}[\boldsymbol{A}, \boldsymbol{b}].\end{aligned}$$

$\Leftarrow$: Suppose that rank $\boldsymbol{A} = \operatorname{rank}[\boldsymbol{A}, \boldsymbol{b}] = r$. Thus, we have $r$ linearly independent columns of $\boldsymbol{A}$. Without loss of generality, let $\boldsymbol{a}_1, \boldsymbol{a}_2, \ldots, \boldsymbol{a}_r$ be these columns. Therefore, $\boldsymbol{a}_1, \boldsymbol{a}_2, \ldots, \boldsymbol{a}_r$ are also linearly independent columns of the matrix $[\boldsymbol{A}, \boldsymbol{b}]$. Since rank $[\boldsymbol{A}, \boldsymbol{b}] = r$, the remaining columns of $[\boldsymbol{A}, \boldsymbol{b}]$ can be expressed as linear combinations of $\boldsymbol{a}_1, \boldsymbol{a}_2, \ldots, \boldsymbol{a}_r$. In particular $\boldsymbol{b}$ can be expressed as a linear combination of these columns. Hence, there exist $x_1, \ldots, x_n$ such that $x_1\boldsymbol{a}_1 + x_2\boldsymbol{a}_2 + \cdots + x_n\boldsymbol{a}_n = \boldsymbol{b}$. ■

Let the symbol $\mathbb{R}^{m \times n}$ denote the set of $m \times n$ matrices whose elements are real numbers.

**Theorem 2.2.** *Consider the equation $\boldsymbol{A}\boldsymbol{x} = \boldsymbol{b}$, where $\boldsymbol{A} \in \mathbb{R}^{m \times n}$, and rank $\boldsymbol{A} = m$. A solution to $\boldsymbol{A}\boldsymbol{x} = \boldsymbol{b}$ can be obtained by assigning arbitrary values for $n - m$ variables and solving for the remaining ones.* □

*Proof.* We have rank $\boldsymbol{A} = m$, and therefore we can find $m$ linearly independent columns of $\boldsymbol{A}$. Without loss of generality, let $\boldsymbol{a}_1, \boldsymbol{a}_2, \ldots, \boldsymbol{a}_m$ be such columns. Rewrite the equation $\boldsymbol{A}\boldsymbol{x} = \boldsymbol{b}$ as

$$x_1\boldsymbol{a}_1 + x_2\boldsymbol{a}_2 + \cdots + x_m\boldsymbol{a}_m = \boldsymbol{b} - x_{m+1}\boldsymbol{a}_{m+1} - \cdots - x_n\boldsymbol{a}_n.$$

Assign to $x_{m+1}, x_{m+2}, \ldots, x_n$ arbitrary values, say

$$x_{m+1} = d_{m+1}, x_{m+2} = d_{m+2}, \ldots, x_n = d_n,$$

and let

$$\boldsymbol{B} = [\boldsymbol{a}_1, \boldsymbol{a}_2, \ldots, \boldsymbol{a}_m] \in \mathbb{R}^{m \times m}.$$

Note that $\det \boldsymbol{B} \neq 0$. We can represent the above system of equations as

$$\boldsymbol{B}\begin{bmatrix} x_1 \\ x_2 \\ \vdots \\ x_m \end{bmatrix} = [\boldsymbol{b} - d_{m+1}\boldsymbol{a}_{m+1} - \cdots - d_n\boldsymbol{a}_n].$$

The matrix $\boldsymbol{B}$ is invertible, and therefore we can solve for $[x_1, x_2, \ldots, x_m]^T$.

Specifically,

$$\begin{bmatrix} x_1 \\ x_2 \\ \vdots \\ x_m \end{bmatrix} = \boldsymbol{B}^{-1}[\boldsymbol{b} - d_{m+1}\boldsymbol{a}_{m+1} - \cdots - d_n\boldsymbol{a}_n]. \qquad \blacksquare$$

## 2.4. INNER PRODUCTS AND NORMS

The absolute value of a real number $a$, denoted $|a|$, is defined as

$$|a| = \begin{cases} a, & \text{if } a \geqslant 0, \\ -a, & \text{if } a < 0. \end{cases}$$

The following formulas hold:

1. $|a| = |-a|$
2. $-|a| \leqslant a \leqslant |a|$
3. $|a+b| \leqslant |a| + |b|$
4. $||a| - |b|| \leqslant |a| - |b| \leqslant |a| + |b|$
5. $|ab| = |a||b|$
6. $|a| \leqslant c$ and $|b| \leqslant d$ imply $|a+b| \leqslant c+d$
7. The inequality $|a| < b$ is equivalent to $-b < a < b$ (i.e., $a < b$ and $-a < b$). The same holds if we replace every occurrence of "$<$" by "$\leqslant$".
8. The inequality $|a| > b$ is equivalent to $a > b$ or $-a > b$. The same holds if we replace every occurrence of "$>$" by "$\geqslant$".

For $\boldsymbol{x}, \boldsymbol{y} \in \mathbb{R}^n$, we define the *Euclidean inner product* by

$$\langle \boldsymbol{x}, \boldsymbol{y} \rangle = \sum_{i=1}^{n} x_i y_i = \boldsymbol{x}^T \boldsymbol{y}.$$

The inner product is a real valued function $\langle \cdot, \cdot \rangle$: $\mathbb{R}^n \times \mathbb{R}^n \to \mathbb{R}$ having the following properties:

1. Positivity: $\langle \boldsymbol{x}, \boldsymbol{x} \rangle \geqslant 0$, $\langle \boldsymbol{x}, \boldsymbol{x} \rangle = 0$ if, and only if, $\boldsymbol{x} = 0$
2. Symmetry: $\langle \boldsymbol{x}, \boldsymbol{y} \rangle = \langle \boldsymbol{y}, \boldsymbol{x} \rangle$
3. Additivity: $\langle \boldsymbol{x} + \boldsymbol{y}, \boldsymbol{z} \rangle = \langle \boldsymbol{x}, \boldsymbol{z} \rangle + \langle \boldsymbol{y}, \boldsymbol{z} \rangle$
4. Homogeneity: $\langle r\boldsymbol{x}, \boldsymbol{y} \rangle = r \langle \boldsymbol{x}, \boldsymbol{y} \rangle$ for every $r \in \mathbb{R}$.

The properties of additivity and homogeneity in the second vector also hold;

that is,

$$\langle \boldsymbol{x}, \boldsymbol{y}+\boldsymbol{z} \rangle = \langle \boldsymbol{x}, \boldsymbol{y} \rangle + \langle \boldsymbol{x}, \boldsymbol{z} \rangle,$$

$$\langle \boldsymbol{x}, r\boldsymbol{y} \rangle = r\langle \boldsymbol{x}, \boldsymbol{y} \rangle \quad \text{for every } r \in \mathbb{R}.$$

The above can be shown using properties 2 to 4. Indeed,

$$\begin{aligned}\langle \boldsymbol{x}, \boldsymbol{y}+\boldsymbol{z} \rangle &= \langle \boldsymbol{y}+\boldsymbol{z}, \boldsymbol{x} \rangle \\ &= \langle \boldsymbol{y}, \boldsymbol{x} \rangle + \langle \boldsymbol{z}, \boldsymbol{x} \rangle \\ &= \langle \boldsymbol{x}, \boldsymbol{y} \rangle + \langle \boldsymbol{x}, \boldsymbol{z} \rangle,\end{aligned}$$

and

$$\langle \boldsymbol{x}, r\boldsymbol{y} \rangle = \langle r\boldsymbol{y}, \boldsymbol{x} \rangle = r\langle \boldsymbol{y}, \boldsymbol{x} \rangle = r\langle \boldsymbol{x}, \boldsymbol{y} \rangle.$$

It is possible to define other real valued functions on $\mathbb{R}^n \times \mathbb{R}^n$ that satisfy properties 1 to 4 above (see Exercise 2.5). Many results involving the Euclidean inner product also hold for these other forms of inner products.

The vectors $\boldsymbol{x}$ and $\boldsymbol{y}$ are said to be *orthogonal* if $\langle \boldsymbol{x}, \boldsymbol{y} \rangle = 0$. The *Euclidean norm* of a vector $\boldsymbol{x}$ is defined as

$$\|\boldsymbol{x}\| = \sqrt{\langle \boldsymbol{x}, \boldsymbol{x} \rangle} = \sqrt{\boldsymbol{x}^T \boldsymbol{x}}.$$

**Theorem 2.3. Cauchy-Schwarz Inequality.** *For any two vectors $\boldsymbol{x}$ and $\boldsymbol{y}$ in $\mathbb{R}^n$, the Cauchy-Schwarz inequality*

$$|\langle \boldsymbol{x}, \boldsymbol{y} \rangle| \leqslant \|\boldsymbol{x}\| \, \|\boldsymbol{y}\|$$

*holds. Furthermore, equality holds if, and only if, $\boldsymbol{x} = \alpha\boldsymbol{y}$ for some $\alpha \in \mathbb{R}$.* □

*Proof.* First assume that $\boldsymbol{x}$ and $\boldsymbol{y}$ are unit vectors, that is, $\|\boldsymbol{x}\| = \|\boldsymbol{y}\| = 1$. Then,

$$\begin{aligned}0 \leqslant \|\boldsymbol{x}-\boldsymbol{y}\|^2 &= \langle \boldsymbol{x}-\boldsymbol{y}, \boldsymbol{x}-\boldsymbol{y} \rangle \\ &= \|\boldsymbol{x}\|^2 - 2\langle \boldsymbol{x}, \boldsymbol{y} \rangle + \|\boldsymbol{y}\|^2 \\ &= 2 - 2\langle \boldsymbol{x}, \boldsymbol{y} \rangle,\end{aligned}$$

or

$$\langle \boldsymbol{x}, \boldsymbol{y} \rangle \leqslant 1,$$

with equality holding if, and only if, $\boldsymbol{x} = \boldsymbol{y}$.

Next, assuming that neither $\boldsymbol{x}$ nor $\boldsymbol{y}$ is zero (for the inequality obviously holds if one of them is zero), we replace $\boldsymbol{x}$ and $\boldsymbol{y}$ by the unit vectors $\boldsymbol{x}/\|\boldsymbol{x}\|$ and $\boldsymbol{y}/\|\boldsymbol{y}\|$. Then, apply property 4 to get

$$\langle \boldsymbol{x}, \boldsymbol{y} \rangle \leqslant \|\boldsymbol{x}\| \, \|\boldsymbol{y}\|.$$

Now replace $\boldsymbol{x}$ by $-\boldsymbol{x}$ and again apply property 4 to get

$$-\langle \boldsymbol{x}, \boldsymbol{y} \rangle \leqslant \| \boldsymbol{x} \| \, \| \boldsymbol{y} \|.$$

The last two inequalities imply the absolute value inequality. Equality holds if, and only if, $\boldsymbol{x}/\|\boldsymbol{x}\| = \pm \boldsymbol{y}/\|\boldsymbol{y}\|$; that is, $\boldsymbol{x} = \alpha \boldsymbol{y}$ for some $\alpha \in \mathbb{R}$. ■

The Euclidean norm of a vector $\|\boldsymbol{x}\|$ has the following properties:

1. Positivity: $\|\boldsymbol{x}\| \geqslant 0$, $\|\boldsymbol{x}\| = 0$ if, and only if, $\boldsymbol{x} = \boldsymbol{0}$
2. Homogeneity: $\|r\boldsymbol{x}\| = |r| \, \|\boldsymbol{x}\|, r \in \mathbb{R}$
3. Triangle inequality: $\|\boldsymbol{x} + \boldsymbol{y}\| \leqslant \|\boldsymbol{x}\| + \|\boldsymbol{y}\|$.

The triangle inequality can be proved using the Cauchy-Schwarz inequality, as follows. We have

$$\|\boldsymbol{x} + \boldsymbol{y}\|^2 = \|\boldsymbol{x}\|^2 + 2\langle \boldsymbol{x}, \boldsymbol{y} \rangle + \|\boldsymbol{y}\|^2.$$

By the Cauchy-Schwarz inequality

$$\begin{aligned} \|\boldsymbol{x} + \boldsymbol{y}\|^2 &\leqslant \|\boldsymbol{x}\|^2 + 2\|\boldsymbol{x}\| \, \|\boldsymbol{y}\| + \|\boldsymbol{y}\|^2 \\ &= (\|\boldsymbol{x}\| + \|\boldsymbol{y}\|)^2, \end{aligned}$$

and therefore

$$\|\boldsymbol{x} + \boldsymbol{y}\| \leqslant \|\boldsymbol{x}\| + \|\boldsymbol{y}\|.$$

Note that if $\boldsymbol{x}$ and $\boldsymbol{y}$ are orthogonal, that is, $\langle \boldsymbol{x}, \boldsymbol{y} \rangle = 0$, then

$$\|\boldsymbol{x} + \boldsymbol{y}\|^2 = \|\boldsymbol{x}\|^2 + \|\boldsymbol{y}\|^2,$$

which is the *Pythagorean theorem* for $\mathbb{R}^n$.

The Euclidean norm is an example of a general *vector norm*, which is any function satisfying the above three properties of positivity, homogeneity, and triangle inequality. Other examples of vector norms on $\mathbb{R}^n$ include the 1-norm, defined by $\|\boldsymbol{x}\|_1 = |x_1| + \cdots + |x_n|$, and the $\infty$-norm, defined by $\|\boldsymbol{x}\|_\infty = \max_i |x_i|$. The Euclidean norm is often referred to as the 2-norm, and denoted $\|\boldsymbol{x}\|_2$. The above norms are special cases of the $p$-norm, given by

$$\|\boldsymbol{x}\|_p = \begin{cases} (|x_1|^p + \cdots + |x_n|^p)^{1/p}, & \text{if } 1 \leqslant p < \infty, \\ \max(|x_1|, \ldots, |x_n|), & \text{if } p = \infty. \end{cases}$$

We can use norms to define the notion of a *continuous function*, as follows. A function $\boldsymbol{f}: \mathbb{R}^n \to \mathbb{R}^m$ is continuous at $\boldsymbol{x}$ if, for all $\varepsilon > 0$, there exists $\delta > 0$ such that

$\|y - x\| < \delta \Rightarrow \|f(y) - f(x)\| < \varepsilon$. If the function $f$ is continuous at every point in $\mathbb{R}^n$, we say that it is continuous on $\mathbb{R}^n$. Note that $f = [f_1, \ldots, f_m]^T$ is continuous if, and only if, each component $f_i$, $i = 1, \ldots, m$, is continuous.

For the complex vector space $\mathbb{C}^n$, we define an inner product $\langle x, y \rangle$ to be $\sum_{i=1}^n x_i \bar{y}_i$, where the bar over $y_i$ denotes complex conjugation. The inner product on $\mathbb{C}^n$ is a complex valued function having the following properties:

1. $\langle x, x \rangle \geqslant 0$, $\langle x, x \rangle = 0$ if, and only if, $x = \mathbf{0}$
2. $\langle x, y \rangle = \overline{\langle y, x \rangle}$
3. $\langle x + y, z \rangle = \langle x, z \rangle + \langle y, z \rangle$
4. $\langle rx, y \rangle = r\langle x, y \rangle$, where $r \in \mathbb{C}$.

From properties 1 to 4, we can deduce other properties, such as

$$\langle x, r_1 y + r_2 z \rangle = \bar{r}_1 \langle x, y \rangle + \bar{r}_2 \langle x, z \rangle,$$

where $r_1$, $r_2 \in \mathbb{C}$. For $\mathbb{C}^n$, the vector norm can similarly be defined by $\|x\|^2 = \langle x, x \rangle$. For more information, consult Gel'fand [Ref. 25].

## EXERCISES

**2.1** Let $A \in \mathbb{R}^{m \times n}$ and rank $A = m$. Show that $m \leqslant n$.

**2.2** Prove that the system $Ax = b$, $A \in \mathbb{R}^{m \times n}$, has a unique solution if, and only if, rank $A$ = rank $[A, b] = n$.

**2.3** (Adapted from Ref. 18.) We know that if $k \geqslant n + 1$, then the vectors $a_1, a_2, \ldots, a_k \in \mathbb{R}^n$ are linearly dependent; that is, there exist scalars $\alpha_1, \ldots, \alpha_k$ such that at least one $\alpha_i \neq 0$ and $\sum_{i=1}^k \alpha_i a_i = \mathbf{0}$. Show that if $k \geqslant n + 2$, then there exist scalars $\alpha_1, \ldots, \alpha_k$ such that at least one $\alpha_i \neq 0$, $\sum_{i=1}^k \alpha_i a_i = \mathbf{0}$, and $\sum_{i=1}^k \alpha_i = 0$. *Hint*: Introduce the vectors $\bar{a}_i = [1, a_i^T]^T \in \mathbb{R}^{n+1}$, $i = 1, \ldots, k$, and use the fact that any $n + 2$ vectors in $\mathbb{R}^{n+1}$ are linearly dependent.

**2.4** Prove the seven properties of the absolute value of a real number.

**2.5** Consider the function $\langle \cdot, \cdot \rangle_2 : \mathbb{R}^2 \times \mathbb{R}^2 \to \mathbb{R}$, defined by $\langle x, y \rangle_2 = 2x_1y_1 + 3x_2y_1 + 3x_1y_2 + 5x_2y_2$, where $x = [x_1, x_2]^T$ and $y = [y_1, y_2]^T$. Show that $\langle \cdot, \cdot \rangle_2$ satisfies conditions 1 to 4 for inner products. *Note*: This is a special case of Exercise 3.14.

**2.6** Show that for any two vectors $x, y \in \mathbb{R}^n$, $|\,\|x\| - \|y\|\,| \leqslant \|x - y\|$. *Hint*: Write $x = (x - y) + y$, and use the triangle inequality. Do the same for $y$.

**2.7** Use Exercise 2.6 to show that the norm $\|\cdot\|$ is a uniformly continuous function; that is, for all $\varepsilon > 0$, there exists $\delta > 0$ such that if $\|x - y\| < \delta$, then $|\,\|x\| - \|y\|\,| < \varepsilon$.

# 3

# Transformations

## 3.1. LINEAR TRANSFORMATIONS

A function $\mathscr{L}:\mathbb{R}^n \to \mathbb{R}^m$ is called a *linear transformation* if

1. $\mathscr{L}(ax) = a\mathscr{L}(x)$ for every $x \in \mathbb{R}^n$ and $a \in \mathbb{R}$
2. $\mathscr{L}(x + y) = \mathscr{L}(x) + \mathscr{L}(y)$ for every $x, y \in \mathbb{R}^n$.

If we fix the bases for $\mathbb{R}^n$ and $\mathbb{R}^m$, then the linear transformation $\mathscr{L}$ can be represented by a matrix. Specifically, there exists $A \in \mathbb{R}^{m \times n}$ such that the following representation holds. Suppose $x \in \mathbb{R}^n$ is a given vector, and $x'$ is the representation of $x$ with respect to the given basis for $\mathbb{R}^n$. If $y = \mathscr{L}(x)$, and $y'$ is the representation of $y$ with respect to the given basis for $\mathbb{R}^m$, then

$$y' = Ax'.$$

We call $A$ the *matrix representation* of $\mathscr{L}$ with respect to the given bases for $\mathbb{R}^n$ and $\mathbb{R}^m$. In the special case where we assume the natural bases for $\mathbb{R}^n$ and $\mathbb{R}^m$, the matrix representation $A$ satisfies

$$\mathscr{L}(x) = Ax.$$

Let $\{e_1, e_2, \ldots, e_n\}$ and $\{e'_1, e'_2, \ldots, e'_n\}$ be two bases for $\mathbb{R}^n$. Define the matrix

$$T = [e'_1, e'_2, \ldots, e'_n]^{-1}[e_1, e_2, \ldots, \mathrm{e}_n].$$

We call $T$ the *transformation matrix* from $\{e_1, e_2, \ldots, e_n\}$ to $\{e'_1, e'_2, \ldots, e'_n\}$. It is clear that

$$[e_1, e_2, \ldots, e_n] = [e'_1, e'_2, \ldots, e'_n]T;$$

that is, the $i$th column of $T$ is the vector of coordinates of $e_i$ with respect to the basis $\{e'_1, e'_2, \ldots, e'_n\}$.

Fix a vector in $\mathbb{R}^n$, and let $x$ be the column of the coordinates of the vector with respect to $\{e_1, e_2, \ldots, e_n\}$, and $x'$ the coordinates of the same vector with respect to $\{e'_1, e'_2, \ldots, e'_n\}$. Then, we can show that $x' = Tx$ (see Exercise 3.1).

Consider a linear transformation

$$\mathscr{L}:\mathbb{R}^n \to \mathbb{R}^n,$$

and let $\boldsymbol{A}$ be its representation with respect to $\{\boldsymbol{e}_1, \boldsymbol{e}_2, \ldots, \boldsymbol{e}_n\}$, and $\boldsymbol{B}$ its representation with respect to $\{\boldsymbol{e}'_1, \boldsymbol{e}'_2, \ldots, \boldsymbol{e}'_n\}$. Let $\boldsymbol{y} = \boldsymbol{A}\boldsymbol{x}$ and $\boldsymbol{y}' = \boldsymbol{B}\boldsymbol{x}'$. Therefore, $\boldsymbol{y}' = \boldsymbol{T}\boldsymbol{y} = \boldsymbol{T}\boldsymbol{A}\boldsymbol{x} = \boldsymbol{B}\boldsymbol{x}' = \boldsymbol{B}\boldsymbol{T}\boldsymbol{x}$, and hence $\boldsymbol{T}\boldsymbol{A} = \boldsymbol{B}\boldsymbol{T}$, or $\boldsymbol{A} = \boldsymbol{T}^{-1}\boldsymbol{B}\boldsymbol{T}$.

Two $n \times n$ matrices $\boldsymbol{A}$ and $\boldsymbol{B}$ are said to be similar if there exists a nonsingular matrix $\boldsymbol{T}$ such that $\boldsymbol{A} = \boldsymbol{T}^{-1}\boldsymbol{B}\boldsymbol{T}$. In conclusion, similar matrices correspond to the same linear transformation with respect to different bases.

## 3.2. EIGENVALUES AND EIGENVECTORS

Let $\boldsymbol{A}$ be an $n \times n$ square matrix. A scalar $\lambda$ (possibly complex) and a nonzero vector $\boldsymbol{v}$ satisfying the equation $\boldsymbol{A}\,\boldsymbol{v} = \lambda\boldsymbol{v}$ are said to be, respectively, an *eigenvalue* and *eigenvector* of $\boldsymbol{A}$. For $\lambda$ to be an eigenvalue, it is necessary and sufficient for the matrix $\lambda\boldsymbol{I} - \boldsymbol{A}$ to be singular, that is, $\det[\lambda\boldsymbol{I} - \boldsymbol{A}] = 0$, where $\boldsymbol{I}$ is the $n \times n$ identity matrix. This leads to an $n$th-order polynomial equation

$$\det[\lambda\boldsymbol{I} - \boldsymbol{A}] = \lambda^n + a_{n-1}\lambda^{n-1} + \cdots + a_1\lambda + a_0 = 0.$$

We call the polynomial $\det[\lambda\boldsymbol{I} - \boldsymbol{A}]$ the *characteristic polynomial* of the matrix $\boldsymbol{A}$. According to the fundamental theorem of algebra, the above equation must have $n$ (possibly nondistinct) roots that are the eigenvalues of $\boldsymbol{A}$. The following theorem states that if $\boldsymbol{A}$ has $n$ distinct eigenvalues, then it also has $n$ linearly independent eigenvectors.

**Theorem 3.1.** *Suppose the characteristic equation* $\det[\lambda\boldsymbol{I} - \boldsymbol{A}] = 0$ *has n distinct roots* $\lambda_1, \lambda_2, \ldots, \lambda_n$. *Then, there exist n linearly independent vectors* $\boldsymbol{v}_1, \boldsymbol{v}_2, \ldots, \boldsymbol{v}_n$ *such that*

$$\boldsymbol{A}\boldsymbol{v}_i = \lambda_i\boldsymbol{v}_i, \quad i = 1, 2, \ldots, n. \qquad \square$$

*Proof.* Since $\det[\lambda_i\boldsymbol{I} - \boldsymbol{A}] = 0$, $i = 1, \ldots, n$, there exist nonzero $\boldsymbol{v}_i$, $i = 1, \ldots, n$, such that $\boldsymbol{A}\boldsymbol{v}_i = \lambda_i\boldsymbol{v}_i$, $i = 1, \ldots, n$. We now prove linear independence of $\{\boldsymbol{v}_1, \boldsymbol{v}_2, \ldots, \boldsymbol{v}_n\}$. To do this, let $c_1, \ldots, c_n$ be scalars such that $\sum_{i=1}^n c_i\boldsymbol{v}_i = \boldsymbol{0}$. We will show that $c_i = 0$, $i = 1, \ldots, n$.

Consider the matrix

$$\boldsymbol{Z} = (\lambda_2\boldsymbol{I} - \boldsymbol{A})(\lambda_3\boldsymbol{I} - \boldsymbol{A})\cdots(\lambda_n\boldsymbol{I} - \boldsymbol{A}).$$

We first show that $c_1 = 0$. Note that.

$$\begin{aligned}\boldsymbol{Z}\boldsymbol{v}_n &= (\lambda_2\boldsymbol{I} - \boldsymbol{A})(\lambda_3\boldsymbol{I} - \boldsymbol{A})\cdots(\lambda_{n-1}\boldsymbol{I} - \boldsymbol{A})(\lambda_n\boldsymbol{I} - \boldsymbol{A})\boldsymbol{v}_n \\ &= (\lambda_2\boldsymbol{I} - \boldsymbol{A})(\lambda_3\boldsymbol{I} - \boldsymbol{A})\cdots(\lambda_{n-1}\boldsymbol{I} - \boldsymbol{A})(\lambda_n\boldsymbol{v}_n - \boldsymbol{A}\boldsymbol{v}_n) \\ &= \boldsymbol{0}\end{aligned}$$

since $\lambda_n\boldsymbol{v}_n - \boldsymbol{A}\boldsymbol{v}_n = \boldsymbol{0}$.

Repeating the above argument, we get

$$\boldsymbol{Z}\boldsymbol{v}_k = \boldsymbol{0}, \quad k = 2, 3, \ldots, n.$$

But

$$\begin{aligned}
\boldsymbol{Z}\boldsymbol{v}_1 &= (\lambda_2\boldsymbol{I} - \boldsymbol{A})(\lambda_3\boldsymbol{I} - \boldsymbol{A})\cdots(\lambda_{n-1}\boldsymbol{I} - \boldsymbol{A})(\lambda_n\boldsymbol{I} - \boldsymbol{A})\boldsymbol{v}_1 \\
&= (\lambda_2\boldsymbol{I} - \boldsymbol{A})(\lambda_3\boldsymbol{I} - \boldsymbol{A})\cdots(\lambda_{n-1}\boldsymbol{v}_1 - \boldsymbol{A}\boldsymbol{v}_1)(\lambda_n - \lambda_1) \\
&\ \vdots \\
&= (\lambda_2\boldsymbol{I} - \boldsymbol{A})(\lambda_3\boldsymbol{I} - \boldsymbol{A})\boldsymbol{v}_1\cdots(\lambda_{n-1} - \lambda_1)(\lambda_n - \lambda_1) \\
&= (\lambda_2 - \lambda_1)(\lambda_3 - \lambda_1)\cdots(\lambda_{n-1} - \lambda_1)(\lambda_n - \lambda_1)\boldsymbol{v}_1.
\end{aligned}$$

Using the above equation, we see that

$$\begin{aligned}
\boldsymbol{Z}\left(\sum_{i=1}^{n} c_i\boldsymbol{v}_i\right) &= \sum_{i=1}^{n} c_i\boldsymbol{Z}\boldsymbol{v}_i \\
&= c_1\boldsymbol{Z}\boldsymbol{v}_1 \\
&= c_1(\lambda_2 - \lambda_1)(\lambda_3 - \lambda_1)\cdots(\lambda_n - \lambda_1)\boldsymbol{v}_1 = \boldsymbol{0}.
\end{aligned}$$

Since the $\lambda_i$ are distinct, it must follow that $c_1 = 0$.

Using similar arguments, we can show that all $c_i$ must vanish, and therefore the set of eigenvectors $\{\boldsymbol{v}_1, \boldsymbol{v}_2, \ldots, \boldsymbol{v}_n\}$ is linearly independent. ■

Consider a basis formed by a linearly independent set of eigenvectors $\{\boldsymbol{v}_1, \boldsymbol{v}_2, \ldots, \boldsymbol{v}_n\}$. With respect to this basis, the matrix $\boldsymbol{A}$ has a diagonal form. Indeed, let

$$\boldsymbol{T} = [\boldsymbol{v}_1, \boldsymbol{v}_2, \ldots, \boldsymbol{v}_n].$$

Then,

$$\begin{aligned}
\boldsymbol{T}^{-1}\boldsymbol{A}\boldsymbol{T} &= \boldsymbol{T}^{-1}\boldsymbol{A}[\boldsymbol{v}_1, \boldsymbol{v}_2, \ldots, \boldsymbol{v}_n] \\
&= \boldsymbol{T}^{-1}[\boldsymbol{A}\boldsymbol{v}_1, \boldsymbol{A}\boldsymbol{v}_2, \ldots, \boldsymbol{A}\boldsymbol{v}_n] \\
&= \boldsymbol{T}^{-1}[\lambda_1\boldsymbol{v}_1, \lambda_2\boldsymbol{v}_2, \ldots, \lambda_n\boldsymbol{v}_n] \\
&= \boldsymbol{T}^{-1}\boldsymbol{T}\begin{bmatrix} \lambda_1 & & & 0 \\ & \lambda_2 & & \\ & & \ddots & \\ 0 & & & \lambda_n \end{bmatrix} \\
&= \begin{bmatrix} \lambda_1 & & & 0 \\ & \lambda_2 & & \\ & & \ddots & \\ 0 & & & \lambda_n \end{bmatrix},
\end{aligned}$$

since $\boldsymbol{T}^{-1}\boldsymbol{T} = \boldsymbol{I}$.

Let us now consider symmetric matrices.

**Theorem 3.2.** *All eigenvalues of a symmetric matrix are real.* □

*Proof.* Let

$$\boldsymbol{Ax} = \lambda \boldsymbol{x},$$

where $\boldsymbol{x} \neq \boldsymbol{0}$. Taking the inner product of $\boldsymbol{Ax}$ with $\boldsymbol{x}$ yields

$$\langle \boldsymbol{Ax}, \boldsymbol{x} \rangle = \langle \lambda \boldsymbol{x}, \boldsymbol{x} \rangle = \lambda \langle \boldsymbol{x}, \boldsymbol{x} \rangle.$$

On the other hand

$$\langle \boldsymbol{Ax}, \boldsymbol{x} \rangle = \langle \boldsymbol{x}, \boldsymbol{A}^T \boldsymbol{x} \rangle = \langle \boldsymbol{x}, \boldsymbol{Ax} \rangle = \langle \boldsymbol{x}, \lambda \boldsymbol{x} \rangle = \bar{\lambda} \langle \boldsymbol{x}, \boldsymbol{x} \rangle.$$

The above follows from the definition of the inner product on $\mathbb{C}^n$. We note that $\langle \boldsymbol{x}, \boldsymbol{x} \rangle$ is real and $\langle \boldsymbol{x}, \boldsymbol{x} \rangle > 0$. Hence,

$$\lambda \langle \boldsymbol{x}, \boldsymbol{x} \rangle = \bar{\lambda} \langle \boldsymbol{x}, \boldsymbol{x} \rangle$$

and

$$(\lambda - \bar{\lambda}) \langle \boldsymbol{x}, \boldsymbol{x} \rangle = 0.$$

Since $\langle \boldsymbol{x}, \boldsymbol{x} \rangle > 0$,

$$\lambda = \bar{\lambda}.$$

Thus, $\lambda$ is real. ■

**Theorem 3.3.** *Any real symmetric $n \times n$ matrix has a set of $n$ eigenvectors that are mutually orthogonal.* □

*Proof.* We prove the result for the case when the $n$ eigenvalues are distinct. For a general proof, see Ref. 35 (p. 104).

Suppose $\boldsymbol{Av}_1 = \lambda_1 \boldsymbol{v}_1, \boldsymbol{Av}_2 = \lambda_2 \boldsymbol{v}_2$, where $\lambda_1 \neq \lambda_2$. Then,

$$\langle \boldsymbol{Av}_1, \boldsymbol{v}_2 \rangle = \langle \lambda_1 \boldsymbol{v}_1, \boldsymbol{v}_2 \rangle = \lambda_1 \langle \boldsymbol{v}_1, \boldsymbol{v}_2 \rangle.$$

Since $\boldsymbol{A} = \boldsymbol{A}^T$

$$\langle \boldsymbol{Av}_1, \boldsymbol{v}_2 \rangle = \langle \boldsymbol{v}_1, \boldsymbol{A}^T \boldsymbol{v}_2 \rangle = \langle \boldsymbol{v}_1, \boldsymbol{Av}_2 \rangle = \lambda_2 \langle \boldsymbol{v}_1, \boldsymbol{v}_2 \rangle.$$

Therefore

$$\lambda_1 \langle \boldsymbol{v}_1, \boldsymbol{v}_2 \rangle = \lambda_2 \langle \boldsymbol{v}_1, \boldsymbol{v}_2 \rangle.$$

Since $\lambda_1 \neq \lambda_2$, it follows that

$$\langle \boldsymbol{v}_1, \boldsymbol{v}_2 \rangle = 0.$$

■

If $A$ is symmetric, then a set of its eigenvectors forms an orthogonal basis for $\mathbb{R}^n$. If the basis $\{v_1, v_2, \ldots, v_n\}$ is normalized so that each element has a norm of unity, then defining the matrix

$$T = [v_1, v_2, \ldots, v_n],$$

we have

$$T^T T = I,$$

and hence

$$T^T = T^{-1}.$$

A matrix whose transpose is its inverse is said to be an *orthogonal* matrix.

## 3.3. ORTHOGONAL PROJECTIONS

We know that a subspace $\mathcal{V}$ of $\mathbb{R}^n$ is a subset that is closed under the operations of vector addition and scalar multiplication. In other words, $\mathcal{V}$ is a subspace of $\mathbb{R}^n$ if $x_1, x_2 \in \mathcal{V} \Rightarrow \alpha x_1 + \beta x_2 \in \mathcal{V}$ for all $\alpha$, $\beta \in \mathbb{R}$. Furthermore, the dimension of a subspace $\mathcal{V}$ is equal to the maximum number of linearly independent vectors in $\mathcal{V}$. If $\mathcal{V}$ is a subspace of $\mathbb{R}^n$, then the *orthogonal complement* of $\mathcal{V}$, denoted $\mathcal{V}^\perp$, consists of all vectors that are orthogonal to every vector in $\mathcal{V}$. Thus,

$$\mathcal{V}^\perp = \{x : v^T x = 0 \text{ for all } v \in \mathcal{V}\}.$$

The orthogonal complement of $\mathcal{V}$ is also a subspace (see Exercise 3.3). Together, $\mathcal{V}$ and $\mathcal{V}^\perp$ span $\mathbb{R}^n$ in the sense that every vector $x \in \mathbb{R}^n$ can be represented uniquely as

$$x = x_1 + x_2,$$

where $x_1 \in \mathcal{V}$ and $x_2 \in \mathcal{V}^\perp$. We call the above representation the *orthogonal decomposition* of $x$ (with respect to $\mathcal{V}$). We say that $x_1$ and $x_2$ are *orthogonal projections* of $x$ onto the subspaces $\mathcal{V}$ and $\mathcal{V}^\perp$, respectively. We write $\mathbb{R}^n = \mathcal{V} \oplus \mathcal{V}^\perp$, and say that $\mathbb{R}^n$ is a direct sum of $\mathcal{V}$ and $\mathcal{V}^\perp$. We say that a linear transformation $P$ is an orthogonal projector onto $\mathcal{V}$ if, for all $x \in \mathbb{R}^n$, we have $Px \in \mathcal{V}$ and $x - Px \in \mathcal{V}^\perp$.

In the subsequent discussion, we use the following notation. Let $A \in \mathbb{R}^{m \times n}$. Let the *range*, or *image*, of $A$ be denoted

$$\mathcal{R}(A) \triangleq \{Ax : x \in \mathbb{R}^n\},$$

and the *nullspace*, or *kernel*, of $A$ be denoted

$$\mathcal{N}(A) \triangleq \{x \in \mathbb{R}^n : Ax = 0\}.$$

Note that $\mathcal{R}(A)$ and $\mathcal{N}(A)$ are subspaces (see Exercise 3.4).

**Theorem 3.4.** *Let $A$ be a given matrix. Then $\mathscr{R}(A)^{\perp} = \mathscr{N}(A^T)$, and $\mathscr{N}(A)^{\perp} = \mathscr{R}(A^T)$.* □

*Proof.* Suppose $x \in \mathscr{R}(A)^{\perp}$. Then $y^T(A^T x) = (Ay)^T x = 0$ for all $y$, so that $A^T x = 0$. Hence, $x \in \mathscr{N}(A^T)$. This implies that $\mathscr{R}(A)^{\perp} \subset \mathscr{N}(A^T)$. If now $x \in \mathscr{N}(A^T)$, then $(Ay)^T x = y^T(A^T x) = 0$ for all $y$, so that $x \in \mathscr{R}(A)^{\perp}$, and consequently $\mathscr{N}(A^T) \subset \mathscr{R}(A)^{\perp}$. Thus $\mathscr{R}(A)^{\perp} = \mathscr{N}(A^T)$. The equation $\mathscr{N}(A)^{\perp} = \mathscr{R}(A^T)$ follows from what we have proved above, and the fact that for any subspace $\mathscr{V}$, we have $(\mathscr{V}^{\perp})^{\perp} = \mathscr{V}$ (see Exercise 3.6). ■

Theorem 3.4 allows us to establish the following necessary and sufficient condition for orthogonal projectors. For this, note that if $P$ is an orthogonal projector onto $\mathscr{V}$, then $Px = x$ for all $x \in \mathscr{V}$, and $\mathscr{R}(P) = \mathscr{V}$ (see Exercise 3.9).

**Theorem 3.5.** *A matrix $P$ is an orthogonal projector [onto the subspace $\mathscr{V} = \mathscr{R}(P)$] if, and only if, $P^2 = P = P^T$.*

*Proof.* $\Rightarrow$: Suppose $P$ is an orthogonal projector onto $\mathscr{V} = \mathscr{R}(P)$. Then, $\mathscr{R}(I - P) \subset \mathscr{R}(P)^{\perp}$. However, $\mathscr{R}(P)^{\perp} = \mathscr{N}(P^T)$ by Theorem 3.4. Therefore, $\mathscr{R}(I - P) \subset \mathscr{N}(P^T)$. Hence, $P^T(I - P)y = 0$ for all $y$, which implies that $P^T(I - P) = O$, where $O$ is the matrix with all entries equal to zero. Therefore, $P^T = P^T P$, and thus $P = P^T = P^2$.

$\Leftarrow$: Suppose $P^2 = P = P^T$. For any $x$, we have $(Py)^T(I - P)x = y^T P^T(I - P)x = y^T P(I - P)x = 0$ for all $y$. Thus, $(I - P)x \in \mathscr{R}(P)^{\perp}$, which means that $P$ is an orthogonal projector. ■

## 3.4. QUADRATIC FORMS

A *quadratic form* $f: \mathbb{R}^n \to \mathbb{R}$ is a function

$$f(x) = x^T Q x,$$

where $Q$ is an $n \times n$ real matrix. There is no loss of generality in assuming $Q$ to be symmetric, that is, $Q = Q^T$. For if the matrix $Q$ is not symmetric, we can always replace it with the symmetric matrix

$$Q_0 = Q_0^T = \tfrac{1}{2}(Q + Q^T).$$

Note that

$$x^T Q x = x^T Q_0 x = x^T(\tfrac{1}{2}Q + \tfrac{1}{2}Q^T)x.$$

A quadratic form $x^T Q x$, $Q = Q^T$, is said to be *positive definite* if $x^T Q x > 0$ for all nonzero vectors $x$. It is *positive semidefinite* if $x^T Q x \geqslant 0$ for all $x$. Similarly, we define the quadratic form to be *negative definite*, or *negative semidefinite*, if $x^T Q x < 0$ for all nonzero vectors $x$, or $x^T Q x \leqslant 0$ for all $x$, respectively.

Recall that the minors of a matrix $\boldsymbol{Q}$ are the determinants of the matrices obtained by successively removing rows and columns from $\boldsymbol{Q}$. The *principal minors* are det $\boldsymbol{Q}$ itself and the determinants of matrices obtained by successively removing an $i$th row and an $i$th column. That is, the principal minors are:

$$\det\begin{bmatrix} q_{i_1i_1} & q_{i_1i_2} & \cdots & q_{i_1i_p} \\ q_{i_2i_1} & q_{i_2i_2} & \cdots & q_{i_2i_p} \\ \vdots & \vdots & & \vdots \\ q_{i_pi_1} & q_{i_pi_2} & \cdots & q_{i_pi_p} \end{bmatrix}, \quad 1 \leqslant i_1 < i_2 < \cdots < i_p \leqslant n, p = 1, 2, \ldots, n.$$

The *leading principal minors* are det $\boldsymbol{Q}$ and the minors obtained by successively removing the last row and the last column. That is, the leading principal minors are

$$\Delta_1 = q_{11}, \quad \Delta_2 = \det\begin{bmatrix} q_{11} & q_{12} \\ q_{21} & q_{22} \end{bmatrix}, \quad \Delta_3 = \det\begin{bmatrix} q_{11} & q_{12} & q_{13} \\ q_{21} & q_{22} & q_{23} \\ q_{31} & q_{32} & q_{33} \end{bmatrix}, \ldots, \Delta_n = \det \boldsymbol{Q}.$$

We now prove *Sylvester's criterion*, which allows us to determine if a quadratic form $\boldsymbol{x}^T\boldsymbol{Q}\boldsymbol{x}$ is positive definite using only the leading principal minors of $\boldsymbol{Q}$.

**Theorem 3.6. Sylvester's Criterion.** *A quadratic form $\boldsymbol{x}^T\boldsymbol{Q}\boldsymbol{x}$, $\boldsymbol{Q} = \boldsymbol{Q}^T$, is positive definite if, and only if, the leading principal minors of $\boldsymbol{Q}$ are positive.* □

*Proof.* The key to the proof of Sylvester's criterion is the fact that a quadratic form whose leading principal minors are nonzero can be expressed in some basis as a sum of squares

$$\frac{\Delta_0}{\Delta_1}\tilde{x}_1^2 + \frac{\Delta_1}{\Delta_2}\tilde{x}_2^2 + \cdots + \frac{\Delta_{n-1}}{\Delta_n}\tilde{x}_n^2,$$

where $\tilde{x}_i$ are the coordinates of the vector $\boldsymbol{x}$ in the new basis, $\Delta_0 \triangleq 1$, and $\Delta_1, \ldots, \Delta_n$ are the leading principal minors of $\boldsymbol{Q}$.

To this end, consider a quadratic form $f(\boldsymbol{x}) = \boldsymbol{x}^T\boldsymbol{Q}\boldsymbol{x}$, where $\boldsymbol{Q} = \boldsymbol{Q}^T$. Let $\{\boldsymbol{e}_1, \boldsymbol{e}_2, \ldots, \boldsymbol{e}_n\}$ be the natural basis for $\mathbb{R}^n$, and let

$$\boldsymbol{x} = x_1\boldsymbol{e}_1 + x_2\boldsymbol{e}_2 + \cdots + x_n\boldsymbol{e}_n$$

be a given vector in $\mathbb{R}^n$. Let $\{\boldsymbol{f}_1, \boldsymbol{f}_2, \ldots, \boldsymbol{f}_n\}$ be another basis for $\mathbb{R}^n$. Then, the vector $\boldsymbol{x}$ is represented in the new basis as $\tilde{\boldsymbol{x}}$, where

$$\boldsymbol{x} = [\boldsymbol{f}_1, \boldsymbol{f}_2, \ldots, \boldsymbol{f}_n]\tilde{\boldsymbol{x}} \triangleq \boldsymbol{F}\tilde{\boldsymbol{x}}.$$

Accordingly, the quadratic form can be written as

$$\boldsymbol{x}^T\boldsymbol{Q}\boldsymbol{x} = \tilde{\boldsymbol{x}}^T\boldsymbol{F}^T\boldsymbol{Q}\boldsymbol{F}\tilde{\boldsymbol{x}} = \tilde{\boldsymbol{x}}^T\tilde{\boldsymbol{Q}}\tilde{\boldsymbol{x}},$$

where

$$\tilde{Q} = F^T Q F = \begin{bmatrix} \tilde{q}_{11} & \cdots & \tilde{q}_{1n} \\ \vdots & \ddots & \vdots \\ \tilde{q}_{n1} & \cdots & \tilde{q}_{nn} \end{bmatrix}.$$

Note that $\tilde{q}_{ij} = \langle f_i, Qf_j \rangle$. Our goal is to determine conditions on the new basis $\{f_1, f_2, \ldots, f_n\}$ such that $\tilde{q}_{ij} = 0$ for $i \neq j$.

We seek the new basis in the form

$$\begin{aligned} f_1 &= \alpha_{11} e_1 \\ f_2 &= \alpha_{21} e_1 + \alpha_{22} e_2 \\ &\vdots \\ f_n &= \alpha_{n1} e_1 + \alpha_{n2} e_2 + \cdots + \alpha_{nn} e_n. \end{aligned}$$

Observe that, for $j = 1, \ldots, i-1$, if

$$\langle f_i, Q e_j \rangle = 0,$$

then

$$\langle f_i, Q f_j \rangle = 0.$$

Our goal therefore is to determine the coefficients $\alpha_{i1}, \alpha_{i2}, \ldots, \alpha_{ii}$, $i = 1, \ldots, n$, such that the vector

$$f_i = \alpha_{i1} e_1 + \alpha_{i2} e_2 + \cdots + \alpha_{ii} e_i$$

satisfies the $i$ relations

$$\begin{aligned} \langle f_i, Q e_j \rangle &= 0, \quad j = 1, \ldots, i-1, \\ \langle e_i, Q f_i \rangle &= 1. \end{aligned}$$

In this case, we get

$$\tilde{Q} = \begin{bmatrix} \alpha_{11} & \cdots & 0 \\ \vdots & \ddots & \vdots \\ 0 & \cdots & \alpha_{nn} \end{bmatrix}.$$

For each $i = 1, \ldots, n$, the above $i$ relations determine the coefficients $\alpha_{i1}, \ldots, \alpha_{ii}$ in a unique way. Indeed, upon substituting the expression for $f_i$ into the above equations, we obtain the set of the equations

$$\begin{aligned} \alpha_{i1} q_{11} + \alpha_{i2} q_{12} + \cdots + \alpha_{ii} q_{ii} &= 0 \\ &\vdots \\ \alpha_{i-11} q_{i-1j} + \alpha_{i-12} q_{i-12} + \cdots + \alpha_{i-1i} q_{i-1i} &= 0 \\ \alpha_{i1} q_{i1} + \alpha_{i2} q_{i2} + \cdots + \alpha_{ii} q_{ii} &= 1. \end{aligned}$$

The above set of equations can be expressed in matrix form as

$$\begin{bmatrix} q_{11} & q_{12} & \cdots & q_{1i} \\ q_{21} & q_{22} & \cdots & q_{2i} \\ \vdots & \vdots & \ddots & \vdots \\ q_{i1} & q_{i2} & \cdots & q_{ii} \end{bmatrix} \begin{bmatrix} \alpha_{i1} \\ \alpha_{i2} \\ \vdots \\ \alpha_{ii} \end{bmatrix} = \begin{bmatrix} 0 \\ 0 \\ \vdots \\ 1 \end{bmatrix}.$$

If the leading principal minors of the matrix $\boldsymbol{Q}$ do not vanish, then the coefficients $\alpha_{ij}$ can be obtained using Cramer's rule. In particular

$$\alpha_{ii} = \frac{1}{\Delta_i} \det \begin{bmatrix} q_{11} & \cdots & q_{1i-1} & 0 \\ \vdots & \vdots & \vdots & 0 \\ q_{i-11} & \cdots & q_{i-1i-1} & 0 \\ q_{i1} & \cdots & q_{ii-1} & 1 \end{bmatrix} = \frac{\Delta_{i-1}}{\Delta_i}.$$

Hence,

$$\tilde{\boldsymbol{Q}} = \begin{bmatrix} \frac{1}{\Delta_1} & & & 0 \\ & \frac{\Delta_1}{\Delta_2} & & \\ & & \ddots & \\ 0 & & & \frac{\Delta_{n-1}}{\Delta_n} \end{bmatrix}.$$

In the new basis, the quadratic form can be expressed as a sum of squares

$$\boldsymbol{x}^T \boldsymbol{Q} \boldsymbol{x} = \tilde{\boldsymbol{x}}^T \tilde{\boldsymbol{Q}} \tilde{\boldsymbol{x}} = \frac{1}{\Delta_1} \tilde{x}_1^2 + \frac{\Delta_1}{\Delta_2} \tilde{x}_2^2 + \cdots + \frac{\Delta_{n-1}}{\Delta_n} \tilde{x}_n^2.$$

We now show that a necessary and sufficient condition for the quadratic form to be positive definite is $\Delta_i > 0$, $i = 1, \ldots, n$.

Sufficiency is clear, because, by the previous argument, if $\Delta_i > 0$, $i = 1, \ldots, n$, then there is a basis such that

$$\boldsymbol{x}^T \boldsymbol{Q} \boldsymbol{x} = \tilde{\boldsymbol{x}}^T \tilde{\boldsymbol{Q}} \tilde{\boldsymbol{x}} > 0$$

for any $\boldsymbol{x} \neq \boldsymbol{0}$ (or, equivalently, any $\tilde{\boldsymbol{x}} \neq \boldsymbol{0}$).

To prove necessity, we first show that for $i = 1, \ldots, n$, we have $\Delta_i \neq 0$. To see this, suppose that $\Delta_k = 0$ for some $k$. Note that $\Delta_k = \det \boldsymbol{Q}_k$,

$$\boldsymbol{Q}_k = \begin{bmatrix} q_{11} & \cdots & q_{1k} \\ \vdots & \ddots & \vdots \\ q_{k1} & \cdots & q_{kk} \end{bmatrix}.$$

Then, there exists a vector $\boldsymbol{v} \in \mathbb{R}^k$, $\boldsymbol{v} \neq \boldsymbol{0}$, such that $\boldsymbol{v}^T \boldsymbol{Q}_k = \boldsymbol{0}^T$. Now, let $\boldsymbol{x} \in \mathbb{R}^n$ be given by $\boldsymbol{x} = [\boldsymbol{v}^T, \boldsymbol{0}^T]^T$. Then,

$$\boldsymbol{x}^T \boldsymbol{Q} \boldsymbol{x} = \boldsymbol{v}^T \boldsymbol{Q}_k \boldsymbol{v} = 0.$$

But $x \neq 0$, which contradicts the fact that the quadratic form $f$ is positive definite. Therefore, if $x^T Q x > 0$, then $\Delta_i \neq 0$, $i = 1, \ldots, n$. Then, using our previous argument, we may write

$$x^T Q x = \tilde{x}^T \tilde{Q} \tilde{x} = \frac{1}{\Delta_1} \tilde{x}_1^2 + \frac{\Delta_1}{\Delta_2} \tilde{x}_2^2 + \cdots + \frac{\Delta_{n-1}}{\Delta_n} \tilde{x}_n^2,$$

where $x = [f_1, \ldots, f_n] \tilde{x}$. Hence, if the quadratic form is positive definite, then all leading principal minors must be positive. ■

Note that if $Q$ is not symmetric, Sylvester's criterion cannot be used to check positive definiteness of the quadratic form $x^T Q x$. To see this, consider an example where

$$Q = \begin{bmatrix} 1 & 0 \\ -4 & 1 \end{bmatrix}.$$

The leading principal minors of $Q$ are $\Delta_1 = 1 > 0$ and $\Delta_2 = \det Q = 1 > 0$. However, if $x = [1, 1]^T$, then $x^T Q x = -2 < 0$, and hence the associated quadratic form is not positive definite. Note that

$$\begin{aligned} x^T Q x = x^T \begin{bmatrix} 1 & 0 \\ -4 & 1 \end{bmatrix} x &= \frac{1}{2} x^T \left( \begin{bmatrix} 1 & 0 \\ -4 & 1 \end{bmatrix} + \begin{bmatrix} 1 & -4 \\ 0 & 1 \end{bmatrix} \right) x \\ &= x^T \begin{bmatrix} 1 & -2 \\ -2 & 1 \end{bmatrix} x = x^T Q_0 x. \end{aligned}$$

The leading principal minors of $Q_0$ are $\Delta_1 = 1 > 0$ and $\Delta_2 = \det Q_0 = -3 < 0$, as expected.

A necessary condition for a real quadratic form to be positive semidefinite is that the leading principal minors be nonnegative. However, this is not a sufficient condition (see Exercise 3.11). In fact, a real quadratic form is positive semidefinite if, and only if, all principal minors are nonnegative.

A symmetric matrix $Q$ is said to be positive definite if the quadratic form $x^T Q x$ is positive definite. If $Q$ is positive definite, we write $Q > 0$. Similarly, we define a symmetric matrix $Q$ to be positive semidefinite ($Q \geqslant 0$), negative definite ($Q < 0$), and negative semidefinite ($Q \leqslant 0$), if the corresponding quadratic forms have the respective properties. The symmetric matrix $Q$ is indefinite if it is neither positive semidefinite nor negative semidefinite. Note that the matrix $Q$ is positive definite (semidefinite) if, and only if, the matrix $-Q$ is negative definite (semidefinite).

Sylvester's criterion provides a way of checking the definiteness of a quadratic form, or equivalently a symmetric matrix. An alternative method involves checking the eigenvalues of $Q$, as stated below.

**Theorem 3.7.** *A symmetric matrix $\boldsymbol{Q}$ is positive definite (or positive semidefinite) if, and only if, all eigenvalues of $\boldsymbol{Q}$ are positive (or nonnegative).* □

*Proof.* For any $\boldsymbol{x}$, let $\boldsymbol{y} = \boldsymbol{T}^{-1}\boldsymbol{x} = \boldsymbol{T}^T\boldsymbol{x}$, where $\boldsymbol{T}$ is an orthogonal matrix whose columns are eigenvectors of $\boldsymbol{Q}$. Then, $\boldsymbol{x}^T\boldsymbol{Q}\boldsymbol{x} = \boldsymbol{y}^T\boldsymbol{T}^T\boldsymbol{Q}\boldsymbol{T}\boldsymbol{y} = \sum_{i=1}^n \lambda_i y_i^2$. From this, the result follows. ■

Through diagonalization, we can show that a symmetric positive semidefinite matrix $\boldsymbol{Q}$ has a positive semidefinite (symmetric) square root $\boldsymbol{Q}^{1/2}$ satisfying $\boldsymbol{Q}^{1/2}\boldsymbol{Q}^{1/2} = \boldsymbol{Q}$. For this, we use $\boldsymbol{T}$ as above and define

$$\boldsymbol{Q}^{1/2} = \boldsymbol{T}\begin{bmatrix} \lambda_1^{1/2} & & & 0 \\ & \lambda_2^{1/2} & & \\ & & \ddots & \\ 0 & & & \lambda_n^{1/2} \end{bmatrix}\boldsymbol{T}^T,$$

which is easily verified to have the desired properties. Note that the quadratic form $\boldsymbol{x}^T\boldsymbol{Q}\boldsymbol{x}$ can be expressed as $\|\boldsymbol{Q}^{1/2}\boldsymbol{x}\|^2$.

In summary, we have presented two tests for definiteness of quadratic forms and symmetric matrices. We point out again that nonnegativity of leading principal minors is a necessary but not a sufficient condition for positive semidefiniteness.

## 3.5. MATRIX NORMS

The norm of a matrix may be chosen in a variety of ways. Since the set of matrices $\mathbb{R}^{m\times n}$ can be viewed as the real vector space $\mathbb{R}^{mn}$, matrix norms should be no different from regular vector norms. Therefore, we define the norm of a matrix $\boldsymbol{A}$, denoted $\|\boldsymbol{A}\|$, to be any function $\|\cdot\|$ that satisfies the conditions:

1. $\|\boldsymbol{A}\| > 0$, if $\boldsymbol{A} \neq \boldsymbol{O}$, and $\|\boldsymbol{O}\| = 0$, where $\boldsymbol{O}$ is the matrix with all entries equal to zero
2. $\|c\boldsymbol{A}\| = |c|\,\|\boldsymbol{A}\|$, for any $c \in \mathbb{R}$
3. $\|\boldsymbol{A}+\boldsymbol{B}\| \leqslant \|\boldsymbol{A}\| + \|\boldsymbol{B}\|$.

An example of a matrix norm is the *Frobenius norm*, defined as

$$\|\boldsymbol{A}\|_F = \left(\sum_{i=1}^{m}\sum_{j=1}^{n}|a_{ij}|^2\right)^{1/2},$$

where $\boldsymbol{A} \in \mathbb{R}^{m\times n}$. Note that the Frobenius norm is equivalent to the Euclidean norm on $\mathbb{R}^{mn}$.

For our purposes, we consider only matrix norms that satisfy the following additional condition:

**4.** $\|AB\| \leqslant \|A\| \, \|B\|$.

It turns out that the Frobenius norm above satisfies condition 4 as well.

In many problems, both matrices and vectors appear simultaneously. Therefore, it is convenient to construct the norm of a matrix in such a way that it will be related with vector norms. To this end, we consider a special class of matrix norms, called *induced* norms. Let $\|\cdot\|_{(n)}$ and $\|\cdot\|_{(m)}$ be vector norms on $\mathbb{R}^n$ and $\mathbb{R}^m$, respectively. We say that the matrix norm is induced by, or is compatible with, the given vector norms if, for any matrix $A \in \mathbb{R}^{m \times n}$ and any vector $x \in \mathbb{R}^n$, the following inequality is satisfied:

$$\|Ax\|_{(m)} \leqslant \|A\| \, \|x\|_{(n)}.$$

We can define an induced matrix norm as

$$\|A\| = \max_{\|x\|_{(n)} = 1} \|Ax\|_{(m)};$$

that is, $\|A\|$ is the maximum of the norms of the vectors $Ax$ where the vector $x$ runs over the set of all vectors with unit norm. When there is no ambiguity, we omit the subscripts $(m)$ and $(n)$ from $\|\cdot\|_{(m)}$ and $\|\cdot\|_{(n)}$.

Because of the continuity of a vector norm (see Exercise 2.7), for each matrix $A$ the maximum

$$\max_{\|x\| = 1} \|Ax\|$$

is attainable; that is, a vector $x_0$ exists such that $\|x_0\| = 1$ and $\|Ax_0\| = \|A\|$. This fact follows from the Theorem of Weierstrass (see Theorem 4.2).

The induced norm satisfies conditions 1–4, and the compatibility condition, as we shall prove below.

*Proof of Condition 1.* Let $A \neq O$. Then, a vector $x$, $\|x\| = 1$, can be found such that $Ax \neq 0$, and thus $\|Ax\| \neq 0$. Hence, $\|A\| = \max_{\|x\|=1} \|Ax\| \neq 0$. If, on the other hand, $A = O$, then $\|A\| = \max_{\|x\|=1} \|Ox\| = 0$. ■

*Proof of Condition 2.* By definition, $\|cA\| = \max_{\|x\|=1} \|cAx\|$. Obviously $\|cAx\| = |c| \, \|Ax\|$, and therefore $\|cA\| = \max_{\|x\|=1} |c| \, \|Ax\| = |c| \max_{\|x\|=1} \|Ax\| = |c| \, \|A\|$. ■

*Proof of Compatibility Condition.* Let $y \neq 0$ be any vector. Then, $x = (1/\|y\|)y$ satisfies the condition $\|x\| = 1$. Consequently $\|Ay\| = \|A(\|y\|x)\| = \|y\| \, \|Ax\| \leqslant \|y\| \, \|A\|$. ■

*Proof of Condition 3.* For the matrix $\boldsymbol{A}+\boldsymbol{B}$, we can find a vector $\boldsymbol{x}_0$ such that $\|\boldsymbol{A}+\boldsymbol{B}\| = \|(\boldsymbol{A}+\boldsymbol{B})\boldsymbol{x}_0\|$ and $\|\boldsymbol{x}_0\| = 1$. Then, we have

$$\begin{aligned}\|\boldsymbol{A}+\boldsymbol{B}\| &= \|(\boldsymbol{A}+\boldsymbol{B})\boldsymbol{x}_0\| \\ &= \|\boldsymbol{A}\boldsymbol{x}_0 + \boldsymbol{B}\boldsymbol{x}_0\| \\ &\leqslant \|\boldsymbol{A}\boldsymbol{x}_0\| + \|\boldsymbol{B}\boldsymbol{x}_0\| \\ &\leqslant \|\boldsymbol{A}\|\,\|\boldsymbol{x}_0\| + \|\boldsymbol{B}\|\,\|\boldsymbol{x}_0\| \\ &= \|\boldsymbol{A}\| + \|\boldsymbol{B}\|.\end{aligned}$$

■

*Proof of Condition 4.* For the matrix $\boldsymbol{AB}$, we can find a vector $\boldsymbol{x}_0$ such that $\|\boldsymbol{x}_0\| = 1$ and $\|\boldsymbol{AB}\boldsymbol{x}_0\| = \|\boldsymbol{AB}\|$. Then, we have

$$\begin{aligned}\|\boldsymbol{AB}\| &= \|\boldsymbol{AB}\boldsymbol{x}_0\| \\ &= \|\boldsymbol{A}(\boldsymbol{B}\boldsymbol{x}_0)\| \\ &\leqslant \|\boldsymbol{A}\|\,\|\boldsymbol{B}\boldsymbol{x}_0\| \\ &\leqslant \|\boldsymbol{A}\|\,\|\boldsymbol{B}\|\,\|\boldsymbol{x}_0\| \\ &= \|\boldsymbol{A}\|\,\|\boldsymbol{B}\|.\end{aligned}$$

■

**Theorem 3.8.** *Let*

$$\|\boldsymbol{x}\| = \left(\sum_{k=1}^{n} |x_k|^2\right)^{1/2} = \sqrt{\langle \boldsymbol{x}, \boldsymbol{x}\rangle}.$$

*The matrix norm induced by this vector norm is*

$$\|\boldsymbol{A}\| = \sqrt{\lambda_1},$$

*where $\lambda_1$ is the largest eigenvalue of the matrix $\boldsymbol{A}^T\boldsymbol{A}$.* □

*Proof.* We have

$$\|\boldsymbol{A}\boldsymbol{x}\|^2 = \langle \boldsymbol{A}\boldsymbol{x}, \boldsymbol{A}\boldsymbol{x}\rangle = \langle \boldsymbol{x}, \boldsymbol{A}^T\boldsymbol{A}\boldsymbol{x}\rangle.$$

The matrix $\boldsymbol{A}^T\boldsymbol{A}$ is symmetric and positive semidefinite. Let $\lambda_1 \geqslant \lambda_2 \geqslant \cdots \geqslant \lambda_n \geqslant 0$ be its eigenvalues and $\boldsymbol{x}_1, \boldsymbol{x}_2, \ldots, \boldsymbol{x}_n$ the orthonormal set of the eigenvectors corresponding to these eigenvalues. Now, we take an arbitrary vector $\boldsymbol{x}$ with $\|\boldsymbol{x}\| = 1$ and represent it as a linear combination of $\boldsymbol{x}_i$, $i = 1, \ldots, n$; that is,

$$\boldsymbol{x} = c_1\boldsymbol{x}_1 + c_2\boldsymbol{x}_2 + \cdots + c_n\boldsymbol{x}_n.$$

Note that

$$\langle \boldsymbol{x}, \boldsymbol{x} \rangle = c_1^2 + c_2^2 + \cdots + c_n^2 = 1.$$

Furthermore

$$\begin{aligned}
\|\boldsymbol{A}\boldsymbol{x}\|^2 &= \langle \boldsymbol{x}, \boldsymbol{A}^T\boldsymbol{A}\boldsymbol{x} \rangle \\
&= \langle c_1\boldsymbol{x}_1 + \cdots + c_n\boldsymbol{x}_n, c_1\lambda_1\boldsymbol{x}_1 + \cdots + c_n\lambda_n\boldsymbol{x}_n \rangle \\
&= \lambda_1 c_1^2 + \cdots + \lambda_n c_n^2 \\
&\leqslant \lambda_1(c_1^2 + \cdots + c_n^2) \\
&= \lambda_1.
\end{aligned}$$

For the eigenvector $\boldsymbol{x}_1$ of $\boldsymbol{A}^T\boldsymbol{A}$ corresponding to the eigenvalue $\lambda_1$, we have

$$\|\boldsymbol{A}\boldsymbol{x}_1\|^2 = \langle \boldsymbol{x}_1, \boldsymbol{A}^T\boldsymbol{A}\boldsymbol{x}_1 \rangle = \langle \boldsymbol{x}_1, \lambda_1\boldsymbol{x}_1 \rangle = \lambda_1,$$

and hence

$$\max_{\|\boldsymbol{x}\|=1} \|\boldsymbol{A}\boldsymbol{x}\| = \sqrt{\lambda_1}.$$

■

Using arguments similar to the above, we can deduce the following important inequality.

***Rayleigh's Inequality.*** If an $n \times n$ matrix $\boldsymbol{P}$ is real symmetric positive definite, then

$$\lambda_{\min}(\boldsymbol{P})\|\boldsymbol{x}\|^2 \leqslant \boldsymbol{x}^T\boldsymbol{P}\boldsymbol{x} \leqslant \lambda_{\max}(\boldsymbol{P})\|\boldsymbol{x}\|^2,$$

where $\lambda_{\min}(\boldsymbol{P})$ denotes the smallest eigenvalue of $\boldsymbol{P}$, and $\lambda_{\max}(\boldsymbol{P})$ denotes the largest eigenvalue of $\boldsymbol{P}$. ■

***Example 3.1.*** Consider the matrix

$$\boldsymbol{A} = \begin{bmatrix} 2 & 1 \\ 1 & 2 \end{bmatrix},$$

and let the norm in $\mathbb{R}^2$ be given by

$$\|\boldsymbol{x}\| = \sqrt{x_1^2 + x_2^2}.$$

Then

$$\boldsymbol{A}^T\boldsymbol{A} = \begin{bmatrix} 5 & 4 \\ 4 & 5 \end{bmatrix},$$

and $\det[\lambda\boldsymbol{I}_2 - \boldsymbol{A}^T\boldsymbol{A}] = \lambda^2 - 10\lambda + 9 = (\lambda - 1)(\lambda - 9)$. Thus, $\|\boldsymbol{A}\| = \sqrt{9} = 3$.

The eigenvector of $A^TA$ corresponding to $\lambda_1 = 9$ is

$$x_1 = \frac{1}{\sqrt{2}}\begin{bmatrix}1\\1\end{bmatrix}.$$

Note that $\|Ax_1\| = \|A\|$. Indeed,

$$\begin{aligned}\|Ax_1\| &= \left\|\frac{1}{\sqrt{2}}\begin{bmatrix}2 & 1\\1 & 2\end{bmatrix}\begin{bmatrix}1\\1\end{bmatrix}\right\| \\ &= \frac{1}{\sqrt{2}}\left\|\begin{bmatrix}3\\3\end{bmatrix}\right\| \\ &= \frac{1}{\sqrt{2}}\sqrt{3^2+3^2} \\ &= 3.\end{aligned}$$

Since $A = A^T$ in this example, we also have $\|A\| = \max_{1 \leqslant i \leqslant n}|\lambda_i(A)|$, where $\lambda_1(A), \ldots, \lambda_n(A)$ are the eigenvalues of $A$ (possibly repeated). ■

*Warning.* In general, $\max_{1 \leqslant i \leqslant n}|\lambda_i(A)| \neq \|A\|$. Instead, we have $\|A\| \geqslant \max_{1 \leqslant i \leqslant n}|\lambda_i(A)|$, as illustrated in the following example (see also Exercise 3.18).

***Example 3.2.*** Let

$$A = \begin{bmatrix}0 & 1\\0 & 0\end{bmatrix},$$

then

$$A^TA = \begin{bmatrix}0 & 0\\0 & 1\end{bmatrix},$$

and

$$\det[\lambda I_2 - A^TA] = \det\begin{bmatrix}\lambda & 0\\0 & \lambda - 1\end{bmatrix} = \lambda(\lambda - 1).$$

Note that 0 is the only eigenvalue of $A$. Thus, for $i = 1, 2$,

$$\|A\| = 1 > |\lambda_i(A)| = 0.$$

■

We end this chapter with a result that will be of use in a later discussion.

**Theorem 3.9.** *Let* $A \in \mathbb{R}^{n \times n}$. *Then,* $\lim_{k\to\infty} A^k = O$ *if, and only if,* $|\lambda_i(A)| < 1$, $i = 1, \ldots, n$. □

*Proof.* To prove this theorem, we use the *Jordan form* [see, for example, Ref. 25]. Specifically, it is well known that any square matrix is similar to the Jordan form; that is, there exists a nonsingular $\boldsymbol{T}$ such that

$$\boldsymbol{T}\boldsymbol{A}\boldsymbol{T}^{-1} = \operatorname{diag}[\boldsymbol{J}_{m_1}(\lambda_1), \ldots, \boldsymbol{J}_{m_s}(\lambda_1), \boldsymbol{J}_{n_1}(\lambda_2), \ldots, \boldsymbol{J}_{t_v}(\lambda_q)] \triangleq \boldsymbol{J},$$

where $\boldsymbol{J}_r(\lambda)$ is the $r \times r$ matrix:

$$\boldsymbol{J}_r(\lambda) = \begin{bmatrix} \lambda & 1 & & 0 \\ & \lambda & \ddots & \\ & & \ddots & 1 \\ & & & \lambda \end{bmatrix}.$$

The $\lambda_1, \ldots, \lambda_q$ above are distinct eigenvalues of $\boldsymbol{A}$, the multiplicity of $\lambda_1$ is $m_1 + \cdots + m_s$, and so on.

We may rewrite the above as $\boldsymbol{A} = \boldsymbol{T}^{-1}\boldsymbol{J}\boldsymbol{T}$. To complete the proof observe that

$$(\boldsymbol{J}_r(\lambda))^k = \begin{bmatrix} \lambda^k & \binom{k}{1}\lambda^{k-1} & \cdots & \binom{k}{k-1}\lambda^{k-r+1} \\ 0 & \lambda^k & \cdots & \binom{k}{k-2}\lambda^{k-r+1} \\ \vdots & \vdots & \ddots & \vdots \\ 0 & 0 & \cdots & \lambda^k \end{bmatrix},$$

where

$$\binom{k}{i} = \frac{k!}{i!(k-i)!}.$$

Furthermore

$$\boldsymbol{A}^k = \boldsymbol{T}^{-1}\boldsymbol{J}^k\boldsymbol{T}.$$

Hence

$$\lim_{k\to\infty} \boldsymbol{A}^k = \boldsymbol{T}^{-1}\left(\lim_{k\to\infty} \boldsymbol{J}^k\right)\boldsymbol{T} = \boldsymbol{O}$$

if, and only if, $|\lambda_i| < 1$, $i = 1, \ldots, n$. ■

For a more complete but still basic treatment of topics in linear algebra as discussed in this and the previous chapter, see Refs. 25, 38, 55, 74. For a treatment of matrices, we refer the reader to Refs. 23 and 35. Numerical aspects of matrix computations are discussed in Refs. 20 and 29.

## EXERCISES

**3.1** Fix a vector in $\mathbb{R}^n$, and let $\boldsymbol{x}$ be the column of the coordinates of the vector with respect to the basis $\{\boldsymbol{e}_1, \boldsymbol{e}_2, \ldots, \boldsymbol{e}_n\}$, and $\boldsymbol{x}'$ the coordinates of the same vector with respect to the basis $\{\boldsymbol{e}'_1, \boldsymbol{e}'_2, \ldots, \boldsymbol{e}'_n\}$. Show that $\boldsymbol{x}' = \boldsymbol{T}\boldsymbol{x}$, where $\boldsymbol{T}$ is the transformation matrix from $\{\boldsymbol{e}_1, \boldsymbol{e}_2, \ldots, \boldsymbol{e}_n\}$ to $\{\boldsymbol{e}'_1, \boldsymbol{e}'_2, \ldots, \boldsymbol{e}'_n\}$.

**3.2** Let $\lambda_1, \ldots, \lambda_n$ be the eigenvalues of the matrix $\boldsymbol{A} \in \mathbb{R}^{n \times n}$. Show that the eigenvalues of the matrix $\boldsymbol{I}_n - \boldsymbol{A}$ are $1 - \lambda_1, \ldots, 1 - \lambda_n$.

**3.3** Let $\mathscr{V}$ be a subspace. Show that $\mathscr{V}^\perp$ is also a subspace.

**3.4** Let $\boldsymbol{A} \in \mathbb{R}^{m \times n}$ be a matrix. Show that $\mathscr{R}(\boldsymbol{A})$ is a subspace of $\mathbb{R}^m$ and $\mathscr{N}(\boldsymbol{A})$ is a subspace of $\mathbb{R}^n$.

**3.5** Prove that if $\boldsymbol{A}$ and $\boldsymbol{B}$ are two matrices with $m$ rows, and $\mathscr{N}(\boldsymbol{A}^T) \subset \mathscr{N}(\boldsymbol{B}^T)$, then $\mathscr{R}(\boldsymbol{B}) \subset \mathscr{R}(\boldsymbol{A})$. *Hint*: Use the fact that for any matrix $\boldsymbol{M}$ with $m$ rows, we have $\dim \mathscr{R}(\boldsymbol{M}) + \dim \mathscr{N}(\boldsymbol{M}^T) = m$ (this is one of the fundamental theorems of linear algebra [see Ref. 74, p.75]).

**3.6** Let $\mathscr{V}$ be a subspace. Show that $(\mathscr{V}^\perp)^\perp = \mathscr{V}$. *Hint*: Use Exercise 3.5.

**3.7** Let $\mathscr{V}$ and $\mathscr{W}$ be subspaces. Show that if $\mathscr{V} \subset \mathscr{W}$, then $\mathscr{W}^\perp \subset \mathscr{V}^\perp$.

**3.8** Let $\mathscr{V}$ be a subspace of $\mathbb{R}^n$. Show that there exist matrices $\boldsymbol{V}$ and $\boldsymbol{U}$ such that $\mathscr{V} = \mathscr{R}(\boldsymbol{V}) = \mathscr{N}(\boldsymbol{U})$.

**3.9** Let $\boldsymbol{P}$ be an orthogonal projector onto a subspace $\mathscr{V}$. Show that

**a.** $\boldsymbol{P}\boldsymbol{x} = \boldsymbol{x}$ for $\boldsymbol{x} \in \mathscr{V}$

**b.** $\mathscr{R}(\boldsymbol{P}) = \mathscr{V}$.

**3.10** Is the quadratic form

$$\boldsymbol{x}^T \begin{bmatrix} 1 & -8 \\ 1 & 1 \end{bmatrix} \boldsymbol{x}$$

positive definite, positive semidefinite, negative definite, negative semidefinite, or indefinite?

**3.11** Let

$$\boldsymbol{A} = \begin{bmatrix} 2 & 2 & 2 \\ 2 & 2 & 2 \\ 2 & 2 & 0 \end{bmatrix}.$$

Show that although all leading principal minors of $\boldsymbol{A}$ are nonnegative, $\boldsymbol{A}$ is not positive semidefinite.

**3.12** Consider the matrix

$$Q=\begin{bmatrix} 0 & 1 & 1 \\ 1 & 0 & 1 \\ 1 & 1 & 0 \end{bmatrix}.$$

**a.** Is this matrix positive definite, negative definite, or indefinite?

**b.** Is this matrix positive definite, negative definite, or indefinite on the subspace

$$\mathcal{M}=\{\boldsymbol{x}:x_1+x_2+x_3=0\}?$$

**3.13** Consider the quadratic form

$$f(x_1,x_2,x_3)=x_1^2+x_2^2+5x_3^2+2\xi x_1x_2-2x_1x_3+4x_2x_3.$$

Find the values of the parameter $\xi$ for which this quadratic form is positive definite.

**3.14** Consider the function $\langle\cdot,\cdot\rangle_Q:\mathbb{R}^n\times\mathbb{R}^n\to\mathbb{R}$, defined by $\langle\boldsymbol{x},\boldsymbol{y}\rangle_Q=\boldsymbol{x}^T\boldsymbol{Q}\boldsymbol{y}$, where $\boldsymbol{x},\boldsymbol{y}\in\mathbb{R}^n$ and $\boldsymbol{Q}\in\mathbb{R}^{n\times n}$ is a symmetric positive definite matrix. Show that $\langle\cdot,\cdot\rangle_Q$ satisfies conditions 1 to 4 for inner products (see Section 2.4).

**3.15** Consider the vector norm $\|\cdot\|_\infty$ on $\mathbb{R}^n$ given by $\|\boldsymbol{x}\|_\infty=\max_i|x_i|$, where $\boldsymbol{x}=[x_1,\ldots,x_n]^T$. Similarly define the norm $\|\cdot\|_\infty$ on $\mathrm{R}^m$. Show that the matrix norm induced by these vector norms is given by

$$\|\boldsymbol{A}\|_\infty=\max_i\sum_{k=1}^{n}|a_{ik}|$$

where $a_{ij}$ is the $(i,j)$th element of $\boldsymbol{A}\in\mathbb{R}^{m\times n}$.

**3.16** Consider the vector norm $\|\cdot\|_1$ on $\mathbb{R}^n$ given by $\|\boldsymbol{x}\|_1=\sum_{i=1}^n|x_i|$, where $\boldsymbol{x}=[x_1,\ldots,x_n]^T$. Similarly define the norm $\|\cdot\|_1$ on $\mathbb{R}^m$. Show that the matrix norm induced by these vector norms is given by

$$\|\boldsymbol{A}\|_1=\max_k\sum_{i=1}^{m}|a_{ik}|$$

where $a_{ij}$ is the $(i,j)$th element of $\boldsymbol{A}\in\mathbb{R}^{m\times n}$.

**3.17** Show that a sufficient condition for $\lim_{k\to\infty}\boldsymbol{A}^k=\boldsymbol{O}$ is $\|\boldsymbol{A}\|<1$.

**3.18** Show that for any matrix $\boldsymbol{A}\in\mathbb{R}^{n\times n}$,

$$\|\boldsymbol{A}\|\geqslant\max_{1\leqslant i\leqslant n}|\lambda_i(\boldsymbol{A})|.$$

*Hint*: Use Exercise 3.17.

# 4

# Concepts from Geometry

## 4.1. LINE SEGMENTS

In the following analysis, we concern ourselves only with $\mathbb{R}^n$. The elements of this space are the $n$-component vectors $\boldsymbol{x} = [x_1, x_2, \ldots, x_n]^T$. The *line segment* between two points $\boldsymbol{x}$ and $\boldsymbol{y}$ in $\mathbb{R}^n$ is the set of points on the straight line joining the points $\boldsymbol{x}$ and $\boldsymbol{y}$ (see Figure 4.1). Note that if $\boldsymbol{z}$ lies on the line segment, then

$$\boldsymbol{z} - \boldsymbol{y} = \alpha(\boldsymbol{x} - \boldsymbol{y}),$$

where $\alpha$ is a real number from the interval $[0, 1]$. The above equation can be rewritten as $\boldsymbol{z} = \alpha\boldsymbol{x} + (1 - \alpha)\boldsymbol{y}$. Hence, the line segment between $\boldsymbol{x}$ and $\boldsymbol{y}$ can be represented as

$$\{\alpha\boldsymbol{x} + (1 - \alpha)\boldsymbol{y} : \alpha \in [0, 1]\}.$$

## 4.2. HYPERPLANES AND LINEAR VARIETIES

Let $u_1, u_2, \ldots, u_n, v \in \mathbb{R}$, where at least one of the $u_i$ is nonzero. The set of all points $\boldsymbol{x} = [x_1, x_2, \ldots, x_n]^T$ that satisfy the linear equation

$$u_1 x_1 + u_2 x_2 + \cdots + u_n x_n = v$$

is called a *hyperplane* of the space $\mathbb{R}^n$. We may describe the hyperplane by

$$\{\boldsymbol{x} \in \mathbb{R}^n : \boldsymbol{u}^T \boldsymbol{x} = v\},$$

where

$$\boldsymbol{u} = [u_1, u_2, \ldots, u_n]^T.$$

A hyperplane is not necessarily a subspace of $\mathbb{R}^n$ since, in general, it does not contain the origin. For $n = 2$, the equation of the hyperplane has the form $u_1 x_1 + u_2 x_2 = v$, which is the equation of a straight line. Thus, straight lines are hyperplanes in $\mathbb{R}^2$. In $\mathbb{R}^3$ (three-dimensional space), hyperplanes are ordinary

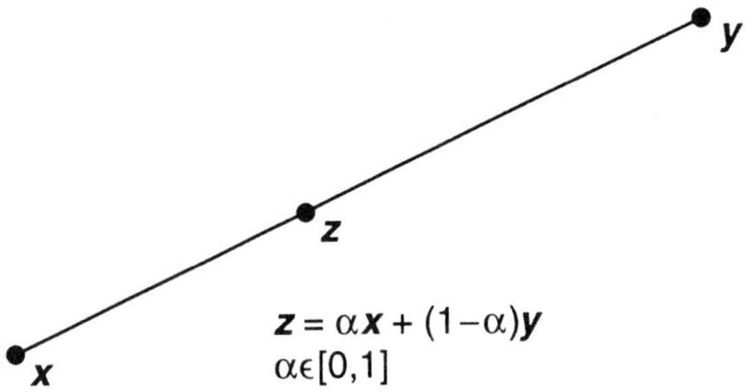


**Figure 4.1.** A line segment

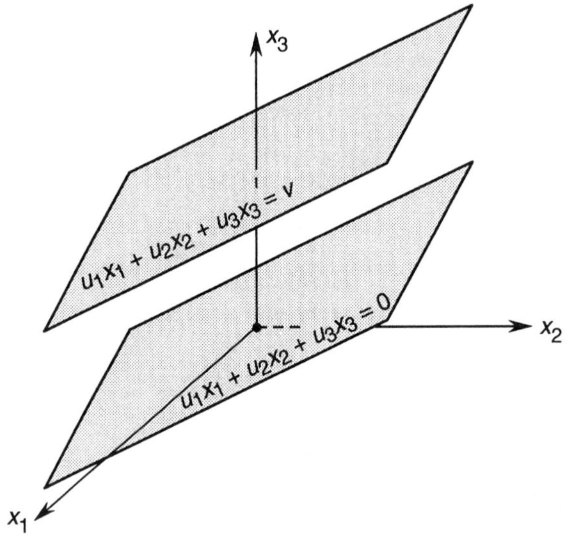


**Figure 4.2.** Translation of a hyperplane

planes. By translating a hyperplane so that it contains the origin of $\mathbb{R}^n$, it becomes a subspace of $\mathbb{R}^n$ (see Figure 4.2). Since the dimension of this subspace is $n-1$, we say that the hyperplane has dimension $n-1$.

The hyperplane $H = \{x: u_1x_1 + \cdots + u_nx_n = v\}$ divides $\mathbb{R}^n$ into two half-spaces. One of these half-spaces consists of the points satisfying the inequality $u_1x_1 + u_2x_2 + \cdots + u_nx_n \geqslant v$, denoted

$$H_+ = \{\boldsymbol{x} \in \mathbb{R}^n : \boldsymbol{u}^T\boldsymbol{x} \geqslant v\},$$

where, as before,

$$\boldsymbol{u} = [u_1, u_2, \ldots, u_n]^T.$$

The other half-space consists of the points satisfying the inequality $u_1x_1 + u_2x_2 + \cdots + u_nx_n \leqslant v$, denoted

$$H_- = \{\boldsymbol{x}\in\mathbb{R}^n : \boldsymbol{u}^T\boldsymbol{x} \leqslant v\}.$$

The half-space $H_+$ is called the *positive half-space*, and the half-space $H_-$ is called the *negative half-space*.

Let $\boldsymbol{a} = [a_1, a_2, \ldots, a_n]^T$ be an arbitrary point of the hyperplane $H$. Thus $\boldsymbol{u}^T\boldsymbol{a} - v = 0$. We can write

$$\begin{aligned} \boldsymbol{u}^T\boldsymbol{x} - v &= \boldsymbol{u}^T\boldsymbol{x} - v - (\boldsymbol{u}^T\boldsymbol{a} - v) \\ &= \boldsymbol{u}^T(\boldsymbol{x} - \boldsymbol{a}) \\ &= u_1(x_1 - a_1) + u_2(x_2 - a_2) + \cdots + u_n(x_n - a_n) = 0. \end{aligned}$$

The numbers $(x_i - a_i)$, $i = 1, \ldots, n$, are the components of the vector $\boldsymbol{x} - \boldsymbol{a}$. Therefore, the hyperplane $H$ consists of the points $\boldsymbol{x}$ for which $\langle \boldsymbol{u}, \boldsymbol{x} - \boldsymbol{a} \rangle = 0$. In other words, the hyperplane $H$ consists of the points $\boldsymbol{x}$ for which the vectors $\boldsymbol{u}$ and $\boldsymbol{x} - \boldsymbol{a}$ are orthogonal (see Figure 4.3). We call the vector $\boldsymbol{u}$ the *normal* to the hyperplane $H$. The set $H_+$ consists of those points $\boldsymbol{x}$ for which $\langle \boldsymbol{u}, \boldsymbol{x} - \boldsymbol{a} \rangle \geqslant 0$, and $H_-$ consists of those points $\boldsymbol{x}$ for which $\langle \boldsymbol{u}, \boldsymbol{x} - \boldsymbol{a} \rangle \leqslant 0$.

A *linear variety* is a set of the form $\{\boldsymbol{x} \in \mathbb{R}^n : \boldsymbol{A}\boldsymbol{x} = \boldsymbol{b}\}$ for some matrix $\boldsymbol{A} \in \mathbb{R}^{m\times n}$ and vector $\boldsymbol{b} \in \mathbb{R}^n$. If $\dim \mathcal{N}(\boldsymbol{A}) = r$, we say that the linear variety has dimension $r$. A linear variety is a subspace if and only if $\boldsymbol{b} = \boldsymbol{0}$. If $\boldsymbol{A} = \boldsymbol{0}$, the linear variety is $\mathbb{R}^n$. If the dimension of the linear variety is less than $n$, then it is the intersection of a finite number of hyperplanes.

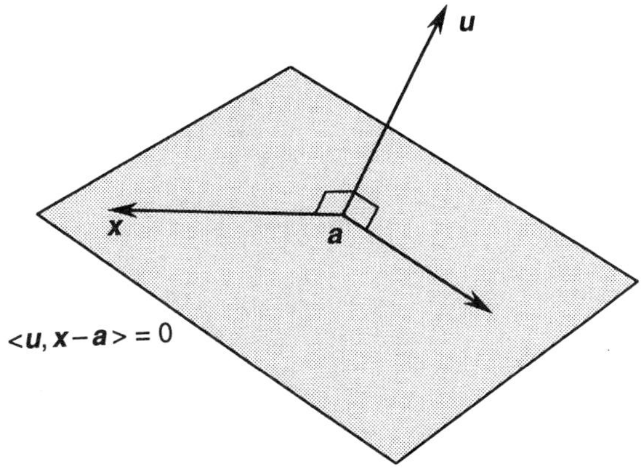


**Figure 4.3.** The hyperplane $H = \{\boldsymbol{x}\in\mathbb{R}^n : \boldsymbol{u}^T(\boldsymbol{x} - \boldsymbol{a}) = 0\}$

## 4.3. CONVEX SETS

Recall that the line segment between two points $\boldsymbol{u}, \boldsymbol{v} \in \mathbb{R}^n$ is the set $\{\boldsymbol{w} \in \mathbb{R}^n : \boldsymbol{w} = \alpha \boldsymbol{u} + (1-\alpha)\boldsymbol{v}, \alpha \in [0,1]\}$.

A set $\Theta \subset \mathbb{R}^n$ is *convex* if, for all $\boldsymbol{u}, \boldsymbol{v} \in \Theta$, the line segment between $\boldsymbol{u}$ and $\boldsymbol{v}$ lies in $\Theta$. Figure 4.4 gives examples of convex sets, whereas Figure 4.5 gives examples of sets that are not convex. Note that $\Theta$ is convex if, and only if, $\alpha \boldsymbol{u} + (1-\alpha)\boldsymbol{v} \in \Theta$ for all $\boldsymbol{u}, \boldsymbol{v} \in \Theta$ and $\alpha \in (0,1)$.

Examples of convex sets include

The empty set
A set consisting of a single point
A line or a line segment
A subspace
A hyperplane
A linear variety
A half-space
$\mathbb{R}^n$

**Theorem 4.1.** *Convex subsets of $\mathbb{R}^n$ have the following properties:*

**a.** *If $\Theta$ is a convex set and $\beta$ is a real number, then the set*

$$\beta\Theta = \{\boldsymbol{x} : \boldsymbol{x} = \beta \boldsymbol{v}, \boldsymbol{v} \in \Theta\}$$

*is also convex.*

**b.** *If $\Theta_1$ and $\Theta_2$ are convex sets, then the set*

$$\Theta_1 + \Theta_2 = \{\boldsymbol{x} : \boldsymbol{x} = \boldsymbol{v}_1 + \boldsymbol{v}_2, \boldsymbol{v}_1 \in \Theta_1, \boldsymbol{v}_2 \in \Theta_2\}$$

*is also convex.*

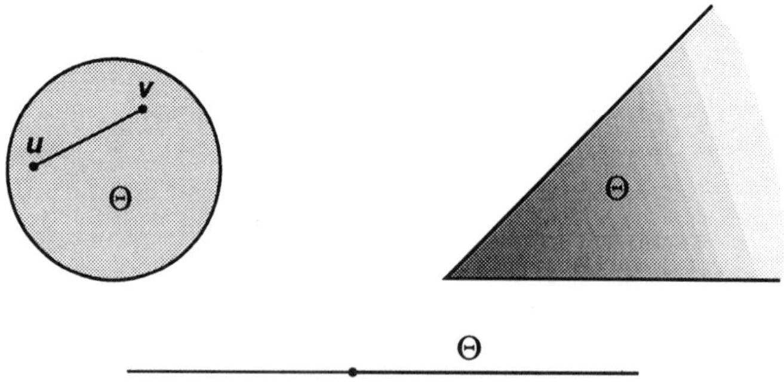


**Figure 4.4.** Convex sets

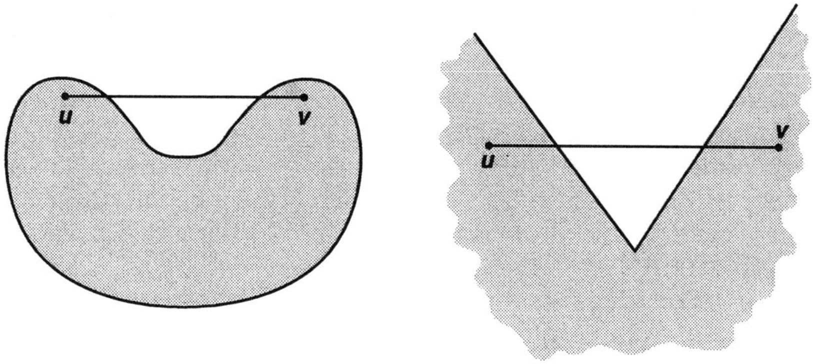


**Figure 4.5.** Sets that are not convex

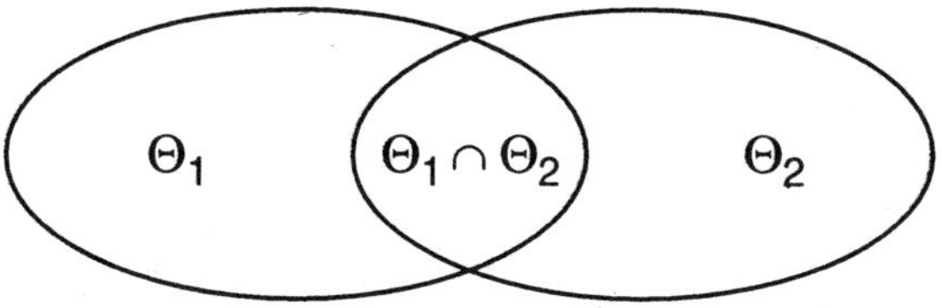


**Figure 4.6.** Intersection of two convex sets

**c.** *The intersection of any collection of convex sets is convex (see Figure 4.6 for an illustration of this result for two sets).* □

*Proof.*

**a.** Let $\beta \boldsymbol{v}_1, \beta \boldsymbol{v}_2 \in \beta\Theta$, where $\boldsymbol{v}_1, \boldsymbol{v}_2 \in \Theta$. Since $\Theta$ is convex, then $\alpha \boldsymbol{v}_1 + (1-\alpha)\boldsymbol{v}_2 \in \Theta$ for any $\alpha \in (0, 1)$. Hence,

$$\alpha\beta \boldsymbol{v}_1 + (1-\alpha)\beta \boldsymbol{v}_2 = \beta(\alpha \boldsymbol{v}_1 + (1-\alpha)\boldsymbol{v}_2) \in \beta\Theta,$$

and thus $\beta\Theta$ is convex.

**b.** Let $\boldsymbol{v}_1, \boldsymbol{v}_2 \in \Theta_1 + \Theta_2$. Then, $\boldsymbol{v}_1 = \boldsymbol{v}_1' + \boldsymbol{v}_1''$, and $\boldsymbol{v}_2 = \boldsymbol{v}_2' + \boldsymbol{v}_2''$, where $\boldsymbol{v}_1', \boldsymbol{v}_2' \in \Theta_1$, and $\boldsymbol{v}_1'', \boldsymbol{v}_2'' \in \Theta_2$. Since $\Theta_1$ and $\Theta_2$ are convex, then for all $\alpha \in (0, 1)$,

$$\boldsymbol{x}_1 = \alpha \boldsymbol{v}_1' + (1-\alpha)\boldsymbol{v}_2' \in \Theta_1$$

and

$$\boldsymbol{x}_2 = \alpha \boldsymbol{v}_1'' + (1-\alpha)\boldsymbol{v}_2'' \in \Theta_2.$$

By definition of $\Theta_1 + \Theta_2$, $\boldsymbol{x}_1 + \boldsymbol{x}_2 \in \Theta_1 + \Theta_2$. Now,

$$\begin{aligned}\alpha \boldsymbol{v}_1 + (1-\alpha)\boldsymbol{v}_2 &= \alpha(\boldsymbol{v}_1' + \boldsymbol{v}_1'') + (1-\alpha)(\boldsymbol{v}_2' + \boldsymbol{v}_2'') \\ &= \boldsymbol{x}_1 + \boldsymbol{x}_2 \in \Theta_1 + \Theta_2.\end{aligned}$$

Hence, $\Theta_1 + \Theta_2$ is convex.

**c.** Let $\boldsymbol{x}_1, \boldsymbol{x}_2 \in \bigcap_i \Theta_i$. Then, $\boldsymbol{x}_1, \boldsymbol{x}_2 \in \Theta_i$ for each $i$. Since each $\Theta_i$ is convex, $\alpha \boldsymbol{x}_1 + (1-\alpha)\boldsymbol{x}_2 \in \Theta_i$ for all $\alpha \in (0,1)$ and each $i$. Thus, $\alpha \boldsymbol{x}_1 + (1-\alpha)\boldsymbol{x}_2 \in \bigcap_i \Theta_i$. ■

A point $\boldsymbol{x}$ in a convex set $\Theta$ is said to be an *extreme point* of $\Theta$ if there are no two distinct points $\boldsymbol{u}$ and $\boldsymbol{v}$ in $\Theta$ such that $\boldsymbol{x} = \alpha\boldsymbol{u} + (1-\alpha)\boldsymbol{v}$ for some $\alpha \in (0,1)$. For example, in Figure 4.4, any point on the boundary of the disk is an extreme point, the vertex (corner) of the set on the right is an extreme point, and the endpoint of the half-line is also an extreme point.

## 4.4. NEIGHBORHOODS

A *neighborhood* of a point $\boldsymbol{x} \in \mathbb{R}^n$ is the set

$$\{\boldsymbol{y} \in \mathbb{R}^n : \|\boldsymbol{y} - \boldsymbol{x}\| < \varepsilon\}$$

where $\varepsilon$ is some positive number. The neighborhood is also called the *ball* with radius $\varepsilon$ and center $\boldsymbol{x}$.

In the plane $\mathbb{R}^2$, a neighborhood of $\boldsymbol{x} = [x_1, x_2]^T$ consists of all the points inside of a disc centered at $\boldsymbol{x}$. In $\mathbb{R}^3$, a neighborhood of $\boldsymbol{x} = [x_1, x_2, x_3]^T$ consists of all the points inside of a sphere centered at $\boldsymbol{x}$ (see Figure 4.7).

A point $\boldsymbol{x} \in S$ is said to be an *interior point* of the set $S$ if the set $S$ contains some neighborhood of $\boldsymbol{x}$, that is, if all points within some neighborhood of $\boldsymbol{x}$ are also in $S$ (see Figure 4.8). The set of all the interior points of $S$ is called the *interior* of $S$.

A point $\boldsymbol{x}$ is said to be a *boundary point* of the set $S$ if every neighborhood of $\boldsymbol{x}$ contains a point in $S$ and a point not in $S$ (see Figure 4.8). Note that a boundary point of $S$ may or may not be an element of $S$. The set of all boundary points of $S$ is called the *boundary* of $S$.

A set $S$ is said to be *open* if it contains a neighborhood of each of its points, that is, if each of its points is an interior point, or equivalently, if $S$ contains no boundary points. A set $S$ is said to be *closed* if it contains its boundary (see Figure 4.9). We can show that a set is closed if, and only if, its complement is open. A set that is contained in a ball of finite radius is called a *bounded set*. A set is *compact* if

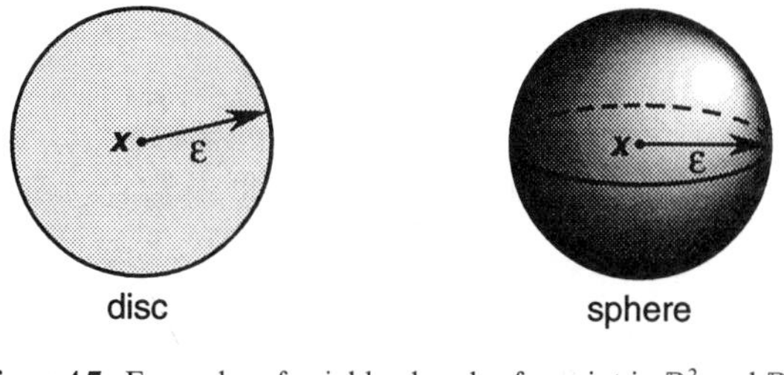


**Figure 4.7.** Examples of neighborhoods of a point in $\mathbb{R}^2$ and $\mathbb{R}^3$

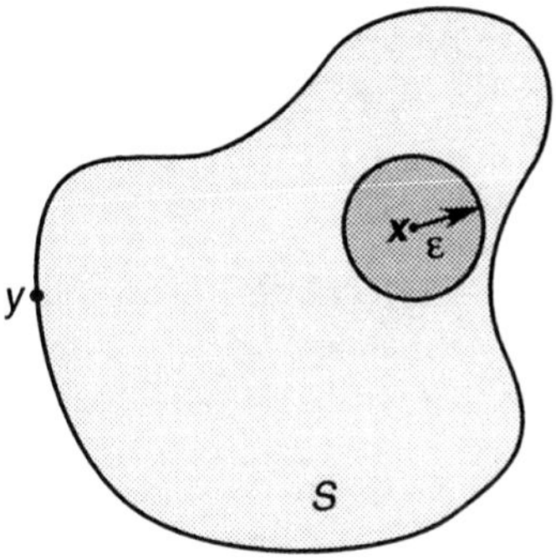


**Figure 4.8.** $x$ is an interior point, while $y$ is a boundary point

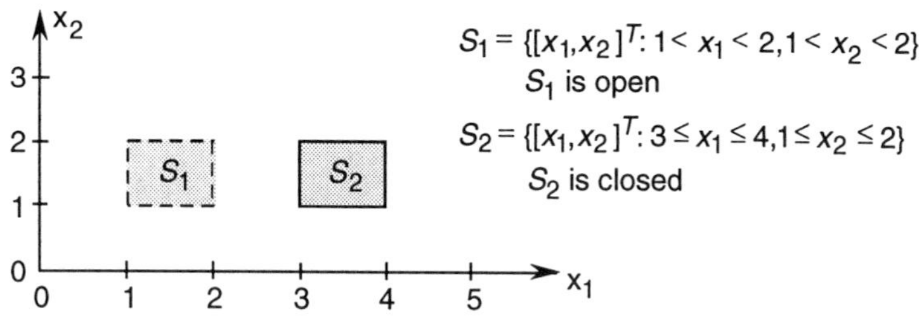


**Figure 4.9.** Open and closed sets

it is both closed and bounded. Compact sets are important in optimization problems for the following reason.

**Theorem 4.2. Theorem of Weierstrass.** *Let $f:\Omega \to \mathbb{R}$ be a continuous function, where $\Omega \subset \mathbb{R}^n$ is a compact set. Then, there exists $\boldsymbol{x}_0 \in \Omega$ such that $f(\boldsymbol{x}_0) \leqslant f(\boldsymbol{x})$ for all $\boldsymbol{x} \in \Omega$. In other words, $f$ achieves its minimum on $\Omega$.* □

*Proof.* See Refs. 65 (p. 89) and 3 (p. 154). ■

## 4.5. POLYTOPES AND POLYHEDRA

Let $\Theta$ be a convex set, and suppose $\boldsymbol{y}$ is a boundary point of $\Theta$. A hyperplane passing through $\boldsymbol{y}$ is called a *hyperplane of support* (or a *supporting hyperplane*) of the set $\Theta$, if the entire set $\Theta$ lies completely in one of the two half-spaces into which this hyperplane divides the space $\mathbb{R}^n$.

Recall that by Theorem 4.1, the intersection of any number of convex sets is convex. In what follows, we are concerned with the intersection of a finite number of half-spaces. Since every half space $H_+$ or $H_-$ is convex in $\mathbb{R}^n$, the intersection of any number of half spaces is a convex set.

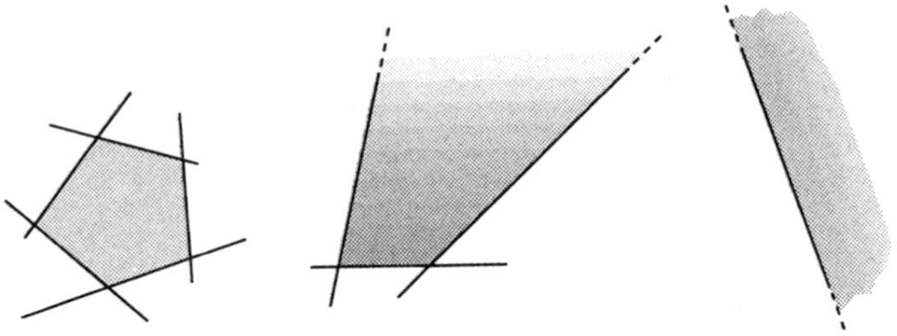

**Figure 4.10.** Polytopes

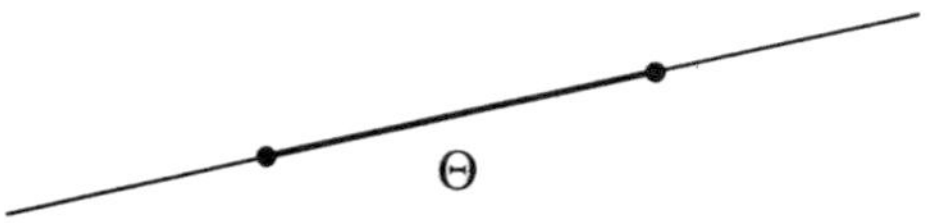


**Figure 4.11.** One-dimensional polyhedron

A set that can be expressed as the intersection of a finite number of half-spaces is called a *convex polytope* (see Figure 4.10). A nonempty bounded polytope is called a *polyhedron* (see Figure 4.11). For every convex polyhedron $\Theta \subset \mathbb{R}^n$, there exists a nonnegative integer $k \leqslant n$ such that $\Theta$ is contained in a linear variety of dimension $k$, but is not entirely contained in any $(k-1)$-dimensional linear variety of $\mathbb{R}^n$. Furthermore, there exists only one $k$-dimensional linear variety containing $\Theta$, called the *carrier* of the polyhedron $\Theta$, and $k$ is called the dimension of $\Theta$. For example, a zero-dimensional polyhedron is a point of $\mathbb{R}^n$, and its carrier is itself. A one-dimensional polyhedron is a segment, and its carrier is the straight line on which it lies. The boundary of any $k$-dimensional polyhedron, $k > 0$, consists of a finite number of $(k-1)$-dimensional polyhedra. For example, the boundary of a one-dimensional polyhedron consists of two points that are the endpoints of the segment.

The $(k-1)$-dimensional polyhedra forming the boundary of a $k$-dimensional polyhedron are called the *faces* of the polyhedron. Each of these faces has in turn $(k-2)$-dimensional faces. Thus, every $k$-dimensional polyhedron has faces of dimensions $k-1, k-2, \ldots, 1, 0$. The zero-dimensional faces of a polyhedron are called its *vertices*, and the one-dimensional faces are called *edges*. A vertex of a polyhedron in $\mathbb{R}^n$ is the intersection of $n$ supporting hyperplanes for the half-spaces defining the polyhedron. Similarly, an edge of a polyhedron in $\mathbb{R}^n$ is the intersection of $n-1$ supporting hyperplanes.

## EXERCISES

**4.1** Show that a set $S \subset \mathbb{R}^n$ is a linear variety if, and only if, for all $\boldsymbol{x}, \boldsymbol{y} \in S$ and $\alpha \in \mathbb{R}$, we have $\alpha \boldsymbol{x} + (1-\alpha)\boldsymbol{y} \in S$.

**4.2** Show that the set $\{\boldsymbol{x} \in \mathbb{R}^n : \|\boldsymbol{x}\| \leqslant r\}$ is convex, where $r > 0$ is a given real number, and $\|\boldsymbol{x}\| = \sqrt{\boldsymbol{x}^T \boldsymbol{x}}$ is the Euclidean norm of $\boldsymbol{x} \in \mathbb{R}^n$.

**4.3** Show that for any matrix $\boldsymbol{A} \in \mathbb{R}^{m \times n}$ and vector $\boldsymbol{b} \in \mathbb{R}^m$, the set (linear variety) $\{\boldsymbol{x} \in \mathbb{R}^n : \boldsymbol{A}\boldsymbol{x} = \boldsymbol{b}\}$ is convex.

**4.4** Show that the set $\{\boldsymbol{x} \in \mathbb{R}^n : \boldsymbol{x} \geqslant \boldsymbol{0}\}$ is convex (where $\boldsymbol{x} \geqslant \boldsymbol{0}$ means that every component of $\boldsymbol{x}$ is nonnegative).

# 5

# Elements of Differential Calculus

## 5.1. DIFFERENTIABILITY

This section follows the exposition in Ref. 79 (pp. 222–226). Differential calculus is based on the idea of approximating an arbitrary function by an *affine function*. To this end, consider a function $\boldsymbol{f}: \mathbb{R}^n \to \mathbb{R}^m$, and a point $\boldsymbol{x}_0 \in \mathbb{R}^n$. A function $\mathscr{A}: \mathbb{R}^n \to \mathbb{R}^m$ is *affine* if there exists a *linear* function $\mathscr{L}: \mathbb{R}^n \to \mathbb{R}^m$ and a vector $\boldsymbol{y} \in \mathbb{R}^m$ such that

$$\mathscr{A}(\boldsymbol{x}) = \mathscr{L}(\boldsymbol{x}) + \boldsymbol{y}$$

for every $\boldsymbol{x} \in \mathbb{R}^n$. We wish to find an affine function $\mathscr{A}$ that approximates $\boldsymbol{f}$ near the point $\boldsymbol{x}_0$. First, it is natural to impose the condition

$$\mathscr{A}(\boldsymbol{x}_0) = \boldsymbol{f}(\boldsymbol{x}_0).$$

Since $\mathscr{A}(\boldsymbol{x}) = \mathscr{L}(\boldsymbol{x}) + \boldsymbol{y}$, we obtain $\boldsymbol{y} = \boldsymbol{f}(\boldsymbol{x}_0) - \mathscr{L}(\boldsymbol{x}_0)$. By the linearity of $\mathscr{L}$,

$$\mathscr{L}(\boldsymbol{x}) + \boldsymbol{y} = \mathscr{L}(\boldsymbol{x}) - \mathscr{L}(\boldsymbol{x}_0) + \boldsymbol{f}(\boldsymbol{x}_0) = \mathscr{L}(\boldsymbol{x} - \boldsymbol{x}_0) + \boldsymbol{f}(\boldsymbol{x}_0).$$

Hence, we may write

$$\mathscr{A}(\boldsymbol{x}) = \mathscr{L}(\boldsymbol{x} - \boldsymbol{x}_0) + \boldsymbol{f}(\boldsymbol{x}_0).$$

Next, we require that $\mathscr{A}(\boldsymbol{x})$ approaches $\boldsymbol{f}(\boldsymbol{x})$ faster than $\boldsymbol{x}$ approaches $\boldsymbol{x}_0$; that is,

$$\lim_{\boldsymbol{x} \to \boldsymbol{x}_0, \boldsymbol{x} \in \Omega} \frac{\|\boldsymbol{f}(\boldsymbol{x}) - \mathscr{A}(\boldsymbol{x})\|}{\|\boldsymbol{x} - \boldsymbol{x}_0\|} = 0.$$

The above conditions on $\mathscr{A}$ ensure that $\mathscr{A}$ approximates $\boldsymbol{f}$ near $\boldsymbol{x}_0$ in the sense that the error in the approximation at a given point is "small" compared with the distance of the point from $\boldsymbol{x}_0$.

In summary, a function $\boldsymbol{f}: \Omega \to \mathbb{R}^m$, $\Omega \subset \mathbb{R}^n$, is said to be *differentiable* at $\boldsymbol{x}_0 \in \Omega$ if there is an affine function that approximates $\boldsymbol{f}$ near $\boldsymbol{x}_0$; that is, there exists a linear function $\mathscr{L}: \mathbb{R}^n \to \mathbb{R}^m$ such that

$$\lim_{\boldsymbol{x} \to \boldsymbol{x}_0, \boldsymbol{x} \in \Omega} \frac{\|\boldsymbol{f}(\boldsymbol{x}) - (\mathscr{L}(\boldsymbol{x} - \boldsymbol{x}_0) + \boldsymbol{f}(\boldsymbol{x}_0))\|}{\|\boldsymbol{x} - \boldsymbol{x}_0\|} = 0.$$

The linear function $\mathscr{L}$ above is uniquely determined by $\boldsymbol{f}$ and $\boldsymbol{x}_0$, and is called the *derivative* of $\boldsymbol{f}$ at $\boldsymbol{x}_0$. The function $\boldsymbol{f}$ is said to be *differentiable* on $\Omega$ if $\boldsymbol{f}$ is differentiable at every point of its domain $\Omega$.

In $\mathbb{R}$, an affine function has the form $ax + b$, with $a, b \in \mathbb{R}$. Hence, a real-valued function $f(x)$ of a real variable $x$ that is differentiable at $x_0$ can be approximated near $x_0$ by a function

$$\mathscr{A}(x) = ax + b.$$

Since $f(x_0) = \mathscr{A}(x_0) = ax_0 + b$, we obtain

$$\mathscr{A}(x) = ax + b = a(x - x_0) + f(x_0).$$

The linear part of $\mathscr{A}(x)$, denoted earlier by $\mathscr{L}(x)$, is in this case just $ax$. The norm of a real number is its absolute value, so the definition of differentiability becomes

$$\lim_{x \to x_0} \frac{|f(x) - (a(x - x_0) + f(x_0))|}{|x - x_0|} = 0,$$

which is equivalent to

$$\lim_{x \to x_0} \frac{f(x) - f(x_0)}{x - x_0} = a.$$

The number $a$ is commonly denoted $f'(x_0)$, and is called the derivative of $f$ at $x_0$. The affine function $\mathscr{A}$ is therefore given by

$$\mathscr{A}(x) = f(x_0) + f'(x_0)(x - x_0).$$

Its graph is the tangent line to the graph of $f$ at $x_0$ (see Figure 5.1).

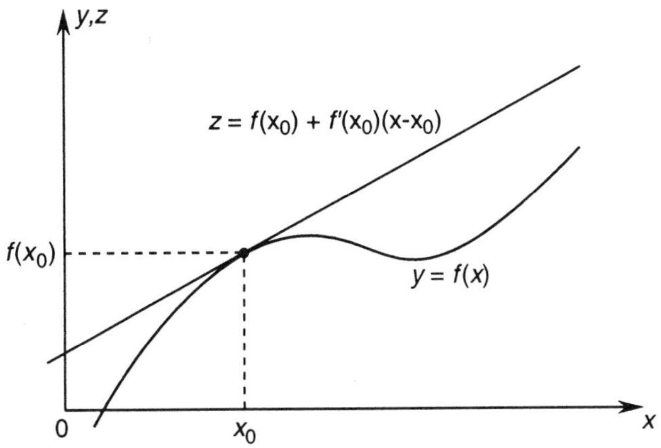


**Figure 5.1.** Illustration of the notion of the derivative

## 5.2. THE DERIVATIVE MATRIX

Any linear transformation from $\mathbb{R}^n$ to $\mathbb{R}^m$, and in particular the derivative $\mathscr{L}$ of $\boldsymbol{f}: \mathbb{R}^n \to \mathbb{R}^m$, can be represented by an $m \times n$ matrix. To find the matrix representation $\boldsymbol{L}$ of the derivative $\mathscr{L}$ of a differentiable function $\boldsymbol{f}: \mathbb{R}^n \to \mathbb{R}^m$, we use the natural basis $\{\boldsymbol{e}_1, \ldots, \boldsymbol{e}_n\}$ for $\mathbb{R}^n$. Consider the vectors

$$\boldsymbol{x}_j = \boldsymbol{x}_0 + t\boldsymbol{e}_j, \quad j = 1, \ldots, n, \quad t \in \mathbb{R}.$$

By the definition of the derivative, we have

$$\lim_{t \to 0} \frac{\boldsymbol{f}(\boldsymbol{x}_j) - (t\boldsymbol{L}\boldsymbol{e}_j + \boldsymbol{f}(\boldsymbol{x}_0))}{t} = \boldsymbol{0}$$

for $j = 1, \ldots, n$. This means that

$$\lim_{t \to 0} \frac{\boldsymbol{f}(\boldsymbol{x}_j) - \boldsymbol{f}(\boldsymbol{x}_0)}{t} = \boldsymbol{L}\boldsymbol{e}_j$$

for $j = 1, \ldots, n$. But $\boldsymbol{L}\boldsymbol{e}_j$ is the $j$th column of the matrix $\boldsymbol{L}$. On the other hand, the vector $\boldsymbol{x}_j$ differs from $\boldsymbol{x}_0$ only in the $j$th coordinate, and in that coordinate the difference is just the number $t$. Therefore, the left side of the last equation is the partial derivative

$$\frac{\partial \boldsymbol{f}}{\partial x_j}(\boldsymbol{x}_0).$$

Since vector limits are computed by taking the limit of each coordinate function, it follows that if

$$\boldsymbol{f}(\boldsymbol{x}) = \begin{bmatrix} f_1(\boldsymbol{x}) \\ \vdots \\ f_m(\boldsymbol{x}) \end{bmatrix},$$

then

$$\frac{\partial \boldsymbol{f}}{\partial x_j}(\boldsymbol{x}_0) = \begin{bmatrix} \dfrac{\partial f_1}{\partial x_j}(\boldsymbol{x}_0) \\ \vdots \\ \dfrac{\partial f_m}{\partial x_j}(\boldsymbol{x}_0) \end{bmatrix},$$

and the matrix $\boldsymbol{L}$ has the form

$$\left[\frac{\partial \boldsymbol{f}}{\partial x_1}(\boldsymbol{x}_0) \cdots \frac{\partial \boldsymbol{f}}{\partial x_n}(\boldsymbol{x}_0)\right] = \begin{bmatrix} \dfrac{\partial f_1}{\partial x_1}(\boldsymbol{x}_0) & \cdots & \dfrac{\partial f_1}{\partial x_n}(\boldsymbol{x}_0) \\ \vdots & & \vdots \\ \dfrac{\partial f_m}{\partial x_1}(\boldsymbol{x}_0) & \cdots & \dfrac{\partial f_m}{\partial x_n}(\boldsymbol{x}_0) \end{bmatrix}.$$

The matrix $\boldsymbol{L}$ is called the *Jacobian matrix*, or *derivative matrix*, of $\boldsymbol{f}$ at $\boldsymbol{x}_0$ and is denoted $D\boldsymbol{f}(\boldsymbol{x}_0)$. For convenience, we often refer to $D\boldsymbol{f}(\boldsymbol{x}_0)$ simply as the derivative of $\boldsymbol{f}$ at $\boldsymbol{x}_0$. We summarize the above discussion in the following theorem.

**Theorem 5.1.** *If a function $\boldsymbol{f}: \mathbb{R}^n \to \mathbb{R}^m$ is differentiable at $\boldsymbol{x}_0$, then the derivative of $\boldsymbol{f}$ at $\boldsymbol{x}_0$ is uniquely determined and is represented by the $m \times n$ derivative matrix $D\boldsymbol{f}(\boldsymbol{x}_0)$. The best affine approximation to $\boldsymbol{f}$ near $\boldsymbol{x}_0$ is then given by*

$$\mathscr{A}(\boldsymbol{x}) = \boldsymbol{f}(\boldsymbol{x}_0) + D\boldsymbol{f}(\boldsymbol{x}_0)(\boldsymbol{x} - \boldsymbol{x}_0).$$

*The columns of the derivative matrix $D\boldsymbol{f}(\boldsymbol{x}_0)$ are vector partial derivatives. The vector*

$$\frac{\partial \boldsymbol{f}}{\partial x_j}(\boldsymbol{x}_0)$$

*is a tangent vector at $\boldsymbol{x}_0$ to the curve $\boldsymbol{f}$ obtained by varying only the* $j$th *coordinate of $\boldsymbol{x}$.* □

If $f: \mathbb{R}^n \to \mathbb{R}$ is differentiable, then the function $\nabla f$ defined by

$$\nabla f(\boldsymbol{x}) = \begin{bmatrix} \dfrac{\partial f}{\partial x_1}(\boldsymbol{x}) \\ \vdots \\ \dfrac{\partial f}{\partial x_n}(\boldsymbol{x}) \end{bmatrix} = Df(\boldsymbol{x})^T$$

is called the *gradient* of $f$. The gradient is a function from $\mathbb{R}^n$ to $\mathbb{R}^n$ and can be pictured as a *vector field*, by drawing the arrow representing $\nabla f(\boldsymbol{x})$ so that its tail starts at $\boldsymbol{x}$.

Given $f: \mathbb{R}^n \to \mathbb{R}$, if $Df$ is differentiable, we say that $f$ is *twice differentiable*, and we write the derivative of $Df$ as

$$D^2 f = \begin{bmatrix} \dfrac{\partial^2 f}{\partial x_1^2} & \dfrac{\partial^2 f}{\partial x_2 \partial x_1} & \cdots & \dfrac{\partial^2 f}{\partial x_n \partial x_1} \\ \dfrac{\partial^2 f}{\partial x_1 \partial x_2} & \dfrac{\partial^2 f}{\partial x_2^2} & \cdots & \dfrac{\partial^2 f}{\partial x_n \partial x_2} \\ \vdots & \vdots & \ddots & \vdots \\ \dfrac{\partial^2 f}{\partial x_1 \partial x_n} & \dfrac{\partial^2 f}{\partial x_2 \partial x_n} & \cdots & \dfrac{\partial^2 f}{\partial x_n^2} \end{bmatrix}$$

The matrix $D^2 f(\boldsymbol{x})$ is called the *Hessian* matrix of $f$ at $\boldsymbol{x}$.

In general, if $D^{m-1} f$ is differentiable, we say that $f$ is $m$ times differentiable.

## 5.3. CHAIN RULE

A function $f:\Omega\to\mathbb{R}^m$, $\Omega\subset\mathbb{R}^n$, is said to be *continuously differentiable* on $\Omega$ if it is differentiable (on $\Omega$), and $Df:\Omega\to\mathbb{R}^{m\times n}$ is continuous, that is, the components of $f$ have continuous partial derivatives. In this case, we write $f\in\mathscr{C}^1$. If $D^p f$ exists and is continuous (i.e., the components of $f$ have continuous partial derivatives of order $p$), then we write $f\in\mathscr{C}^p$. Note that the Hessian matrix of $f$ at $x$ is symmetric if $f$ is twice continuously differentiable at $x$.

We now prove the Chain Rule for differentiating the composition $g(f(t))$, of a function $f:\mathbb{R}\to\mathbb{R}^n$ and a function $g:\mathbb{R}^n\to\mathbb{R}$.

**Theorem 5.2.** *Let $g:\mathscr{D}\to\mathbb{R}$ be continuously differentiable on an open set $\mathscr{D}\subset\mathbb{R}^n$, and let $f:(a,b)\to\mathscr{D}$ be differentiable on $(a,b)$. Then, the composite function $F:(a,b)\to\mathbb{R}$ given by $F(t)=g(f(t))$ is differentiable on $(a,b)$, and*

$$F'(t)=Dg(f(t))Df(t)=\nabla g(f(t))^T\begin{bmatrix}f_1'(t)\\ \vdots\\ f_n'(t)\end{bmatrix}.$$

□

*Proof.* By definition,

$$F'(t)=\lim_{h\to 0}\frac{F(t+h)-F(t)}{h}=\lim_{h\to 0}\frac{g(f(t+h))-g(f(t))}{h}$$

if the limit exists. Since $f$ is differentiable, it is continuous. We now apply the mean-value theorem to $g$, and obtain

$$g(y)-g(x)=Dg(x_0)(y-x),$$

where $x_0$ is some point on the segment joining $y$ and $x$. Letting $x=f(t)$ and $y=f(t+h)$, we have

$$\frac{F(t+h)-F(t)}{h}=Dg(x_0)\frac{f(t+h)-f(t)}{h}.$$

By continuity of $Dg$, $Dg(x_0)\to Dg(f(t))$ as $h\to 0$. Thus,

$$F'(t)=\lim_{h\to 0}Dg(x_0)\frac{f(t+h)-f(t)}{h}=Dg(f(t))Df(t).$$

■

## 5.4. LEVEL SETS AND GRADIENTS

The *level set* of a function $f: \mathbb{R}^n \to \mathbb{R}$ at level $c$ is the set of points

$$S = \{x: f(x) = c\}.$$

For $f: \mathbb{R}^2 \to \mathbb{R}$, we are usually interested in $S$ when it is a curve. For $f: \mathbb{R}^3 \to \mathbb{R}$, the sets $S$ most often considered are surfaces.

***Example 5.1.*** Consider the following real-valued function on $\mathbb{R}^2$:

$$f(x) = 100(x_2 - x_1^2)^2 + (1 - x_1)^2, \quad x = [x_1, x_2]^T.$$

The above function is called *Rosenbrock's function.* A graph of the function $f$ is illustrated in Figure 5.2. The level sets of $f$ at levels 0.7, 7, 70, 200, and 700 are depicted in Figure 5.3. These level sets have a particular shape resembling bananas. For this reason, Rosenbrock's function is also called the banana function. ■

To say that a point $x_0$ is on the level set $S$ at level $c$ means $f(x_0) = c$. Now suppose that there is a curve $\gamma$ lying in $S$ and parameterized by a continuously differentiable function $g: \mathbb{R} \to \mathbb{R}^n$. Suppose also that $g(t_0) = x_0$ and $Dg(t_0) = v \neq 0$, so that $v$ is a tangent vector to $\gamma$ at $x_0$ (see Figure 5.4). Applying the chain rule to

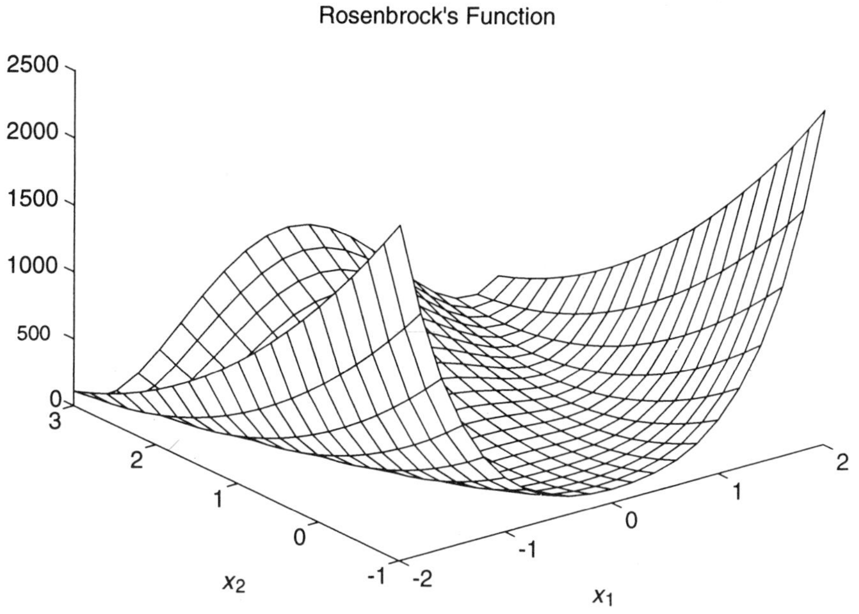


**Figure 5.2.** Graph of Rosenbrock's function

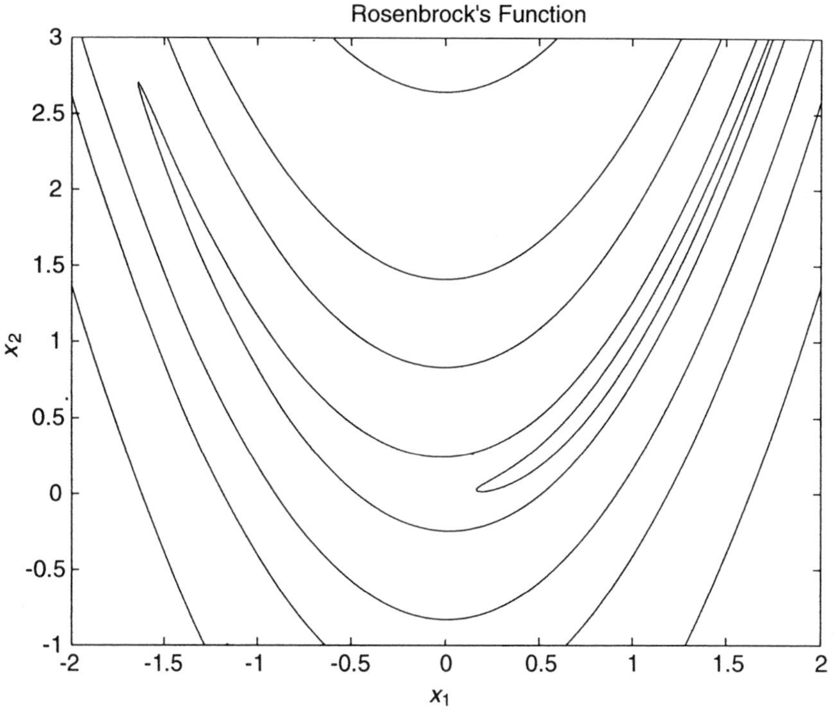


**Figure 5.3.** Level sets of Rosenbrock's (banana) function

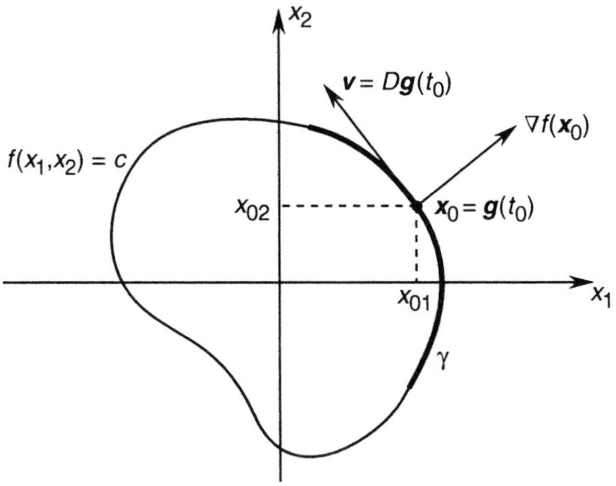


**Figure 5.4.** Orthogonality of the gradient to the level set

the function $h(t)=f(\boldsymbol{g}(t))$ at $t_0$, gives

$$h'(t_0)=Df(\boldsymbol{g}(t_0))D\boldsymbol{g}(t_0)=Df(\boldsymbol{x}_0)\boldsymbol{v}.$$

But since $\gamma$ lies on $S$, we have

$$h(t)=f(\boldsymbol{g}(t))=c,$$

that is, $h$ is constant. Thus, $h'(t_0)=0$ and

$$Df(\boldsymbol{x}_0)\boldsymbol{v}=\nabla f(\boldsymbol{x}_0)^T\boldsymbol{v}=0.$$

Hence, we have proved, assuming $f$ continuously differentiable, the following theorem (see Figure 5.4).

**Theorem 5.3.** *The vector $\nabla f(\boldsymbol{x}_0)$ is orthogonal to the tangent vector to an arbitrary smooth curve passing through $\boldsymbol{x}_0$ on the level set determined by $f(\boldsymbol{x})=f(\boldsymbol{x}_0)$.* □

It is natural to say that $\nabla f(\boldsymbol{x}_0)$ is orthogonal or normal to the level set $S$ corresponding to $\boldsymbol{x}_0$, and to take as the tangent plane (or line) to $S$ at $\boldsymbol{x}_0$ the set of all points satisfying

$$\nabla f(\boldsymbol{x}_0)^T(\boldsymbol{x}-\boldsymbol{x}_0)=0, \quad \text{if } \nabla f(\boldsymbol{x}_0)\neq\mathbf{0}.$$

As we shall see later, $\nabla f(\boldsymbol{x}_0)$ is the direction of *maximum rate of increase* of $f$ at $\boldsymbol{x}_0$. Since $\nabla f(\boldsymbol{x}_0)$ is orthogonal to the level set through $\boldsymbol{x}_0$ determined by $f(\boldsymbol{x})=f(\boldsymbol{x}_0)$, we deduce the following fact: the direction of maximum rate of increase of a real-valued differentiable function at a point is orthogonal to the level set of the function through that point.

Figure 5.5 illustrates the above discussion for the case $f:\mathbb{R}^2\to\mathbb{R}$. The curve on the graph in Figure 5.5 running from bottom to top has the property that its projection onto the $(x_1,x_2)$ plane is always orthogonal to the level curves, and is called a *path of steepest ascent*, because it always heads in the direction of maximum rate of increase for $f$.

The graph of $f:\mathbb{R}^n\to\mathbb{R}$ is the set $\{[\boldsymbol{x}^T,f(\boldsymbol{x})]^T:\boldsymbol{x}\in\mathbb{R}^n\}\subset\mathbb{R}^{n+1}$. The notion of the gradient of a function has an alternative useful interpretation in terms of the tangent hyperplane to its graph. To proceed, let $\boldsymbol{x}_0\in\mathbb{R}^n$ and $z_0=f(\boldsymbol{x}_0)$. The point $[\boldsymbol{x}_0^T,z_0]^T\in\mathbb{R}^{n+1}$ is a point on the graph of $f$. If $f$ is differentiable at $\xi$, then the graph admits a nonvertical tangent hyperplane at $\xi=[\boldsymbol{x}_0^T,z_0]^T$. The hyperplane through $\xi$ is the set of all points $[x_1,\ldots,x_n,z]^T\in\mathbb{R}^{n+1}$ satisfying the equation

$$u_1(x_1-x_{01})+\cdots+u_n(x_n-x_{0n})+v(z-z_0)=0,$$

where the vector $[u_1,\ldots,u_n,v]^T\in\mathbb{R}^{n+1}$ is normal to the hyperplane. Assuming

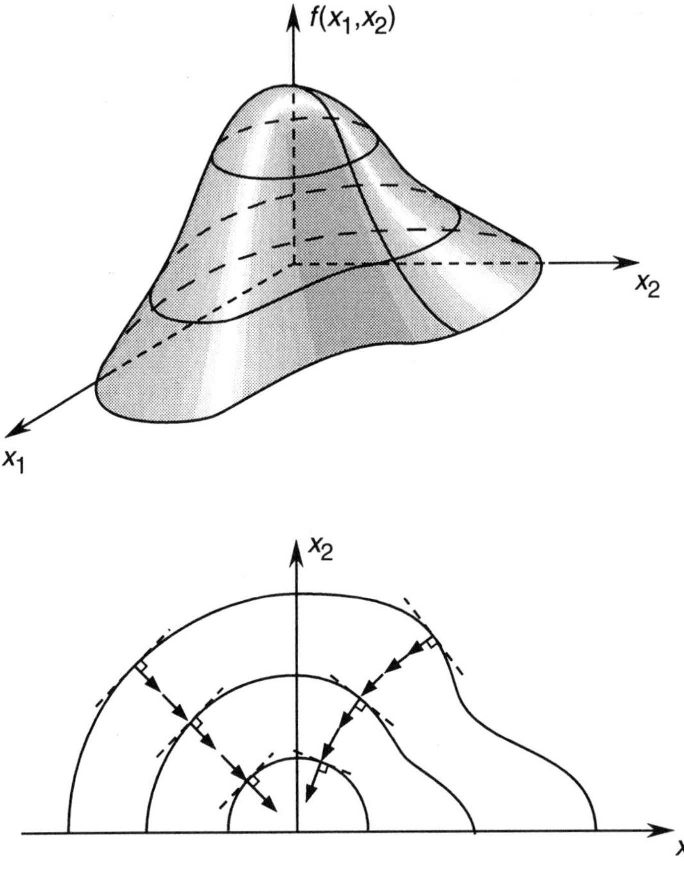


**Figure 5.5.** Illustration of a path of steepest ascent

that this hyperplane is nonvertical, that is, $v \neq 0$, let

$$d_i = -\frac{u_i}{v}.$$

Thus, we can rewrite the hyperplane equation above as

$$z = d_1(x_1 - x_{01}) + \cdots + d_n(x_n - x_{0n}) + z_0.$$

We can think of the right side of the above equation as a function $z: \mathbb{R}^n \to \mathbb{R}$. Observe that for the hyperplane to be tangent to the graph of $f$, the functions $f$ and $z$ must have the same partial derivatives at the point $\boldsymbol{x}_0$. Hence, if $f$ is differentiable at $\boldsymbol{x}_0$, its tangent hyperplane can be written in terms of its gradient, as given by the equation

$$z - z_0 = Df(\boldsymbol{x}_0)(\boldsymbol{x} - \boldsymbol{x}_0) = (\boldsymbol{x} - \boldsymbol{x}_0)^T \nabla f(\boldsymbol{x}_0).$$

## 5.5. TAYLOR SERIES

The basis for many numerical methods and models for optimization is Taylor's formula.

**Theorem 5.4. Taylor's Theorem.** *Assume that a function $f: \mathbb{R} \to \mathbb{R}$ is $m$ times continuously differentiable (i.e., $f \in \mathscr{C}^m$) on an interval $[a, b]$. Denote $h = b - a$. Then,*

$$f(b) = f(a) + \frac{h}{1!} f^{(1)}(a) + \frac{h^2}{2!} f^{(2)}(a) + \cdots + \frac{h^{m-1}}{(m-1)!} f^{(m-1)}(a) + R_m,$$

*where $f^{(i)}$ is the* ith *derivative of $f$, and*

$$R_m = \frac{h^m(1-\theta)^{m-1}}{(m-1)!} f^{(m)}(a + \theta h) = \frac{h^m}{m!} f^{(m)}(a + \theta' h),$$

*with $\theta, \theta' \in (0, 1)$.* □

*Proof.*

We have

$$R_m = f(b) - f(a) - \frac{h}{1!} f^{(1)}(a) - \frac{h^2}{2!} f^{(2)}(a) - \cdots - \frac{h^{m-1}}{(m-1)!} f^{(m-1)}(a).$$

Denote by $g_m(x)$ an auxiliary function obtained from $R_m$ by replacing $a$ by $x$. Hence,

$$g_m(x) = f(b) - f(x) - \frac{b-x}{1!} f^{(1)}(x) - \frac{(b-x)^2}{2!} f^{(2)}(x) - \cdots - \frac{(b-x)^{m-1}}{(m-1)!} f^{(m-1)}(x).$$

Differentiating $g_m(x)$ yields

$$\begin{aligned} g_m^{(1)}(x) &= -f^{(1)}(x) + \left[ f^{(1)}(x) - \frac{b-x}{1!} f^{(2)}(x) \right] \\ &\quad + \left[ 2\frac{b-x}{2!} f^{(2)}(x) - \frac{(b-x)^2}{2!} f^{(3)}(x) \right] + \cdots \\ &\quad + \left[ (m-1)\frac{(b-x)^{m-2}}{(m-1)!} f^{(m-1)}(x) - \frac{(b-x)^{m-1}}{(m-1)!} f^{(m)}(x) \right] \\ &= -\frac{(b-x)^{m-1}}{(m-1)!} f^{(m)}(x). \end{aligned}$$

Observe that $g_m(b) = 0$ and $g_m(a) = R_m$. Applying the mean-value theorem yields

$$\frac{g_m(b) - g_m(a)}{b - a} = g_m^{(1)}(a + \theta h),$$

where $\theta \in (0, 1)$. The above equation is equivalent to

$$-\frac{R_m}{h} = -\frac{(b - a - \theta h)^{m-1}}{(m-1)!} f^{(m)}(a + \theta h) = -\frac{h^{m-1}(1-\theta)^{m-1}}{(m-1)!} f^{(m)}(a + \theta h).$$

Hence

$$R_m = \frac{h^m(1-\theta)^{m-1}}{(n-1)!} f^{(m)}(a + \theta h).$$

To prove the formula

$$R_m = \frac{h^m}{m!} f^{(m)}(a + \theta' h),$$

see, e.g., Refs. 49 and 50. ■

An important property of Taylor's theorem arises from the form of the remainder $R_m$. To further discuss this property, we introduce the so-called *order symbols*, $O$ (big-oh) and $o$ (little-oh).

Let $g$ be a real-valued function defined in some neighborhood of $\mathbf{0} \in \mathbb{R}^n$, with $g(\boldsymbol{x}) > 0$ if $\boldsymbol{x} \neq \mathbf{0}$. Let $\boldsymbol{f}: \Omega \to \mathbb{R}^m$ be defined in a domain $\Omega \subset \mathbb{R}^n$ that includes $\mathbf{0}$. Then we write

1. $\boldsymbol{f}(\boldsymbol{x}) = O(g(\boldsymbol{x}))$ to mean that the quotient $\|\boldsymbol{f}(\boldsymbol{x})\|/g(\boldsymbol{x})$ is bounded near $\mathbf{0}$, i.e., there exist numbers $K > 0$ and $\delta > 0$ such that if $\|\boldsymbol{x}\| < \delta$, $\boldsymbol{x} \in \Omega$, then $\|\boldsymbol{f}(\boldsymbol{x})\|/g(\boldsymbol{x}) \leqslant K$.
2. $\boldsymbol{f}(\boldsymbol{x}) = o(g(\boldsymbol{x}))$ to mean that

$$\lim_{\boldsymbol{x} \to \mathbf{0}, \boldsymbol{x} \in \Omega} \frac{\|\boldsymbol{f}(\boldsymbol{x})\|}{g(\boldsymbol{x})} = 0.$$

Note that $o(\alpha)$ is a function that goes to zero "faster" than $\alpha$, that is, $\lim_{\alpha \to 0} o(\alpha)/\alpha = 0$. Examples of such a function are

1. $o(\alpha) = \alpha^2$
2. $o(\alpha) = \begin{bmatrix} \alpha^3 \\ \alpha^2 + \alpha^4 \end{bmatrix}$.

Suppose $f \in \mathscr{C}^m$. Recall that the remainder term has the form

$$R_m = \frac{h^m}{m!} f^{(m)}(a + \theta h),$$

where $\theta \in (0, 1)$. Substituting the above into Taylor's formula, we get

$$f(b) = f(a) + \frac{h}{1!} f^{(1)}(a) + \frac{h^2}{2!} f^{(2)}(a) + \cdots + \frac{h^{m-1}}{(m-1)!} f^{(m-1)}(a) + \frac{h^m}{m!} f^{(m)}(a + \theta h).$$

By the continuity of $f^{(m)}$, we have $f^{(m)}(a + \theta h) \to f^{(m)}(a)$ as $h \to 0$, that is, $f^{(m)}(a + \theta h) = f^{(m)}(a) + o(1)$. Therefore,

$$\frac{h^m}{m!} f^{(m)}(a + \theta h) = \frac{h^m}{m!} f^{(m)}(a) + o(h^m),$$

since $h^m o(1) = o(h^m)$. We may then write Taylor's formula as

$$f(b) = f(a) + \frac{h}{1!} f^{(1)}(a) + \frac{h^2}{2!} f^{(2)}(a) + \cdots + \frac{h^m}{m!} f^{(m)}(a) + o(h^m).$$

If, in addition, we assume that $f \in \mathscr{C}^{m+1}$, we may replace the term $o(h^m)$ above by $O(h^{m+1})$. To see this, we first write Taylor's formula with $R_{m+1}$:

$$f(b) = f(a) + \frac{h}{1!} f^{(1)}(a) + \frac{h^2}{2!} f^{(2)}(a) + \cdots + \frac{h^m}{m!} f^{(m)}(a) + R_{m+1},$$

where

$$R_{m+1} = \frac{h^{m+1}}{(m+1)!} f^{(m+1)}(a + \theta' h),$$

with $\theta' \in (0, 1)$. Since $f^{(m+1)}$ is bounded on $[a, b]$ (by Theorem 4.2),

$$R_{m+1} = O(h^{m+1}).$$

Therefore, if $f \in \mathscr{C}^{m+1}$, we may write Taylor's formula as

$$f(b) = f(a) + \frac{h}{1!} f^{(1)}(a) + \frac{h^2}{2!} f^{(2)}(a) + \cdots + \frac{h^m}{m!} f^{(m)}(a) + O(h^{m+1}).$$

We now turn to the Taylor series expansion of a real-valued function $f: \mathbb{R}^n \to \mathbb{R}$ about the point $\boldsymbol{x}_0 \in \mathbb{R}^n$. Suppose $f \in \mathscr{C}^2$. Let $\boldsymbol{x}$ and $\boldsymbol{x}_0$ be points in $\mathbb{R}^n$, and let $\boldsymbol{z}(\alpha) = \boldsymbol{x}_0 + \alpha(\boldsymbol{x} - \boldsymbol{x}_0)/\|\boldsymbol{x} - \boldsymbol{x}_0\|$. Define $g: \mathbb{R} \to \mathbb{R}$ by

$$g(\alpha) = f(\boldsymbol{z}(\alpha)) = f(\boldsymbol{x}_0 + \alpha(\boldsymbol{x} - \boldsymbol{x}_0)/\|\boldsymbol{x} - \boldsymbol{x}_0\|).$$

Using the chain rule, we obtain

$$\begin{aligned} g'(\alpha) &= \frac{dg}{d\alpha}(\alpha) \\ &= Df(\boldsymbol{z}(\alpha)) D\boldsymbol{z}(\alpha) = Df(\boldsymbol{z}(\alpha)) \frac{(\boldsymbol{x} - \boldsymbol{x}_0)}{\|\boldsymbol{x} - \boldsymbol{x}_0\|} \\ &= \frac{1}{\|\boldsymbol{x} - \boldsymbol{x}_0\|} (\boldsymbol{x} - \boldsymbol{x}_0)^T Df(\boldsymbol{z}(\alpha))^T, \end{aligned}$$

and

$$\begin{aligned} g''(\alpha) &= \frac{d^2 g}{d\alpha^2}(\alpha) \\ &= \frac{d}{d\alpha}\left(\frac{dg}{d\alpha}\right)(\alpha) \\ &= \frac{(\boldsymbol{x}-\boldsymbol{x}_0)^T}{\|\boldsymbol{x}-\boldsymbol{x}_0\|}\frac{d}{d\alpha} Df(z(\alpha))^T \\ &= \frac{(\boldsymbol{x}-\boldsymbol{x}_0)^T}{\|\boldsymbol{x}-\boldsymbol{x}_0\|} D(Df)(z(\alpha))^T \frac{dz}{d\alpha}(\alpha) \\ &= \frac{1}{\|\boldsymbol{x}-\boldsymbol{x}_0\|^2}(\boldsymbol{x}-\boldsymbol{x}_0)^T D^2 f(z(\alpha))^T(\boldsymbol{x}-\boldsymbol{x}_0) \\ &= \frac{1}{\|\boldsymbol{x}-\boldsymbol{x}_0\|^2}(\boldsymbol{x}-\boldsymbol{x}_0)^T D^2 f(z(\alpha))(\boldsymbol{x}-\boldsymbol{x}_0), \end{aligned}$$

where we recall that

$$D^2 f = \begin{bmatrix} \dfrac{\partial^2 f}{\partial x_1^2} & \dfrac{\partial^2 f}{\partial x_2 \partial x_1} & \cdots & \dfrac{\partial^2 f}{\partial x_n \partial x_1} \\ \dfrac{\partial^2 f}{\partial x_1 \partial x_2} & \dfrac{\partial^2 f}{\partial x_2^2} & \cdots & \dfrac{\partial^2 f}{\partial x_n \partial x_2} \\ \vdots & \vdots & \ddots & \vdots \\ \dfrac{\partial^2 f}{\partial x_1 \partial x_n} & \dfrac{\partial^2 f}{\partial x_2 \partial x_n} & \cdots & \dfrac{\partial^2 f}{\partial x_n^2} \end{bmatrix},$$

and $D^2 f = (D^2 f)^T$ since $f \in \mathscr{C}^2$. Observe that

$$\begin{aligned} f(\boldsymbol{x}) &= g(\|\boldsymbol{x}-\boldsymbol{x}_0\|) \\ &= g(0) + \frac{\|\boldsymbol{x}-\boldsymbol{x}_0\|}{1!} g'(0) + \frac{\|\boldsymbol{x}-\boldsymbol{x}_0\|^2}{2!} g''(0) + o(\|\boldsymbol{x}-\boldsymbol{x}_0\|^2). \end{aligned}$$

Hence,

$$\begin{aligned} f(\boldsymbol{x}) = f(\boldsymbol{x}_0) &+ \frac{1}{1!} Df(\boldsymbol{x}_0)(\boldsymbol{x}-\boldsymbol{x}_0) \\ &+ \frac{1}{2!}(\boldsymbol{x}-\boldsymbol{x}_0)^T D^2 f(\boldsymbol{x}_0)(\boldsymbol{x}-\boldsymbol{x}_0) + o(\|\boldsymbol{x}-\boldsymbol{x}_0\|^2). \end{aligned}$$

If we assume that $f \in \mathscr{C}^3$, we may use the formula for the remainder term $R_3$ to conclude that

$$f(x) = f(x_0) + \frac{1}{1!} Df(x_0)(x - x_0)$$
$$+ \frac{1}{2!}(x - x_0)^T D^2 f(x_0)(x - x_0) + O(\|x - x_0\|^3).$$

For further reading in calculus, consult Refs. 9, 49, 50, 68, 70, and 79. A basic treatment of real analysis can be found in Refs. 3 and 65, whereas a more advanced treatment is provided in Refs. 53 and 64.

## EXERCISES

**5.1** Consider $f(x) = x_1^2/6 + x_2^2/4$, $g(t) = [3t + 5, 2t - 6]^T$. Evaluate $(d/dt)f(g(t))$ using the chain rule.

**5.2** Consider $f(x) = x_1 x_2/2$, $g(s, t) = [4s + 3t, 2s + t]^T$. Evaluate $(\partial/\partial s)f(g(s, t))$ and $(\partial/\partial t)f(g(s, t))$ using the chain rule.

**5.3** Let $x(t) = [e^t + t^3, t^2, t + 1]^T$, $t \in \mathbb{R}$, and $f(x) = x_1^3 x_2 x_3^2 + x_1 x_2 + x_3$, $x = [x_1, x_2, x_3]^T \in \mathbb{R}^3$. Find $(d/dt)f(x(t))$ in terms of $t$.

**5.4** Suppose that $f(x) = o(g(x))$. Show that for any given $\varepsilon > 0$, there exists $\delta > 0$ such that if $\|x\| < \delta$, then $\|f(x)\| < \varepsilon |g(x)|$.

**5.5** Use Exercise 5.4 to show that if functions $f: \mathbb{R}^n \to \mathbb{R}$ and $g: \mathbb{R}^n \to \mathbb{R}$ satisfy $f(x) = -g(x) + o(g(x))$ and $g(x) > 0$ for all $x \neq 0$, then for all $x \neq 0$ sufficiently small, we have $f(x) < 0$.

**5.6** Let

$$f_1(x_1, x_2) = x_1^2 - x_2^2;$$
$$f_2(x_1, x_2) = 2x_1 x_2.$$

Sketch the level sets associated with $f_1(x_1, x_2) = 12$ and $f_2(x_1, x_2) = 16$ on the same diagram. Indicate on the diagram the values of $x = [x_1, x_2]^T$ for which $f(x) = [f_1(x_1, x_2), f_2(x_1, x_2)]^T = [12, 16]^T$.

**5.7** Write down the Taylor series expansion of the following functions about the given points $x_0$. Neglect terms of order three or higher.

**a.** $f(x) = x_1 e^{-x_2} + x_2 + 1, x_0 = [1, 0]^T$
**b.** $f(x) = x_1^4 + 2x_1^2 x_2^2 + x_2^4, x_0 = [1, 1]^T$
**c.** $f(x) = e^{x_1 - x_2} + e^{x_1 + x_2} + x_1 + x_2 + 1, x_0 = [1, 0]^T$

# Part II

# UNCONSTRAINED OPTIMIZATION

# 6

# Basics of Unconstrained Optimization

## 6.1. INTRODUCTION

In this chapter, we consider the optimization problem

$$\begin{array}{ll} \text{minimize} & f(\boldsymbol{x}) \\ \text{subject to} & \boldsymbol{x} \in \Omega. \end{array}$$

The function $f: \mathbb{R}^n \to \mathbb{R}$ that we wish to minimize is a real-valued function, and is called the *objective function*, or *cost function*. The vector $\boldsymbol{x}$ is an $n$-vector of independent variables, that is, $\boldsymbol{x} = [x_1, x_2, \ldots, x_n]^T \in \mathbb{R}^n$. The variables $x_1, \ldots, x_n$ are often referred to as *decision variables*. The set $\Omega$ is a subset of $\mathbb{R}^n$, called the *constraint set* or *feasible set*.

The optimization problem above can be viewed as a decision problem that involves finding the "best" vector $\boldsymbol{x}$ of the decision variables over all possible vectors in $\Omega$. By the "best" vector, we mean the one that results in the smallest value of the objective function. This vector is called the *minimizer* of $f$ over $\Omega$. It is possible that there may be many minimizers. In this case, finding any of the minimizers will suffice.

There are also optimization problems that require maximization of the objective function. These problems, however, can be represented in the above form since maximizing $f$ is equivalent to minimizing $-f$. Therefore, we can confine our attention to minimization problems without loss of generality.

The above problem is a general form of a *constrained* optimization problem, since the decision variables are constrained to be in the constraint set $\Omega$. If $\Omega = \mathbb{R}^n$, then we refer to the problem as an *unconstrained* optimization problem. In this chapter, we discuss basic properties of the general optimization problem above, which includes the unconstrained case. In the remaining chapters of this part, we deal with iterative algorithms for solving unconstrained optimization problems.

The constraint "$\boldsymbol{x} \in \Omega$" is called a *set constraint*. Often, the constraint set $\Omega$ takes the form $\Omega = \{\boldsymbol{x} : \boldsymbol{h}(\boldsymbol{x}) = \boldsymbol{0}, \boldsymbol{g}(\boldsymbol{x}) \leqslant \boldsymbol{0}\}$, where $\boldsymbol{h}$ and $\boldsymbol{g}$ are given functions. We refer to such constraints as *functional constraints*. The remainder of this chapter

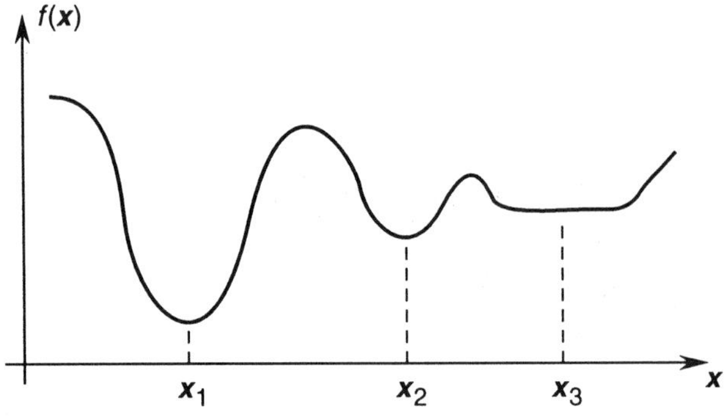


**Figure 6.1.** Examples of minimizers: $x_1$: strict global minimizer; $x_2$: strict local minimizer; $x_3$: local (not strict) minimizer

deals with general set constraints. In Parts III and IV, we consider constrained optimization problems with functional constraints.

In considering the general optimization problem above, we distinguish between two kinds of minimizers, as specified by the following definitions.

***Definition 6.1.*** ***Local minimizer.*** Suppose $f:\mathbb{R}^n \to \mathbb{R}$ is a real-valued function defined on some set $\Omega \subset \mathbb{R}^n$. A point $x^* \in \Omega$ is a local minimizer of $f$ over $\Omega$ if there exists $\varepsilon > 0$ such that $f(x) \geqslant f(x^*)$ for all $x \in \Omega \backslash \{x^*\}$ and $\| x - x^* \| < \varepsilon$.

***Global minimizer.*** A point $x^* \in \Omega$ is a global minimizer of $f$ over $\Omega$ if $f(x) \geqslant f(x^*)$ for all $x \in \Omega \backslash \{x^*\}$. ■

If, in the above definitions, we replace "$\geqslant$" with "$>$", then we have a *strict local minimizer* and a *strict global minimizer*, respectively. In Figure 6.1, we graphically illustrate the above definitions for $n = 1$.

Strictly speaking, an optimization problem is solved only when a global minimizer is found. However, global minimizers are in general difficult to find. Therefore, in practice, we often have to be satisfied with finding local minimizers.

## 6.2. CONDITIONS FOR LOCAL MINIMA

In this section, we derive conditions for a point $x^*$ to be a local minimizer. We use derivatives of a function $f:\mathbb{R}^n \to \mathbb{R}$. Recall that the first-order derivative of $f$, denoted by $Df$, is

$$Df \triangleq \left[\frac{\partial f}{\partial x_1}, \frac{\partial f}{\partial x_2}, \ldots, \frac{\partial f}{\partial x_n}\right].$$

Note that the gradient $\nabla f$ is just the transpose of $Df$; that is, $\nabla f = (Df)^T$. The second derivative of $f:\mathbb{R}^n \to \mathbb{R}$ (also called the *Hessian* of $f$) is

$$F(x) \triangleq D^2 f(x) = \begin{bmatrix} \frac{\partial^2 f}{\partial x_1^2}(x) & \cdots & \frac{\partial^2 f}{\partial x_n \partial x_1}(x) \\ \vdots & & \vdots \\ \frac{\partial^2 f}{\partial x_1 \partial x_n}(x) & \cdots & \frac{\partial^2 f}{\partial x_n^2}(x) \end{bmatrix}.$$

***Example 6.1.*** Let $f(x_1, x_2) = 5x_1 + 8x_2 + x_1 x_2 - x_1^2 - 2x_2^2$. Then

$$Df(x) = (\nabla f(x))^T = \left[\frac{\partial f}{\partial x_1}(x), \frac{\partial f}{\partial x_2}(x)\right] = [5 + x_2 - 2x_1, 8 + x_1 - 4x_2],$$

and

$$F(x) = D^2 f(x) = \begin{bmatrix} \frac{\partial^2 f}{\partial x_1^2}(x) & \frac{\partial^2 f}{\partial x_2 \partial x_1}(x) \\ \frac{\partial^2 f}{\partial x_1 \partial x_2}(x) & \frac{\partial^2 f}{\partial x_2^2}(x) \end{bmatrix} = \begin{bmatrix} -2 & 1 \\ 1 & -4 \end{bmatrix}.$$ ■

Given an optimization problem with constraint set $\Omega$, a minimizer may lie either in the interior or on the boundary of $\Omega$. To study the case when it lies on the boundary, we need the notion of feasible directions.

***Definition 6.2. Feasible direction.*** A vector $d \in \mathbb{R}^n$, $d \neq 0$, is a feasible direction at $x \in \Omega$ if there exists $\alpha_0 > 0$ such that $x + \alpha d \in \Omega$ for all $\alpha \in [0, \alpha_0]$. ■

Figure 6.2 illustrates the notion of feasible directions geometrically.

Let $f:\mathbb{R}^n \to \mathbb{R}$ be a real-valued function and let $d$ be a feasible direction at $x \in \Omega$. The *directional derivative* of $f$ in the direction $d$, denoted by $\partial f/\partial d$, is the

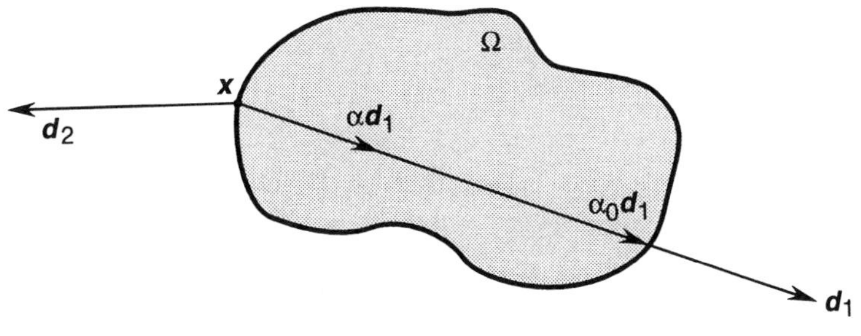


**Figure 6.2.** Two-dimensional illustration of feasible directions; $\mathbf{d}_1$ is a feasible direction, $d_2$ is not a feasible direction

real-valued function defined by

$$\frac{\partial f}{\partial \boldsymbol{d}}(\boldsymbol{x}) = \lim_{\alpha \to 0} \frac{f(\boldsymbol{x} + \alpha \boldsymbol{d}) - f(\boldsymbol{x})}{\alpha}.$$

If $\|\boldsymbol{d}\| = 1$, then $\partial f / \partial \boldsymbol{d}$ is the rate of increase of $f$ at $\boldsymbol{x}$ in the direction $\boldsymbol{d}$. To compute the above directional derivative, suppose that $\boldsymbol{x}$ and $\boldsymbol{d}$ are given. Then, $f(\boldsymbol{x} + \alpha \boldsymbol{d})$ is a function of $\alpha$, and

$$\frac{\partial f}{\partial \boldsymbol{d}}(\boldsymbol{x}) = \frac{d}{d\alpha} f(\boldsymbol{x} + \alpha \boldsymbol{d}) \bigg|_{\alpha = 0}$$

Applying the chain rule yields

$$\frac{\partial f}{\partial \boldsymbol{d}}(\boldsymbol{x}) = \frac{d}{d\alpha} f(\boldsymbol{x} + \alpha \boldsymbol{d}) \bigg|_{\alpha = 0} = \nabla f(\boldsymbol{x})^T \boldsymbol{d} = \langle \nabla f(\boldsymbol{x}), \boldsymbol{d} \rangle = \boldsymbol{d}^T \nabla f(\boldsymbol{x}).$$

In summary, if $\boldsymbol{d}$ is a unit vector (that is, $\|\boldsymbol{d}\| = 1$), then $\langle \nabla f(\boldsymbol{x}), \boldsymbol{d} \rangle$ is the rate of increase of $f$ at the point $\boldsymbol{x}$ in the direction $\boldsymbol{d}$.

***Example 6.2.*** Define $f: \mathbb{R}^n \to \mathbb{R}$ by $f(\boldsymbol{x}) = x_1 x_2 x_3$, and let

$$\boldsymbol{d} = \left[\frac{1}{2}, \frac{1}{2}, \frac{1}{\sqrt{2}}\right]^T.$$

The directional derivative of $f$ in the direction $\boldsymbol{d}$ is

$$\frac{\partial f}{\partial \boldsymbol{d}}(\boldsymbol{x}) = \nabla f(\boldsymbol{x})^T \boldsymbol{d} = [x_2 x_3, x_1 x_3, x_1 x_2] \begin{bmatrix} \frac{1}{2} \\ \frac{1}{2} \\ \frac{1}{\sqrt{2}} \end{bmatrix} = \frac{x_2 x_3 + x_1 x_3 + \sqrt{2} x_1 x_2}{2}.$$

Note that since $\|\boldsymbol{d}\| = 1$, the above is also the rate of increase of $f$ at $\boldsymbol{x}$ in the direction $\boldsymbol{d}$. ■

We are now ready to state and prove the following theorem.

**Theorem 6.1. First-Order Necessary Condition (FONC).** *Let $\Omega$ be a subset of $\mathbb{R}^n$ and $f \in \mathscr{C}^1$ a real-valued function on $\Omega$. If $\boldsymbol{x}^*$ is a local minimizer of $f$ over $\Omega$, then for any feasible direction $\boldsymbol{d}$ at $\boldsymbol{x}^*$, we have*

$$\boldsymbol{d}^T \nabla f(\boldsymbol{x}^*) \geqslant 0.$$

□

*Proof.* Define

$$x(\alpha) = x^* + \alpha d \in \Omega.$$

Note that $x(0) = x^*$. Let

$$\phi(\alpha) = f(x(\alpha)).$$

Then, by Taylor's theorem

$$f(x^* + \alpha d) - f(x^*) = \phi(\alpha) - \phi(0) = \phi'(0)\alpha + o(\alpha) = \alpha d^T \nabla f(x(0)) + o(\alpha),$$

where $\alpha \geqslant 0$. Thus, if $\phi(\alpha) \geqslant \phi(0)$, that is, $f(x^* + \alpha d) \geqslant f(x^*)$ for sufficiently small values of $\alpha > 0$ ($x^*$ is a local minimizer), then we have to have $d^T \nabla f(x^*) \geqslant 0$ (see Exercise 5.5). ■

An alternative way to express the FONC is

$$\frac{\partial f}{\partial d}(x^*) \geqslant 0$$

for all feasible directions $d$. We thus conclude that if $x^*$ is a local minimizer then the rate of increase of $f$ at $x^*$ in any feasible direction $d$ in $\Omega$ is nonnegative. The above theorem is graphically illustrated in Figure 6.3.

A special case of interest is when $x^*$ is an interior point of $\Omega$ (see Section 4.4). In this case, any direction is feasible, and we have the following result.

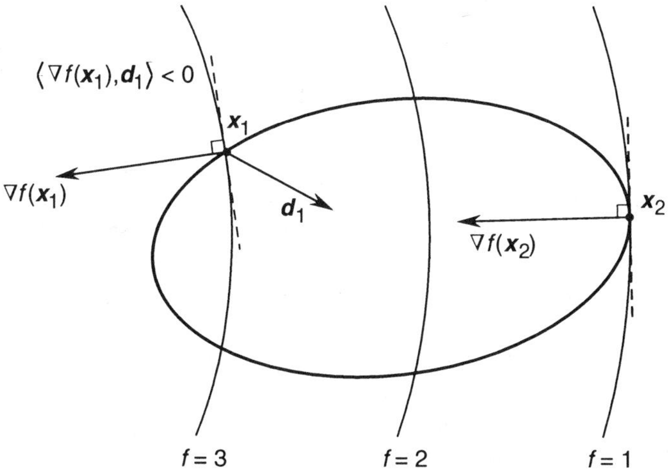


**Figure 6.3.** Illustration of the first-order necessary condition (FONC) for the constrained case; $x_1$ does not satisfy the FONC, $x_2$ satisfies the FONC

**Corollary 6.1. Interior case.** *Let $\Omega$ be a subset of $\mathbb{R}^n$ and $f \in \mathscr{C}^1$ be a real-valued function on $\Omega$. If $x^*$ is a local minimizer of $f$ over $\Omega$ and if $x^*$ is an interior point of $\Omega$, then*

$$\nabla f(x^*) = \mathbf{0}.$$

□

*Proof.* Suppose $f$ has a local minimizer $x^*$ that is an interior point of $\Omega$. Since $x^*$ is an interior point of $\Omega$, the set of feasible directions at $x^*$ is the whole $\mathbb{R}^n$. Thus, for any $d \in \mathbb{R}^n$, $d^T \nabla f(x^*) \geqslant 0$ and $-d^T \nabla f(x^*) \geqslant 0$. Therefore,

$$\nabla f(x^*) = \mathbf{0}.$$

■

***Example 6.3.*** Consider the problem

$$\begin{array}{ll} \text{minimize} & x_1^2 + 0.5x_2^2 + 3x_2 + 4.5 \\ \text{subject to} & x_1, x_2 \geqslant 0. \end{array}$$

Questions:

1. Is the first-order necessary condition (FONC) for a local minimizer satisfied at $x = [1,3]^T$?
2. Is the FONC for a local minimizer satisfied at $x = [0,3]^T$?
3. Is the FONC for a local minimizer satisfied at $x = [1,0]^T$?
4. Is the FONC for a local minimizer satisfied at $x = [0,0]^T$?

Answers: First, let $f: \mathbb{R}^2 \to \mathbb{R}$ be defined by $f(x) = x_1^2 + 0.5x_2^2 + 3x_2 + 4.5$, where $x = [x_1, x_2]^T$.

1. At $x = [1,3]^T$, we have $\nabla f(x) = [2x_1, x_2 + 3]^T = [2,6]^T$. The point $x = [1,3]^T$ is an interior point of $\Omega = \{x : x_1 \geqslant 0, x_2 \geqslant 0\}$. Hence, the FONC requires $\nabla f(x) = \mathbf{0}$. The point $x = [1,3]^T$ does not satisfy the FONC for a local minimum.
2. At $x = [0,3]^T$, we have $\nabla f(x) = [0,6]^T$, and hence $d^T \nabla f(x) = 6d_2$, where $d = [d_1, d_2]^T$. For $d$ to be feasible at $x$, we need $d_1 \geqslant 0$, and $d_2$ can take an arbitrary value in $\mathbb{R}$. The point $x = [0,3]^T$ does not satisfy the FONC for a minimum because $d_2$ is allowed to be less than zero. For example, $d = [1, -1]^T$ is a feasible direction, but $d^T \nabla f(x) = -6 < 0$.
3. At $x = [1,0]^T$, we have $\nabla f(x) = [2,3]^T$, and hence $d^T \nabla f(x) = 2d_1 + 3d_2$. For $d$ to be feasible, we need $d_2 \geqslant 0$, and $d_1$ can take an arbitrary value in $\mathbb{R}$. For example, $d = [-5,1]^T$ is a feasible direction. But $d^T \nabla f(x) = -7 < 0$. Thus, $x = [1,0]^T$ does not satisfy the FONC for a local minimum.
4. At $x = [0,0]^T$, we have $\nabla f(x) = [0,3]^T$, and hence $d^T \nabla f(x) = 3d_2$. For $d$ to be feasible, we need $d_2 \geqslant 0$ and $d_1 \geqslant 0$. Hence, $x = [0,0]^T$ satisfies the FONC for a local minimum.

■

We now derive a second-order necessary condition that is satisfied by a local minimizer.

**Theorem 6.2. Second-Order Necessary Condition (SONC).** *Let $\Omega \subset \mathbb{R}^n$, $f \in \mathscr{C}^2$ a function on $\Omega$, $\boldsymbol{x}^*$ a local minimizer of $f$ over $\Omega$, and $\boldsymbol{d}$ a feasible direction at $\boldsymbol{x}^*$. If $\boldsymbol{d}^T \nabla f(\boldsymbol{x}^*) = 0$, then*

$$\boldsymbol{d}^T \boldsymbol{F}(\boldsymbol{x}^*) \boldsymbol{d} \geqslant 0,$$

*where $\boldsymbol{F}$ is the Hessian of $f$.* □

*Proof.* We prove the result by contradiction. Suppose there is a feasible direction $\boldsymbol{d}$ at $\boldsymbol{x}^*$ such that $\boldsymbol{d}^T \nabla f(\boldsymbol{x}^*) = 0$ and $\boldsymbol{d}^T \boldsymbol{F}(\boldsymbol{x}^*) \boldsymbol{d} < 0$. Let $\boldsymbol{x}(\alpha) = \boldsymbol{x}^* + \alpha \boldsymbol{d}$ and $\phi(\alpha) = f(\boldsymbol{x}^* + \alpha \boldsymbol{d}) = f(\boldsymbol{x}(\alpha))$. Then, by Taylor's formula

$$\phi(\alpha) = \phi(0) + \phi''(0) \frac{\alpha^2}{2} + o(\alpha^2),$$

where by assumption $\phi'(0) = \boldsymbol{d}^T \nabla f(\boldsymbol{x}^*) = 0$, and $\phi''(0) = \boldsymbol{d}^T \boldsymbol{F}(\boldsymbol{x}^*) \boldsymbol{d} < 0$. For sufficiently small $\alpha$,

$$\phi(\alpha) - \phi(0) = \phi''(0) \frac{\alpha^2}{2} + o(\alpha^2) < 0,$$

that is,

$$f(\boldsymbol{x}^* + \alpha \boldsymbol{d}) < f(\boldsymbol{x}^*),$$

which contradicts the assumption that $\boldsymbol{x}^*$ is a local minimizer. Thus,

$$\phi''(0) = \boldsymbol{d}^T \boldsymbol{F}(\boldsymbol{x}^*) \boldsymbol{d} \geqslant 0.$$ ■

**Corollary 6.2. Interior Case.** *Let $\boldsymbol{x}^*$ be an interior point of $\Omega \subset \mathbb{R}^n$. If $\boldsymbol{x}^*$ is a local minimizer of $f: \Omega \to \mathbb{R}$, $f \in \mathscr{C}^2$, then*

$$\nabla f(\boldsymbol{x}^*) = \boldsymbol{0},$$

*and $\boldsymbol{F}(\boldsymbol{x}^*)$ is positive semidefinite ($\boldsymbol{F}(\boldsymbol{x}^*) \geqslant 0$), that is, for all $\boldsymbol{d} \in \mathbb{R}^n$,*

$$\boldsymbol{d}^T \boldsymbol{F}(\boldsymbol{x}^*) \boldsymbol{d} \geqslant 0.$$ □

*Proof.* If $\boldsymbol{x}^*$ is an interior point then all directions are feasible. The result then follows from Corollary 6.1 and Theorem 6.2. ■

In the examples below, we show that the necessary conditions are not sufficient.

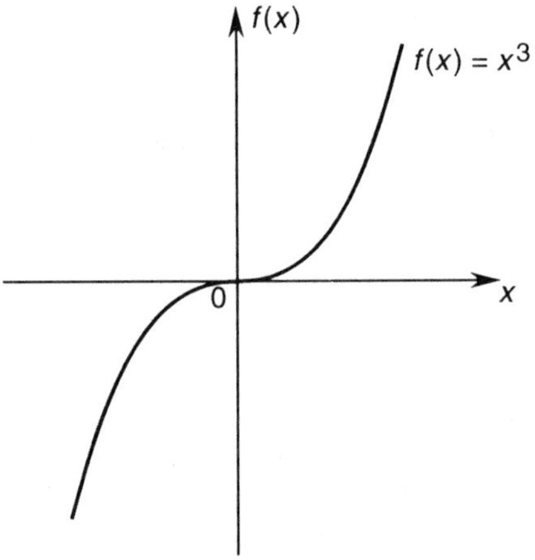


**Figure 6.4.** The point **0** satisfies the first-order necessary condition (FONC) and the second-order necessary condition (SONC), but is not a minimizer

***Example 6.4.*** Consider a function of one variable $f(x) = x^3, f: \mathbb{R} \to \mathbb{R}$. Since $f'(0) = 0$, and $f''(0) = 0$, the point $x = 0$ satisfies both the FONC and SONC. However, $x = 0$ is not a minimizer (see Figure 6.4). ■

***Example 6.5.*** Consider a function $f: \mathbb{R}^2 \to \mathbb{R}$, where $f(\boldsymbol{x}) = x_1^2 - x_2^2$. The FONC requires that $\nabla f(\boldsymbol{x}) = [2x_1, -2x_2]^T = \mathbf{0}$. Thus, $\boldsymbol{x} = [0, 0]^T$ satisfies the FONC. The Hessian matrix of $f$ is

$$\boldsymbol{F}(\boldsymbol{x}) = \begin{bmatrix} 2 & 0 \\ 0 & -2 \end{bmatrix}.$$

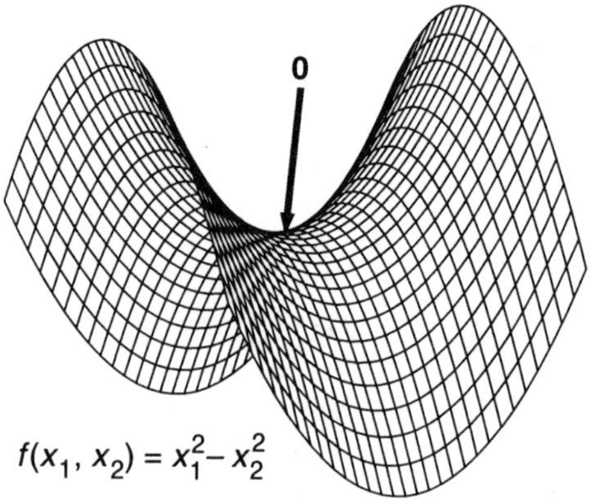


**Figure 6.5.** The point **0** satisfies the first- but not the second-order necessary condition; this point is not a minimizer

The Hessian matrix is indefinite, that is, for some $\boldsymbol{d}_1 \in \mathbb{R}^2$, we have $\boldsymbol{d}_1^T \boldsymbol{F} \boldsymbol{d}_1 > 0$ (e.g., $\boldsymbol{d}_1 = [1, 0]^T$), and, for some $\boldsymbol{d}_2$, we have $\boldsymbol{d}_2^T \boldsymbol{F} \boldsymbol{d}_2 < 0$ (e.g., $\boldsymbol{d}_2 = [0, 1]^T$). Thus, $\boldsymbol{x} = [0, 0]^T$ does not satisfy the SONC, and hence it is not a minimizer. The graph of $f = x_1^2 - x_2^2$ is shown in Figure 6.5. ■

We now derive sufficient conditions that imply that $\boldsymbol{x}^*$ is a local minimizer.

**Theorem 6.3. Second-Order Sufficient Condition (SOSC) Interior Case.** *Let $f \in \mathscr{C}^2$ be defined on a region in which $\boldsymbol{x}^*$ is an interior point. Suppose that*

1. $\nabla f(\boldsymbol{x}^*) = \boldsymbol{0}$; *and*
2. $\boldsymbol{F}(\boldsymbol{x}^*) > 0$.

*Then, $\boldsymbol{x}^*$ is a strict local minimizer of $f$.* □

*Proof.* Since $f \in \mathscr{C}^2$, $\boldsymbol{F}(\boldsymbol{x}^*) = \boldsymbol{F}^T(\boldsymbol{x}^*)$. Using assumption 2 and Rayleigh's inequality it follows that if $\boldsymbol{d} \neq \boldsymbol{0}$, then $0 < \lambda_{\min}(\boldsymbol{F}(\boldsymbol{x}^*)) \| \boldsymbol{d} \|^2 \leqslant \boldsymbol{d}^T \boldsymbol{F}(\boldsymbol{x}^*) \boldsymbol{d}$. From Taylor's theorem

$$f(\boldsymbol{x}^* + \boldsymbol{d}) - f(\boldsymbol{x}^*) = \frac{1}{2} \boldsymbol{d}^T \boldsymbol{F}(\boldsymbol{x}^*) \boldsymbol{d} + o(\| \boldsymbol{d} \|^2) \geqslant \frac{\lambda_{\min}(\boldsymbol{F}(\boldsymbol{x}^*))}{2} \| \boldsymbol{d} \|^2 + o(\| \boldsymbol{d} \|^2),$$

since by assumption 1, $\nabla f(\boldsymbol{x}^*) = \boldsymbol{0}$. For all $\boldsymbol{d}$ such that $\| \boldsymbol{d} \|$ is sufficiently small,

$$f(\boldsymbol{x}^* + \boldsymbol{d}) > f(\boldsymbol{x}^*),$$

and the proof is completed. ■

***Example 6.6.*** Let $f(\boldsymbol{x}) = x_1^2 + x_2^2$. We have $\nabla f(\boldsymbol{x}) = [2x_1, 2x_2]^T = \boldsymbol{0}$ if and only if $\boldsymbol{x} = [0, 0]^T$. For all $\boldsymbol{x} \in \mathbb{R}^2$, we have

$$\boldsymbol{F}(\boldsymbol{x}) = \begin{bmatrix} 2 & 0 \\ 0 & 2 \end{bmatrix} > 0.$$

The point $\boldsymbol{x} = [0, 0]^T$ satisfies the FONC, SONC, and SOSC. It is a strict local minimizer. Actually $\boldsymbol{x} = [0, 0]^T$ is a strict global minimizer. Figure 6.6 shows the graph of $f(\boldsymbol{x}) = x_1^2 + x_2^2$.

In this chapter, we presented a theoretical basis for the solution of nonlinear unconstrained problems. In the following chapters, we are concerned with iterative methods for solving such problems. Such methods are of great importance in practice. Indeed, suppose one is confronted with a highly nonlinear function of 20 variables. Then, the FONC requires the solution of 20 nonlinear simultaneous equations of 20 variables. These equations, being nonlinear, will

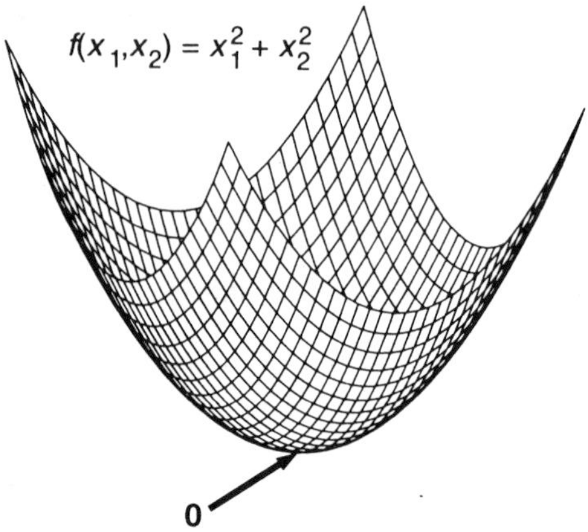


**Figure 6.6.** Graph of $f(\boldsymbol{x}) = x_1^2 + x_2^2$

normally have multiple solutions. In addition, we would have to compute 210 second derivatives (provided $f \in \mathscr{C}^2$) to use the SONC or SOSC. We begin our discussion of iterative methods in the next chapter with search methods for functions of one variable.

## EXERCISES

**6.1** Show that, if $\boldsymbol{x}^*$ is a global minimizer of $f$ over $\Omega$, and $\boldsymbol{x}^* \in \Omega' \subset \Omega$, then $\boldsymbol{x}^*$ is a global minimizer of $f$ over $\Omega'$.

**6.2** Suppose that $\boldsymbol{x}^*$ is a local minimizer of $f$ over $\Omega$, and $\Omega \subset \Omega'$. Show that, if $\boldsymbol{x}^*$ is an interior point of $\Omega$, then $\boldsymbol{x}^*$ is a local minimizer of $f$ over $\Omega'$. Show that the same conclusion cannot be made if $\boldsymbol{x}^*$ is not an interior point of $\Omega$.

**6.3** Consider the function $f: \mathbb{R}^2 \to \mathbb{R}$ given below:

$$f(\boldsymbol{x}) = \boldsymbol{x}^T \begin{bmatrix} 1 & 2 \\ 4 & 7 \end{bmatrix} \boldsymbol{x} + \boldsymbol{x}^T \begin{bmatrix} 3 \\ 5 \end{bmatrix} + 6$$

**a.** Find the gradient and Hessian of $f$ at the point $[1, 1]^T$.

**b.** Find the directional derivative of $f$ at $[1, 1]^T$ with respect to a unit vector in the direction of maximal rate of increase.

**c.** Find a point that satisfies the FONC (interior case) for $f$. Does this point satisfy the SONC (for a minimizer)?

**6.4** Suppose we are given $n$ real numbers, $x_1, \ldots, x_n$. Find the number $\bar{x} \in \mathbb{R}$ such that the sum of the squared difference between $\bar{x}$ and the above numbers is minimized (assuming the solution $\bar{x}$ exists).

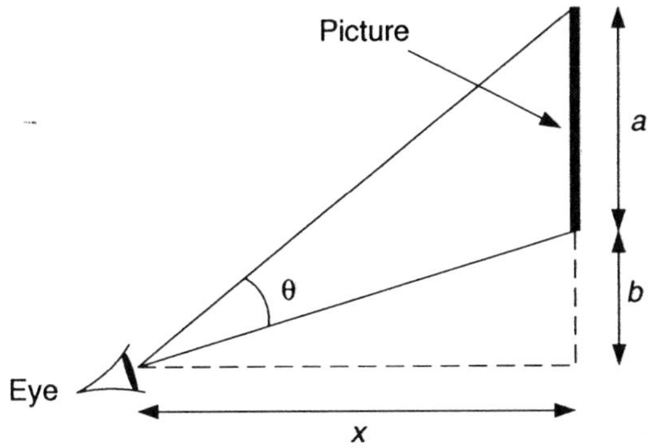


**Figure 6.7.** Exercise 6.5

**6.5** An art collector stands at distance of $x$ feet from the wall where a piece of art (picture) of height $a$ feet is hung, $b$ feet above his eyes, as shown above. Find the distance from the wall for which the angle $\theta$ subtended by the eye to the picture is maximized. *Hint*: (1) Maximizing $\theta$ is equivalent to maximizing $\tan(\theta)$. (2) If $\theta = \theta_2 - \theta_1$, then $\tan(\theta) = (\tan(\theta_2) - \tan(\theta_1))/(1 + \tan(\theta_2)\tan(\theta_1))$.

**6.6** Let $f: \mathbb{R}^2 \to \mathbb{R}$ be defined by

$$f(\boldsymbol{x}) = (x_1 - x_2)^4 + x_1^2 - x_2^2 - 2x_1 + 2x_2 + 1,$$

where $\boldsymbol{x} = [x_1, x_2]^T$. Suppose we wish to minimize $f$ over $\mathbb{R}^2$. Find all points satisfying the FONC. Do these points satisfy the SONC?

**6.7** Show that if $\boldsymbol{d}$ is a feasible direction at a point $\boldsymbol{x} \in \Omega$, then for all $\beta > 0$, the vector $\beta\boldsymbol{d}$ is also a feasible direction at $\boldsymbol{x}$.

**6.8** Let $f: \mathbb{R}^2 \to \mathbb{R}$. Consider the problem

$$\begin{array}{ll} \text{minimize} & f(\boldsymbol{x}) \\ \text{subject to} & x_1, x_2 \geqslant 0, \end{array}$$

where $\boldsymbol{x} = [x_1, x_2]^T$. Suppose $\nabla f(\mathbf{0}) \neq \mathbf{0}$, and

$$\frac{\partial f}{\partial x_1}(\mathbf{0}) \leqslant 0, \quad \frac{\partial f}{\partial x_2}(\mathbf{0}) \leqslant 0.$$

Show that $\mathbf{0}$ cannot be a minimizer for the above problem.

**6.9** Let $\boldsymbol{c} \in \mathbb{R}^n$, $\boldsymbol{c} \neq \mathbf{0}$, and consider the problem of minimizing the function

$f(\boldsymbol{x}) = \boldsymbol{c}^T\boldsymbol{x}$ over a constraint set $\Omega \subset \mathbb{R}^n$. Show that we cannot have a solution lying in the interior of $\Omega$.

**6.10** Consider the problem:

$$\begin{aligned} \text{maximize} \quad & c_1x_1 + c_2x_2 \\ \text{subject to} \quad & x_1 + x_2 \leqslant 1 \\ & x_1x_2 \geqslant 0 \end{aligned}$$

where $c_1$ and $c_2$ are constants such that $c_1 > c_2 \geqslant 0$. The above is a linear programming problem (see Part III). Assuming that the problem has an optimal feasible solution, use the FONC to show that the unique optimal feasible solution $\boldsymbol{x}^*$ is $[1, 0]^T$. *Hint*: First show that $\boldsymbol{x}^*$ cannot lie in the interior of the constraint set. Then, show that $\boldsymbol{x}^*$ cannot lie on the line segments $\boldsymbol{L}_1 = \{\boldsymbol{x}: x_1 = 0, 0 \leqslant x_2 < 1\}$, $\boldsymbol{L}_2 = \{\boldsymbol{x}: 0 \leqslant x_1 < 1,\ x_2 = 0\}$, $\boldsymbol{L}_3 = \{\boldsymbol{x}: 0 \leqslant x_1 < 1, x_2 = 1 - x_1\}$.

**6.11** *Linear Regression.* Let $[x_1, y_1]^T, \ldots, [x_n, y_n]^T, n \geqslant 2$, be points on the $\mathbb{R}^2$ plane (each $x_i, y_i \in \mathbb{R}$). We wish to find the straight line of "best fit" through these points ("best" in the sense that the average squared error is minimized), that is, we wish to find $a, b \in \mathbb{R}$ to minimize

$$f(a, b) = \frac{1}{n} \sum_{i=1}^{n} (ax_i + b - y_i)^2.$$

**a.** Let

$$\overline{X} = \frac{1}{n} \sum_{i=1}^{n} x_i$$

$$\overline{Y} = \frac{1}{n} \sum_{i=1}^{n} y_i$$

$$\overline{X^2} = \frac{1}{n} \sum_{i=1}^{n} x_i^2$$

$$\overline{Y^2} = \frac{1}{n} \sum_{i=1}^{n} y_i^2$$

$$\overline{XY} = \frac{1}{n} \sum_{i=1}^{n} x_iy_i.$$

Show that $f(a, b)$ can be written in the form $\boldsymbol{z}^T\boldsymbol{Q}\boldsymbol{z} - 2\boldsymbol{c}^T\boldsymbol{z} + d$, where $\boldsymbol{z} = [a, b]^T$, $\boldsymbol{Q} = \boldsymbol{Q}^T \in \mathbb{R}^{2\times 2}, \boldsymbol{c} \in \mathbb{R}^2$, and $d \in \mathbb{R}$, and find expressions for $\boldsymbol{Q}, \boldsymbol{c}$, and $d$ in terms of $\overline{X}, \overline{Y}$, $\overline{X^2}$, $\overline{Y^2}$, and $\overline{XY}$.

**b.** Assume that the $x_i, i = 1, \ldots, n$, are not all equal. Find the parameters $a^*$ and $b^*$ for the line of best fit in terms of $\overline{X}$, $\overline{Y}$, $\overline{X^2}$, $\overline{Y^2}$, and $\overline{XY}$. Show that the point $[a^*, b^*]^T$ is the only local minimizer of $f$. *Hint*: $\overline{X^2} - (\overline{X})^2 = (1/n)\sum_{i=1}^{n}(x_i - \overline{X})^2$.

**c.** Show that if $a^*$ and $b^*$ are the parameters of the line of best fit, then $\overline{Y} = a^*\overline{X} + b^*$ (and hence once we have computed $a^*$, we can compute $b^*$ using the formula $b^* = \overline{Y} - a^*\overline{X}$).

**6.12** Suppose we are given a set of vectors $\{\boldsymbol{x}^{(1)}, \ldots, \boldsymbol{x}^{(p)}\}$, $\boldsymbol{x}^{(i)} \in \mathbb{R}^n$, $i = 1, \ldots, p$. Find the vector $\bar{\boldsymbol{x}} \in \mathbb{R}^n$ such that the average squared distance (norm) between $\bar{\boldsymbol{x}}$ and $\boldsymbol{x}^{(1)}, \ldots, \boldsymbol{x}^{(p)}$,

$$\frac{1}{p}\sum_{i=1}^{p} \| \bar{\boldsymbol{x}} - \boldsymbol{x}^{(i)} \|^2,$$

is minimized. Use the SOSC to prove that the vector $\bar{\boldsymbol{x}}$ found above is a strict local minimizer.

**6.13** Prove the following generalization of the SOSC:

**Theorem.** *Let $\Omega$ be a convex subsest, $\mathbb{R}^n$, $f \in \mathscr{C}^2$ a real-valued function on $\Omega$, and $\boldsymbol{x}^*$ a point in $\Omega$. Suppose that there exists $c \in \mathbb{R}$, $c > 0$, such that for all feasible directions $\boldsymbol{d}$ at $\boldsymbol{x}^*$ ($\boldsymbol{d} \neq \boldsymbol{0}$), the following hold:*

**a.** $\boldsymbol{d}^T \nabla f(\boldsymbol{x}^*) \geqslant 0$; *and*

**b.** $\boldsymbol{d}^T \boldsymbol{F}(\boldsymbol{x}^*)\boldsymbol{d} \geqslant c \| \boldsymbol{d} \|^2$.

*Then, $\boldsymbol{x}^*$ is a strict local minimizer of $f$.* □

**6.14** Consider the quadratic function $f : \mathbb{R}^n \to \mathbb{R}$ given by

$$f(\boldsymbol{x}) = \tfrac{1}{2}\boldsymbol{x}^T \boldsymbol{Q} \boldsymbol{x} - \boldsymbol{x}^T \boldsymbol{b},$$

where $\boldsymbol{Q} = \boldsymbol{Q}^T > 0$. Show that $\boldsymbol{x}^*$ minimizes $f$ if, and only if, $\boldsymbol{x}^*$ satisfies the FONC.

**6.15** Consider the linear system $x_{k+1} = a x_k + b u_{x+1}$, $k \geqslant 0$, where $x_i \in \mathbb{R}$, $u_i \in \mathbb{R}$, and the initial condition is $x_0 = 0$. Find the values of the control inputs $u_1, \ldots, u_n$ to minimize

$$-q x_n + r \sum_{i=1}^{n} u_i^2,$$

where $q, r > 0$ are given constants. The above can be interpreted as desiring to make $x_n$ as large as possible, but at the same time desiring to make the total input energy $\sum_{i=1}^{n} u_i^2$ as small as possible. The constants $q$ and $r$ reflect the relative weights of the above two desires.

# 7

# One-Dimensional Search Methods

## 7.1. GOLDEN SECTION SEARCH

The search methods we discuss in this and the next section allow us to determine the minimizer of a function $f: \mathbb{R} \to \mathbb{R}$ over a closed interval, say $[a_0, b_0]$. The only property that we assume of the objective function $f$ is that it is *unimodal*, which means that $f$ has only one local minimizer. An example of such a function is depicted in Figure 7.1

The methods we discuss are based on evaluating the objective function at different points in the interval $[a_0, b_0]$. We choose these points in such a way that an approximation to the minimizer of $f$ may be achieved in as few evaluations as possible. Our goal is to progressively narrow the range until the minimizer is boxed in with sufficient accuracy.

Consider a unimodal function $f$ of one variable and the interval $[a_0, b_0]$. If we evaluate $f$ at only one intermediate point of the interval, we cannot narrow the range within which we know the minimizer is located. We have to evaluate $f$ at two intermediate points, as illustrated in Figure 7.2. We choose the intermediate points in such a way that the reduction in the range is symmetric, in the sense that

$$a_1 - a_0 = b_0 - b_1 = \rho(b_0 - a_0),$$

where

$$\rho < \tfrac{1}{2}.$$

We then evaluate $f$ at the intermediate points. If $f(a_1) < f(b_1)$, then the minimizer must lie in the range $[a_0, b_1]$. If, on the other hand, $f(a_1) \geqslant f(b_1)$, then the minimizer is located in the range $[a_1, b_0]$ (see Figure 7.3).

Starting with the reduced range of uncertainty we can repeat the process and similarly find two new points, say $a_2$ and $b_2$, using the same value of $\rho < \frac{1}{2}$ as before. However, we would like to minimize the number of the objective function evaluations while reducing the width of the uncertainty range. Suppose, for example, that $f(a_1) < f(b_1)$, as in Figure 7.3. Then, we know that $x^* \in [a_0, b_1]$. Since $a_1$ is already in the uncertainty range and $f(a_1)$ is already known, we can make $a_1$ coincide with $b_2$. Thus, only one new evaluation of $f$ at $a_2$ would be

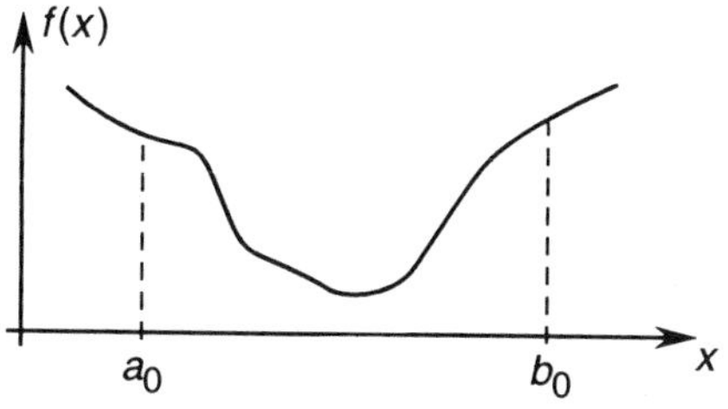


**Figure 7.1.** A unimodal function

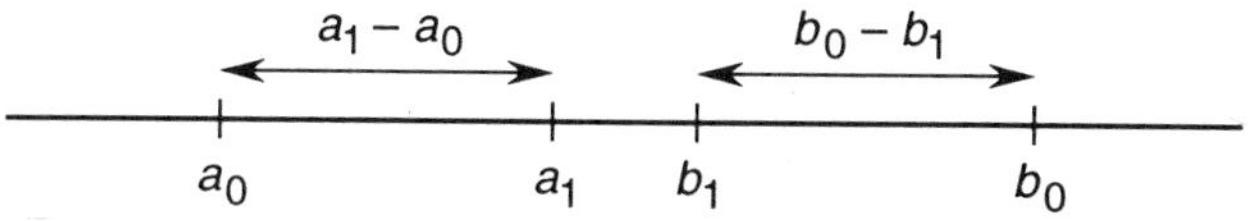


**Figure 7.2.** Evaluating the objective function at two intermediate points

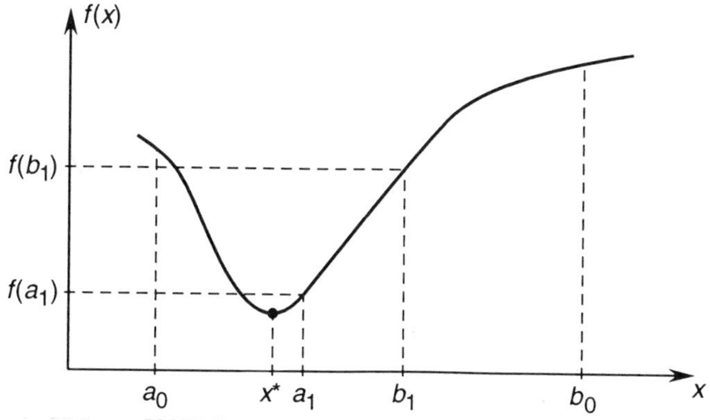


**Figure 7.3.** The case when $f(a_1) < f(b_1)$; the minimizer $x^* \in [a_0, b_1]$

necessary. To find the value of $\rho$ that results in only one new evaluation of $f$, see Figure 7.4. Without loss of generality, imagine that the original range $[a_0, b_0]$ is of unit length. Then, to have only one new evaluation of $f$ it is enough to choose $\rho$ so that

$$\rho(b_1 - a_0) = b_1 - b_2.$$

Since $b_1 - a_0 = 1 - \rho$ and $b_1 - b_2 = 1 - 2\rho$, we have

$$\rho(1 - \rho) = 1 - 2\rho.$$

We write the above quadratic function of $\rho$ as

$$\rho^2 - 3\rho + 1 = 0.$$

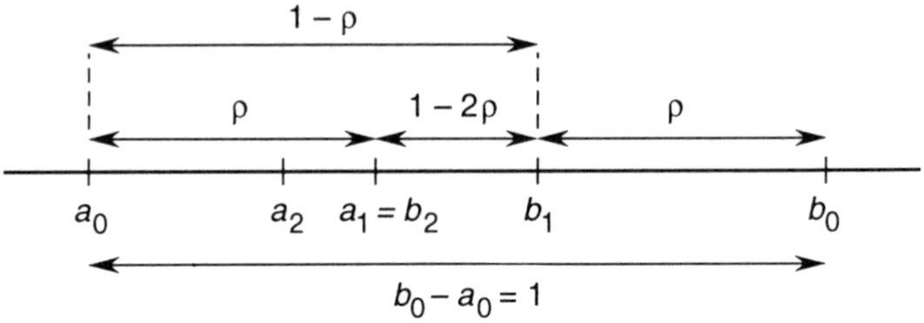


**Figure 7.4.** Finding value of $\rho$ resulting in only one new evaluation of $f$

The solutions are

$$\rho_1 = \frac{3+\sqrt{5}}{2}, \quad \rho_2 = \frac{3-\sqrt{5}}{2}.$$

Since we require $\rho < \frac{1}{2}$, we take

$$\rho = \frac{3-\sqrt{5}}{2} \approx 0.382.$$

Observe that

$$1-\rho = \frac{\sqrt{5}-1}{2},$$

and

$$\frac{\rho}{1-\rho} = \frac{3-\sqrt{5}}{\sqrt{5}-1} = \frac{\sqrt{5}-1}{2} = \frac{1-\rho}{1},$$

that is,

$$\frac{\rho}{1-\rho} = \frac{1-\rho}{1}.$$

Thus, dividing a range in the ratio of $\rho$ to $1-\rho$ has the effect that the ratio of the shorter segment to the longer equals the ratio of the longer to the sum of the two. This rule was referred to by ancient Greek geometers as the *Golden Section.*

Using the Golden Section rule means that at every stage of the uncertainty range reduction (except the first one), the objective function $f$ need only be evaluated at one new point. The uncertainty range is reduced by the ratio $1-\rho \approx 0.61803$ at every stage. Hence, $N$ steps of reduction using the Golden Section method reduce the range by the factor

$$(1-\rho)^N \approx (0.61803)^N.$$

***Example 7.1.*** Use the Golden Section search to find the value of $x$ that minimizes

$$f(x) = x^4 - 14x^3 + 60x^2 - 70x$$

in the range [0, 2] (this function comes from an example in Ref. 13). Locate this value of $x$ to within a 0.3 range.

After $N$ stages, the range [0, 2] is reduced by $(0.61803)^N$. So, we choose $N$ so that

$$(0.61803)^N \leqslant 0.3/2.$$

Four stages of reduction will do, i.e., $N = 4$.

*Iteration 1.* We evaluate $f$ at two intermediate points $a_1$ and $b_1$. We have

$$a_1 = a_0 + \rho(b_0 - a_0) = 0.7639,$$
$$b_1 = a_0 + (1 - \rho)(b_0 - a_0) = 1.236,$$

where $\rho = (3 - \sqrt{5})/2$. We compute

$$f(a_1) = -24.36,$$
$$f(b_1) = -18.96.$$

Thus, $f(a_1) < f(b_1)$, and so the uncertainty interval is reduced to

$$[a_0, b_1] = [0, 1.236].$$

*Iteration 2.* We choose $b_2$ to coincide with $a_1$, and $f$ need only be evaluated at one new point

$$a_2 = a_0 + \rho(b_1 - a_0) = 0.4721.$$

We have

$$f(a_2) = -21.10,$$
$$f(b_2) = f(a_1) = -24.36.$$

Now, $f(b_2) < f(a_2)$, so the uncertainty interval is reduced to

$$[a_2, b_1] = [0.4721, 1.236].$$

*Iteration 3.* We set $a_3 = b_2$, and compute $b_3$:

$$b_3 = a_2 + (1 - \rho)(b_1 - a_2) = 0.9443.$$

We have

$$f(a_3) = f(b_2) = -24.36,$$
$$f(b_3) = -23.59.$$

So $f(b_3) > f(a_3)$. Hence, the uncertainty interval is further reduced to

$$[a_2, b_3] = [0.4721, 0.9443].$$

*Iteration 4*. We set $b_4 = a_3$, and

$$a_4 = a_2 + \rho(b_3 - a_2) = 0.6525.$$

We have

$$f(a_4) = -23.84,$$
$$f(b_4) = f(a_3) = -24.36.$$

Hence, $f(a_4) > f(b_4)$. Thus, the value of $x$ that minimizes $f$ is located in the interval

$$[a_4, b_3] = [0.6525, 0.9443].$$

Note that $b_3 - a_4 = 0.292 < 0.3$. ■

## 7.2. FIBONACCI SEARCH

Recall that the Golden Section method uses the same value of $\rho$ throughout. Suppose now that we are allowed to vary the value $\rho$ from stage to stage, so that at the $k$th stage in the reduction process we use a value $\rho_k$; at the next stage we use a value $\rho_{k+1}$, and so on.

As in the Golden Section search, our goal is to select successive values of $\rho_k$, $0 \leqslant \rho_k \leqslant \frac{1}{2}$, such that only one new function evaluation is required at each stage. To derive the strategy for selecting evaluation points, consider Figure 7.5. From Figure 7.5, we see that it is sufficient to choose the $\rho_k$ such that

$$\rho_{k+1}(1 - \rho_k) = 1 - 2\rho_k.$$

After some manipulations, we obtain

$$\rho_{k+1} = 1 - \frac{\rho_k}{1 - \rho_k}.$$

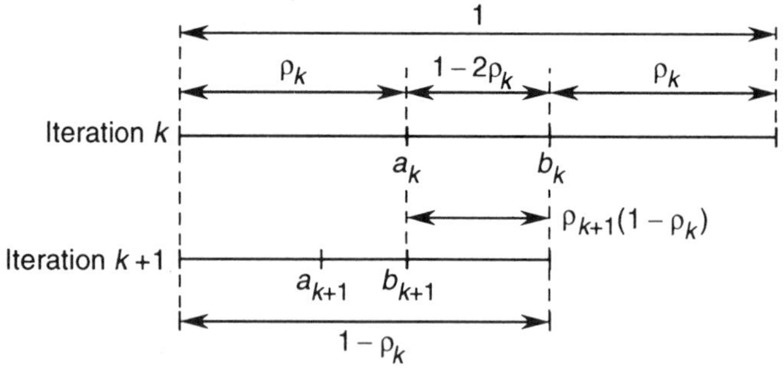


**Figure 7.5.** Selecting evaluation points

There are many sequences $\rho_1, \rho_2, \ldots$ that satisfy the above law of formation, and the condition that $0 \leqslant \rho_k \leqslant \frac{1}{2}$. For example, the sequence $\rho_1 = \rho_2 = \rho_3 = \cdots = (3 - \sqrt{5})/2$ satisfies the above conditions, and gives rise to the Golden Section method.

Suppose we are given a sequence $\rho_1, \rho_2, \ldots$ satisfying the above conditions, and we use this sequence in our search algorithm. Then, after $N$ iterations of the algorithm, the uncertainty range is reduced by a factor of

$$(1-\rho_1)(1-\rho_2)\cdots(1-\rho_N).$$

Depending on the sequence $\rho_1, \rho_2, \ldots$, we get a different reduction factor. The natural question is as follows: What sequence $\rho_1, \rho_2, \ldots$ minimizes the above reduction factor? This problem is a constrained optimization problem that can be formally stated:

$$\begin{aligned} &\text{minimize} \quad (1-\rho_1)(1-\rho_2)\cdots(1-\rho_N) \\ &\text{subject to} \quad \rho_{k+1} = 1 - \frac{\rho_k}{1-\rho_k}, \quad k = 1, \ldots, N-1 \\ &\qquad\qquad\quad 0 \leqslant \rho_k \leqslant \tfrac{1}{2}, \quad k = 1, \ldots, N. \end{aligned}$$

Before we give the solution to the above optimization problem, we first need to introduce the *Fibonacci sequence*, $F_1, F_2, F_3, \ldots$. This sequence is defined as follows. First, let $F_{-1} = 0$ and $F_0 = 1$ by convention. Then, for $k \geqslant 0$,

$$F_{k+1} = F_k + F_{k-1}.$$

Some values of elements in the Fibonacci sequence are as follows:

| $F_1$ | $F_2$ | $F_3$ | $F_4$ | $F_5$ | $F_6$ | $F_7$ | $F_8$ |
|---|---|---|---|---|---|---|---|
| 1 | 2 | 3 | 5 | 8 | 13 | 21 | 34 |

It turns out that the solution to the above optimization problem is:

$$\begin{aligned} \rho_1 &= 1 - \frac{F_N}{F_{N+1}}, \\ \rho_2 &= 1 - \frac{F_{N-1}}{F_N}, \\ &\vdots \\ \rho_k &= 1 - \frac{F_{N-k+1}}{F_{N-k+2}}, \\ &\vdots \\ \rho_N &= 1 - \frac{F_1}{F_2}, \end{aligned}$$

where the $F_k$ are the elements of the Fibonacci sequence. The resulting algorithm is called the *Fibonacci search method*. We present a proof for the optimality of the Fibonacci search method later in this section.

In the Fibonacci search method, the uncertainty range is reduced by the factor

$$(1-\rho_1)(1-\rho_2)\cdots(1-\rho_N)=\frac{F_N}{F_{N+1}}\frac{F_{N-1}}{F_N}\cdots\frac{F_1}{F_2}=\frac{F_1}{F_{N+1}}=\frac{1}{F_{N+1}}.$$

Since the Fibonacci method uses the optimal values of $\rho_1, \rho_2, \ldots$, the above reduction factor is less than that of the Golden Section method. In other words, the Fibonacci method is better than the Golden Section method in that it gives a smaller final uncertainty range.

We point out that there is an anomaly in the final iteration of the Fibonacci search method, because

$$\rho_N = 1 - \frac{F_1}{F_2} = \frac{1}{2}.$$

Recall that we need two intermediate points at each stage, one that comes from a previous iteration and another that is a new evaluation point. However, with $\rho_N = 1/2$, the two intermediate points coincide in the middle of the uncertainty interval, and, therefore, we cannot further reduce the uncertainty range. To get around this problem, we perform the new evaluation for the last iteration using $\rho_N = 1/2 - \varepsilon$, where $\varepsilon$ is a small number. In other words, the new evaluation point is just to the left or right of the midpoint of the uncertainty interval. This modification to the Fibonacci method is, of course, of no significant practical consequence.

As a result of the above modification, the reduction in the uncertainty range at the last iteration may be either

$$1 - \rho_N = \frac{1}{2},$$

or

$$1 - (\rho_N - \varepsilon) = \frac{1}{2} + \varepsilon = \frac{1+2\varepsilon}{2},$$

depending on which of the two points has the smaller objective function value. Therefore, in the worst case, the reduction factor in the uncertainty range for the Fibonacci method is

$$\frac{1+2\varepsilon}{F_{N+1}}.$$

***Example 7.2.*** Consider the function

$$f(x) = x^4 - 14x^3 + 60x^2 - 70x.$$

Use the Fibonacci search method to find the value of $x$ that minimizes $f$ over the range $[0, 2]$. Locate this value of $x$ to within a range 0.3.

After $N$ steps, the range is reduced by $(1 + 2\varepsilon)/F_{N+1}$ in the worst case. We need to choose $N$ such that

$$\frac{1+2\varepsilon}{F_{N+1}} \leqslant \frac{\text{final range}}{\text{initial range}} = \frac{0.3}{2} = 0.15.$$

Thus, we need

$$F_{N+1} \geqslant \frac{1+2\varepsilon}{0.15}.$$

If we choose $\varepsilon \leqslant 0.1$, then $N = 4$ will do.

*Iteration 1*. We start with

$$1 - \rho_1 = \frac{F_4}{F_5} = \frac{5}{8}.$$

We then compute

$$\begin{aligned}
a_1 &= a_0 + \rho_1(b_0 - a_0) = \tfrac{3}{4}\\
b_1 &= a_0 + (1 - \rho_1)(b_0 - a_0) = \tfrac{5}{4}\\
f(a_1) &= -24.34\\
f(b_1) &= -18.65\\
f(a_1) &< f(b_1).
\end{aligned}$$

The range is reduced to

$$[a_0, b_1] = [0, \tfrac{5}{4}].$$

*Iteration 2*. We have

$$\begin{aligned}
1 - \rho_2 &= \frac{F_3}{F_4} = \frac{3}{5}\\
a_2 &= a_0 + \rho_2(b_1 - a_0) = \tfrac{1}{2}\\
b_2 &= a_1 = \frac{3}{4}\\
f(a_2) &= -21.69\\
f(b_2) &= f(a_1) = -24.34\\
f(a_2) &> f(b_2).
\end{aligned}$$

So the range is reduced to

$$[a_2, b_1] = [\tfrac{1}{2}, \tfrac{5}{4}].$$

*Iteration 3.* We compute

$$\begin{aligned}
1 - \rho_3 &= \frac{F_2}{F_3} = \frac{2}{3} \\
a_3 &= b_2 = \tfrac{3}{4} \\
b_3 &= a_2 + (1 - \rho_3)(b_1 - a_2) = 1 \\
f(a_3) &= f(b_2) = -24.34 \\
f(b_3) &= -23 \\
f(a_3) &< f(b_3).
\end{aligned}$$

The range is reduced to

$$[a_2, b_3] = [\tfrac{1}{2}, 1].$$

*Iteration 4.* We choose $\varepsilon = 0.05$. We have

$$\begin{aligned}
1 - \rho_4 &= \frac{F_1}{F_2} = \frac{1}{2} \\
a_4 &= a_2 + (\rho_4 - \varepsilon)(b_3 - a_2) = 0.725 \\
b_4 &= a_3 = \tfrac{3}{4} \\
f(a_4) &= -24.27 \\
f(b_4) &= f(a_3) = -24.34 \\
f(a_4) &> f(b_4).
\end{aligned}$$

The range is reduced to

$$[a_4, b_3] = [0.725, 1].$$

Note $b_3 - a_4 = 0.275 < 0.3$. ∎

For the diligent reader, we now turn to a proof of the optimality of the Fibonacci search method. Skipping the rest of this section will not affect the continuity of the presentation.

To begin, recall that we wish to prove that the values of $\rho_1, \rho_2, \ldots, \rho_N$ used in the Fibonacci method, where $\rho_k = 1 - F_{N-k+1}/F_{N-k+2}$, solve the optimization problem:

$$\begin{aligned} \text{minimize} \quad & (1-\rho_1)(1-\rho_2)\cdots(1-\rho_N) \\ \text{subject to} \quad & \rho_{k+1} = 1 - \frac{\rho_k}{1-\rho_k}, \quad k = 1, \ldots, N-1 \\ & 0 \leqslant \rho_k \leqslant \tfrac{1}{2}, \ k = 1, \ldots, N. \end{aligned}$$

It is easy to check that the values of $\rho_1, \rho_2, \ldots$ above for the Fibonacci search method satisfy the feasibility conditions in the optimization problem above (see Exercise 7.4). Recall that the Fibonacci method has an overall reduction factor of $(1-\rho_1)\cdots(1-\rho_N) = 1/F_{N+1}$. To prove that the Fibonacci search method is optimal, we will show that for any feasible values of $\rho_1, \ldots, \rho_N$, we have $(1-\rho_1)\cdots(1-\rho_N) \geqslant 1/F_{N+1}$.

It will be more convenient to work with $r_k = 1 - \rho_k$ rather than $\rho_k$. The optimization problem stated in terms of $r_k$ is:

$$\begin{aligned} \text{minimize} \quad & r_1 \cdots r_N \\ \text{subject to} \quad & r_{k+1} = \frac{1}{r_k} - 1, \ k = 1, \ldots, N-1 \\ & \tfrac{1}{2} \leqslant r_k \leqslant 1, \ k = 1, \ldots, N. \end{aligned}$$

Note that, if $r_1, r_2, \ldots$ satisfy $r_{k+1} = (1/r_k) - 1$, then $r_k \geqslant \frac{1}{2}$ if, and only if, $r_{k+1} \leqslant 1$. Also, $r_k \geqslant \frac{1}{2}$ if, and only if, $r_{k-1} \leqslant \frac{2}{3} \leqslant 1$. Therefore, in the above constraints, we may remove the constraint $r_k \leqslant 1$, since it is implicitly implied by $r_k \geqslant \frac{1}{2}$ and the other constraints. Therefore, the above constraints reduce to

$$\begin{aligned} r_{k+1} &= \frac{1}{r_k} - 1, \quad && k = 1, \ldots, N-1 \\ r_k &\geqslant \tfrac{1}{2}, && k = 1, \ldots, N. \end{aligned}$$

To proceed, we need the following technical lemmas. In the statements of the lemmas, we assume that $r_1, r_2, \ldots$ is a sequence that satisfies

$$r_{k+1} = \frac{1}{r_k} - 1, \quad r_k \geqslant \tfrac{1}{2}, \quad k = 1, 2, \ldots.$$

**Lemma 7.1.** *For* $k \geqslant 2$,

$$r_k = -\frac{F_{k-2} - F_{k-1} r_1}{F_{k-3} - F_{k-2} r_1}. \qquad \square$$

*Proof.* We proceed by induction. For $k = 2$, we have

$$r_2 = \frac{1}{r_1} - 1 = \frac{1 - r_1}{r_1} = -\frac{F_0 - F_1 r_1}{F_{-1} - F_0 r_1},$$

and hence the lemma holds for $k = 2$. Suppose now that the lemma holds for $k \geqslant 2$. We show that it also holds for $k + 1$. We have

$$\begin{aligned} r_{k+1} &= \frac{1}{r_k} - 1 \\ &= \frac{-F_{k-3} + F_{k-2} r_1}{F_{k-2} - F_{k-1} r_1} - \frac{F_{k-2} - F_{k-1} r_1}{F_{k-2} - F_{k-1} r_1} \\ &= -\frac{F_{k-2} + F_{k-3} - (F_{k-1} + F_{k-2}) r_1}{F_{k-2} - F_{k-1} r_1} \\ &= -\frac{F_{k-1} - F_k r_1}{F_{k-2} - F_{k-1} r_1}, \end{aligned}$$

where we used the formation law for the Fibonacci sequence. ■

**Lemma 7.2.** *For* $k \geqslant 2$,

$$(-1)^k (F_{k-2} - F_{k-1} r_1) > 0. \qquad \square$$

*Proof.* We proceed by induction. For $k = 2$, we have

$$(-1)^2 (F_0 - F_1 r_1) = 1 - r_1.$$

But $r_1 = 1/(1 + r_2) \leqslant \frac{2}{3}$, and hence $1 - r_1 > 0$. Therefore, the result holds for $k = 2$. Suppose now that the lemma holds for $k \geqslant 2$. We show that it also holds for $k + 1$. We have

$$(-1)^{k+1} (F_{k-1} - F_k r_1) = (-1)^{k+1} r_{k+1} \frac{1}{r_{k+1}} (F_{k-1} - F_k r_1).$$

By Lemma 7.1,

$$r_{k+1} = -\frac{F_{k-1} - F_k r_1}{F_{k-2} - F_{k-1} r_1}.$$

Substituting for $1/r_{k+1}$, we obtain

$$(-1)^{k+1} (F_{k-1} - F_k r_1) = r_{k+1} (-1)^k (F_{k-2} - F_{k-1} r_1) > 0,$$

and the proof is completed. ■

**Lemma 7.3.** *For $k \geqslant 2$,*

$$(-1)^{k+1} r_1 \geqslant (-1)^{k+1} \frac{F_k}{F_{k+1}}. \qquad \square$$

*Proof.* Since $r_{k+1} = (1/r_k) - 1$ and $r_k \geqslant \frac{1}{2}$, then $r_{k+1} \leqslant 1$. Substituting for $r_{k+1}$ from Lemma 7.1, we get

$$-\frac{F_{k-1} - F_k r_1}{F_{k-2} - F_{k-1} r_1} \leqslant 1.$$

Multiplying the numerator and denominator by $(-1)^k$ yields

$$\frac{(-1)^{k+1}(F_{k-1} - F_k r_1)}{(-1)^k (F_{k-2} - F_{k-1} r_1)} \leqslant 1.$$

By Lemma 7.2, $(-1)^k (F_{k-2} - F_{k-1} r_1) > 0$, and therefore we can multiply both sides of the above inequality by $(-1)^k (F_{k-2} - F_{k-1} r_1)$ to obtain

$$(-1)^{k+1}(F_{k-1} - F_k r_1) \leqslant (-1)^k (F_{k-2} - F_{k-1} r_1).$$

Rearranging the above yields

$$(-1)^{k+1}(F_{k-1} + F_k) r_1 \geqslant (-1)^{k+1}(F_{k-2} + F_{k-1}).$$

Using the law of formation of the Fibonacci sequence, we get

$$(-1)^{k+1} F_{k+1} r_1 \geqslant (-1)^{k+1} F_k$$

which upon dividing by $F_{k+1}$ on both sides gives the desired result. ■

We are now ready to prove the optimality of the Fibonacci search method, and the uniqueness of this optimal solution.

**Theorem 7.1.** *Let $r_1, \ldots, r_N$, $N \geqslant 2$, satisfy the constraints*

$$r_{k+1} = \frac{1}{r_k} - 1, \quad k = 1, \ldots, N-1$$

$$r_k \geqslant \tfrac{1}{2}, \quad k = 1, \ldots, N.$$

*Then,*

$$r_1 \cdots r_N \geqslant \frac{1}{F_{N+1}}.$$

*Furthermore,*

$$r_1 \cdots r_N = \frac{1}{F_{N+1}}$$

*if, and only if,* $r_k = F_{N-k+1}/F_{N-k+2}$, $k = 1,\ldots,N$. *In other words, the values of* $r_1,\ldots,r_N$ *used in the Fibonacci search method form the unique solution to the optimization problem.* □

*Proof.* By substituting expressions for $r_1,\ldots,r_N$ from Lemma 7.1 and performing the appropriate cancellations, we obtain

$$r_1\cdots r_N = (-1)^N(F_{N-2} - F_{N-1}r_1) = (-1)^N F_{N-2} + F_{N-1}(-1)^{N+1}r_1.$$

Using Lemma 7.3,

$$\begin{aligned} r_1\cdots r_N &\geqslant (-1)^N F_{N-2} + F_{N-1}(-1)^{N+1}\frac{F_N}{F_{N+1}} \\ &= (-1)^N(F_{N-2}F_{N+1} - F_{N-1}F_N)\frac{1}{F_{N+1}}. \end{aligned}$$

By Exercise 7.5, $(-1)^N(F_{N-2}F_{N+1} - F_{N-1}F_N) = 1$. Hence,

$$r_1\cdots r_N \geqslant \frac{1}{F_{N+1}}.$$

From the above, we see that

$$r_1\cdots r_N = \frac{1}{F_{N+1}}$$

if, and only if,

$$r_1 = \frac{F_N}{F_{N+1}}.$$

The above is simply the value of $r_1$ for the Fibonacci search method. Note that fixing $r_1$ uniquely determines $r_2,\ldots,r_N$. ■

For further discussion on the Fibonacci search method and its variants, see Ref. 78.

## 7.3. NEWTON'S METHOD

Suppose again that we are confronted with the problem of minimizing a function $f$ of a single real variable $x$. We assume now that at each measurement point $x^{(k)}$ we can calculate $f(x^{(k)})$, $f'(x^{(k)})$, and $f''(x^{(k)})$. We can fit a quadratic function through $x^{(k)}$ that matches its first and second derivatives with that of the function

$f$. This quadratic has the form

$$q(x) = f(x^{(k)}) + f'(x^{(k)})(x - x^{(k)}) + \tfrac{1}{2} f''(x^{(k)})(x - x^{(k)})^2.$$

Note that $q(x^{(k)}) = f(x^{(k)}), q'(x^{(k)}) = f'(x^{(k)})$, and $q''(x^{(k)}) = f''(x^{(k)})$. Then, instead of minimizing $f$, we minimize its approximation $q$. The first-order necessary condition for a minimum of $q$ yields

$$0 = q'(x) = f'(x^{(k)}) + f''(x^{(k)})(x - x^{(k)}).$$

Setting $x = x^{(k+1)}$, we find

$$x^{(k+1)} = x^{(k)} - \frac{f'(x^{(k)})}{f''(x^{(k)})}.$$

***Example*** *7.3.* Using Newton's method, find the minimum of

$$f(x) = \tfrac{1}{2}x^2 - \sin x.$$

(the above function is taken from Ref. 13). The initial value is $x^{(0)} = 0.5$. The required accuracy is $\varepsilon = 10^{-5}$, in the sense that we stop when $|x^{(k+1)} - x^{(k)}| < \varepsilon$.

We compute

$$f'(x) = x - \cos x, \quad f''(x) = 1 + \sin x.$$

Hence

$$\begin{aligned} x^{(1)} &= 0.5 - \left[\frac{0.5 - \cos 0.5}{1 + \sin 0.5}\right] \\ &= 0.5 - \left[\frac{-0.3775}{1.479}\right] \\ &= 0.7552. \end{aligned}$$

Proceeding in a similar manner, we obtain

$$x^{(2)} = x^{(1)} - \frac{f'(x^{(1)})}{f''(x^{(1)})} = x^{(1)} - \frac{0.02710}{1.685} = 0.7391,$$

$$x^{(3)} = x^{(2)} - \frac{f'(x^{(2)})}{f''(x^{(2)})} = x^{(2)} - \frac{9.461 \times 10^{-5}}{1.673} = 0.7390,$$

$$x^{(4)} = x^{(3)} - \frac{f'(x^{(3)})}{f''(x^{(3)})} = x^{(3)} - \frac{1.17 \times 10^{-9}}{1.673} = 0.7390.$$

Note that $|x^{(4)} - x^{(3)}| < \varepsilon = 10^{-5}$. Furthermore $f'(x^{(4)}) = -8.6 \times 10^{-6} \approx 0$. Observe that $f''(x^{(4)}) = 1.673 > 0$, so we can assume that $x^* \approx x^{(4)}$ is a strict minimizer. ■

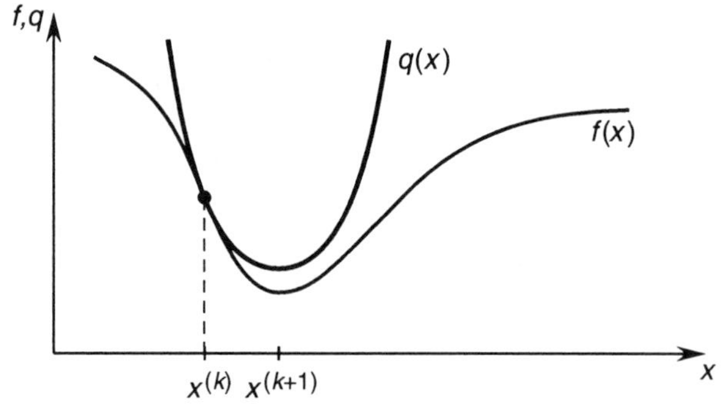


**Figure 7.6.** Newton's algorithm with $f''(x) > 0$

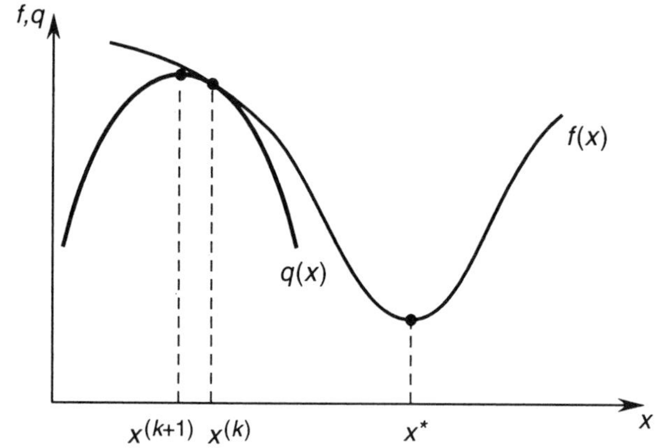


**Figure 7.7.** Newton's algorithm with $f''(x) < 0$

Newton's method works well if $f''(x) > 0$ everywhere (see Figure 7.6). However, if $f''(x) < 0$ for some $x$, Newton's method may fail to converge to the minimizer (see Figure 7.7).

Newton's method can also be viewed as a way to drive the first derivative of $f$ to zero. Indeed, if we set $g(x) = f'(x)$, then we obtain a formula for iterative solution of the equation $g(x) = 0$:

$$x^{(k+1)} = x^{(k)} - \frac{g(x^{(k)})}{g'(x^{(k)})}.$$

***Example 7.4.*** We apply Newton's method to improve a first approximation, $x^{(0)} = 12$, to the root of the equation

$$g(x) = x^3 - 12.2x^2 + 7.45x + 42 = 0.$$

We have $g'(x) = 3x^2 - 24.4x + 7.45$.

Performing two iterations yields

$$x^{(1)} = 12 - \frac{102.6}{146.65} = 11.33$$

$$x^{(2)} = 11.33 - \frac{14.73}{116.11} = 11.21.$$

■

Newton's method for solving equations of the form $g(x) = 0$ is also referred to as *Newton's method of tangents.* This name is easily justified if we look at a geometric interpretation of the method when applied to the solution of the equation $g(x) = 0$ (see Figure 7.8).

If we draw a tangent to $g(x)$ at the given point $x^{(k)}$, then the tangent line intersects the $x$ axis at the point $x^{(k+1)}$, which we expect to be closer to the root $x^*$ of $g(x) = 0$. Note that the slope of $g(x)$ at $x^{(k)}$ is

$$g'(x^{(k)}) = \frac{g(x^{(k)})}{x^{(k)} - x^{(k+1)}}.$$

Hence

$$x^{(k+1)} = x^{(k)} - \frac{g(x^{(k)})}{g'(x^{(k)})}.$$

Newton's method of tangents may fail if the first approximation to the root is such that the ratio $g(x^{(0)})/g'(x^{(0)})$ is not small enough (see Figure 7.9). Thus, an initial approximation to the root is very important.

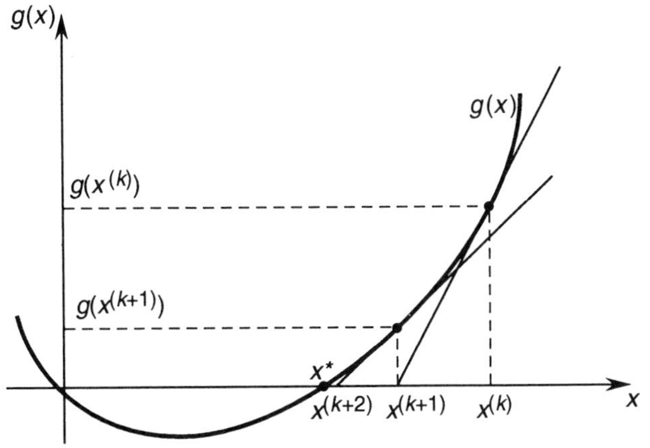


**Figure 7.8.** Newton's method of tangents

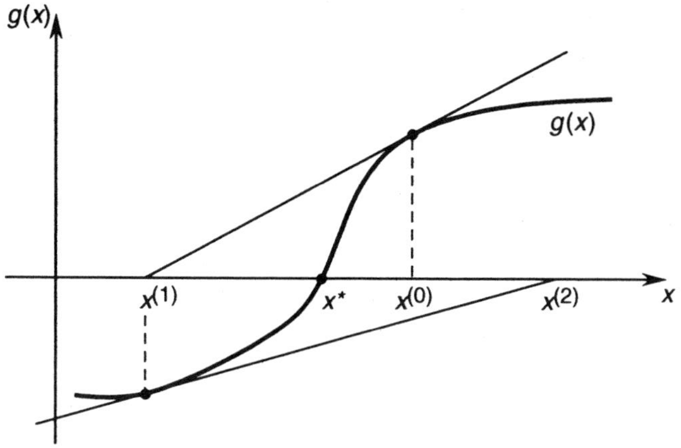


**Figure 7.9.** Example where Newton's method of tangents fails to converge to the root $x^*$ of $g(x) = 0$

## 7.4. SECANT METHOD

Newton's method for minimizing $f$ uses second derivatives of $f$:

$$x^{(k+1)} = x^{(k)} - \frac{f'(x^{(k)})}{f''(x^{(k)})}.$$

If the second derivative is not available, we may attempt to approximate it using first derivative information. In particular, we may approximate $f''(x^{(k)})$ above with

$$\frac{f'(x^{(k)}) - f'(x^{(k-1)})}{x^{(k)} - x^{(k-1)}}.$$

Using the above approximation of the second derivative, we obtain the algorithm

$$x^{(k+1)} = x^{(k)} - \frac{x^{(k)} - x^{(k-1)}}{f'(x^{(k)}) - f'(x^{(k-1)})} f'(x^{(k)}).$$

The above algorithm is called the *secant method*. Note that the algorithm requires two initial points to start it, which we will denote $x^{(-1)}$ and $x^{(0)}$. The secant algorithm can be represented in the following equivalent form:

$$x^{(k+1)} = \frac{f'(x^{(k)})x^{(k-1)} - f'(x^{(k-1)})x^{(k)}}{f'(x^{(k)}) - f'(x^{(k-1)})}.$$

Observe that, like Newton's method, the secant method does not directly involve values of $f(x^{(k)})$. Instead, it tries to drive the derivative $f'$ to zero. In fact, as we did for Newton's method, we can interpret the secant method as an algorithm for solving equations of the form $g(x)=0$. Specifically, the secant algorithm for finding a root of the equation $g(x)=0$ takes the form

$$x^{(k+1)}=x^{(k)}-\frac{x^{(k)}-x^{(k-1)}}{g(x^{(k)})-g(x^{(k-1)})}g(x^{(k)}).$$

or, equivalently,

$$x^{(k+1)}=\frac{g(x^{(k)})x^{(k-1)}-g(x^{(k-1)})x^{(k)}}{g(x^{(k)})-g(x^{(k-1)})}.$$

The secant method for root finding is illustrated in Figure 7.10 (compare this with Figure 7.8). Unlike Newton's method, which uses the slope of $g$ to determine the next point, the secant method uses the "secant" between the $(k-1)$st and $k$th points to determine the $(k+1)$st point.

***Example 7.5.*** We apply the secant method to find the root of the equation

$$g(x)=x^3-12.2x^2+7.45x+42=0.$$

We perform two iterations, with starting points $x^{(-1)}=13$ and $x^{(0)}=12$. We obtain

$$x^{(1)}=11.40$$

$$x^{(2)}=11.25.$$ ■

For further reading on the secant method, see Ref. 15.

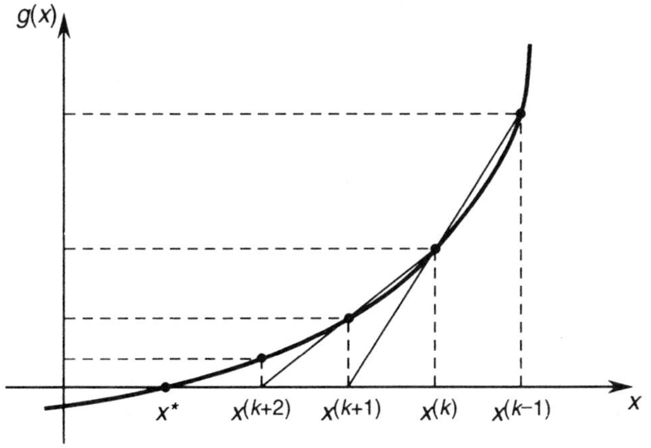


**Figure 7.10.** Secant method for root finding

## 7.5. REMARKS ON LINE SEARCH METHODS

One-dimensional search methods play an important role in multidimensional optimization problems. In particular, iterative algorithms for solving such optimization problems (to be discussed in the following chapters) typically involve a "line search" at every iteration. To be specific, let $f:\mathbb{R}^n \to \mathbb{R}$ be a function that we wish to minimize. Iterative algorithms for finding a minimizer of $f$ are of the form

$$\boldsymbol{x}^{(k+1)} = \boldsymbol{x}^{(k)} + \alpha_k \boldsymbol{d}^{(k)},$$

where $\boldsymbol{x}^{(0)}$ is a given initial point, and $\alpha_k \geqslant 0$ is chosen to minimize $\phi_k(\alpha) = f(\boldsymbol{x}^{(k)} + \alpha_k \boldsymbol{d}^{(k)})$. The vector $\boldsymbol{d}^{(k)}$ is called the search direction. Note that choice of $\alpha_k$ involves a one-dimensional minimization. This choice ensures that, under appropriate conditions,

$$f(\boldsymbol{x}^{(k+1)}) < f(\boldsymbol{x}^{(k)}).$$

We may, for example, use the secant method to find $\alpha_k$. In this case, we will need the derivative of $\phi_k$, which is

$$\phi_k'(\alpha) = \boldsymbol{d}^{(k)T} \nabla f(\boldsymbol{x}^{(k)} + \alpha \boldsymbol{d}^{(k)}).$$

The above is obtained using the chain rule. Therefore, applying the secant method for the line search requires the gradient $\nabla f$, the initial line search point $\boldsymbol{x}^{(k)}$, and the search direction $\boldsymbol{d}^{(k)}$ (see Exercise 7.8). Of course, other one-dimensional search methods may be used for line search (see, e.g., Refs. 51 and 22).

Line search algorithms used in practice are much more involved than the one-dimensional search methods presented in this chapter. The reason for this stems from several practical considerations. First, determining the value of $\alpha_k$ that exactly minimizes $\phi_k$ may be computationally demanding; even worse, the minimizer of $\phi_k$ may not even exist. Second, practical experience suggests that it is better to allocate more computational time on iterating the optimization algorithm rather than performing exact line searches. These considerations led to the development of conditions for terminating line search algorithms that would result in low-accuracy line searches while still securing a decrease in the value of $f$ from one iteration to the next. For more information on practical line search methods, we refer the reader to Refs. 22 (pp. 26–40), 26, and 27.[1]

## EXERCISES

**7.1** Suppose we have a unimodal function over the interval $[5, 8]$. Give an example of a desired final uncertainty range where the Golden Section

[1] We thank Dennis M. Goodman for furnishing us with Refs. 26 and 27.

method requires at least four iterations, whereas the Fibonacci method requires only three. You may choose an arbitrarily small value of $\varepsilon$ for the Fibonacci method.

**7.2** Let $f(x) = x^2 + 4\cos x$, $x \in \mathbb{R}$. We wish to find the minimizer $x^*$ of $f$ over the interval $[1, 2]$. (Calculator users: Note that in $\cos x$, $x$ is in radians.)

**a.** Plot $f(x)$ versus $x$ over the interval $[1, 2]$.

**b.** Use the Golden Section method to locate $x^*$ to within an uncertainty of 0.2. Display all intermediate steps using a table as follows:

| Iteration $k$ | $a_k$ | $b_k$ | $f(a_k)$ | $f(b_k)$ | New Uncertainty Interval |
|---|---|---|---|---|---|
| 1 | ? | ? | ? | ? | [?, ?] |
| 2 | ? | ? | ? | ? | [?, ?] |
| ⋮ | ⋮ | ⋮ | ⋮ | ⋮ | ⋮ |

**c.** Repeat b using the Fibonacci method, with $\varepsilon = 0.05$. Display all intermediate steps using a table as follows:

| Iteration $k$ | $\rho_k$ | $a_k$ | $b_k$ | $f(a_k)$ | $f(b_k)$ | New Uncertainty Interval |
|---|---|---|---|---|---|---|
| 1 | ? | ? | ? | ? | ? | [?, ?] |
| 2 | ? | ? | ? | ? | ? | [?, ?] |
| ⋮ | ⋮ | ⋮ | ⋮ | ⋮ | ⋮ | ⋮ |

**d.** Apply Newton's method, using the same number of iterations as in b, with $x^{(0)} = 1$.

**7.3** Let $f(x) = 8e^{1-x} + 7\log(x)$, where $\log(\cdot)$ represents the natural logarithm function.

**a.** Use MATLAB to plot $f(x)$ versus $x$ over the interval $[1, 2]$, and verify that $f$ is unimodal over $[1, 2]$.

**b.** Write a simple MATLAB routine to implement the Golden Section method that locates the minimizer of $f$ over $[1, 2]$ to within an uncertainty of 0.23. Display all intermediate steps using a table as in Exercise 7.2.

**c.** Repeat b using the Fibonacci method, with $\varepsilon = 0.05$. Display all intermediate steps using a table as in Exercise 7.2.

**7.4** Suppose $\rho_1, \ldots, \rho_N$ are the values used in the Fibonacci search method. Show that for each $k = 1, \ldots, N$, $0 \leqslant \rho_k \leqslant \frac{1}{2}$, and for each $k = 1, \ldots, N-1$,

$$\rho_{k+1} = 1 - \frac{\rho_k}{1 - \rho_k}.$$

**7.5** Show that if $F_0, F_1, \ldots$ is the Fibonacci sequence, then for each $k = 2, 3, \ldots,$

$$F_{k-2}F_{k+1} - F_{k-1}F_k = (-1)^k.$$

**7.6** Suppose we have an efficient way of calculating exponentials. Based on this, use Newton's method to devise a method to approximate log(2) (where log(·) is the natural logarithm function). Use an initial point of $x^{(0)} = 1$, and perform two iterations.

**7.7** **a.** Write a simple MATLAB routine to implement the secant method to locate the root of the equation $g(x) = 0$. For the stopping criterion, use the condition $|x^{(k+1)} - x^{(k)}| < |x^{(k)}|\varepsilon$, where $\varepsilon > 0$ is a given constant.

**b.** Let $g(x) = (2x - 1)^2 + 4(4 - 1024x)^4$. Find the root of $g(x) = 0$ using the secant method with $x^{(-1)} = 0$, $x^{(0)} = 1$, and $\varepsilon = 10^{-5}$. Also determine the value of $g$ at the obtained solution.

**7.8** Write a MATLAB function that implements a line search using the secant method. The arguments to this function are the name of the M-file for the gradient, the current point, and the search direction. For example, the function may be called linesearch_secant, and used by the function call alpha = linesearch_secant ('grad', x, d), where grad.m is the M-file containing the gradient, x is the starting line search point, d is the search direction, and alpha is the value returned by the function (which we will use in the following chapters as the "step-size" for iterative algorithms; see, e.g., Exercises 8.9 and 10.6).

*Note*: In the solutions manual, we used the stopping criterion $|\boldsymbol{d}^T\nabla f(\boldsymbol{x} + \alpha\boldsymbol{d})| \leqslant \varepsilon|\boldsymbol{d}^T\nabla f(\boldsymbol{x})|$, where $\varepsilon > 0$ is a prespecified number, $\nabla f$ is the gradient, $\boldsymbol{x}$ is the starting line search point, and $\boldsymbol{d}$ is the search direction. The rationale for the above stopping criterion is that we want to reduce the directional derivative of $f$ in the direction $\boldsymbol{d}$ by the specified fraction $\varepsilon$. We used a value of $\varepsilon = 10^{-4}$, and initial conditions of 0 and 0.001.

# 8

# Gradient Methods

## 8.1. INTRODUCTION

In this chapter, we consider a class of search methods for real-valued functions on $\mathbb{R}^n$. These methods use the gradient of the given function as well as the function values. In our discussion, we use terms like level sets, normal vectors, tangent vectors, and so on. These notions were discussed in some detail in Part I.

Recall that a level set of a function $f:\mathbb{R}^n \to \mathbb{R}$ is the set of points $\boldsymbol{x}$ satisfying $f(\boldsymbol{x}) = c$ for some constant $c$. Thus, a point $\boldsymbol{x}_0 \in \mathbb{R}^n$ is on the level set corresponding to level $c$ if $f(\boldsymbol{x}_0) = c$. In the case of functions of two real variables, $f:\mathbb{R}^2 \to \mathbb{R}$, the notion of the level set is illustrated in Figure 8.1.

The gradient of $f$ at $\boldsymbol{x}_0$, denoted by $\nabla f(\boldsymbol{x}_0)$, if it is not a zero vector, is orthogonal to the tangent vector to an arbitrary smooth curve passing through $\boldsymbol{x}_0$ on the level set $f(\boldsymbol{x}) = c$. Thus, the direction of maximum rate of increase of a real-valued differentiable function at a point is orthogonal to the level set of the function through that point. In other words, the gradient acts in such a direction that for a given small displacement, the function $f$ increases more in the direction of the gradient than in any other direction. To prove this statement, recall that $\langle \nabla f(\boldsymbol{x}), \boldsymbol{d} \rangle$, $\|\boldsymbol{d}\| = 1$, is the rate of increase of $f$ in the direction $\boldsymbol{d}$ at the point $\boldsymbol{x}$. By the Cauchy–Schwarz inequality,

$$\langle \nabla f(\boldsymbol{x}), \boldsymbol{d} \rangle \leqslant \|\nabla f(\boldsymbol{x})\|$$

since $\|\boldsymbol{d}\| = 1$. But, if $\boldsymbol{d} = \nabla f(\boldsymbol{x}) / \|\nabla f(\boldsymbol{x})\|$, then

$$\left\langle \nabla f(\boldsymbol{x}), \frac{\nabla f(\boldsymbol{x})}{\|\nabla f(\boldsymbol{x})\|} \right\rangle = \|\nabla f(\boldsymbol{x})\|.$$

Thus, the direction in which $\nabla f(\boldsymbol{x})$ points is the direction of maximum rate of increase of $f$ at $\boldsymbol{x}$. The direction in which $-\nabla f(\boldsymbol{x})$ points is the direction of maximum rate of decrease of $f$ at $\boldsymbol{x}$. Hence, the direction of negative gradient is a good direction to search if we want to find a function minimizer.

We proceed as follows. Let $\boldsymbol{x}^{(0)}$ be a starting point, and consider the point $\boldsymbol{x}^{(0)} - \alpha \nabla f(\boldsymbol{x}^{(0)})$. Then, from Taylor's formula we obtain

$$f(\boldsymbol{x}^{(0)} - \alpha \nabla f(\boldsymbol{x}^{(0)})) = f(\boldsymbol{x}^{(0)}) - \alpha \|\nabla f(\boldsymbol{x}^{(0)})\|^2 + o(\alpha).$$

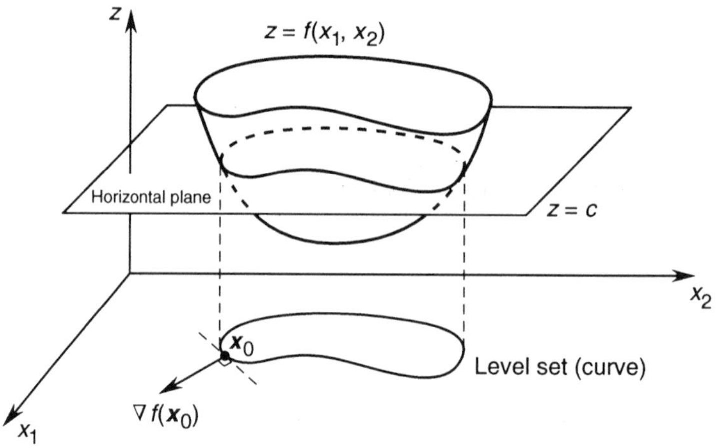


**Figure 8.1.** Constructing a level set corresponding to level $c$ for $f$

Thus, if $\nabla f(\boldsymbol{x}^{(0)}) \neq \mathbf{0}$, then for sufficiently small $\alpha > 0$, we have

$$f(\boldsymbol{x}^{(0)} - \alpha \nabla f(\boldsymbol{x}^{(0)})) < f(\boldsymbol{x}^{(0)}).$$

This means that the point $\boldsymbol{x}^{(0)} - \alpha \nabla f(\boldsymbol{x}^{(0)})$ is an improvement over the point $\boldsymbol{x}^{(0)}$ if we are searching for a minimizer.

To formulate an algorithm that implements the above idea, suppose we are given a point $\boldsymbol{x}^{(k)}$. To find the next point $\boldsymbol{x}^{(k+1)}$, we start at $\boldsymbol{x}^{(k)}$ and move by an amount $-\alpha_k \nabla f(\boldsymbol{x}^{(k)})$, where $\alpha_k$ is a positive scalar called the *step size*. The above procedure leads to the following iterative algorithm:

$$\boldsymbol{x}^{(k+1)} = \boldsymbol{x}^{(k)} - \alpha_k \nabla f(\boldsymbol{x}^{(k)}).$$

We refer to the above as a *gradient descent algorithm* (or simply a *gradient algorithm*). The gradient varies as the search proceeds, tending to zero as we approach the minimizer. We have the option of either taking very small steps and reevaluating the gradient at every step, or we can take large steps each time. The first approach results in a laborious method of reaching the minimizer, whereas the second approach may result in a more zigzag path to the minimizer. The advantage of the second approach is that it may require fewer gradient evaluations. Among the many different methods that use the above philosophy, the most popular is the method of steepest descent, which we discuss next.

## 8.2. STEEPEST DESCENT METHOD

The *method of steepest descent* is a gradient algorithm where the step size $\alpha_k$ is chosen to achieve the maximum amount of decrease of the objective function at

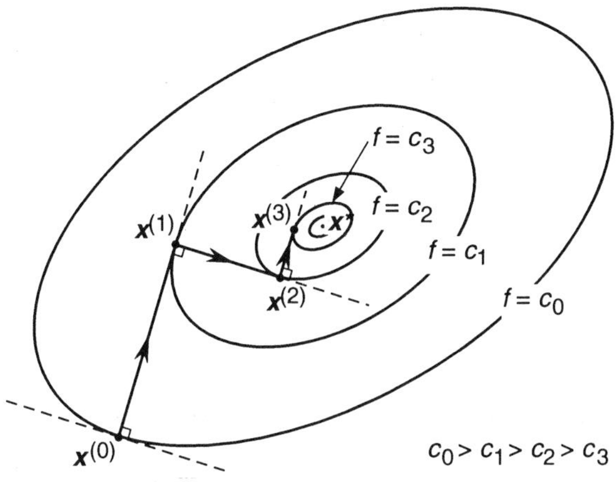


**Figure 8.2.** Typical sequence resulting from the method of steepest descent

each individual step. Specifically, $\alpha_k$ is chosen to minimize $\phi_k(\alpha) \triangleq f(\boldsymbol{x}^{(k)} - \alpha \nabla f(\boldsymbol{x}^{(k)}))$. In other words,

$$\alpha_k = \arg\min_{\alpha \geq 0} f(\boldsymbol{x}^{(k)} - \alpha \nabla f(\boldsymbol{x}^{(k)})).$$

To summarize, the steepest descent algorithm proceeds as follows: at each step, starting from the point $\boldsymbol{x}^{(k)}$, we conduct a line search in the direction $-\nabla f(\boldsymbol{x}^{(k)})$ until a minimizer, $\boldsymbol{x}^{(k+1)}$, is found. A typical sequence resulting from the method of steepest descent is depicted in Figure 8.2.

Observe that the method of steepest descent moves in orthogonal steps, as stated in the following proposition.

**Proposition 8.1.** *If* $\{\boldsymbol{x}^{(k)}\}_{k=0}^{\infty}$ *is a steepest descent sequence for a given function* $f: \mathbb{R}^n \to \mathbb{R}$, *then for each* $k$ *the vector* $\boldsymbol{x}^{(k+1)} - \boldsymbol{x}^{(k)}$ *is orthogonal to the vector* $\boldsymbol{x}^{(k+2)} - \boldsymbol{x}^{(k+1)}$. □

*Proof.* From the iterative formula of the method of steepest descent, it follows that

$$\langle \boldsymbol{x}^{(k+1)} - \boldsymbol{x}^{(k)}, \boldsymbol{x}^{(k+2)} - \boldsymbol{x}^{(k+1)} \rangle = \alpha_k \alpha_{k+1} \langle \nabla f(\boldsymbol{x}^{(k)}), \nabla f(\boldsymbol{x}^{(k+1)}) \rangle.$$

To complete the proof it is enough to show that

$$\langle \nabla f(\boldsymbol{x}^{(k)}), \nabla f(\boldsymbol{x}^{(k+1)}) \rangle = 0.$$

To this end, observe that $\alpha_k$ is a nonnegative scalar that minimizes $\phi_k(\alpha) \triangleq$

$f(\boldsymbol{x}^{(k)} - \alpha \nabla f(\boldsymbol{x}^{(k)}))$. Hence, using the FONC and the chain rule,

$$\begin{aligned}
0 &= \phi_k'(\alpha_k) \\
&= \frac{d\phi_k}{d\alpha}(\alpha_k) \\
&= \nabla f(\boldsymbol{x}^{(k)} - \alpha_k \nabla f(\boldsymbol{x}^{(k)}))^T(-\nabla f(\boldsymbol{x}^{(k)})) \\
&= -\langle \nabla f(\boldsymbol{x}^{(k+1)}), \nabla f(\boldsymbol{x}^{(k)}) \rangle
\end{aligned}$$

and the proof is completed. ■

The above proposition implies that $\nabla f(\boldsymbol{x}^{(k)})$ is parallel to the tangent plane to the level set $\{f(\boldsymbol{x}) = f(\boldsymbol{x}^{(k+1)})\}$ at $\boldsymbol{x}^{(k+1)}$. Note that as each new point is generated by the steepest descent algorithm, the corresponding value of the function $f$ decreases in value, as stated below.

**Proposition 8.2.** *If $\{\boldsymbol{x}^{(k)}\}_{k=0}^{\infty}$ is the steepest descent sequence for $f: \mathbb{R}^n \to \mathbb{R}$ and if $\nabla f(\boldsymbol{x}^{(k)}) \neq \mathbf{0}$ then $f(\boldsymbol{x}^{(k+1)}) < f(\boldsymbol{x}^{(k)})$.*

*Proof.* First recall that

$$\boldsymbol{x}^{(k+1)} = \boldsymbol{x}^{(k)} - \alpha_k \nabla f(\boldsymbol{x}^{(k)}),$$

where $\alpha_k \geqslant 0$ is the minimizer of

$$\phi_k(\alpha) = f(\boldsymbol{x}^{(k)} - \alpha \nabla f(\boldsymbol{x}^{(k)}))$$

over all $\alpha \geqslant 0$. Thus, for $\alpha \geqslant 0$, we have

$$\phi_k(\alpha_k) \leqslant \phi_k(\alpha).$$

By the chain rule,

$$\phi_k'(0) = \frac{d\phi_k}{d\alpha}(0) = -(\nabla f(\boldsymbol{x}^{(k)} - 0\nabla f(\boldsymbol{x}^{(k)})))^T \nabla f(\boldsymbol{x}^{(k)}) = -\|\nabla f(\boldsymbol{x}^{(k)})\|^2 < 0,$$

since $\nabla f(\boldsymbol{x}^{(k)}) \neq \mathbf{0}$ by assumption. Thus, $\phi_k'(0) < 0$, and this implies that there is an $\bar{\alpha} > 0$ such that $\phi_k(0) > \phi_k(\alpha)$ for all $\alpha \in (0, \bar{\alpha}]$. Hence

$$f(\boldsymbol{x}^{(k+1)}) = \phi_k(\alpha_k) \leqslant \phi_k(\bar{\alpha}) < \phi_k(0) = f(\boldsymbol{x}^{(k)})$$

and the proof of the statement is completed. ■

In the above theorem, we used the assumption that $\nabla f(\boldsymbol{x}^{(k)}) \neq \mathbf{0}$ to prove that the algorithm possesses a descent property, that is, $f(\boldsymbol{x}^{(k+1)}) < f(\boldsymbol{x}^{(k)})$ if

$\nabla f(\boldsymbol{x}^{(k)}) \neq \mathbf{0}$. If for some $k$, we have $\nabla f(\boldsymbol{x}^{(k)}) = \mathbf{0}$, then the point $\boldsymbol{x}^{(k)}$ satisfies the FONC. In this case, $\boldsymbol{x}^{(k+1)} = \boldsymbol{x}^{(k)}$. We can use the above as the basis for a stopping (termination) criterion for the algorithm.

The condition $\nabla f(\boldsymbol{x}^{(k+1)}) = \mathbf{0}$, however, is not directly suitable as a practical stopping criterion, since the numerical computation of the gradient will rarely be identically equal to zero. A practical stopping criterion is to check if the norm $\|\nabla f(\boldsymbol{x}^{(k)})\|$ of the gradient is less than a prespecified threshold, in which case we stop. Alternatively, we may compute the absolute difference $|f(\boldsymbol{x}^{(k+1)}) - f(\boldsymbol{x}^{(k)})|$ between objective function values for every two successive iterations, and if the difference is less than some prespecified threshold, then we stop. Yet another alternative is to compute the norm $\|\boldsymbol{x}^{(k+1)} - \boldsymbol{x}^{(k)}\|$ of the difference between two successive iterates, and we stop if the norm is less than a prespecified threshold. The above stopping criteria are relevant to all the iterative algorithms we discuss in this part.

***Example 8.1.*** We use the method of steepest descent to find the minimizer of

$$f(x_1, x_2, x_3) = (x_1 - 4)^4 + (x_2 - 3)^2 + 4(x_3 + 5)^4.$$

The initial point is $\boldsymbol{x}^{(0)} = [4, 2, -1]^T$. We perform three iterations.

We find

$$\nabla f(\boldsymbol{x}) = [4(x_1 - 4)^3, 2(x_2 - 3), 16(x_3 + 5)^3]^T.$$

Hence

$$\nabla f(\boldsymbol{x}^{(0)}) = [0, -2, 1024]^T.$$

To compute $\boldsymbol{x}^{(1)}$, we need

$$\begin{aligned}\alpha_0 &= \arg\min_{\alpha \geqslant 0} f(\boldsymbol{x}^{(0)} - \alpha \nabla f(\boldsymbol{x}^{(0)})) \\ &= \arg\min_{\alpha \geqslant 0} (0 + (2 + 2\alpha - 3)^2 + 4(-1 - 1024\alpha + 5)^4) \\ &= \arg\min_{\alpha \geqslant 0} \phi_0(\alpha).\end{aligned}$$

Using the secant method from the previous chapter, we obtain

$$\alpha_0 = 3.967 \times 10^{-3}.$$

For illustrative purpose, we show a plot of $\phi_0(\alpha)$ versus $\alpha$ in Figure 8.3, obtained using MATLAB.

Thus

$$\boldsymbol{x}^{(1)} = \boldsymbol{x}^{(0)} - \alpha_0 \nabla f(\boldsymbol{x}^{(0)}) = [4.000, 2.008, -5.062]^T.$$

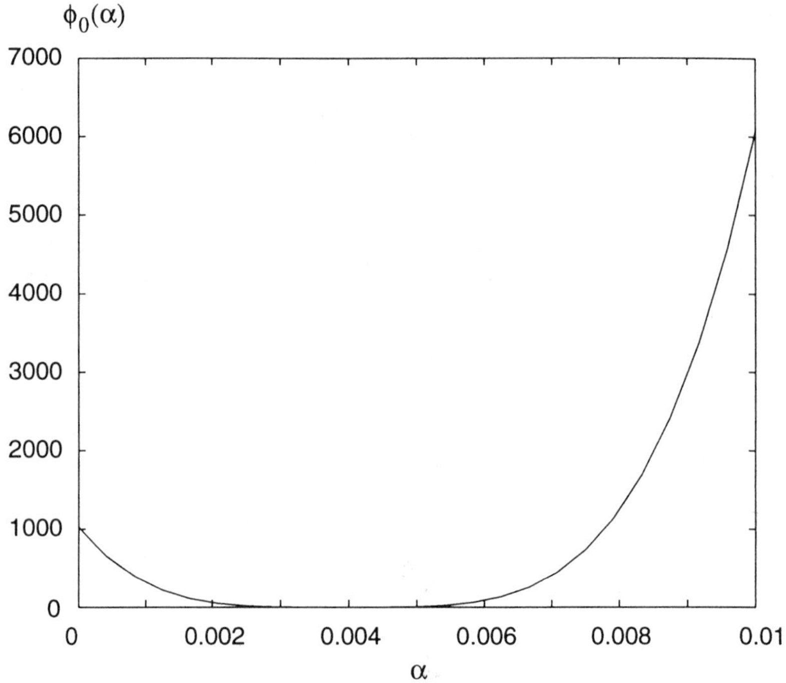


**Figure 8.3.** Plot of $\phi_0(\alpha)$ versus $\alpha$

To find $x^{(2)}$, we first determine

$$\nabla f(x^{(1)}) = [0.000,\ -1.984,\ -0.003875]^T.$$

Next, we find $\alpha_1$, where

$$\begin{aligned}\alpha_1 &= \arg\min_{\alpha \geqslant 0} (0 + (2.008 + 1.984\alpha - 3)^2 + 4(-5.062 + 0.003875\alpha + 5)^4) \\ &= \arg\min_{\alpha \geqslant 0} \phi_1(\alpha).\end{aligned}$$

Using the secant method again, we obtain $\alpha_1 = 0.5000$. Figure 8.4 depicts a plot of $\phi_1(\alpha)$ versus $\alpha$.

Thus

$$x^{(2)} = x^{(1)} - \alpha_1 \nabla f(x^{(1)}) = [4.000, 3.000, -5.060]^T,$$

To find $x^{(3)}$, we first determine

$$\nabla f(x^{(2)}) = [0.000, 0.000, -0.003525]^T$$

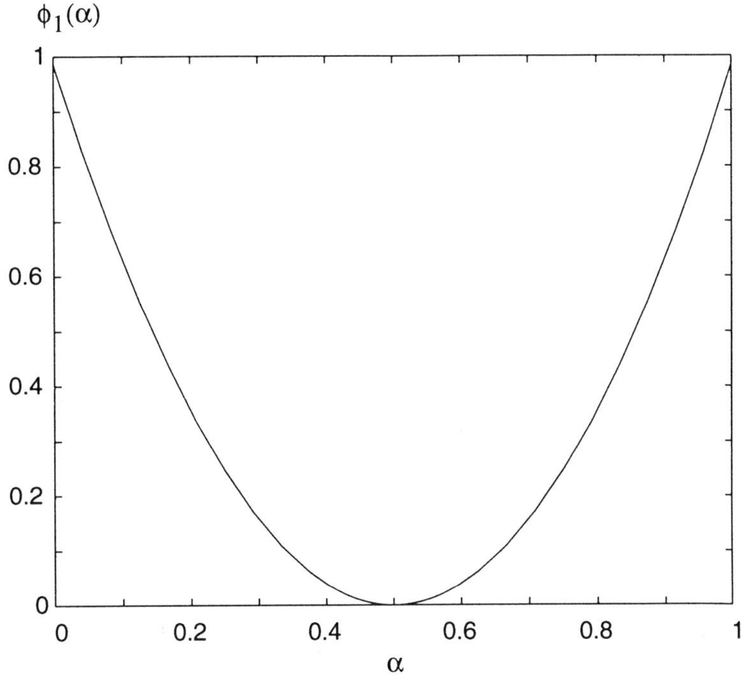


**Figure 8.4.** Plot of $\phi_1(\alpha)$ versus $\alpha$

and

$$\begin{aligned}\alpha_2 &= \arg\min_{\alpha \geqslant 0} (0.000 + 0.000 + 4(-5.060 + 0.003525\alpha + 5)^4) \\ &= \arg\min_{\alpha \geqslant 0} \phi_2(\alpha).\end{aligned}$$

We proceed as in the previous iterations to obtain $\alpha_2 = 16.29$. A plot of $\phi_2(\alpha)$ versus $\alpha$ is shown in Figure 8.5.

The value of $\boldsymbol{x}^{(3)}$ is

$$\boldsymbol{x}^{(3)} = [4.000, 3.000, -5.002]^T.$$

Note that the minimizer of $f$ is $[4, 3, -5]^T$, and hence it appears that we have arrived at the minimizer in only three iterations. The reader should be cautioned not to draw any conclusions from this example about the number of iterations required to arrive at a solution in general.

It goes without saying that numerical computations, such as those in this example, are performed in practice using a computer (rather than by hand). The above calculations were written out explicitly, step by step, for the purpose of illustrating the operations involved in the steepest descent algorithm. The

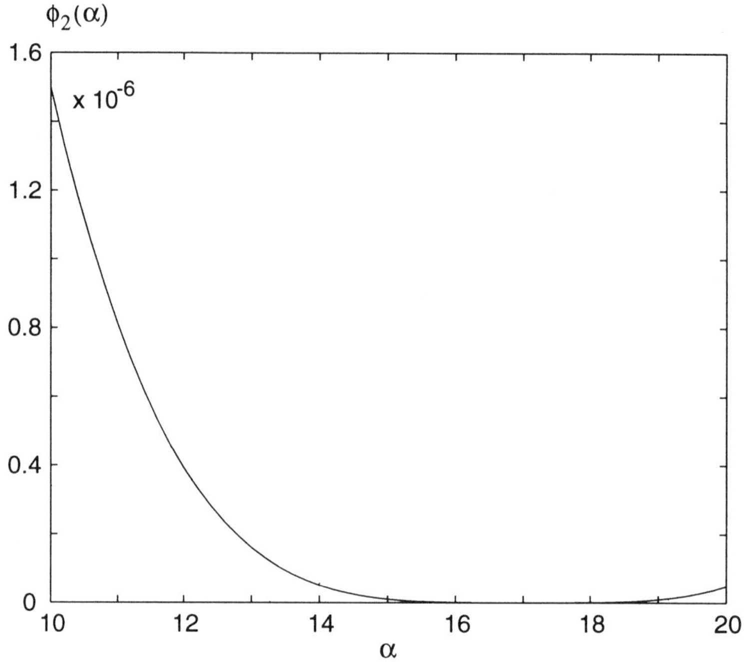


**Figure 8.5.** Plot of $\phi_2(\alpha)$ versus $\alpha$

computations themselves were in fact carried out using a MATLAB routine (see Exercise 8.9). ■

Let us now see what the method of steepest descent does with a quadratic function of the form

$$f(\boldsymbol{x}) = \tfrac{1}{2}\boldsymbol{x}^T\boldsymbol{Q}\boldsymbol{x} - \boldsymbol{b}^T\boldsymbol{x},$$

where $\boldsymbol{Q} \in \mathbb{R}^{n \times n}$ is a symmetric positive definite matrix, $\boldsymbol{b} \in \mathbb{R}^n$, and $\boldsymbol{x} \in \mathbb{R}^n$. The unique minimizer of $f$ can be found by setting the gradient of $f$ to zero, where

$$\nabla f(\boldsymbol{x}) = \boldsymbol{Q}\boldsymbol{x} - \boldsymbol{b},$$

since $D(\frac{1}{2}\boldsymbol{x}^T\boldsymbol{Q}\boldsymbol{x}) = \frac{1}{2}\boldsymbol{x}^T(\boldsymbol{Q} + \boldsymbol{Q}^T) = \boldsymbol{x}^T\boldsymbol{Q}$, and $D(\boldsymbol{b}^T\boldsymbol{x}) = \boldsymbol{b}^T$. There is no loss of generality in assuming $\boldsymbol{Q}$ to be a symmetric matrix. For, if we are given a quadratic form $\boldsymbol{x}^T\boldsymbol{A}\boldsymbol{x}$ and $\boldsymbol{A} \neq \boldsymbol{A}^T$, then, since the transposition of a scalar equals itself, we obtain

$$(\boldsymbol{x}^T\boldsymbol{A}\boldsymbol{x})^T = \boldsymbol{x}^T\boldsymbol{A}^T\boldsymbol{x} = \boldsymbol{x}^T\boldsymbol{A}\boldsymbol{x}.$$

Hence

$$\begin{aligned} \boldsymbol{x}^T\boldsymbol{A}\boldsymbol{x} &= \tfrac{1}{2}\boldsymbol{x}^T\boldsymbol{A}\boldsymbol{x} + \tfrac{1}{2}\boldsymbol{x}^T\boldsymbol{A}^T\boldsymbol{x} \\ &= \tfrac{1}{2}\boldsymbol{x}^T(\boldsymbol{A}+\boldsymbol{A}^T)\boldsymbol{x} \\ &\triangleq \tfrac{1}{2}\boldsymbol{x}^T\boldsymbol{Q}\boldsymbol{x}. \end{aligned}$$

Note that

$$(\boldsymbol{A}+\boldsymbol{A}^T)^T = \boldsymbol{Q}^T = \boldsymbol{A}+\boldsymbol{A}^T = \boldsymbol{Q}.$$

The Hessian of $f$ is $\boldsymbol{F}(\boldsymbol{x}) = \boldsymbol{Q} = \boldsymbol{Q}^T > 0$. To simplify the notation, we write $\boldsymbol{g}^{(k)} = \nabla f(\boldsymbol{x}^{(k)})$. Then, the steepest descent algorithm for the quadratic function can be represented as

$$\boldsymbol{x}^{(k+1)} = \boldsymbol{x}^{(k)} - \alpha_k \boldsymbol{g}^{(k)},$$

where

$$\begin{aligned} \alpha_k &= \arg\min_{\alpha\geqslant 0} f(\boldsymbol{x}^{(k)} - \alpha\boldsymbol{g}^{(k)}) \\ &= \arg\min_{\alpha\geqslant 0} \left(\tfrac{1}{2}(\boldsymbol{x}^{(k)} - \alpha\boldsymbol{g}^{(k)})^T\boldsymbol{Q}(\boldsymbol{x}^{(k)} - \alpha\boldsymbol{g}^{(k)}) - (\boldsymbol{x}^{(k)} - \alpha\boldsymbol{g}^{(k)})^T\boldsymbol{b}\right). \end{aligned}$$

In the quadratic case, we can find an explicit formula for $\alpha_k$. We proceed as follows. Since $\alpha_k \geqslant 0$ is a minimizer of $\phi_k(\alpha) = f(\boldsymbol{x}^{(k)} - \alpha\boldsymbol{g}^{(k)})$, we apply the FONC to $\phi_k(\alpha)$ to obtain

$$\phi_k'(\alpha) = (\boldsymbol{x}^{(k)} - \alpha\boldsymbol{g}^{(k)})^T\boldsymbol{Q}(-\boldsymbol{g}^{(k)}) - \boldsymbol{b}^T(-\boldsymbol{g}^{(k)}).$$

Therefore, $\phi_k'(\alpha) = 0$ if $\alpha\boldsymbol{g}^{(k)T}\boldsymbol{Q}\boldsymbol{g}^{(k)} = (\boldsymbol{x}^{(k)T}\boldsymbol{Q} - \boldsymbol{b}^T)\boldsymbol{g}^{(k)}$. But

$$\boldsymbol{x}^{(k)T}\boldsymbol{Q} - \boldsymbol{b}^T = \boldsymbol{g}^{(k)T}.$$

Hence

$$\alpha_k = \frac{\boldsymbol{g}^{(k)T}\boldsymbol{g}^{(k)}}{\boldsymbol{g}^{(k)T}\boldsymbol{Q}\boldsymbol{g}^{(k)}}.$$

In summary, the method of steepest descent for the quadratic takes the form

$$\boldsymbol{x}^{(k+1)} = \boldsymbol{x}^{(k)} - \left(\frac{\boldsymbol{g}^{(k)T}\boldsymbol{g}^{(k)}}{\boldsymbol{g}^{(k)T}\boldsymbol{Q}\boldsymbol{g}^{(k)}}\right)\boldsymbol{g}^{(k)},$$

where

$$\boldsymbol{g}^{(k)} = \nabla f(\boldsymbol{x}^{(k)}) = \boldsymbol{Q}\boldsymbol{x}^{(k)} - \boldsymbol{b}.$$

***Example 8.2.*** Let

$$f(x_1, x_2) = x_1^2 + x_2^2.$$

Then, starting from an arbitrary initial point $\boldsymbol{x}^{(0)} \in \mathbb{R}^2$ we arrive at the solution $\boldsymbol{x}^* = \boldsymbol{0} \in \mathbb{R}^2$ in only one step. See Figure 8.6.

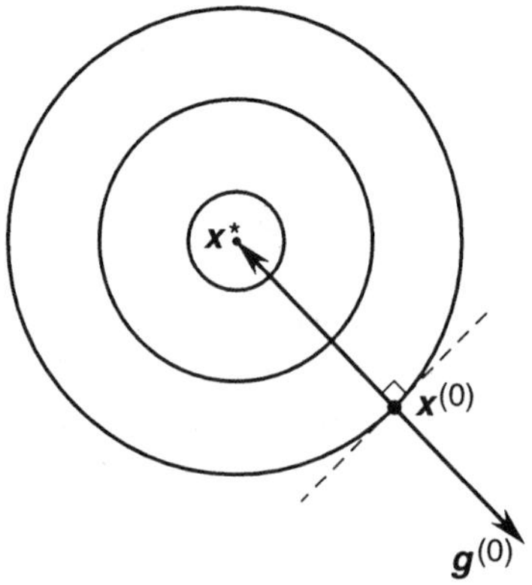


**Figure 8.6.** Steepest descent method applied to $f(x_1, x_2) = x_1^2 + x_2^2$

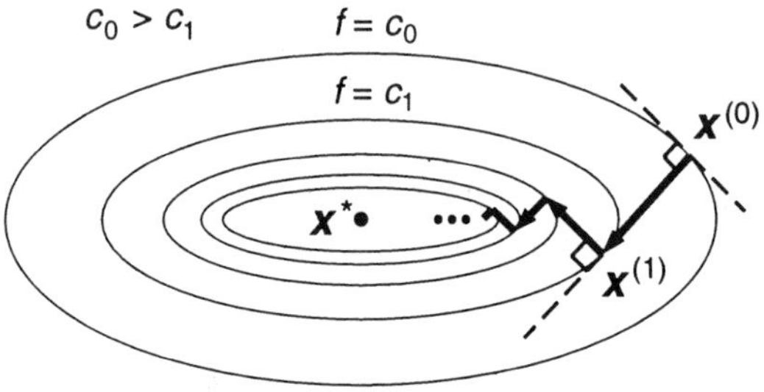


**Figure 8.7.** Steepest descent method in search for minimizer in a narrow valley

However, if

$$f(x_1, x_2) = \frac{x_1^2}{5} + x_2^2,$$

then the method of steepest descent shuffles ineffectively back and forth when searching for the minimizer in a narrow valley (see Figure 8.7). This example illustrates a major drawback in the steepest-descent method. More sophisticated methods that alleviate this problem are discussed in subsequent chapters. ■

To better understand the method of steepest descent, we examine its convergence properties in the next section.

## 8.3. ANALYSIS OF GRADIENT METHODS

The method of steepest descent is an example of an iterative algorithm. This means that the algorithm generates a sequence of points, each calculated on the basis of the points preceding it. The method is a *descent method*, because as each new point is generated by the algorithm, the corresponding value of the objective function decreases in value (i.e., the algorithm possesses a descent property).

We say that an iterative algorithm is *globally convergent* if for any arbitrary starting point the algorithm is guaranteed to generate a sequence of points converging to a point that satisfies the FONC for a minimum. When the algorithm is not globally convergent, it may still generate a sequence that converges to a point satisfying the FONC, provided that the initial point is sufficiently close to the point. In this case, we say that the algorithm is *locally convergent*. How close we need to start to a solution point for the algorithm to converge depends on the local convergence properties of the algorithm. A related issue of interest pertaining to a given locally or globally convergent algorithm is the *rate of convergence*, that is, how fast the algorithm converges to a solution point.

In this section, we analyze the convergence properties of descent gradient methods, including the method of steepest descent and gradient methods with fixed step size. We can investigate important convergence characteristics of a gradient method by applying the method to quadratic problems. The convergence analysis is more convenient if, instead of working with $f$, we deal with

$$V(\boldsymbol{x}) = f(\boldsymbol{x}) + \tfrac{1}{2}\boldsymbol{x}^{*T}\boldsymbol{Q}\boldsymbol{x}^* = \tfrac{1}{2}(\boldsymbol{x}-\boldsymbol{x}^*)^T\boldsymbol{Q}(\boldsymbol{x}-\boldsymbol{x}^*),$$

where $\boldsymbol{Q} = \boldsymbol{Q}^T > 0$. The solution point $\boldsymbol{x}^*$ is obtained by solving $\boldsymbol{Q}\boldsymbol{x} = \boldsymbol{b}$, that is, $\boldsymbol{x}^* = \boldsymbol{Q}^{-1}\boldsymbol{b}$. The function $V$ differs from $f$ only by a constant $\frac{1}{2}\boldsymbol{x}^{*T}\boldsymbol{Q}\boldsymbol{x}^*$. We begin our analysis with the following useful lemma that applies to a general gradient algorithm.

**Lemma 8.1.** *The iterative algorithm*

$$\boldsymbol{x}^{(k+1)} = \boldsymbol{x}^{(k)} - \alpha_k \boldsymbol{g}^{(k)}$$

*with* $\boldsymbol{g}^{(k)} = \boldsymbol{Q}\boldsymbol{x}^{(k)} - \boldsymbol{b}$ *satisfies*

$$V(\boldsymbol{x}^{(k+1)}) = (1-\gamma_k)V(\boldsymbol{x}^{(k)}),$$

*where, if* $\boldsymbol{g}^{(k)} = \boldsymbol{0}$ *then* $\gamma_k = 1$, *and if* $\boldsymbol{g}^{(k)} \neq \boldsymbol{0}$, *then*

$$\gamma_k = \alpha_k \frac{\boldsymbol{g}^{(k)T}\boldsymbol{Q}\boldsymbol{g}^{(k)}}{\boldsymbol{g}^{(k)T}\boldsymbol{Q}^{-1}\boldsymbol{g}^{(k)}}\left(2\frac{\boldsymbol{g}^{(k)T}\boldsymbol{g}^{(k)}}{\boldsymbol{g}^{(k)T}\boldsymbol{Q}\boldsymbol{g}^{(k)}} - \alpha_k\right).$$

□

*Proof.* The proof is by direct computation. We first evaluate the expression

$$\frac{V(\boldsymbol{x}^{(k)}) - V(\boldsymbol{x}^{(k+1)})}{V(\boldsymbol{x}^{(k)})}.$$

To facilitate computations, let $\boldsymbol{y}^{(k)} = \boldsymbol{x}^{(k)} - \boldsymbol{x}^*$. Then, $V(\boldsymbol{x}^{(k)}) = \frac{1}{2}\boldsymbol{y}^{(k)T}\boldsymbol{Q}\boldsymbol{y}^{(k)}$. Hence

$$\begin{aligned} V(\boldsymbol{x}^{(k+1)}) &= \tfrac{1}{2}(\boldsymbol{x}^{(k+1)} - \boldsymbol{x}^*)^T\boldsymbol{Q}(\boldsymbol{x}^{(k+1)} - \boldsymbol{x}^*) \\ &= \tfrac{1}{2}(\boldsymbol{x}^{(k)} - \boldsymbol{x}^* - \alpha_k\boldsymbol{g}^{(k)})^T\boldsymbol{Q}(\boldsymbol{x}^{(k)} - \boldsymbol{x}^* - \alpha_k\boldsymbol{g}^{(k)}) \\ &= \tfrac{1}{2}\boldsymbol{y}^{(k)T}\boldsymbol{Q}\boldsymbol{y}^{(k)} - \alpha_k\boldsymbol{g}^{(k)T}\boldsymbol{Q}\boldsymbol{y}^{(k)} + \tfrac{1}{2}\alpha_k^2\boldsymbol{g}^{(k)T}\boldsymbol{Q}\boldsymbol{g}^{(k)}. \end{aligned}$$

Therefore

$$\frac{V(\boldsymbol{x}^{(k)}) - V(\boldsymbol{x}^{(k+1)})}{V(\boldsymbol{x}^{(k)})} = \frac{2\alpha_k\boldsymbol{g}^{(k)T}\boldsymbol{Q}\boldsymbol{y}^{(k)} - \alpha_k^2\boldsymbol{g}^{(k)T}\boldsymbol{Q}\boldsymbol{g}^{(k)}}{\boldsymbol{y}^{(k)T}\boldsymbol{Q}\boldsymbol{y}^{(k)}}.$$

Since

$$\boldsymbol{g}^{(k)} = \boldsymbol{Q}\boldsymbol{x}^{(k)} - \boldsymbol{b} = \boldsymbol{Q}\boldsymbol{x}^{(k)} - \boldsymbol{Q}\boldsymbol{x}^* = \boldsymbol{Q}\boldsymbol{y}^{(k)},$$

we have

$$\begin{aligned} \boldsymbol{y}^{(k)T}\boldsymbol{Q}\boldsymbol{y}^{(k)} &= \boldsymbol{g}^{(k)T}\boldsymbol{Q}^{-1}\boldsymbol{g}^{(k)}, \\ \boldsymbol{g}^{(k)T}\boldsymbol{Q}\boldsymbol{y}^{(k)} &= \boldsymbol{g}^{(k)T}\boldsymbol{g}^{(k)}. \end{aligned}$$

Therefore, substituting the above, we get

$$\frac{V(\boldsymbol{x}^{(k)}) - V(\boldsymbol{x}^{(k+1)})}{V(\boldsymbol{x}^{(k)})} = \alpha_k\frac{\boldsymbol{g}^{(k)T}\boldsymbol{Q}\boldsymbol{g}^{(k)}}{\boldsymbol{g}^{(k)T}\boldsymbol{Q}^{-1}\boldsymbol{g}^{(k)}}\left(2\frac{\boldsymbol{g}^{(k)T}\boldsymbol{g}^{(k)}}{\boldsymbol{g}^{(k)T}\boldsymbol{Q}\boldsymbol{g}^{(k)}} - \alpha_k\right) = \gamma_k.$$

■

We are now ready to state and prove our key convergence theorem for gradient methods. The theorem gives a necessary and sufficient condition for the sequence $\{\boldsymbol{x}^{(k)}\}$ generated by a gradient method to converge to $\boldsymbol{x}^*$, that is, $\boldsymbol{x}^{(k)} \to \boldsymbol{x}^*$, or $\lim_{k\to\infty}\boldsymbol{x}^{(k)} = \boldsymbol{x}^*$.

**Theorem 8.1.** *Let $\{\boldsymbol{x}^{(k)}\}$ be the sequence resulting from a gradient algorithm $\boldsymbol{x}^{(k+1)} = \boldsymbol{x}^{(k)} - \alpha_k\boldsymbol{g}^{(k)}$. Let $\gamma_k$ be as defined in Lemma 8.1, and suppose $\gamma_k > 0$ for all k. Then, $\{\boldsymbol{x}^{(k)}\}$ converges to $\boldsymbol{x}^*$ for any initial condition $\boldsymbol{x}^{(0)}$ if, and only if,*

$$\sum_{k=0}^{\infty}\gamma_k = \infty.$$

□

*Proof.* From Lemma 8.1, we have $V(\boldsymbol{x}^{(k+1)}) = (1 - \gamma_k)V(\boldsymbol{x}^{(k)})$, from which we obtain

$$V(\boldsymbol{x}^{(k)}) = \left(\prod_{i=0}^{k-1}(1 - \gamma_i)\right)V(\boldsymbol{x}^{(0)}).$$

Assume that $\gamma_k < 1$ for all $k$, for otherwise the result holds trivially. Note that $\boldsymbol{x}^{(k)} \to \boldsymbol{x}^*$ if, and only if, $V(\boldsymbol{x}^{(k)}) \to 0$. By the above equation, we see that this occurs if, and only if, $\prod_{i=0}^{\infty}(1 - \gamma_i) = 0$, which in turn holds if, and only if, $\sum_{i=0}^{\infty} -\log(1-\gamma_i) = \infty$ (we get this simply by taking logs). Therefore, it remains to be shown that $\sum_{i=0}^{\infty} -\log(1-\gamma_i) = \infty$ if, and only if,

$$\sum_{i=0}^{\infty} \gamma_i = \infty.$$

We first show that $\sum_{i=0}^{\infty} \gamma_i = \infty$ implies that $\sum_{i=0}^{\infty} -\log(1-\gamma_i) = \infty$. For this, first observe that for any $x \in \mathbb{R}, x > 0$, we have $\log(x) \leqslant x - 1$ [this is easy to see simply by plotting $\log(x)$ and $x - 1$ versus $x$]. Therefore, $\log(1-\gamma_i) \leqslant 1 - \gamma_i - 1 = -\gamma_i$, and hence $-\log(1-\gamma_i) \geqslant \gamma_i$. Thus, if $\sum_{i=0}^{\infty} \gamma_i = \infty$, then clearly $\sum_{i=0}^{\infty} -\log(1-\gamma_i) = \infty$.

Finally, we show that $\sum_{i=0}^{\infty} -\log(1-\gamma_i) = \infty$ implies that $\sum_{i=0}^{\infty} \gamma_i = \infty$. We proceed by contraposition. Suppose $\sum_{i=0}^{\infty} \gamma_i < \infty$. Then, it must be that $\gamma_i \to 0$. Now, observe that, for $x \in \mathbb{R}$, $x < 1$, and $x$ sufficiently close to 1, we have $\log(x) \geqslant 2(x-1)$ [as before, this is easy to see simply by plotting $\log(x)$ and $2(x-1)$ versus $x$]. Therefore, for sufficiently large $i$, $\log(1-\gamma_i) \geqslant 2(1-\gamma_i-1) = -2\gamma_i$, which implies that $-\log(1-\gamma_i) \leqslant 2\gamma_i$. Hence, $\sum_{i=0}^{\infty} \gamma_i < \infty$ implies that $\sum_{i=0}^{\infty} -\log(1-\gamma_i) < \infty$. This completes the proof. ■

The assumption in the above theorem that $\gamma_k \geqslant 0$ for all $k$ is significant in that it corresponds to the algorithm having the descent property (see Exercise 8.6). Furthermore, the result of the theorem does not hold in general if we do not assume that $\gamma_k \geqslant 0$ for all $k$ (see Exercise 8.7).

Using the above general theorem, we can now establish the covergence of specific cases of the gradient algorithm, including the steepest descent algorithm and algorithms with fixed step size. In the analysis to follow, we use Rayleigh's inequality, which states that for any $\boldsymbol{Q} = \boldsymbol{Q}^T > 0$, we have

$$\lambda_{\min}(\boldsymbol{Q})\|\boldsymbol{x}\|^2 \leqslant \boldsymbol{x}^T\boldsymbol{Q}\boldsymbol{x} \leqslant \lambda_{\max}(\boldsymbol{Q})\|\boldsymbol{x}\|^2,$$

where $\lambda_{\min}(\boldsymbol{Q})$ denotes the minimal eigenvalue of $\boldsymbol{Q}$, and $\lambda_{\max}(\boldsymbol{Q})$ denotes the maximal eigenvalue of $\boldsymbol{Q}$. For $\boldsymbol{Q} = \boldsymbol{Q}^T > 0$, we also have

$$\lambda_{\min}(\boldsymbol{Q}^{-1}) = \frac{1}{\lambda_{\max}(\boldsymbol{Q})},$$

$$\lambda_{\max}(\boldsymbol{Q}^{-1}) = \frac{1}{\lambda_{\min}(\boldsymbol{Q})},$$

and

$$\lambda_{\min}(\boldsymbol{Q}^{-1})\|\boldsymbol{x}\|^2 \leqslant \boldsymbol{x}^T\boldsymbol{Q}^{-1}\boldsymbol{x} \leqslant \lambda_{\max}(\boldsymbol{Q}^{-1})\|\boldsymbol{x}\|^2.$$

**Lemma 8.2.** *Let $\boldsymbol{Q} = \boldsymbol{Q}^T > 0$ be an $n \times n$ real symmetric positive definite matrix. Then, for any $\boldsymbol{x} \in \mathbb{R}^n$, we have*

$$\frac{\lambda_{\min}(\boldsymbol{Q})}{\lambda_{\max}(\boldsymbol{Q})} \leqslant \frac{(\boldsymbol{x}^T\boldsymbol{x})^2}{(\boldsymbol{x}^T\boldsymbol{Q}\boldsymbol{x})(\boldsymbol{x}^T\boldsymbol{Q}^{-1}\boldsymbol{x})} \leqslant \frac{\lambda_{\max}(\boldsymbol{Q})}{\lambda_{\min}(\boldsymbol{Q})}.$$

□

*Proof.* Applying Rayleigh's inequality and using the previously listed properties of symmetric positive definite matrices, we get

$$\frac{(x^T x)^2}{(x^T Q x)(x^T Q^{-1} x)} \leqslant \frac{\|x\|^4}{\lambda_{\min}(Q)\|x\|^2 \lambda_{\min}(Q^{-1})\|x\|^2} = \frac{\lambda_{\max}(Q)}{\lambda_{\min}(Q)}$$

and

$$\frac{(x^T x)^2}{(x^T Q x)(x^T Q^{-1} x)} \geqslant \frac{\|x\|^4}{\lambda_{\max}(Q)\|x\|^2 \lambda_{\max}(Q^{-1})\|x\|^2} = \frac{\lambda_{\min}(Q)}{\lambda_{\max}(Q)}.$$

■

We are now ready to establish the convergence of the steepest descent method.

**Theorem 8.2.** *In the steepest descent algorithm, we have $x^{(k)} \to x^*$ for any $x^{(0)}$.* □

*Proof.* Recall that for the steepest descent algorithm,

$$\alpha_k = \frac{g^{(k)T} g^{(k)}}{g^{(k)T} Q g^{(k)}}.$$

Substituting the above expression for $\alpha_k$ in the formula for $\gamma_k$ yields

$$\gamma_k = \frac{(g^{(k)T} g^{(k)})^2}{(g^{(k)T} Q g^{(k)})(g^{(k)T} Q^{-1} g^{(k)})}.$$

Note that in this case, $\gamma_k > 0$ for all k. Furthermore, by Lemma 8.2, we have $\gamma_k \geqslant (\lambda_{\min}(Q)/\lambda_{\max}(Q)) > 0$. Therefore, we have $\sum_{k=0}^{\infty} \gamma_k = \infty$, and hence by Theorem 8.1, we conclude that $x^{(k)} \to x^*$. ■

Consider now a gradient method with fixed step size, that is, $\alpha_k = \alpha \in \mathbb{R}$ for all $k$. The resulting algorithm is of the form

$$x^{(k+1)} = x^{(k)} - \alpha g^{(k)}.$$

We refer to the above algorithm as a fixed step size gradient algorithm. The algorithm is of practical interest because of its simplicity. In particular, the algorithm does not require a line search at each step to determine $\alpha_k$, because the same step size $\alpha$ is used at each step. Clearly, the convergence of the algorithm depends on the choice of $\alpha$, and we would not expect the algorithm to work for arbitrary $\alpha$. The following theorem gives a necessary and sufficient condition on $\alpha$ for convergence of the algorithm.

**Theorem 8.3.** *For the fixed step-size gradient algorithm, $x^{(k)} \to x^*$ for any $x^{(0)}$ if, and only if,*

$$0 < \alpha < \frac{2}{\lambda_{\max}(Q)}.$$

□

*Proof.* $\Leftarrow$: By Rayleigh's inequality, we have

$$\lambda_{\min}(\boldsymbol{Q})\boldsymbol{g}^{(k)T}\boldsymbol{g}^{(k)} \leqslant \boldsymbol{g}^{(k)T}\boldsymbol{Q}\boldsymbol{g}^{(k)} \leqslant \lambda_{\max}(\boldsymbol{Q})\boldsymbol{g}^{(k)T}\boldsymbol{g}^{(k)}$$

and

$$\boldsymbol{g}^{(k)T}\boldsymbol{Q}^{-1}\boldsymbol{g}^{(k)} \leqslant \frac{1}{\lambda_{\min}(\boldsymbol{Q})}\boldsymbol{g}^{(k)T}\boldsymbol{g}^{(k)}.$$

Therefore, substituting the above into the formula for $\gamma_k$, we get

$$\gamma_k \geqslant \alpha(\lambda_{\min}(\boldsymbol{Q}))^2\left(\frac{2}{\lambda_{\max}(\boldsymbol{Q})} - \alpha\right) > 0.$$

Therefore, $\gamma_k > 0$ for all k, and $\Sigma_{k=0}^{\infty}\gamma_k = \infty$. Hence, by Theorem 8.1, we conclude that $\boldsymbol{x}^{(k)} \to \boldsymbol{x}^*$.

$\Rightarrow$: We use contraposition. Suppose either $\alpha \leqslant 0$ or $\alpha \geqslant 2/\lambda_{\max}(\boldsymbol{Q})$. Let $\boldsymbol{x}^{(0)}$ be chosen such that $\boldsymbol{x}^{(0)} - \boldsymbol{x}^*$ is an eigenvector of $\boldsymbol{Q}$ corresponding to the eigenvalue $\lambda_{\max}(\boldsymbol{Q})$. Since

$$\boldsymbol{x}^{(k+1)} = \boldsymbol{x}^{(k)} - \alpha(\boldsymbol{Q}\boldsymbol{x}^{(k)} - \boldsymbol{b}) = \boldsymbol{x}^{(k)} - \alpha(\boldsymbol{Q}\boldsymbol{x}^{(k)} - \boldsymbol{Q}\boldsymbol{x}^*)$$

then

$$\begin{aligned}\boldsymbol{x}^{(k+1)} - \boldsymbol{x}^* &= \boldsymbol{x}^{(k)} - \boldsymbol{x}^* - \alpha(\boldsymbol{Q}\boldsymbol{x}^{(k)} - \boldsymbol{Q}\boldsymbol{x}^*)\\ &= (\boldsymbol{I}_n - \alpha\boldsymbol{Q})(\boldsymbol{x}^{(k)} - \boldsymbol{x}^*)\\ &= (\boldsymbol{I}_n - \alpha\boldsymbol{Q})^{k+1}(\boldsymbol{x}^{(0)} - \boldsymbol{x}^*)\\ &= (1 - \alpha\lambda_{\max}(\boldsymbol{Q}))^{k+1}(\boldsymbol{x}^{(0)} - \boldsymbol{x}^*)\end{aligned}$$

where, in the last line, we used the property that $\boldsymbol{x}^{(0)} - \boldsymbol{x}^*$ is an eigenvector of $\boldsymbol{Q}$. Taking norms on both sides, we get

$$\|\boldsymbol{x}^{(k+1)} - \boldsymbol{x}^*\| = |1 - \alpha\lambda_{\max}(\boldsymbol{Q})|^{k+1}\|\boldsymbol{x}^{(0)} - \boldsymbol{x}^*\|.$$

Since $\alpha \leqslant 0$ or $\alpha \geqslant 2/\lambda_{\max}(\boldsymbol{Q})$, then

$$|1 - \alpha\lambda_{\max}(\boldsymbol{Q})| \geqslant 1.$$

Hence, $\|\boldsymbol{x}^{(k+1)} - \boldsymbol{x}^*\|$ cannot converge to 0, and thus the sequence $\{\boldsymbol{x}^{(k)}\}$ does not converge to $\boldsymbol{x}^*$. ■

***Example 8.3.*** Let the function $f$ be given by

$$f(\boldsymbol{x}) = \boldsymbol{x}^T\begin{bmatrix}4 & 2\sqrt{2}\\ 0 & 5\end{bmatrix}\boldsymbol{x} + \boldsymbol{x}^T\begin{bmatrix}3\\ 6\end{bmatrix} + 24.$$

We wish to find the minimizer of $f$ using a fixed step-size gradient algorithm

$$x^{(k+1)} = x^{(k)} - \alpha \nabla f(x^{(k)})$$

where $\alpha \in \mathbb{R}$ is a fixed step size.

To apply Theorem 8.3, we first symmetrize the matrix in the quadratic term of $f$ to get

$$f(x) = \frac{1}{2} x^T \begin{bmatrix} 8 & 2\sqrt{2} \\ 2\sqrt{2} & 10 \end{bmatrix} x + x^T \begin{bmatrix} 3 \\ 6 \end{bmatrix} + 24.$$

The eigenvalues of the matrix in the quadratic term are 6 and 12. Hence, using Theorem 8.3, the above algorithm converges to the minimizer for all $x^{(0)}$ if, and only if, $\alpha$ lies in the range $0 < \alpha < 2/12$. ■

We now turn our attention to the issue of convergence rates of gradient algorithms. In particular, we focus on the steepest descent algorithm. We first present the following theorem.

**Theorem 8.4.** *In the method of steepest descent applied to the quadratic function, at every step k, we have*

$$V(x^{(k+1)}) \leqslant \left( \frac{\lambda_{\max}(Q) - \lambda_{\min}(Q)}{\lambda_{\max}(Q)} \right) V(x^{(k)}).$$

□

*Proof.* In the proof of Theorem 8.2, we showed that $\gamma_k \geqslant (\lambda_{\min}(Q)/\lambda_{\max}(Q))$. Therefore,

$$\frac{V(x^{(k)}) - V(x^{(k+1)})}{V(x^{(k)})} = \gamma_k \geqslant \frac{\lambda_{\min}(Q)}{\lambda_{\max}(Q)},$$

and the result follows. ■

The above theorem is relevant to our consideration of the convergence rate of the steepest descent algorithm as follows. Let

$$r = \frac{\lambda_{\max}(Q)}{\lambda_{\min}(Q)} = \|Q\| \, \|Q^{-1}\|,$$

the so-called *condition number* of $Q$. Then, it follows from Theorem 8.4 that

$$V(x^{(k+1)}) \leqslant \left(1 - \frac{1}{r}\right) V(x^{(k)}).$$

The term $(1-1/r)$ plays an important role in the convergence of $\{V(\boldsymbol{x}^{(k)})\}$ to 0 [and hence of $\{\boldsymbol{x}^{(k)}\}$ to $\boldsymbol{x}^*$]. We refer to $(1-1/r)$ as the *convergence ratio.* Specifically, we see that the smaller the value of $(1-1/r)$, the smaller $V(\boldsymbol{x}^{(k+1)})$ will be relative to $V(\boldsymbol{x}^{(k)})$, and hence the "faster" $V(\boldsymbol{x}^{(k)})$ converges to 0, as indicated by the inequality above. The convergence ratio $(1-1/r)$ decreases as $r$ decreases. If $r=1$, then $\lambda_{\max}(\boldsymbol{Q})=\lambda_{\min}(\boldsymbol{Q})$, and this corresponds to circular contours of $f$ (see Figure 8.6). In this case, the algorithm converges in a single step to the minimizer. As $r$ increases, the speed of convergence of $\{V(\boldsymbol{x}^{(k)})\}$ [and hence of $\{\boldsymbol{x}^{(k)}\}$] decreases. The increase in $r$ reflects the fact that the contours of $f$ are more eccentric (see, for example, Figure 8.7). We refer the reader to Ref. 51 (pp. 218 and 219) for an alternative approach to the above analysis.

To further investigate the convergence properties of $\{\boldsymbol{x}^{(k)}\}$, we need the following definition.

***Definition 8.1.*** Given a sequence $\{\boldsymbol{x}^{(k)}\}$ that converges to $\boldsymbol{x}^*$, that is, $\lim_{k\to\infty}\|\boldsymbol{x}^{(k)}-\boldsymbol{x}^*\|=0$, we say that the order of convergence is $p$, where $p\in\mathbb{R}$, if

$$0<\lim_{k\to\infty}\frac{\|\boldsymbol{x}^{(k+1)}-\boldsymbol{x}^*\|}{\|\boldsymbol{x}^{(k)}-\boldsymbol{x}^*\|^p}<\infty.$$

If for all $p>0$,

$$\lim_{k\to\infty}\frac{\|\boldsymbol{x}^{(k+1)}-\boldsymbol{x}^*\|}{\|\boldsymbol{x}^{(k)}-\boldsymbol{x}^*\|^p}=0,$$

then we say that the order of convergence is $\infty$. ■

Note that in the above definition, 0/0 should be understood to be 0.

If $p=1$ (first-order convergence), we say that the convergence is *linear*. If $p=2$ (second-order convergence), we say that the convergence is *quadratic*. The above definition can be interpreted using the notion of the order symbol $O$ as follows. Recall that $a=O(h)$ if there exists a constant $c$ such that $|a|\leqslant ch$ for sufficiently small $h$. Then, the order of convergence is at least $p$ if

$$\|\boldsymbol{x}^{(k+1)}-\boldsymbol{x}^*\|=O(\|\boldsymbol{x}^{(k)}-\boldsymbol{x}^*\|^p)$$

(see Exercise 8.1). In particular, the order of convergence is at least 1 if

$$\|\boldsymbol{x}^{(k+1)}-\boldsymbol{x}^*\|=O(\|\boldsymbol{x}^{(k)}-\boldsymbol{x}^*\|),$$

and the order of convergence is at least 2 if

$$\|\boldsymbol{x}^{(k+1)}-\boldsymbol{x}^*\|=O(\|\boldsymbol{x}^{(k)}-\boldsymbol{x}^*\|^2).$$

*Example 8.4.*

1. Suppose that $x^{(k)} = 1/k$ and thus $x^{(k)} \to 0$. Then,

$$\frac{|x^{(k+1)}|}{|x^{(k)}|^p} = \frac{1/(k+1)}{1/k^p} = \frac{k^p}{k+1}.$$

If $p < 1$, the above sequence converges to 0, whereas, if $p > 1$, it grows to $\infty$. If $p = 1$, the sequence converges to 1. Hence, the order of convergence is 1.

2. Suppose $x^{(k)} = \gamma^k$, where $0 < \gamma < 1$, and thus $x^{(k)} \to 0$. Then,

$$\frac{|x^{(k+1)}|}{|x^{(k)}|^p} = \frac{\gamma^{k+1}}{(\gamma^k)^p} = \gamma^{k+1-kp} = \gamma^{k(1-p)+1}.$$

If $p < 1$, the above sequence converges to 0, whereas, if $p > 1$, it grows to $\infty$. If $p = 1$, the sequence converges to $\gamma$ (in fact, remains constant at $\gamma$). Hence, the order of convergence is 1.

3. Suppose $x^{(k)} = \gamma^{(q^k)}$, where $q > 1$ and $0 < \gamma < 1$, and thus $x^{(k)} \to 0$. Then,

$$\frac{|x^{(k+1)}|}{|x^{(k)}|^p} = \frac{\gamma^{(q^{k+1})}}{(\gamma^{(q^k)})^p} = \gamma^{(q^{k+1} - pq^k)} = \gamma^{(q-p)q^k}.$$

If $p < q$, the above sequence converges to 0, whereas, if $p > q$, it grows to $\infty$. If $p = q$, the sequence converges to 1 (in fact, remains constant at 1). Hence, the order of convergence is $q$.

4. Suppose $x^{(k)} = 1$ for all $k$, and thus $x^{(k)} \to 1$ trivially. Then,

$$\frac{|x^{(k+1)} - 1|}{|x^{(k)} - 1|^p} = \frac{0}{0^p} = 0,$$

for all $p$. Hence, the order of convergence is $\infty$.

■

With the above definition of the order of convergence, we can prove the following result.

**Theorem 8.5.** *Consider the steepest descent algorithm applied to a quadratic function, and let $\{\boldsymbol{x}^{(k)}\}$ be the sequence of iterates. Then the order of convergence of $\{\boldsymbol{x}^{(k)}\}$ is at least 1.* □

*Proof.* From Rayleigh's inequality, we have

$$V(\boldsymbol{x}^{(k+1)}) \geqslant \frac{\lambda_{\min}(\boldsymbol{Q})}{2} \|\boldsymbol{x}^{(k+1)} - \boldsymbol{x}^*\|^2,$$

$$V(\boldsymbol{x}^{(k)}) \leqslant \frac{\lambda_{\max}(\boldsymbol{Q})}{2} \|\boldsymbol{x}^{(k)} - \boldsymbol{x}^*\|^2.$$

Substituting the above into the inequality in Theorem 8.4, yields

$$\lambda_{\min}(\boldsymbol{Q})\|\boldsymbol{x}^{(k+1)}-\boldsymbol{x}^*\|^2 \leqslant (\lambda_{\max}(\boldsymbol{Q})-\lambda_{\min}(\boldsymbol{Q}))\|\boldsymbol{x}^{(k)}-\boldsymbol{x}^*\|^2.$$

Rearranging terms, we get

$$\frac{\|\boldsymbol{x}^{(k+1)}-\boldsymbol{x}^*\|}{\|\boldsymbol{x}^{(k)}-\boldsymbol{x}^*\|} \leqslant \sqrt{\frac{\lambda_{\max}(\boldsymbol{Q})-\lambda_{\min}(\boldsymbol{Q})}{\lambda_{\min}(\boldsymbol{Q})}} \geqslant 0.$$

Hence, $\|\boldsymbol{x}^{(k+1)}-\boldsymbol{x}^*\| = O(\|\boldsymbol{x}^{(k)}-\boldsymbol{x}^*\|)$, and the order of convergence is at least 1. ■

In the next chapter, we discuss Newton's method, which has an order of convergence at least 2 if the initial guess is near the solution.

## EXERCISES

**8.1** Let $\{\boldsymbol{x}^{(k)}\}$ be a sequence that converges to $\boldsymbol{x}^*$. Show that if

$$\|\boldsymbol{x}^{(k+1)}-\boldsymbol{x}^*\| = O(\|\boldsymbol{x}^{(k)}-\boldsymbol{x}^*\|^p),$$

then the order of convergence is at least $p$.

**8.2** Suppose we use the Golden Section algorithm to find the minimizer of a function. Let $u_k$ be the uncertainty range at the $k$th iteration. Find the order of convergence of $\{u_k\}$.

**8.3** Suppose we wish to minimize a function $f:\mathbb{R}\to\mathbb{R}$ that has a derivative $f'$. A simple line search method, called the *derivative descent search* (DDS), is described as follows: given that we are at a point $x^{(k)}$, we move in the direction of the negative derivative with step-size $\alpha$; that is, $x^{(k+1)} = x^{(k)} - \alpha f'(x^{(k)})$, where $\alpha > 0$ is a constant.

In the following parts, assume that $f$ is quadratic: $f(x) = \frac{1}{2}ax^2 - bx + c$ (where $a$, $b$, and $c$ are constants, $a > 0$).

**a.** Write down the value of $x^*$ (in terms of $a$, $b$, and $c$) that minimizes $f$.

**b.** Write down the recursive equation for the DDS algorithm explicitly for this quadratic $f$.

**c.** Assuming that the DDS algorithm converges, show that it converges to the optimal value $x^*$ (found in part a).

**d.** Find the order of convergence of the algorithm, assuming that it does converge.

**e.** Find the range of values of $\alpha$ for which the algorithm converges (for this particular $f$) for all starting points $x^{(0)}$.

**8.4** Let $f:\mathbb{R}^n \to \mathbb{R}$ be given by $f(\boldsymbol{x}) = \frac{1}{2}\boldsymbol{x}^T\boldsymbol{Q}\boldsymbol{x} - \boldsymbol{x}^T\boldsymbol{b}$, where $\boldsymbol{b} \in \mathbb{R}^n$, and $\boldsymbol{Q}$ is a real symmetric positive definite $n \times n$ matrix. Suppose we apply the steepest descent method to this function, with $\boldsymbol{x}^{(0)} \neq \boldsymbol{Q}^{-1}\boldsymbol{b}$. Show that the method converges in one step, that is, $\boldsymbol{x}^{(1)} = \boldsymbol{Q}^{-1}\boldsymbol{b}$, if, and only if, $\boldsymbol{x}^{(0)}$ is chosen such that $\boldsymbol{g}^{(0)} = \boldsymbol{Q}\boldsymbol{x}^{(0)} - \boldsymbol{b}$ is an eigenvector of $\boldsymbol{Q}$.

**8.5** Let $f:\mathbb{R}^n \to \mathbb{R}$ be given by $f(\boldsymbol{x}) = \frac{1}{2}\boldsymbol{x}^T\boldsymbol{Q}\boldsymbol{x} - \boldsymbol{x}^T\boldsymbol{b}$, where $\boldsymbol{b} \in \mathbb{R}^n$, and $\boldsymbol{Q}$ is a real symmetric positive definite $n \times n$ matrix. Consider the algorithm

$$\boldsymbol{x}^{(k+1)} = \boldsymbol{x}^{(k)} - \beta\alpha_k \boldsymbol{g}^{(k)},$$

where $\boldsymbol{g}^{(k)} = \boldsymbol{Q}\boldsymbol{x}^{(k)} - \boldsymbol{b}$, $\alpha_k = \boldsymbol{g}^{(k)T}\boldsymbol{g}^{(k)}/\boldsymbol{g}^{(k)T}\boldsymbol{Q}\boldsymbol{g}^{(k)}$, and $\beta \in \mathbb{R}$ is a given constant. (Note that the above reduces to the steepest descent algorithm if $\beta = 1$.)

Show that $\{\boldsymbol{x}^{(k)}\}$ converges to $\boldsymbol{x}^* = \boldsymbol{Q}^{-1}\boldsymbol{b}$ for any initial condition $\boldsymbol{x}^{(0)}$ if, and only if, $0 < \beta < 2$.

**8.6** Let $f:\mathbb{R}^n \to \mathbb{R}$ be given by $f(\boldsymbol{x}) = \frac{1}{2}\boldsymbol{x}^T\boldsymbol{Q}\boldsymbol{x} - \boldsymbol{x}^T\boldsymbol{b}$, where $\boldsymbol{b} \in \mathbb{R}^n$, and $\boldsymbol{Q}$ is a real symmetric positive definite $n \times n$ matrix. Consider a gradient algorithm

$$\boldsymbol{x}^{(k+1)} = \boldsymbol{x}^{(k)} - \alpha_k \boldsymbol{g}^{(k)},$$

where $\boldsymbol{g}^{(k)} = \boldsymbol{Q}\boldsymbol{x}^{(k)} - \boldsymbol{b}$ is the gradient of $f$ at $\boldsymbol{x}^{(k)}$, and $\alpha_k$ is some step size.

Show that the above algorithm has a descent property (that is, $f(\boldsymbol{x}^{(k+1)}) < f(\boldsymbol{x}^{(k)})$ whenever $\boldsymbol{g}^{(k)} \neq 0$) if, and only if, $\gamma_k > 0$ for all $k$.

**8.7** Show, using a counterexample, that the assumption that $\gamma_k > 0$ in Theorem 8.1 is necessary for the result of the theorem to hold.

**8.8** Given $f:\mathbb{R}^n \to \mathbb{R}$, consider the algorithm

$$\boldsymbol{x}^{(k+1)} = \boldsymbol{x}^{(k)} + \alpha_k \boldsymbol{d}^{(k)}$$

where $\boldsymbol{d}^{(1)}, \boldsymbol{d}^{(2)}, \ldots$ are given vectors in $\mathbb{R}^n$, and $\alpha_k$ is chosen to minimize $f(\boldsymbol{x}^{(k)} + \alpha\boldsymbol{d}^{(k)})$; that is

$$\alpha_k = \arg\min f(\boldsymbol{x}^{(k)} + \alpha\boldsymbol{d}^{(k)}).$$

Show that for each $k$, the vector $\boldsymbol{x}^{(k+1)} - \boldsymbol{x}^{(k)}$ is orthogonal to $\nabla f(\boldsymbol{x}^{(k+1)})$.

**8.9** Write a simple MATLAB routine for implementing the steepest descent algorithm using the secant method for the line search (for example, the MATLAB function of Exercise 7.8). For the stopping criterion, use the condition $\|\boldsymbol{g}^{(k)}\| \leqslant \varepsilon$, where $\varepsilon = 10^{-6}$. Test your routine by comparing the output with the numbers in Example 8.1. Also test your routine using an initial condition of $[-4, 5, 1]^T$, and determine the number of iterations required to satisfy the above stopping criterion. Evaluate the objective function at the final point to see how close it is to 0.

**8.10** Apply the MATLAB routine from Exercise 8.9 to Rosenbrock's function:

$$f(\boldsymbol{x}) = 100(x_2 - x_1^2)^2 + (1 - x_1)^2.$$

Use an initial condition of $\boldsymbol{x}^{(0)} = [-2, 2]^T$. Terminate the algorithm when the norm of the gradient of $f$ is less than $10^{-4}$.

# 9

# Newton's Method

## 9.1. INTRODUCTION

Recall that the method of steepest descent uses only first derivatives (gradients) in selecting a suitable search direction. This strategy is not always the most effective. If higher derivatives are used, the resulting iterative algorithm may perform better than the steepest descent method. Newton's method (sometimes called the Newton–Raphson method) uses first and second derivatives and indeed does perform better than the steepest descent method if the initial point is close to the minimizer. The idea behind this method is as follows. Given a starting point, we construct a quadratic approximation to the objective function that matches the first- and second-derivative values at that point. We then minimize the approximate (quadratic) function instead of the original objective function. We use the minimizer of the approximate function as the starting point in the next step and repeat the procedure iteratively. If the objective function is quadratic, then the approximation is exact, and the method yields the true minimizer in one step. If on the other hand the objective function is not quadratic, then the approximation will provide only an estimate of the position of the true minimizer. Figure 9.1 illustrates the above idea.

We can obtain a quadratic approximation to the given twice continuously differentiable objection function $f: \mathbb{R}^n \to \mathbb{R}$ using the Taylor series expansion of $f$ about the current point $\boldsymbol{x}^{(k)}$, neglecting terms of order three and higher. We obtain

$$f(\boldsymbol{x}) \approx f(\boldsymbol{x}^{(k)}) + (\boldsymbol{x} - \boldsymbol{x}^{(k)})^T \boldsymbol{g}^{(k)} + \tfrac{1}{2}(\boldsymbol{x} - \boldsymbol{x}^{(k)})^T \boldsymbol{F}(\boldsymbol{x}^{(k)})(\boldsymbol{x} - \boldsymbol{x}^{(k)}) \triangleq q(\boldsymbol{x}),$$

where, for simplicity, we use the notation $\boldsymbol{g}^{(k)} = \nabla f(\boldsymbol{x}^{(k)})$. Applying the FONC to $q$ yields

$$\boldsymbol{0} = \nabla q(\boldsymbol{x}) = \boldsymbol{g}^{(k)} + \boldsymbol{F}(\boldsymbol{x}^{(k)})(\boldsymbol{x} - \boldsymbol{x}^{(k)}).$$

If $\boldsymbol{F}(\boldsymbol{x}^{(k)}) > 0$, then $q$ achieves a minimum at

$$\boldsymbol{x}^{(k+1)} = \boldsymbol{x}^{(k)} - \boldsymbol{F}(\boldsymbol{x}^{(k)})^{-1}\boldsymbol{g}^{(k)}.$$

This recursive formula represents Newton's method.

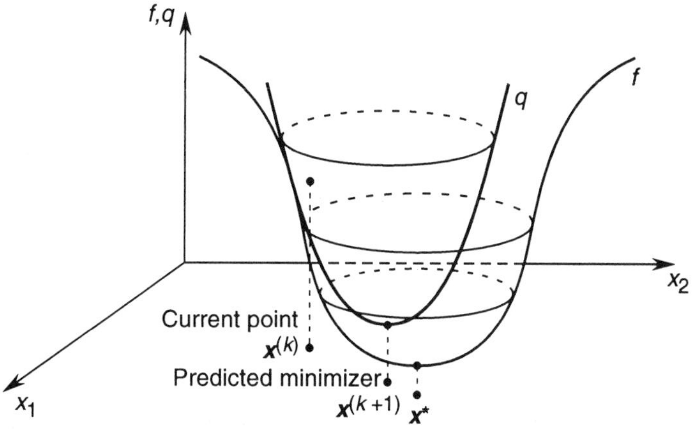


**Figure 9.1.** Quadratic approximation to the objective function using first and second derivatives

***Example 9.1.*** Use Newton's method to minimize the Powell function:

$$f(x_1,x_2,x_3,x_4)=(x_1+10x_2)^2+5(x_3-x_4)^2+(x_2-2x_3)^4+10(x_1-x_4)^4.$$

Use as the starting point $\boldsymbol{x}^{(0)}=[3,\ -1,\ 0,\ 1]^T$. Perform three iterations.

Note that $f(\boldsymbol{x}^{(0)})=215$. We have

$$\nabla f(\boldsymbol{x})=\begin{bmatrix} 2(x_1+10x_2)+40(x_1-x_4)^3 \\ 20(x_1+10x_2)+4(x_2-2x_3)^3 \\ 10(x_3-x_4)-8(x_2-2x_3)^3 \\ -10(x_3-x_4)-40(x_1-x_4)^3 \end{bmatrix},$$

and $\boldsymbol{F}(\boldsymbol{x})$ is given by

$$\begin{bmatrix} 2+120(x_1-x_4)^2 & 20 & 0 & -120(x_1-x_4)^2 \\ 20 & 200+12(x_2-2x_3)^2 & -24(x_2-2x_3)^2 & 0 \\ 0 & -24(x_2-2x_3)^2 & 10+48(x_2-2x_3)^2 & -10 \\ -120(x_1-x_4)^2 & 0 & -10 & 10+120(x_1-x_4)^2 \end{bmatrix}.$$

*Iteration 1.*

$$\boldsymbol{g}^{(0)}=[306,\ -144,\ -2,\ -310]^T,$$

$$\boldsymbol{F}(\boldsymbol{x}^{(0)})=\begin{bmatrix} 482 & 20 & 0 & -480 \\ 20 & 212 & -24 & 0 \\ 0 & -24 & 58 & -10 \\ -480 & 0 & -10 & 490 \end{bmatrix},$$

$$F(x^{(0)})^{-1} = \begin{bmatrix} 0.1126 & -0.0089 & 0.0154 & 0.1106 \\ -0.0089 & 0.0057 & 0.0008 & -0.0087 \\ 0.0154 & 0.0008 & 0.0203 & 0.0155 \\ 0.1106 & -0.0087 & 0.0155 & 0.1107 \end{bmatrix},$$

$$F(x^{(0)})^{-1} g^{(0)} = [1.4127, -0.8413, -0.2540, 0.7460]^T.$$

Hence

$$x^{(1)} = x^{(0)} - F(x^{(0)})^{-1} g^{(0)} = [1.5873, -0.1587, 0.2540, 0.2540]^T,$$

$$f(x^{(1)}) = 31.8.$$

*Iteration 2.*

$$g^{(1)} = [94.81, -1.179, 2.371, -94.81]^T,$$

$$F(x^{(1)}) = \begin{bmatrix} 215.3 & 20 & 0 & -213.3 \\ 20 & 205.3 & -10.67 & 0 \\ 0 & -10.67 & 31.34 & -10 \\ -213.3 & 0 & -10 & 223.3 \end{bmatrix},$$

$$F(x^{(1)})^{-1} g^{(1)} = [0.5291, -0.0529, 0.0846, 0.0846]^T.$$

Hence

$$x^{(2)} = x^{(1)} - F(x^{(1)})^{-1} g^{(1)} = [1.0582, -0.1058, 0.1694, 0.1694]^T,$$

$$f(x^{(2)}) = 6.28.$$

*Iteration 3.*

$$g^{(2)} = [28.09, -0.3475, 0.7031, -28.08]^T,$$

$$F(x^{(2)}) = \begin{bmatrix} 96.80 & 20 & 0 & -94.80 \\ 20 & 202.4 & -4.744 & 0 \\ 0 & -4.744 & 19.49 & -10 \\ -94.80 & 0 & -10 & 104.80 \end{bmatrix},$$

$$x^{(3)} = [0.7037, -0.0704, 0.1121, 0.1111]^T,$$

$$f(x^{(3)}) = 1.24.$$

■

Observe that the $k$th iteration of Newton's method can be written in two steps as

1. Solve $F(x^{(k)})d^{(k)} = -g^{(k)}$ for $d^{(k)}$
2. Set $x^{(k+1)} = x^{(k)} + d^{(k)}$

Step 1 requires the solution of an $n \times n$ system of linear equations. Thus, an efficient method for solving systems of linear equations is essential when using Newton's method.

As in the one variable case, Newton's method can also be viewed as a technique for iteratively solving the equation

$$g(x) = 0,$$

where $x \in \mathbb{R}^n$, and $g: \mathbb{R}^n \to \mathbb{R}^n$. In this case, $F(x)$ is the Jacobian matrix of $g$ at $x$, that is, $F(x)$ is the $n \times n$ matrix whose $(i, j)$ entry is $(\partial g_i/\partial x_j)(x)$, $i, j = 1, 2, \ldots, n$.

## 9.2. ANALYSIS OF NEWTON'S METHOD

As in the one variable case, there is no guarantee that Newton's algorithm heads in the direction of decreasing values of the objective function if $F(x^{(k)})$ is not positive definite (recall Figure 7.7 illustrating Newton's method for functions of one variable when $f'' < 0$). Moreover, even if $F(x^{(k)}) > 0$, Newton's method may not be a descent method; that is, it is possible that $f(x^{(k+1)}) \geqslant f(x^{(k)})$. For example, this may occur if our starting point $x^{(0)}$ is far away from the solution. See the end of this section for a possible remedy to this problem. Despite the above drawbacks, Newton's method has superior convergence properties when the starting point is near the solution, as we shall see in the remainder of this section.

The convergence analysis of Newton's method when $f$ is a quadratic function is straightforward. In fact, Newton's method reaches the point $x^*$ such that $\nabla f(x^*) = 0$ in just one step starting from any initial point $x^{(0)}$. To see this, suppose $Q = Q^T$ is invertible, and

$$f(x) = \tfrac{1}{2}x^T Q x - x^T b.$$

Then

$$g(x) = \nabla f(x) = Qx - b,$$

and

$$F(x) = Q.$$

Hence, given any initial point $x^{(0)}$, by Newton's algorithm

$$\begin{aligned} x^{(1)} &= x^{(0)} - F(x^{(0)})^{-1} g^{(0)} \\ &= x^{(0)} - Q^{-1}[Qx^{(0)} - b] \\ &= Q^{-1}b \\ &= x^*. \end{aligned}$$

Therefore, for the quadratic case, the order of convergence of Newton's algorithm is $\infty$ for any initial point $x^{(0)}$ (compare the above with Exercise 8.4, which deals with the steepest descent algorithm).

To analyze the convergence of Newton's method in the general case we need the following background material.

***Definition 9.1.*** We say that a sequence $\{A_k\}$ of $m \times n$ matrices converges to the $m \times n$ matrix $A$ if

$$\lim_{k \to \infty} \|A - A_k\| = 0. \qquad \blacksquare$$

**Lemma 9.1.** *The series of $n \times n$ matrices*

$$I_n + A + A^2 + \cdots + A^k + \cdots$$

*converges if, and only if, $\lim_{k \to \infty} A^k = O$. In this case the sum of the series equals $(I_n - A)^{-1}$.* □

*Proof.* The necessity of the condition is obvious. To prove the sufficiency note that

$$\lim_{k \to \infty} A^k = O \Leftrightarrow |\lambda_i(A)| < 1, \quad i = 1, \ldots, n.$$

The above implies that $\det(I_n - A) \neq 0$, and hence $(I_n - A)^{-1}$ exists. Consider now the following relation

$$(I_n + A + A^2 + \cdots + A^k)(I_n - A) = I_n - A^{k+1}.$$

Postmultiplying the above equation by $(I_n - A)^{-1}$ yields

$$I_n + A + A^2 + \cdots + A^k = (I_n - A)^{-1} - A^{k+1}(I_n - A)^{-1}.$$

Hence

$$\lim_{k \to \infty} \sum_{j=0}^{k} A^j = (I_n - A)^{-1}$$

since $\lim_{k \to \infty} A^{k+1} = O$. Thus

$$\sum_{j=0}^{\infty} A^j = (I_n - A)^{-1} \qquad \blacksquare$$

***Definition 9.2.*** A matrix-valued function $A: \mathbb{R}^r \to \mathbb{R}^{n \times n}$ is continuous at a point $\xi_0 \in \mathbb{R}^r$ if

$$\lim_{\|\xi - \xi_0\| \to 0} \|A(\xi) - A(\xi_0)\| = 0. \qquad \blacksquare$$

We will need the following lemma, taken from Ref. 67.

**Lemma 9.2.** *Let $A: \mathbb{R}^r \to \mathbb{R}^{n \times n}$ be an $n \times n$ matrix-valued function that is continuous at $\xi_0$. If $A(\xi_0)^{-1}$ exists, then $A(\xi)^{-1}$ exists for $\xi$ sufficiently close to $\xi_0$ and $A(\cdot)^{-1}$ is continuous at $\xi_0$.* □

*Proof.* We first prove the existence of $A(\xi)$ for all $\xi$ sufficiently close to $\xi_0$. We have

$$A(\xi) = A(\xi_0) - A(\xi_0) + A(\xi) = A(\xi_0)(I_n - K(\xi)),$$

where

$$K(\xi) = A(\xi_0)^{-1}(A(\xi_0) - A(\xi)).$$

Thus

$$\| K(\xi) \| \leqslant \| A(\xi_0)^{-1} \| \, \| A(\xi_0) - A(\xi) \|$$

and

$$\lim_{\|\xi - \xi_0\| \to 0} \| K(\xi) \| = 0.$$

Since $A$ is continuous at $\xi_0$, for all $\xi$ close enough to $\xi_0$, we have

$$\| A(\xi_0) - A(\xi) \| \leqslant \frac{\theta}{\| A(\xi_0)^{-1} \|},$$

where $\theta \in (0, 1)$. Then

$$\| K(\xi) \| \leqslant \theta < 1$$

and

$$(I_n - K(\xi))^{-1}$$

exists. But then

$$A(\xi)^{-1} = (A(\xi_0)(I_n - K(\xi)))^{-1} = (I_n - K(\xi))^{-1} A(\xi_0)^{-1},$$

which means that $A(\xi)^{-1}$ exists for $\xi$ sufficiently close to $\xi_0$.

To prove the continuity of $A(\cdot)^{-1}$ note that

$$\| A(\xi_0)^{-1} - A(\xi)^{-1} \| = \| A(\xi)^{-1} - A(\xi_0)^{-1} \| = \| ((I_n - K(\xi))^{-1} - I_n) A(\xi_0)^{-1} \|.$$

However, since $\| K(\xi) \| < 1$, it follows from Lemma 9.1 that

$$(I_n - K(\xi))^{-1} - I_n = K(\xi) + K^2(\xi) + \cdots = K(\xi)(I_n + K(\xi) + \cdots).$$

Hence

$$\begin{aligned} \| (I_n - K(\xi))^{-1} - I_n \| &\leqslant \| K(\xi) \| (1 + \| K(\xi) \| + \| K(\xi) \|^2 + \cdots) \\ &= \frac{\| K(\xi) \|}{1 - \| K(\xi) \|}, \end{aligned}$$

where $\| K(\xi) \| < 1$. Therefore

$$\| A(\xi)^{-1} - A(\xi_0)^{-1} \| \leqslant \left( \frac{\| K(\xi) \|}{1 - \| K(\xi) \|} \right) \| A(\xi_0)^{-1} \|.$$

Since

$$\lim_{\|\xi - \xi_0\| \to 0} \| K(\xi) \| = 0$$

we obtain

$$\lim_{\|\xi-\xi_0\|\to 0} \|A(\xi)^{-1} - A(\xi_0)^{-1}\| = 0,$$

which completes the proof. ■

With the above background, we are now ready to analyze the convergence properties of Newton's method. Let $\{x^{(k)}\}$ be the Newton's method sequence for minimizing a function $f$. We will show that $\{x^{(k)}\}$ converges to the minimizer $x^*$ with order of convergence at least two.

**Theorem 9.1.** *Suppose $f \in \mathscr{C}^3$, and $x^* \in \mathbb{R}^n$ is a point such that $\nabla f(x^*) = \mathbf{0}$ and $F(x^*)$ is invertible. Then, for all $x^{(0)}$ sufficiently close to $x^*$, Newton's method is well defined for all $k$, and converges to $x^*$ with order of convergence at least two.* □

*Proof.* The Taylor expansion of $\nabla f$ about $x^{(0)}$ yields

$$\nabla f(x) - \nabla f(x^{(0)}) - F(x^{(0)})(x - x^{(0)}) = O(\|x - x^{(0)}\|^2).$$

Because by assumption $f \in \mathscr{C}^3$ and $F(x^*)$ is invertible, there exist constants $\varepsilon > 0$, $c_1 > 0$, and $c_2 > 0$ such that, if $x^{(0)}, x \in \{x : \|x - x^*\| \leqslant \varepsilon\}$, we have

$$\|\nabla f(x) - \nabla f(x^{(0)}) - F(x^{(0)})(x - x^{(0)})\| \leqslant c_1 \|x - x^{(0)}\|^2$$

and, by Lemma 9.2, $F(x)^{-1}$ exists and satisfies

$$\|F(x)^{-1}\| \leqslant c_2.$$

The first inequality above holds because the remainder term in the Taylor expansion contains third derivatives of $f$ that are continuous and hence bounded on $\{x : \|x - x^*\| \leqslant \varepsilon\}$.

Suppose $x^{(0)} \in \{x : \|x - x^*\| \leqslant \varepsilon\}$. Then, substituting $x = x^*$ in the above inequality and using the assumption that $\nabla f(x^*) = \mathbf{0}$, we get

$$\|F(x^{(0)})(x^{(0)} - x^*) - \nabla f(x^{(0)})\| \leqslant c_1 \|x^{(0)} - x^*\|^2.$$

Now, subtracting $x^*$ from both sides of Newton's algorithm and taking norms yields

$$\begin{aligned}\|x^{(1)} - x^*\| &= \|x^{(0)} - x^* - F(x^{(0)})^{-1}\nabla f(x^{(0)})\| \\ &= \|F(x^{(0)})^{-1}(F(x^{(0)})(x^{(0)} - x^*) - \nabla f(x^{(0)}))\| \\ &\leqslant \|F(x^{(0)})^{-1}\| \, \|F(x^{(0)})(x^{(0)} - x^*) - \nabla f(x^{(0)})\|.\end{aligned}$$

Applying the above inequalities involving the constants $c_1$ and $c_2$ gives

$$\|x^{(1)} - x^*\| \leqslant c_1 c_2 \|x^{(0)} - x^*\|^2.$$

Suppose that $\boldsymbol{x}^{(0)}$ is such that

$$\|\boldsymbol{x}^{(0)} - \boldsymbol{x}^*\| \leqslant \frac{\alpha}{c_1 c_2},$$

where $\alpha \in (0, 1)$. Then,

$$\|\boldsymbol{x}^{(1)} - \boldsymbol{x}^*\| \leqslant \alpha \|\boldsymbol{x}^{(0)} - \boldsymbol{x}^*\|.$$

By induction, we obtain

$$\|\boldsymbol{x}^{(k+1)} - \boldsymbol{x}^*\| \leqslant c_1 c_2 \|\boldsymbol{x}^{(k)} - \boldsymbol{x}^*\|^2,$$

$$\|\boldsymbol{x}^{(k+1)} - \boldsymbol{x}^*\| \leqslant \alpha \|\boldsymbol{x}^{(k)} - \boldsymbol{x}^*\|.$$

Hence

$$\lim_{k \to \infty} \|\boldsymbol{x}^{(k)} - \boldsymbol{x}^*\| = 0,$$

and therefore the sequence $\{\boldsymbol{x}^{(k)}\}$ converges to $\boldsymbol{x}^*$. The order of convergence is at least two, because $\|\boldsymbol{x}^{(k+1)} - \boldsymbol{x}^*\| \leqslant c_1 c_2 \|\boldsymbol{x}^{(k)} - \boldsymbol{x}^*\|^2$, that is, $\|\boldsymbol{x}^{(k+1)} - \boldsymbol{x}^*\| = O(\|\boldsymbol{x}^{(k)} - \boldsymbol{x}^*\|^2)$. ■

As stated in the above theorem, Newton's method has superior convergence properties if the starting point is near the solution. However, the method is not guaranteed to converge to the solution if we start far away from it (in fact, it may not even be well defined because the Hessian may be singular). In particular, the method may not be a descent method, that is, it is possible that $f(\boldsymbol{x}^{(k+1)}) \geqslant f(\boldsymbol{x}^{(k)})$. Fortunately, it is possible to modify the algorithm such that the descent property holds. To see this, we will need the following result.

**Theorem 9.2.** *Let $\{\boldsymbol{x}^{(k)}\}$ be the sequence generated by Newton's method for minimizing a given objective function $f(\boldsymbol{x})$. If the Hessian $\boldsymbol{F}(\boldsymbol{x}^{(k)}) > 0$ and $\boldsymbol{g}^{(k)} = \nabla f(\boldsymbol{x}^{(k)}) \neq \boldsymbol{0}$, then the direction*

$$\boldsymbol{d}^{(k)} = -\boldsymbol{F}(\boldsymbol{x}^{(k)})^{-1} \boldsymbol{g}^{(k)} = \boldsymbol{x}^{(k+1)} - \boldsymbol{x}^{(k)}$$

*from $\boldsymbol{x}^{(k)}$ to $\boldsymbol{x}^{(k+1)}$ is a descent direction for $f$ in the sense that there exists an $\bar{\alpha} > 0$ such that for all $\alpha \in (0, \bar{\alpha})$,*

$$f(\boldsymbol{x}^{(k)} + \alpha \boldsymbol{d}^{(k)}) < f(\boldsymbol{x}^{(k)}).$$ □

*Proof.* Let

$$\phi(\alpha) = f(\boldsymbol{x}^{(k)} + \alpha \boldsymbol{d}^{(k)}).$$

Then, using the chain rule, we obtain

$$\phi'(\alpha) = \nabla f(\boldsymbol{x}^{(k)} + \alpha \boldsymbol{d}^{(k)})^T \boldsymbol{d}^{(k)}.$$

Hence

$$\phi'(0) = \nabla f(\boldsymbol{x}^{(k)})^T \boldsymbol{d}^{(k)} = -\boldsymbol{g}^{(k)T} \boldsymbol{F}(\boldsymbol{x}^{(k)})^{-1} \boldsymbol{g}^{(k)} < 0,$$

because $\boldsymbol{F}(\boldsymbol{x}^{(k)})^{-1} > 0$ and $\boldsymbol{g}^{(k)} \neq \boldsymbol{0}$. Thus, there exists an $\bar{\alpha} > 0$ so that for all $\alpha \in (0, \bar{\alpha})$, $\phi(\alpha) < \phi(0)$. This implies that for all $\alpha \in (0, \bar{\alpha})$,

$$f(\boldsymbol{x}^{(k)} + \alpha \boldsymbol{d}^{(k)}) < f(\boldsymbol{x}^{(k)})$$

and the proof is completed. ■

The above theorem motivates the following modification of Newton's method:

$$\boldsymbol{x}^{(k+1)} = \boldsymbol{x}^{(k)} - \alpha_k \boldsymbol{F}(\boldsymbol{x}^{(k)})^{-1} \boldsymbol{g}^{(k)},$$

where

$$\alpha_k = \arg\min_{\alpha \geq 0} f(\boldsymbol{x}^{(k)} - \alpha \boldsymbol{F}(\boldsymbol{x}^{(k)})^{-1} \boldsymbol{g}^{(k)}),$$

that is, at each iteration, we perform a line search in the direction $-\boldsymbol{F}(\boldsymbol{x}^{(k)})^{-1} \boldsymbol{g}^{(k)}$. By Theorem 9.2, we conclude that the above modified Newton's method has a descent property, that is, whenever $\boldsymbol{g}^{(k)} \neq \boldsymbol{0}$,

$$f(\boldsymbol{x}^{(k+1)}) < f(\boldsymbol{x}^{(k)}).$$

A drawback of Newton's method is that evaluation of $\boldsymbol{F}(\boldsymbol{x}^{(k)})$ for large $n$ can be computationally expensive. Furthermore, we have to solve the set of $n$ linear equations $\boldsymbol{F}(\boldsymbol{x}^{(k)}) \boldsymbol{d}^{(k)} = -\boldsymbol{g}^{(k)}$. In the following chapters, we discuss methods that alleviate this difficulty.

## EXERCISES

**9.1** Let $f: \mathbb{R} \to \mathbb{R}$ be given by $f(x) = (x - x_0)^4$, where $x_0 \in \mathbb{R}$ is a constant. Suppose we apply Newton's method to the problem of minimizing $f$.

**a.** Write down the update equation for Newton's method applied to the problem.

**b.** Let $y^{(k)} = |x^{(k)} - x_0|$. Show that the sequence $\{y^{(k)}\}$ satisfies $y^{(k+1)} = \frac{2}{3} y^{(k)}$.

**c.** Show that $x^{(k)} \to x_0$ for any initial guess $x^{(0)}$.

**d.** Show that the order of convergence of the sequence $\{x^{(k)}\}$ in step b is 1.

**e.** Theorem 9.1 states that under certain conditions, the order of convergence of Newton's method is at least two. Why does that theorem not hold in this particular problem?

**9.2** Consider the problem of minimizing $f(x) = x^{4/3} = (\sqrt[3]{x})^4$, $x \in \mathbb{R}$. Note that 0 is the global minimizer of $f$.

**a.** Write down the algorithm for Newton's method applied to this problem.

**b.** Show that as long as the starting point is not 0, the algorithm in part a does not converge to 0 (no matter how close to 0 we start).

**9.3** Consider Rosenbrock's function: $f(x) = 100(x_2 - x_1^2)^2 + (1 - x_1)^2$, where $x = [x_1, x_2]^T$ (known to be a "nasty" function—often used as a benchmark for testing algorithms). This function is also known as the *banana function* because of the shape of its level sets.

**a.** Prove that $[1, 1]^T$ is the unique global minimizer of $f$ over $\mathbb{R}^2$.

**b.** With a starting point of $[0, 0]^T$, apply two iterations of Newton's method. *Hint*:

$$\begin{bmatrix} a & b \\ c & d \end{bmatrix}^{-1} = \frac{1}{ad - bc}\begin{bmatrix} d & -b \\ -c & a \end{bmatrix}.$$

c. Repeat part b using a gradient algorithm with a fixed step size of $\alpha_k = 0.05$ at each iteration.

**9.4** Consider the modified Newton's algorithm

$$\boldsymbol{x}^{(k+1)} = \boldsymbol{x}^{(k)} - \alpha_k \boldsymbol{F}(\boldsymbol{x}^{(k)})^{-1}\boldsymbol{g}^{(k)},$$

where $\alpha_k$ arg $\min_{\alpha \geqslant 0} f(\boldsymbol{x}^{(k)} - \alpha \boldsymbol{F}(\boldsymbol{x}^{(k)})^{-1} \boldsymbol{g}^{(k)})$. Suppose that we apply the algorithm to a quadratic function $f(\boldsymbol{x}) = \frac{1}{2}\boldsymbol{x}^T\boldsymbol{Q}\boldsymbol{x} - \boldsymbol{x}^T\boldsymbol{b}$, where $\boldsymbol{Q} = \boldsymbol{Q}^T > 0$. Recall that the standard Newton's method reaches the point $\boldsymbol{x}^*$ such that $\nabla f(\boldsymbol{x}^*) = \boldsymbol{0}$ in just one step starting from any initial point $\boldsymbol{x}^{(0)}$. Does the above modified Newton's algorithm possess the same property? Justify your answer.

# 10

# Conjugate Direction Methods

## 10.1. INTRODUCTION

The class of *conjugate direction methods* can be viewed as being intermediate between the method of steepest descent and Newton's method. The conjugate direction methods have the following properties:

1. Solve quadratics of $n$ variables in $n$ steps
2. The usual implementation, the *conjugate gradient algorithm*, requires no Hessian matrix evaluations
3. No matrix inversion and no storage of an $n \times n$ matrix required.

The conjugate direction methods typically perform better than the method of steepest descent, but not as well as Newton's method. As we saw from the method of steepest descent and Newton's method, the crucial factor in the efficiency of an iterative search method is the direction of search at each iteration. For a quadratic function of $n$ variables $f(\boldsymbol{x}) = \frac{1}{2}\boldsymbol{x}^T\boldsymbol{Q}\boldsymbol{x} - \boldsymbol{x}^T\boldsymbol{b}$, $\boldsymbol{x} \in \mathbb{R}^n$, $\boldsymbol{Q} = \boldsymbol{Q}^T > 0$, the best direction of search, as we shall see, is in the so-called $\boldsymbol{Q}$-conjugate direction. Basically, two directions $\boldsymbol{d}^{(1)}$ and $\boldsymbol{d}^{(2)}$ in $\mathbb{R}^n$ are said to be $\boldsymbol{Q}$-conjugate if $\boldsymbol{d}^{(1)T}\boldsymbol{Q}\boldsymbol{d}^{(2)} = 0$. In general, we have the following definition.

***Definition 10.1.*** Let $\boldsymbol{Q}$ be a real symmetric $n \times n$ matrix. The directions $\boldsymbol{d}^{(0)}, \boldsymbol{d}^{(1)}, \boldsymbol{d}^{(2)}, \ldots, \boldsymbol{d}^{(m)}$ are *$\boldsymbol{Q}$-conjugate* if, for all $i \neq j$,

$$\boldsymbol{d}^{(i)T}\boldsymbol{Q}\boldsymbol{d}^{(j)} = 0. \qquad \blacksquare$$

**Lemma 10.1.** *Let $\boldsymbol{Q}$ be a symmetric positive definite $n \times n$ matrix. If the directions $\boldsymbol{d}^{(0)}, \boldsymbol{d}^{(1)}, \ldots, \boldsymbol{d}^{(k)} \in \mathbb{R}^n$, $k \leqslant n-1$, are nonzero and $\boldsymbol{Q}$-conjugate, then they are linearly independent.* $\square$

*Proof.* Let $\alpha_0, \ldots, \alpha_k$ be scalars such that

$$\alpha_0\boldsymbol{d}^{(0)} + \alpha_1\boldsymbol{d}^{(1)} + \cdots + \alpha_k\boldsymbol{d}^{(k)} = \boldsymbol{0}.$$

Premultiplying the above equality by $\boldsymbol{d}^{(j)T}\boldsymbol{Q}$, $0 \leqslant j \leqslant k$, yields

$$\alpha_j \boldsymbol{d}^{(j)T}\boldsymbol{Q}\boldsymbol{d}^{(j)} = 0$$

because all other terms $\boldsymbol{d}^{(j)T}\boldsymbol{Q}\boldsymbol{d}^{(i)} = 0$, $i \neq j$, by $\boldsymbol{Q}$-conjugacy. But $\boldsymbol{Q} = \boldsymbol{Q}^T > 0$ and $\boldsymbol{d}^{(j)} \neq \boldsymbol{0}$; hence $\alpha_j = 0$, $j = 0, 1, \ldots, k$. Therefore, $\boldsymbol{d}^{(0)}, \boldsymbol{d}^{(1)}, \ldots, \boldsymbol{d}^{(k)}$, $k \leqslant n-1$, are linearly independent. ■

***Example 10.1.*** Let

$$\boldsymbol{Q} = \begin{bmatrix} 3 & 0 & 1 \\ 0 & 4 & 2 \\ 1 & 2 & 3 \end{bmatrix}.$$

Note that $\boldsymbol{Q} = \boldsymbol{Q}^T > 0$. The matrix $\boldsymbol{Q}$ is positive definite, because all its leading principal minors are positive:

$$\Delta_1 = 3 > 0, \quad \Delta_2 = \det\begin{bmatrix} 3 & 0 \\ 0 & 4 \end{bmatrix} = 12 > 0, \quad \Delta_3 = \det \boldsymbol{Q} = 20 > 0.$$

Our goal is to construct a set of $\boldsymbol{Q}$-conjugate vectors $\boldsymbol{d}^{(0)}, \boldsymbol{d}^{(1)}, \boldsymbol{d}^{(2)}$.

Let $\boldsymbol{d}^{(0)} = [1, 0, 0]^T$, $\boldsymbol{d}^{(1)} = [d_1^{(1)}, d_2^{(1)}, d_3^{(1)}]^T$, $\boldsymbol{d}^{(2)} = [d_1^{(2)}, d_2^{(2)}, d_3^{(2)}]^T$. We require $\boldsymbol{d}^{(0)T}\boldsymbol{Q}\boldsymbol{d}^{(1)} = 0$. We have

$$\boldsymbol{d}^{(0)T}\boldsymbol{Q}\boldsymbol{d}^{(1)} = [1, 0, 0]\begin{bmatrix} 3 & 0 & 1 \\ 0 & 4 & 2 \\ 1 & 2 & 3 \end{bmatrix}\begin{bmatrix} d_1^{(1)} \\ d_2^{(1)} \\ d_3^{(1)} \end{bmatrix} = 3d_1^{(1)} + d_3^{(1)}.$$

Let $d_1^{(1)} = 1$, $d_2^{(1)} = 0$, $d_3^{(1)} = -3$. Then, $\boldsymbol{d}^{(1)} = [1, 0, -3]^T$, and thus $\boldsymbol{d}^{(0)T}\boldsymbol{Q}\boldsymbol{d}^{(1)} = 0$.

To find the third vector $\boldsymbol{d}^{(2)}$, which would be $\boldsymbol{Q}$-conjugate with $\boldsymbol{d}^{(0)}$ and $\boldsymbol{d}^{(1)}$, we require $\boldsymbol{d}^{(0)T}\boldsymbol{Q}\boldsymbol{d}^{(2)} = 0$ and $\boldsymbol{d}^{(1)T}\boldsymbol{Q}\boldsymbol{d}^{(2)} = 0$. We have

$$\boldsymbol{d}^{(0)T}\boldsymbol{Q}\boldsymbol{d}^{(2)} = 3d_1^{(2)} + d_3^{(2)} = 0,$$
$$\boldsymbol{d}^{(1)T}\boldsymbol{Q}\boldsymbol{d}^{(2)} = -6d_2^{(2)} - 8d_3^{(2)} = 0.$$

If we take $\boldsymbol{d}^{(2)} = [1, 4, -3]^T$, then the resulting set of vectors is mutually conjugate. ■

The above method of finding $\boldsymbol{Q}$-conjugate vectors is inefficient. A systematic procedure for finding $\boldsymbol{Q}$-conjugate vectors can be devised using the idea underlying the Gram–Schmidt process of transforming a given basis of $\mathbb{R}^n$ into an orthonormal basis of $\mathbb{R}^n$ (see Exercise 10.1).

## 10.2. THE CONJUGATE DIRECTION ALGORITHM

We now present the conjugate direction algorithm for minimizing the quadratic function of $n$ variables

$$f(x) = \tfrac{1}{2}\boldsymbol{x}^T\boldsymbol{Q}\boldsymbol{x} - \boldsymbol{x}^T\boldsymbol{b},$$

where $\boldsymbol{Q} = \boldsymbol{Q}^T > 0$, $\boldsymbol{x} \in \mathbb{R}^n$. Note that since $\boldsymbol{Q} > 0$, the function $f$ has a global minimizer that can be found by solving $\boldsymbol{Q}\boldsymbol{x} = \boldsymbol{b}$.

***Basic Conjugate Direction Algorithm.*** Given a starting point $\boldsymbol{x}^{(0)}$, and $\boldsymbol{Q}$-conjugate directions $\boldsymbol{d}^{(0)}, \boldsymbol{d}^{(1)}, \ldots, \boldsymbol{d}^{(n-1)}$; for $k \geqslant 0$,

$$\boldsymbol{g}^{(k)} = \nabla f(\boldsymbol{x}^{(k)}) = \boldsymbol{Q}\boldsymbol{x}^{(k)} - \boldsymbol{b},$$

$$\alpha_k = -\frac{\boldsymbol{g}^{(k)T}\boldsymbol{d}^{(k)}}{\boldsymbol{d}^{(k)T}\boldsymbol{Q}\boldsymbol{d}^{(k)}},$$

$$\boldsymbol{x}^{(k+1)} = \boldsymbol{x}^{(k)} + \alpha_k\boldsymbol{d}^{(k)}. \qquad \blacksquare$$

**Theorem 10.1.** *For any starting point $\boldsymbol{x}^{(0)}$, the basic conjugate direction algorithm converges to the unique $\boldsymbol{x}^*$ (that solves $\boldsymbol{Q}\boldsymbol{x} = \boldsymbol{b}$) in $n$ steps; that is, $\boldsymbol{x}^{(n)} = \boldsymbol{x}^*$.* □

*Proof.* Consider $\boldsymbol{x}^* - \boldsymbol{x}^{(0)} \in \mathbb{R}^n$. Because the $\boldsymbol{d}^{(i)}$ are linearly independent, there exist constants $\beta_i$, $i = 0, \ldots, n-1$, such that

$$\boldsymbol{x}^* - \boldsymbol{x}^{(0)} = \beta_0\boldsymbol{d}^{(0)} + \cdots + \beta_{n-1}\boldsymbol{d}^{(n-1)}.$$

Now premultiply both sides of the above equation by $\boldsymbol{d}^{(k)T}\boldsymbol{Q}$, $0 \leqslant k < n$, to obtain

$$\boldsymbol{d}^{(k)T}\boldsymbol{Q}(\boldsymbol{x}^* - \boldsymbol{x}^{(0)}) = \beta_k\boldsymbol{d}^{(k)T}\boldsymbol{Q}\boldsymbol{d}^{(k)},$$

where the terms $\boldsymbol{d}^{(k)T}\boldsymbol{Q}\boldsymbol{d}^{(i)} = 0$, $k \neq i$, by the $\boldsymbol{Q}$-conjugate property. Hence

$$\beta_k = \frac{\boldsymbol{d}^{(k)T}\boldsymbol{Q}(\boldsymbol{x}^* - \boldsymbol{x}^{(0)})}{\boldsymbol{d}^{(k)T}\boldsymbol{Q}\boldsymbol{d}^{(k)}}.$$

Now, we can write

$$\boldsymbol{x}^{(k)} = \boldsymbol{x}^{(0)} + \alpha_0\boldsymbol{d}^{(0)} + \cdots + \alpha_{k-1}\boldsymbol{d}^{(k-1)}.$$

Therefore,

$$\boldsymbol{x}^{(k)} - \boldsymbol{x}^{(0)} = \alpha_0\boldsymbol{d}^{(0)} + \cdots + \alpha_{k-1}\boldsymbol{d}^{(k-1)}.$$

So writing

$$x^* - x^{(0)} = (x^* - x^{(k)}) + (x^{(k)} - x^{(0)})$$

and premultiplying the above by $d^{(k)T}Q$ we obtain

$$d^{(k)T}Q(x^* - x^{(0)}) = d^{(k)T}Q(x^* - x^{(k)}) = -d^{(k)T}g^{(k)}$$

since $g^{(k)} = Qx^{(k)} - b$ and $Qx^* = b$. Thus

$$\beta_k = -\frac{d^{(k)T}g^{(k)}}{d^{(k)T}Qd^{(k)}} = \alpha_k$$

and $x^* = x^{(n)}$, which completes the proof. ■

***Example 10.2.*** Find the minimizer of

$$f(x_1, x_2) = \tfrac{1}{2}x^T\begin{bmatrix}4 & 2\\ 2 & 2\end{bmatrix}x - x^T\begin{bmatrix}-1\\ 1\end{bmatrix}, \quad x \in \mathbb{R}^2,$$

using the conjugate direction method with the initial point $x^{(0)} = [0, 0]^T$, and $Q$-conjugate directions $d^{(0)} = [1, 0]^T$ and $d^{(1)} = [-\frac{3}{8}, \frac{3}{4}]^T$.

We have

$$g^{(0)} = -b = [1, -1]^T,$$

and hence

$$\alpha_0 = -\frac{g^{(0)T}d^{(0)}}{d^{(0)T}Qd^{(0)}} = -\frac{[1, -1]\begin{bmatrix}1\\ 0\end{bmatrix}}{[1, 0]\begin{bmatrix}4 & 2\\ 2 & 2\end{bmatrix}\begin{bmatrix}1\\ 0\end{bmatrix}} = -\frac{1}{4}.$$

Thus

$$x^{(1)} = x^{(0)} + \alpha_0 d^{(0)} = \begin{bmatrix}0\\ 0\end{bmatrix} - \frac{1}{4}\begin{bmatrix}1\\ 0\end{bmatrix} = \begin{bmatrix}-\frac{1}{4}\\ 0\end{bmatrix}.$$

To find $x^{(2)}$, we compute

$$g^{(1)} = Qx^{(1)} - b = \begin{bmatrix}4 & 2\\ 2 & 2\end{bmatrix}\begin{bmatrix}-\frac{1}{4}\\ 0\end{bmatrix} - \begin{bmatrix}-1\\ 1\end{bmatrix} = \begin{bmatrix}0\\ -\frac{3}{2}\end{bmatrix},$$

and

$$\alpha_1 = -\frac{g^{(1)T}d^{(1)}}{d^{(1)T}Qd^{(1)}} = -\frac{[0, -\frac{3}{2}]\begin{bmatrix}-\frac{3}{8}\\ \frac{3}{4}\end{bmatrix}}{[-\frac{3}{8}, \frac{3}{4}]\begin{bmatrix}4 & 2\\ 2 & 2\end{bmatrix}\begin{bmatrix}-\frac{3}{8}\\ \frac{3}{4}\end{bmatrix}} = 2.$$

Therefore

$$x^{(2)} = x^{(1)} + \alpha_1 d^{(1)} = \begin{bmatrix} -\frac{1}{4} \\ 0 \end{bmatrix} + 2\begin{bmatrix} -\frac{3}{8} \\ \frac{3}{4} \end{bmatrix} = \begin{bmatrix} -1 \\ \frac{3}{2} \end{bmatrix}.$$

Since $f$ is a quadratic function in two variables, $x^{(2)} = x^*$.

For a quadratic function of $n$ variables, the conjugate direction method reaches the solution after $n$ steps. As we shall see below, the method also possesses a certain desirable property in the intermediate steps. To see this, suppose we start at $x^{(0)}$ and search in the direction $d^{(0)}$ to obtain

$$x^{(1)} = x^{(0)} - \left(\frac{g^{(0)T}d^{(0)}}{d^{(0)T}Qd^{(0)}}\right)d^{(0)}.$$

We claim that

$$g^{(1)T}d^{(0)} = 0.$$

Indeed,

$$\begin{aligned} g^{(1)T}d^{(0)} &= (Qx^{(1)} - b)^T d^{(0)} \\ &= x^{(0)T}Qd^{(0)} - \left(\frac{g^{(0)T}d^{(0)}}{d^{(0)T}Qd^{(0)}}\right)d^{(0)T}Qd^{(0)} - b^T d^{(0)} \\ &= g^{(0)T}d^{(0)} - g^{(0)T}d^{(0)} = 0. \end{aligned}$$

The equation $g^{(1)T}d^{(0)} = 0$ implies that $\alpha_0$ has the property that $\alpha_0 = \arg\min \phi_0(\alpha)$, where $\phi_0(\alpha) = f(x^{(0)} + \alpha d^{(0)})$. To see this, apply the chain rule to get

$$\frac{d\phi_0}{d\alpha}(\alpha) = \nabla f(x^{(0)} + \alpha d^{(0)})^T d^{(0)}.$$

Evaluating the above at $\alpha = \alpha_0$, we get

$$\frac{d\phi_0}{d\alpha}(\alpha_0) = g^{(1)T}d^{(0)} = 0.$$

Since $\phi_0$ is a quadratic function of $\alpha$, and the coefficient of the $\alpha^2$ term in $\phi_0$ is $d^{(0)T}Qd^{(0)} > 0$, the above implies that $\alpha_0 = \arg\min_{\alpha\in\mathbb{R}} \phi_0(\alpha)$.

Using a similar argument, we can show that for all $k$,

$$g^{(k+1)T}d^{(k)} = 0$$

and hence

$$\alpha_k = \arg\min f(x^{(k)} + \alpha d^{(k)}).$$

In fact, an even stronger condition holds, as given by the following lemma.

**Lemma 10.2.** *In the conjugate direction algorithm,*

$$\boldsymbol{g}^{(k+1)T}\boldsymbol{d}^{(i)} = 0$$

*for all* $k$, $0 \leqslant k \leqslant n-1$, *and* $0 \leqslant i \leqslant k$. □

*Proof.* Note that

$$\boldsymbol{Q}(\boldsymbol{x}^{(k+1)} - \boldsymbol{x}^{(k)}) = \boldsymbol{Q}\boldsymbol{x}^{(k+1)} - \boldsymbol{b} - (\boldsymbol{Q}\boldsymbol{x}^{(k)} - \boldsymbol{b}) = \boldsymbol{g}^{(k+1)} - \boldsymbol{g}^{(k)}$$

since $\boldsymbol{g}^{(k)} = \boldsymbol{Q}\boldsymbol{x}^{(k)} - \boldsymbol{b}$. Thus

$$\boldsymbol{g}^{(k+1)} = \boldsymbol{g}^{(k)} + \alpha_k \boldsymbol{Q}\boldsymbol{d}^{(k)}.$$

We prove the lemma by induction. The result is true for $k = 0$ because $\boldsymbol{g}^{(1)T}\boldsymbol{d}^{(0)} = 0$, as shown before. We now show that if the result is true for $k-1$ [that is $\boldsymbol{g}^{(k)T}\boldsymbol{d}^{(i)} = 0$, $i \leqslant k-1$] then it is true for $k$ [that is $\boldsymbol{g}^{(k+1)T}\boldsymbol{d}^{(i)} = 0$, $i \leqslant k$]. Fix $k > 0$ and $0 \leqslant i < k$. By the induction hypothesis, $\boldsymbol{g}^{(k)T}\boldsymbol{d}^{(i)} = 0$. Since

$$\boldsymbol{g}^{(k+1)} = \boldsymbol{g}^{(k)} + \alpha_k \boldsymbol{Q}\boldsymbol{d}^{(k)},$$

then by $\boldsymbol{Q}$-conjugacy, $\boldsymbol{d}^{(k)T}\boldsymbol{Q}\boldsymbol{d}^{(i)} = 0$, and hence

$$\boldsymbol{g}^{(k+1)T}\boldsymbol{d}^{(i)} = \boldsymbol{g}^{(k)T}\boldsymbol{d}^{(i)} + \alpha_k \boldsymbol{d}^{(k)T}\boldsymbol{Q}\boldsymbol{d}^{(i)} = 0.$$

It remains to be shown that

$$\boldsymbol{g}^{(k+1)T}\boldsymbol{d}^{(k)} = 0.$$

Indeed,

$$\begin{aligned}
\boldsymbol{g}^{(k+1)T}\boldsymbol{d}^{(k)} &= (\boldsymbol{Q}\boldsymbol{x}^{(k+1)} - \boldsymbol{b})^T\boldsymbol{d}^{(k)} \\
&= \left(\boldsymbol{x}^{(k)} - \left(\frac{\boldsymbol{g}^{(k)T}\boldsymbol{d}^{(k)}}{\boldsymbol{d}^{(k)T}\boldsymbol{Q}\boldsymbol{d}^{(k)}}\right)\boldsymbol{d}^{(k)}\right)^T \boldsymbol{Q}\boldsymbol{d}^{(k)} - \boldsymbol{b}^T\boldsymbol{d}^{(k)} \\
&= (\boldsymbol{Q}\boldsymbol{x}^{(k)} - \boldsymbol{b})^T\boldsymbol{d}^{(k)} - \boldsymbol{g}^{(k)T}\boldsymbol{d}^{(k)} \\
&= 0
\end{aligned}$$

since $\boldsymbol{Q}\boldsymbol{x}^{(k)} - \boldsymbol{b} = \boldsymbol{g}^{(k)}$.

Therefore, by induction, for all $0 \leqslant k \leqslant n-1$ and $0 \leqslant i \leqslant k$,

$$\boldsymbol{g}^{(k+1)T}\boldsymbol{d}^{(i)} = 0.$$ ■

By the above lemma, we see that $\boldsymbol{g}^{(k+1)}$ is orthogonal to any vector from the subspace spanned by $\boldsymbol{d}^{(0)}, \boldsymbol{d}^{(1)}, \ldots, \boldsymbol{d}^{(k)}$. Figure 10.1 illustrates this statement.

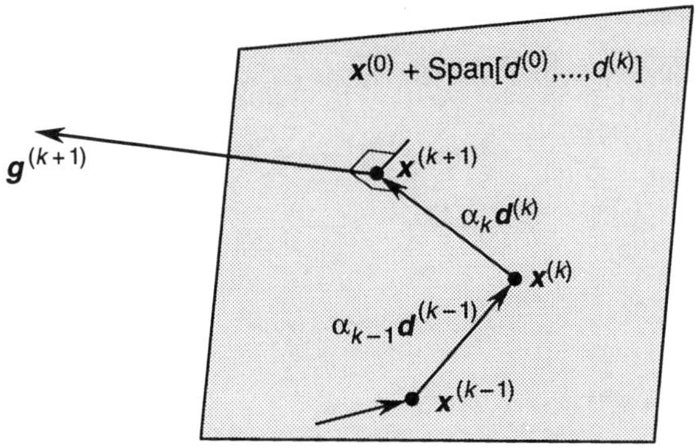


**Figure 10.1.** Illustration of Lemma 10.2

The above lemma can be used to show an interesting optimal property of the conjugate direction algorithm. Specifically, we now show that not only does $f(\boldsymbol{x}^{(k+1)})$ satisfy $f(\boldsymbol{x}^{(k+1)}) = \min_\alpha f(\boldsymbol{x}^{(k)} + \alpha \boldsymbol{d}^{(k)})$, as indicated before, but also,

$$f(\boldsymbol{x}^{(k+1)}) = \min_{a_0,\ldots,a_k} f\left( \boldsymbol{x}^{(0)} + \sum_{i=0}^{k} a_i \boldsymbol{d}^{(i)} \right).$$

In other words, $f(\boldsymbol{x}^{(k+1)}) = \min_{\boldsymbol{x} \in \mathcal{V}_k} f(\boldsymbol{x})$, where $\mathcal{V}_k = \boldsymbol{x}^{(0)} + \operatorname{span}[\boldsymbol{d}^{(0)}, \boldsymbol{d}^{(1)}, \ldots, \boldsymbol{d}^{(k)}]$.

To show the above, define the matrix $\boldsymbol{D}^{(k)}$ by

$$\boldsymbol{D}^{(k)} = [\boldsymbol{d}^{(0)}, \ldots, \boldsymbol{d}^{(k)}],$$

that is, $\boldsymbol{d}^{(i)}$ is the $i$th column of $\boldsymbol{D}^{(k)}$. Note that $\boldsymbol{x}^{(0)} + \mathcal{R}(\boldsymbol{D}^{(k)}) = \mathcal{V}_k$. Also,

$$\begin{aligned} \boldsymbol{x}^{(k+1)} &= \boldsymbol{x}^{(0)} + \sum_{i=0}^{k} \alpha_i \boldsymbol{d}^{(i)} \\ &= \boldsymbol{x}^{(0)} + \boldsymbol{D}^{(k)} \boldsymbol{\alpha} \end{aligned}$$

where $\boldsymbol{\alpha} = [\alpha_0, \ldots, \alpha_k]^T$. Hence,

$$\boldsymbol{x}^{(k+1)} \in \boldsymbol{x}^{(0)} + \mathcal{R}(\boldsymbol{D}^{(k)}) = \mathcal{V}_k.$$

Now, consider any vector $\boldsymbol{x} \in \mathcal{V}_k$. There exists a vector $\boldsymbol{a}$ such that $\boldsymbol{x} = \boldsymbol{x}^{(0)} + \boldsymbol{D}^{(k)} \boldsymbol{a}$. Let $\phi_k(\boldsymbol{a}) = f(\boldsymbol{x}^{(0)} + \boldsymbol{D}^{(k)} \boldsymbol{a})$. Note that $\phi_k$ is a quadratic function and has a unique minimizer that satisfies the FONC (see Exercise 6.14 and Exercise 10.4). By the chain rule,

$$D\phi_k(\boldsymbol{a}) = \nabla f(\boldsymbol{x}^{(0)} + \boldsymbol{D}^{(k)} \boldsymbol{a})^T \boldsymbol{D}^{(k)}.$$

Therefore,

$$\begin{aligned} D\phi_k(\boldsymbol{\alpha}) &= \nabla f(\boldsymbol{x}^{(0)} + \boldsymbol{D}^{(k)}\boldsymbol{\alpha})^T \boldsymbol{D}^{(k)} \\ &= \nabla f(\boldsymbol{x}^{(k+1)})^T \boldsymbol{D}^{(k)} \\ &= \boldsymbol{g}^{(k+1)T} \boldsymbol{D}^{(k)}. \end{aligned}$$

By Lemma 10.2, $\boldsymbol{g}^{(k+1)T}\boldsymbol{D}^{(k)} = \boldsymbol{0}^T$. Therefore, $\boldsymbol{\alpha}$ satisfies the FONC for the quadratic function $\phi_k$, and hence $\boldsymbol{\alpha}$ is the minimizer of $\phi_k$; that is,

$$f(\boldsymbol{x}^{(k+1)}) = \min_{\boldsymbol{a}} f(\boldsymbol{x}^{(0)} + \boldsymbol{D}^{(k)}\boldsymbol{a}) = \min_{x \in \mathcal{V}_k} f(\boldsymbol{x}),$$

and the proof of our claim is completed.

The conjugate direction algorithm is very effective. However, to use the algorithm, we need to specify beforehand the $\boldsymbol{Q}$-conjugate directions. Fortunately there is a way to generate $\boldsymbol{Q}$-conjugate directions as we perform iterations. In the next section, we discuss an algorithm that incorporates the generation of $\boldsymbol{Q}$-conjugate directions.

## 10.3. THE CONJUGATE GRADIENT ALGORITHM

The conjugate gradient algorithm does not use prespecified conjugate directions, but instead computes the directions as the algorithm progresses. At each stage of the algorithm, the direction is calculated as a linear combination of the previous direction and the current gradient, in such a way that all the directions are mutually $\boldsymbol{Q}$-conjugate—hence the name conjugate gradient algorithm. This calculation exploits the fact that for a quadratic function of $n$ variables, we can locate the function minimizer by performing $n$ searches along mutually conjugate directions.

As before, we consider the quadratic function

$$f(\boldsymbol{x}) = \tfrac{1}{2}\boldsymbol{x}^T\boldsymbol{Q}\boldsymbol{x} - \boldsymbol{x}^T\boldsymbol{b}, \quad \boldsymbol{x} \in \mathbb{R}^n,$$

where $\boldsymbol{Q} = \boldsymbol{Q}^T > 0$. Our first search direction from an initial point $\boldsymbol{x}^{(0)}$ is in the direction of steepest descent, that is,

$$\boldsymbol{d}^{(0)} = -\boldsymbol{g}^{(0)}.$$

Thus

$$\boldsymbol{x}^{(1)} = \boldsymbol{x}^{(0)} + \alpha_0 \boldsymbol{d}^{(0)},$$

where

$$\alpha_0 = \arg\min_{\alpha \geqslant 0} f(\boldsymbol{x}^{(0)} + \alpha \boldsymbol{d}^{(0)}) = -\frac{\boldsymbol{g}^{(0)T}\boldsymbol{d}^{(0)}}{\boldsymbol{d}^{(0)T}\boldsymbol{Q}\boldsymbol{d}^{(0)}}.$$

In the next stage, we search in a direction $\boldsymbol{d}^{(1)}$ that is $\boldsymbol{Q}$-conjugate to $\boldsymbol{d}^{(0)}$. We choose $\boldsymbol{d}^{(1)}$ as a linear combination of $\boldsymbol{g}^{(1)}$ and $\boldsymbol{d}^{(0)}$. In general, at the $(k+1)$st step, we choose $\boldsymbol{d}^{(k+1)}$ to be a linear combination of $\boldsymbol{g}^{(k+1)}$ and $\boldsymbol{d}^{(k)}$. Specifically, we choose

$$\boldsymbol{d}^{(k+1)} = -\boldsymbol{g}^{(k+1)} + \beta_k \boldsymbol{d}^{(k)}, \quad k = 0, 1, 2, \ldots.$$

The coefficients $\beta_k$, $k = 1, 2, \ldots$, are chosen in such a way that $\boldsymbol{d}^{(k+1)}$ is $\boldsymbol{Q}$-conjugate to $\boldsymbol{d}^{(0)}, \boldsymbol{d}^{(1)}, \ldots, \boldsymbol{d}^{(k)}$. This is accomplished by choosing $\beta_k$ to be

$$\beta_k = \frac{\boldsymbol{g}^{(k+1)T} \boldsymbol{Q} \boldsymbol{d}^{(k)}}{\boldsymbol{d}^{(k)T} \boldsymbol{Q} \boldsymbol{d}^{(k)}}.$$

The conjugate gradient algorithm is summarized below.

1. Set $k := 0$; select the initial point $\boldsymbol{x}^{(0)}$.
2. $\boldsymbol{g}^{(0)} = \nabla f(\boldsymbol{x}^{(0)})$. If $\boldsymbol{g}^{(0)} = \boldsymbol{0}$, stop, else set $\boldsymbol{d}^{(0)} = -\boldsymbol{g}^{(0)}$.
3. $\alpha_k = -\dfrac{\boldsymbol{g}^{(k)T} \boldsymbol{d}^{(k)}}{\boldsymbol{d}^{(k)T} \boldsymbol{Q} \boldsymbol{d}^{(k)}}$.
4. $\boldsymbol{x}^{(k+1)} = \boldsymbol{x}^{(k)} + \alpha_k \boldsymbol{d}^{(k)}$.
5. $\boldsymbol{g}^{(k+1)} = \nabla f(\boldsymbol{x}^{(k+1)})$. If $\boldsymbol{g}^{(k+1)} = \boldsymbol{0}$, stop.
6. $\beta_k = \boldsymbol{g}^{(k+1)T} \boldsymbol{Q} \boldsymbol{d}^{(k)} / \boldsymbol{d}^{(k)T} \boldsymbol{Q} \boldsymbol{d}^{(k)}$.
7. $\boldsymbol{d}^{(k+1)} = -\boldsymbol{g}^{(k+1)} + \beta_k \boldsymbol{d}^{(k)}$.
8. Set $k := k + 1$; go to 3.

**Proposition 10.1.** *In the conjugate gradient algorithm, the directions $\boldsymbol{d}^{(0)}, \boldsymbol{d}^{(1)}, \ldots, \boldsymbol{d}^{(n-1)}$ are $\boldsymbol{Q}$-conjugate.* □

*Proof.* We use induction. We first show $\boldsymbol{d}^{(0)T} \boldsymbol{Q} \boldsymbol{d}^{(1)} = 0$. To this end, we write

$$\boldsymbol{d}^{(0)T} \boldsymbol{Q} \boldsymbol{d}^{(1)} = \boldsymbol{d}^{(0)T} \boldsymbol{Q}(-\boldsymbol{g}^{(1)} + \beta_0 \boldsymbol{d}^{(0)}).$$

Substituting for

$$\beta_0 = \frac{\boldsymbol{g}^{(1)T} \boldsymbol{Q} \boldsymbol{d}^{(0)}}{\boldsymbol{d}^{(0)T} \boldsymbol{Q} \boldsymbol{d}^{(0)}}$$

in the above equation, we see that $\boldsymbol{d}^{(0)T} \boldsymbol{Q} \boldsymbol{d}^{(1)} = 0$.

We now assume that $\boldsymbol{d}^{(0)}, \boldsymbol{d}^{(1)}, \ldots, \boldsymbol{d}^{(k)}$, $k < n - 1$, are $\boldsymbol{Q}$-conjugate directions. From Lemma 10.2, we have $\boldsymbol{g}^{(k+1)T} \boldsymbol{d}^{(j)} = 0$, $j = 0, 1, \ldots, k$. Thus, $\boldsymbol{g}^{(k+1)}$ is orthogonal to each of the directions $\boldsymbol{d}^{(0)}, \boldsymbol{d}^{(1)}, \ldots, \boldsymbol{d}^{(k)}$. We now show that

$$\boldsymbol{g}^{(k+1)T} \boldsymbol{g}^{(j)} = 0, \quad j = 0, 1, \ldots, k.$$

Fix $j \in \{0, \ldots, k\}$. We have

$$\boldsymbol{d}^{(j)} = -\boldsymbol{g}^{(j)} + \beta_{j-1} \boldsymbol{d}^{(j-1)}.$$

Substituting this equation into the previous one yields

$$g^{(k+1)T}d^{(j)} = 0 = -g^{(k+1)T}g^{(j)} + \beta_{j-1}g^{(k+1)T}d^{(j-1)}.$$

Because $g^{(k+1)T}d^{(j-1)} = 0$, it follows that $g^{(k+1)T}g^{(j)} = 0$.

We are now ready to show that $d^{(k+1)T}Qd^{(j)} = 0$, $j = 0, \ldots, k$. We have

$$d^{(k+1)T}Qd^{(j)} = (-g^{(k+1)} + \beta_k d^{(k)})^T Qd^{(j)}.$$

If $j < k$, then $d^{(k)T}Qd^{(j)} = 0$ by virtue of the inductive hypothesis. Hence, we have

$$d^{(k+1)T}Qd^{(j)} = -g^{(k+1)T}Qd^{(j)}.$$

But $g^{(j+1)} = g^{(j)} + \alpha_j Qd^{(j)}$. Since $g^{(k+1)T}g^{(i)} = 0$, $i = 0, \ldots, k$, then

$$d^{(k+1)T}Qd^{(j)} = -g^{(k+1)T}\frac{(g^{(j+1)} - g^{(j)})}{\alpha_j} = 0.$$

Thus

$$d^{(k+1)T}Qd^{(j)} = 0, \quad j = 0, \ldots, k-1.$$

It remains to be shown that $d^{(k+1)T}Qd^{(k)} = 0$. We have

$$d^{(k+1)T}Qd^{(k)} = (-g^{(k+1)} + \beta_k d^{(k)})^T Qd^{(k)}.$$

Using the expression for $\beta_k$, we get $d^{(k+1)T}Qd^{(k)} = 0$, which completes the proof. ■

***Example 10.3.*** Consider the quadratic function

$$f(x_1, x_2, x_3) = \tfrac{3}{2}x_1^2 + 2x_2^2 + \tfrac{3}{2}x_3^2 + x_1x_3 + 2x_2x_3 - 3x_1 - x_3.$$

We find the minimizer using the conjugate gradient algorithm, using the starting point $x^{(0)} = [0, 0, 0]^T$.

We can represent $f$ as

$$f(x) = \tfrac{1}{2}x^TQx - x^Tb,$$

where

$$Q = \begin{bmatrix} 3 & 0 & 1 \\ 0 & 4 & 2 \\ 1 & 2 & 3 \end{bmatrix}, \quad b = \begin{bmatrix} 3 \\ 0 \\ 1 \end{bmatrix}.$$

We have

$$g(x) = \nabla f(x) = Qx - b = [3x_1 + x_3 - 3, 4x_2 + 2x_3, x_1 + 2x_2 + 3x_3 - 1]^T.$$

Hence

$$g^{(0)} = [-3, 0, -1]^T$$
$$d^{(0)} = -g^{(0)}$$
$$\alpha_0 = -\frac{g^{(0)T} d^{(0)}}{d^{(0)T} Q d^{(0)}} = \frac{10}{36} = 0.2778,$$

and

$$x^{(1)} = x^{(0)} + \alpha_0 d^{(0)} = [0.8333, 0, 0.2778]^T.$$

The next stage yields

$$g^{(1)} = \nabla f(x^{(1)}) = [-0.2222, 0.5556, 0.6667]^T,$$
$$\beta_0 = \frac{g^{(1)T} Q d^{(0)}}{d^{(0)T} Q d^{(0)}} = 0.08025.$$

We can now compute

$$d^{(1)} = -g^{(1)} + \beta_0 d^{(0)} = [0.4630, -0.5556, -0.5864]^T.$$

Hence

$$\alpha_1 = -\frac{g^{(1)T} d^{(1)}}{d^{(1)T} Q d^{(1)}} = 0.2187,$$

and

$$x^{(2)} = x^{(1)} + \alpha_1 d^{(1)} = [0.9346, -0.1215, 0.1495]^T.$$

To perform the third iteration, we compute

$$g^{(2)} = \nabla f(x^{(2)}) = [-0.04673, -0.1869, 0.1402]^T,$$
$$\beta_1 = \frac{g^{(2)T} Q d^{(1)}}{d^{(1)T} Q d^{(1)}} = 0.07075,$$
$$d^{(2)} = -g^{(2)} + \beta_1 d^{(1)} = [0.07948, 0.1476, -0.1817]^T.$$

Hence

$$\alpha_2 = -\frac{g^{(2)T} d^{(2)}}{d^{(2)T} Q d^{(2)}} = 0.8231,$$

and

$$x^{(3)} = x^{(2)} + \alpha_2 d^{(2)} = [1.000, 0.000, 0.000]^T.$$

Note that

$$g^{(3)} = \nabla f(x^{(3)}) = \mathbf{0},$$

as expected since $f$ is a quadratic function of three variables. Hence, $x^* = x^{(3)}$.

## 10.4. THE CONJUGATE GRADIENT ALGORITHM FOR NONQUADRATIC PROBLEMS

In the previous section, we showed that the conjugate gradient algorithm is a conjugate direction method, and therefore minimizes a positive definite quadratic function of $n$ variables in $n$ steps. The algorithm can be extended to general nonlinear functions by interpreting $f(\boldsymbol{x})=\frac{1}{2}\boldsymbol{x}^T\boldsymbol{Q}\boldsymbol{x}-\boldsymbol{x}^T\boldsymbol{b}$ as a second-order Taylor series approximation of the objective function. Near the solution, such functions behave approximately as quadratics, as suggested by the Taylor series expansion. For a quadratic, the matrix $\boldsymbol{Q}$, the Hessian of the quadratic, is constant. However, for a general nonlinear function the Hessian is a matrix that has to be reevaluated at each iteration of the algorithm. This can be very computationally expensive. Thus, an efficient implementation of the conjugate gradient algorithm that eliminates the Hessian evaluation at each step is desirable.

Observe that $\boldsymbol{Q}$ appears only in the computation of the scalars $\alpha_k$ and $\beta_k$. Since

$$\alpha_k = \arg\min_{\alpha\geqslant 0} f(\boldsymbol{x}^{(k)}+\alpha\boldsymbol{d}^{(k)})$$

then the closed-form formula for $\alpha_k$ in the algorithm can be replaced by a numerical line-search procedure. Therefore, we only need to concern ourselves with the formula for $\beta_k$. Fortunately, elimination of $\boldsymbol{Q}$ from the formula is possible and results in algorithms that depend only on the function and gradient values at each iteration. We now discuss modifications of the conjugate gradient algorithm for a quadratic function for the case in which the Hessian is unknown but in which objective function values and gradients are available. The modifications are all based on algebraically manipulating the formula $\beta_k$ in such a way that $\boldsymbol{Q}$ is eliminated. We discuss three well-known modifications.

*The Hestenes–Stiefel formula.* Recall that

$$\beta_k=\frac{\boldsymbol{g}^{(k+1)T}\boldsymbol{Q}\boldsymbol{d}^{(k)}}{\boldsymbol{d}^{(k)T}\boldsymbol{Q}\boldsymbol{d}^{(k)}}$$

The Hestenes–Stiefel formula is based on replacing the term $\boldsymbol{Q}\boldsymbol{d}^{(k)}$ by the term $(\boldsymbol{g}^{(k+1)}-\boldsymbol{g}^{(k)})/\alpha_k$. The two terms are equal in the quadratic case, as we now show. Now, $\boldsymbol{x}^{(k+1)}=\boldsymbol{x}^{(k)}+\alpha_k\boldsymbol{d}^{(k)}$. Premultiplying both sides by $\boldsymbol{Q}$, and recognizing that $\boldsymbol{g}^{(k)}=\boldsymbol{Q}\boldsymbol{x}^{(k)}-\boldsymbol{b}$, we get $\boldsymbol{g}^{(k+1)}=\boldsymbol{g}^{(k)}+\alpha_k\boldsymbol{Q}\boldsymbol{d}^{(k)}$, which we can rewrite as $\boldsymbol{Q}\boldsymbol{d}^{(k)}=(\boldsymbol{g}^{(k+1)}-\boldsymbol{g}^{(k)})/\alpha_k$. Substituting this into the original equation for $\beta_k$ gives

$$\beta_k=\frac{\boldsymbol{g}^{(k+1)T}[\boldsymbol{g}^{(k+1)}-\boldsymbol{g}^{(k)}]}{\boldsymbol{d}^{(k)T}[\boldsymbol{g}^{(k+1)}-\boldsymbol{g}^{(k)}]}$$

which is called the Hestenes–Stiefel formula.

*The Polak–Ribiere formula.* Starting from the Hestenes–Stiefel formula, we multiply out the denominator to get

$$\beta_k = \frac{\boldsymbol{g}^{(k+1)T}[\boldsymbol{g}^{(k+1)} - \boldsymbol{g}^{(k)}]}{\boldsymbol{d}^{(k)T}\boldsymbol{g}^{(k+1)} - \boldsymbol{d}^{(k)T}\boldsymbol{g}^{(k)}}.$$

By Lemma 10.2, $\boldsymbol{d}^{(k)T}\boldsymbol{g}^{(k+1)} = 0$. Also, since $\boldsymbol{d}^{(k)} = -\boldsymbol{g}^{(k)} + \beta_{k-1}\boldsymbol{d}^{(k-1)}$, and premultiplying this by $\boldsymbol{g}^{(k)T}$, we get

$$\boldsymbol{g}^{(k)T}\boldsymbol{d}^{(k)} = -\boldsymbol{g}^{(k)T}\boldsymbol{g}^{(k)} + \beta_{k-1}\boldsymbol{g}^{(k)T}\boldsymbol{d}^{(k-1)} = -\boldsymbol{g}^{(k)T}\boldsymbol{g}^{(k)}$$

where once again we used Lemma 10.2. Hence, we get

$$\beta_k = \frac{\boldsymbol{g}^{(k+1)T}[\boldsymbol{g}^{(k+1)} - \boldsymbol{g}^{(k)}]}{\boldsymbol{g}^{(k)T}\boldsymbol{g}^{(k)}}.$$

This expression for $\beta_k$ is known as the Polak–Ribiere formula.

*The Fletcher–Reeves formula.* Starting with the Polak–Ribiere formula, we multiply out the numerator to get

$$\beta_k = \frac{\boldsymbol{g}^{(k+1)T}\boldsymbol{g}^{(k+1)} - \boldsymbol{g}^{(k+1)T}\boldsymbol{g}^{(k)}}{\boldsymbol{g}^{(k)T}\boldsymbol{g}^{(k)}}.$$

We now use the fact that $\boldsymbol{g}^{(k+1)T}\boldsymbol{g}^{(k)} = 0$, which we get by using the equation

$$\boldsymbol{g}^{(k+1)T}\boldsymbol{d}^{(k)} = -\boldsymbol{g}^{(k+1)T}\boldsymbol{g}^{(k)} + \beta_{k-1}\boldsymbol{g}^{(k+1)T}\boldsymbol{d}^{(k-1)}$$

and applying Lemma 10.2. This leads to

$$\beta_k = \frac{\boldsymbol{g}^{(k+1)T}\boldsymbol{g}^{(k+1)}}{\boldsymbol{g}^{(k)T}\boldsymbol{g}^{(k)}},$$

which is called the Fletcher–Reeves formula.

The above formulas give us conjugate gradient algorithms that do not require explicit knowledge of the Hessian matrix $\boldsymbol{Q}$. All we need are the objective function and gradient values at each iteration. For the quadratic case, the three expressions for $\beta_k$ are exactly equal. However, this is not the case for a general nonlinear objective function.

We will need a few more slight modifications to apply the algorithm to general nonlinear functions in practice. First, as mentioned in our discussion of the steepest descent algorithm (Section 8.2), the termination criterion $\nabla f(\boldsymbol{x}^{(k+1)}) = \boldsymbol{0}$ is not practical. A suitable practical stopping criterion, such as those discussed in Section 8.2, needs to be used.

For nonquadratic problems, the algorithm will not usually converge in $n$ steps, and as the algorithm progresses, the "$\boldsymbol{Q}$-conjugacy" of the direction vectors will tend to deteriorate. Thus, a common practice is to reinitialize the direction vector to the negative gradient after every few iterations (for example, $n$ or $n+1$), and continue until the algorithm satisfies the stopping criterion.

A very important issue in minimization problems of nonquadratic functions is the line search. The purpose of the line search is to minimize $\phi_k(\alpha) = f(\boldsymbol{x}^{(k)} + \alpha \boldsymbol{d}^{(k)})$ with respect to $\alpha \geqslant 0$. A typical approach is to bracket or box in the minimum and then estimate it. The accuracy of the line search is a critical factor in the performance of the conjugate gradient algorithm. If the line search is known to be inaccurate, the Hestenes–Stiefel formula for $\beta_k$ is recommended [Ref. 40].

In general, the choice of which formula for $\beta_k$ to use depends on the objective function. Other variants are also possible. For example, another formula for $\beta_k$ suggested by Powell [Ref. 60] is

$$\beta_k = \max\left[0, \frac{\boldsymbol{g}^{(k+1)T}[\boldsymbol{g}^{(k+1)} - \boldsymbol{g}^{(k)}]}{\boldsymbol{g}^{(k)T}\boldsymbol{g}^{(k)}}\right].$$

For general results on the convergence of conjugate gradient methods, we refer the reader to Ref. 80.

## EXERCISES

**10.1** Adopted from Ref. 51 [Exercise 8.8(1)]. Let $\boldsymbol{Q}$ be a real symmetric positive definite $n \times n$ matrix. Given an arbitrary set of linearly independent vectors $\{\boldsymbol{p}^{(0)}, \ldots, \boldsymbol{p}^{(n-1)}\}$ in $\mathbb{R}^n$, the *Gram–Schmidt* procedure generates a set of vectors $\{\boldsymbol{d}^{(0)}, \ldots, \boldsymbol{d}^{(n-1)}\}$ as follows:

$$\boldsymbol{d}^{(0)} = \boldsymbol{p}^{(0)}$$

$$\boldsymbol{d}^{(k+1)} = \boldsymbol{p}^{(k+1)} - \sum_{i=0}^{k} \frac{\boldsymbol{p}^{(k+1)T}\boldsymbol{Q}\boldsymbol{d}^{(i)}}{\boldsymbol{d}^{(i)T}\boldsymbol{Q}\boldsymbol{d}^{(i)}}\boldsymbol{d}^{(i)}$$

Show that the vectors $\boldsymbol{d}^{(0)}, \ldots, \boldsymbol{d}^{(n-1)}$ are $\boldsymbol{Q}$-conjugate.

**10.2** Let $f: \mathbb{R}^n \to \mathbb{R}$ be given by $f(\boldsymbol{x}) = \frac{1}{2}\boldsymbol{x}^T\boldsymbol{Q}\boldsymbol{x} - \boldsymbol{x}^T\boldsymbol{b}$, where $\boldsymbol{b} \in \mathbb{R}^n$, and $\boldsymbol{Q}$ is a real symmetric positive definite $n \times n$ matrix. Show that in the conjugate gradient method for this $f$, $\boldsymbol{d}^{(k)T}\boldsymbol{Q}\boldsymbol{d}^{(k)} = -\boldsymbol{d}^{(k)T}\boldsymbol{Q}\boldsymbol{g}^{(k)}$.

**10.3** Let $\boldsymbol{Q}$ be a real $n \times n$ symmetric matrix.

**a.** show that there exists a $\boldsymbol{Q}$-conjugate set $\{\boldsymbol{d}^{(1)}, \ldots, \boldsymbol{d}^{(n)}\}$ such that each $\boldsymbol{d}^{(i)}$ $(i = 1, \ldots, n)$ is an eigenvector of $\boldsymbol{Q}$. *Hint*: Use the fact that for any real symmetric $n \times n$ matrix, there exists a set $\{\boldsymbol{v}_1, \ldots, \boldsymbol{v}_n\}$ of its eigenvectors such that $\boldsymbol{v}_i^T\boldsymbol{v}_j = 0$ for all $i, j = 1, \ldots, n, i \neq j$.

**b.** Suppose $\boldsymbol{Q}$ is positive definite. Show that, if $\{\boldsymbol{d}^{(1)},\ldots,\boldsymbol{d}^{(n)}\}$ is a $\boldsymbol{Q}$-conjugate set that is also orthogonal [that is, $\boldsymbol{d}^{(i)T}\boldsymbol{d}^{(j)}=0$ for all $i,j=1,\ldots,n,\ i\neq j$] and if $\boldsymbol{d}^{(i)}\neq\boldsymbol{0},\ i=1,\ldots,n$, then each $\boldsymbol{d}^{(i)},\ i=1,\ldots,n$, is an eigenvector of $\boldsymbol{Q}$.

**10.4** Consider the quadratic function $f:\mathbb{R}^n\to\mathbb{R}$ given by

$$f(\boldsymbol{x})=\tfrac{1}{2}\boldsymbol{x}^T\boldsymbol{Q}\boldsymbol{x}-\boldsymbol{x}^T\boldsymbol{b},$$

where $\boldsymbol{Q}=\boldsymbol{Q}^T>0$. Let $\boldsymbol{D}\in\mathbb{R}^{n\times r}$ be of rank $r$, and $\boldsymbol{x}_0\in\mathbb{R}^n$. Define the function $\phi:\mathbb{R}^r\to\mathbb{R}$ by

$$\phi(\boldsymbol{a})=f(\boldsymbol{x}_0+\boldsymbol{D}\boldsymbol{a}).$$

Show that $\phi$ is a quadratic function with a positive definite quadratic term.

**10.5** Let $f(\boldsymbol{x}),\boldsymbol{x}=[x_1,x_2]^T\in\mathbb{R}^2$, be given by

$$f(\boldsymbol{x})=\tfrac{5}{2}x_1^2+\tfrac{1}{2}x_2^2+2x_1x_2-3x_1-x_2$$

**a.** Express $f(\boldsymbol{x})$ in the form of $f(\boldsymbol{x})=\frac{1}{2}\boldsymbol{x}^T\boldsymbol{Q}\boldsymbol{x}-\boldsymbol{x}^T\boldsymbol{b}$.

**b.** Find the minimizer of $f$ using the conjugate gradient algorithm. Use a starting point of $\boldsymbol{x}^{(0)}=[0,0]^T$.

**c.** Calculate the minimizer of $f$ explicitly from $\boldsymbol{Q}$ and $\boldsymbol{b}$, and check it with your answer in part b.

**10.6** Write a MATLAB routine to implement the conjugate gradient algorithm for general functions. Use the secant method for the line search (e.g., the MATLAB function of Exercise 7.8). Test the different formulas for $\beta_k$ on Rosenbrock's function (see Exercise 9.3), with an initial condition $\boldsymbol{x}^{(0)}=[-2,2]^T$. For this exercise, reinitialize the update direction to the negative gradient every six iterations.

# 11

# Quasi-Newton Methods

## 11.1. INTRODUCTION

Newton's method is one of the more successful algorithms for optimization. If it converges, it has at least a quadratic order of convergence. However, as pointed out before, for a general nonlinear objective function, convergence to a solution cannot be guaranteed from an arbitrary initial point $\boldsymbol{x}^{(0)}$. In general, if the initial point is not sufficiently close to the solution, the algorithm may not be a descent algorithm [that is, $f(\boldsymbol{x}^{(k+1)}) \not< f(\boldsymbol{x}^{(k)})$].

Recall that the idea behind Newton's method is to locally approximate the function $f$ being minimized, at every iteration, by a quadratic function. The minimizer for the quadratic approximation is used as the starting point for the next iteration. This leads to Newton's recursive algorithm

$$\boldsymbol{x}^{(k+1)} = \boldsymbol{x}^{(k)} - \boldsymbol{F}(\boldsymbol{x}^{(k)})^{-1}\boldsymbol{g}^{(k)}.$$

We may try to guarantee that the algorithm is a descent algorithm by modifying the original algorithm as follows:

$$\boldsymbol{x}^{(k+1)} = \boldsymbol{x}^{(k)} - \alpha_k \boldsymbol{F}(\boldsymbol{x}^{(k)})^{-1}\boldsymbol{g}^{(k)},$$

where $\alpha_k$ is chosen to ensure that

$$f(\boldsymbol{x}^{(k+1)}) < f(\boldsymbol{x}^{(k)}).$$

For example, we may choose $\alpha_k = \arg\min_{\alpha \geqslant 0} f(\boldsymbol{x}^{(k)} + \alpha \boldsymbol{F}(\boldsymbol{x}^{(k)})^{-1}\boldsymbol{g}^{(k)})$ (see Theorem 9.2). We can then determine an appropriate value of $\alpha_k$ by performing a line search in the direction $\boldsymbol{F}(\boldsymbol{x}^{(k)})^{-1}\boldsymbol{g}^{(k)}$. Note that although the line search is simply the minimization of the real variable function $\phi_k(\alpha) = f(\boldsymbol{x}^{(k)} - \alpha \boldsymbol{F}(\boldsymbol{x}^{(k)})^{-1}\boldsymbol{g}^{(k)})$, it is not a trivial problem to solve.

A computational drawback of Newton's method is the need to evaluate $\boldsymbol{F}(\boldsymbol{x}^{(k)})$ and solve the equation $\boldsymbol{F}(\boldsymbol{x}^{(k)})\boldsymbol{d}^{(k)} = -\boldsymbol{g}^{(k)}$ (i.e., compute $\boldsymbol{d}^{(k)} = -\boldsymbol{F}(\boldsymbol{x}^{(k)})^{-1}\boldsymbol{g}^{(k)}$). To avoid the computation of $\boldsymbol{F}(\boldsymbol{x}^{(k)})^{-1}$, the quasi-Newton methods use an approximation to $\boldsymbol{F}(\boldsymbol{x}^{(k)})^{-1}$ in place of the true inverse. This approximation is updated at every stage so that it exhibits at least some properties of $\boldsymbol{F}(\boldsymbol{x}^{(k)})^{-1}$. To get some idea about the properties that an approximation to $\boldsymbol{F}(\boldsymbol{x}^{(k)})^{-1}$ should satisfy, consider the formula

$$\boldsymbol{x}^{(k+1)} = \boldsymbol{x}^{(k)} - \alpha \boldsymbol{H}_k \boldsymbol{g}^{(k)},$$

where $\boldsymbol{H}_k$ is an $n \times n$ real matrix, and $\alpha$ is a positive search parameter. Expanding $f$ about $\boldsymbol{x}^{(k)}$ yields

$$\begin{aligned} f(\boldsymbol{x}^{(k+1)}) &= f(\boldsymbol{x}^{(k)}) + \boldsymbol{g}^{(k)T}(\boldsymbol{x}^{(k+1)} - \boldsymbol{x}^{(k)}) + o(\| \boldsymbol{x}^{(k+1)} - \boldsymbol{x}^{(k)} \|) \\ &= f(\boldsymbol{x}^{(k)}) - \alpha \boldsymbol{g}^{(k)T}\boldsymbol{H}_k \boldsymbol{g}^{(k)} + o(\| \boldsymbol{H}_k \boldsymbol{g}^{(k)} \| \alpha). \end{aligned}$$

As $\alpha$ tends to zero, the second term on the right-hand side of the above equation dominates the third. Thus, to guarantee a decrease in $f$ for small $\alpha$, we have to have

$$\boldsymbol{g}^{(k)T}\boldsymbol{H}_k \boldsymbol{g}^{(k)} > 0.$$

A simple way to ensure this is to require that $\boldsymbol{H}_k$ be positive definite. We have proved the following result.

**Proposition 11.1.** *Let $f \in \mathscr{C}^1$, $\boldsymbol{x}^{(k)} \in \mathbb{R}^n$, $\boldsymbol{g}^{(k)} = \nabla f(\boldsymbol{x}^{(k)}) \neq \boldsymbol{0}$, and $\boldsymbol{H}_k$ an $n \times n$ real symmetric positive definite matrix. If we set $\boldsymbol{x}^{(k+1)} = \boldsymbol{x}^{(k)} - \alpha_k \boldsymbol{H}_k \boldsymbol{g}^{(k)}$, where $\alpha^k = \arg\min_{\alpha \geqslant 0} f(\boldsymbol{x}^{(k)} - \alpha \boldsymbol{H}_k \boldsymbol{g}^{(k)})$, then $\alpha_k > 0$, and $f(\boldsymbol{x}^{(k+1)}) < f(\boldsymbol{x}^{(k)})$.* □

In constructing an approximation to the inverse of the Hessian matrix, we should use only objective function and gradient values. Thus, if we can find a suitable method of choosing $\boldsymbol{H}_k$, the iteration may be carried out without any evaluation of the Hessian and without the solution of any set of linear equations.

## 11.2. APPROXIMATING THE INVERSE HESSIAN

Let $\boldsymbol{H}_0, \boldsymbol{H}_1, \boldsymbol{H}_2, \ldots$ be successive approximations of the inverse $\boldsymbol{F}(\boldsymbol{x}^{(k)})^{-1}$ of the Hessian. We now derive a condition that the approximations should satisfy, which forms the starting point for our subsequent discussion of quasi-Newton algorithms. To begin, suppose first that the Hessian matrix $\boldsymbol{F}(\boldsymbol{x})$ of the objective function $f$ is constant and independent of $\boldsymbol{x}$. In other words, the objective function is quadratic, with Hessian $\boldsymbol{F}(\boldsymbol{x}) = \boldsymbol{Q}$ for all $\boldsymbol{x}$, where $\boldsymbol{Q} = \boldsymbol{Q}^T$. Then,

$$\boldsymbol{g}^{(k+1)} - \boldsymbol{g}^{(k)} = \boldsymbol{Q}(\boldsymbol{x}^{(k+1)} - \boldsymbol{x}^{(k)}).$$

Let

$$\Delta \boldsymbol{g}^{(k)} \triangleq \boldsymbol{g}^{(k+1)} - \boldsymbol{g}^{(k)},$$

and

$$\Delta \boldsymbol{x}^{(k)} \triangleq \boldsymbol{x}^{(k+1)} - \boldsymbol{x}^{(k)}.$$

Then, we may write

$$\Delta \boldsymbol{g}^{(k)} = \boldsymbol{Q}\Delta \boldsymbol{x}^{(k)}.$$

We start with a real symmetric positive definite matrix $\boldsymbol{H}_0$. Note that given $k$, the matrix $\boldsymbol{Q}^{-1}$ satisfies

$$\boldsymbol{Q}^{-1}\Delta \boldsymbol{g}^{(i)} = \Delta \boldsymbol{x}^{(i)}, \quad 0 \leqslant i \leqslant k.$$

Therefore, we will also impose the requirement that the approximation $\boldsymbol{H}_{k+1}$ of the inverse Hessian satisfy

$$\boldsymbol{H}_{k+1}\Delta \boldsymbol{g}^{(i)} = \Delta \boldsymbol{x}^{(i)}, \quad 0 \leqslant i \leqslant k.$$

If $n$ steps are involved, then moving in $n$ directions $\Delta \boldsymbol{x}^{(0)}, \Delta \boldsymbol{x}^{(1)}, \ldots, \Delta \boldsymbol{x}^{(n-1)}$ yields

$$\begin{aligned}
\boldsymbol{H}_n \Delta \boldsymbol{g}^{(0)} &= \Delta \boldsymbol{x}^{(0)}, \\
\boldsymbol{H}_n \Delta \boldsymbol{g}^{(1)} &= \Delta \boldsymbol{x}^{(1)}, \\
&\vdots \\
\boldsymbol{H}_n \Delta \boldsymbol{g}^{(n-1)} &= \Delta \boldsymbol{x}^{(n-1)}.
\end{aligned}$$

The above set of equations can be represented as

$$\boldsymbol{H}_n[\Delta \boldsymbol{g}^{(0)}, \Delta \boldsymbol{g}^{(1)}, \ldots, \Delta \boldsymbol{g}^{(n-1)}] = [\Delta \boldsymbol{x}^{(0)}, \Delta \boldsymbol{x}^{(1)}, \ldots, \Delta \boldsymbol{x}^{(n-1)}].$$

Note that $\boldsymbol{Q}$ satisfies

$$\boldsymbol{Q}[\Delta \boldsymbol{x}^{(0)}, \Delta \boldsymbol{x}^{(1)}, \ldots, \Delta \boldsymbol{x}^{(n-1)}] = [\Delta \boldsymbol{g}^{(0)}, \Delta \boldsymbol{g}^{(1)}, \ldots, \Delta \boldsymbol{g}^{(n-1)}]$$

and

$$\boldsymbol{Q}^{-1}[\Delta \boldsymbol{g}^{(0)}, \Delta \boldsymbol{g}^{(1)}, \ldots, \Delta \boldsymbol{g}^{(n-1)}] = [\Delta \boldsymbol{x}^{(0)}, \Delta \boldsymbol{g}^{(1)}, \ldots, \Delta \boldsymbol{x}^{(n-1)}].$$

Therefore, if $[\Delta \boldsymbol{g}^{(0)}, \Delta \boldsymbol{g}^{(1)}, \ldots, \Delta \boldsymbol{g}^{(n-1)}]$ is nonsingular, then $\boldsymbol{Q}^{-1}$ is determined uniquely after $n$ steps, by

$$\boldsymbol{Q}^{-1} = \boldsymbol{H}_n = [\Delta \boldsymbol{x}^{(0)}, \Delta \boldsymbol{x}^{(1)}, \ldots, \Delta \boldsymbol{x}^{(n-1)}][\Delta \boldsymbol{g}^{(0)}, \Delta \boldsymbol{g}^{(1)}, \ldots, \Delta \boldsymbol{g}^{(n-1)}]^{-1}.$$

As a consequence, we conclude that if $\boldsymbol{H}_n$ satisfies the equations $\boldsymbol{H}_n \Delta \boldsymbol{g}^{(i)} = \Delta \boldsymbol{x}^{(i)}$, $0 \leqslant i \leqslant n-1$, then the algorithm $\boldsymbol{x}^{(k+1)} = \boldsymbol{x}^{(k)} - \alpha_k \boldsymbol{H}_k \boldsymbol{g}^{(k)}$, $\alpha_k = \arg\min_{\alpha \geqslant 0} f(\boldsymbol{x}^{(k)} - \alpha \boldsymbol{H}_k \boldsymbol{g}^{(k)})$, is guaranteed to solve problems with quadratic objective functions in $n+1$ steps, since the update $\boldsymbol{x}^{(n+1)} = \boldsymbol{x}^{(n)} - \alpha_n \boldsymbol{H}_n \boldsymbol{g}^{(n)}$ is equivalent to Newton's algorithm. In fact, as we shall see below, such algorithms solve quadratic problems in at most $n$ steps.

The above considerations illustrate the basic idea behind the quasi-Newton methods. Specifically, quasi-Newton algorithms have the form

$$\begin{aligned} \boldsymbol{d}^{(k)} &= -\boldsymbol{H}_k \boldsymbol{g}^{(k)} \\ \alpha_k &= \arg\min_{\alpha \geqslant 0} f(\boldsymbol{x}^{(k)} + \alpha \boldsymbol{d}^{(k)}) \\ \boldsymbol{x}^{(k+1)} &= \boldsymbol{x}^{(k)} + \alpha_k \boldsymbol{d}^{(k)}, \end{aligned}$$

where the matrices $\boldsymbol{H}_0, \boldsymbol{H}_1, \ldots$ are symmetric. In the quadratic case, the above matrices are required to satisfy

$$\boldsymbol{H}_{k+1} \Delta \boldsymbol{g}^{(i)} = \Delta \boldsymbol{x}^{(i)}, \quad 0 \leqslant i \leqslant k,$$

where $\Delta \boldsymbol{x}^{(i)} = \boldsymbol{x}^{(i+1)} - \boldsymbol{x}^{(i)} = \alpha_k \boldsymbol{d}^{(i)}$, and $\Delta \boldsymbol{g}^{(i)} = \boldsymbol{g}^{(i+1)} - \boldsymbol{g}^{(i)} = \boldsymbol{Q} \Delta \boldsymbol{x}^{(i)}$. It turns out that quasi-Newton methods are also conjugate direction methods, as stated in the following theorem.

**Theorem 11.1.** *Consider a quasi-Newton algorithm applied to a quadratic function with Hessian $\boldsymbol{Q} = \boldsymbol{Q}^T$, such that for $0 \leqslant k < n-1$,*

$$\boldsymbol{H}_{k+1} \Delta \boldsymbol{g}^{(i)} = \Delta \boldsymbol{x}^{(i)}, \quad 0 \leqslant i \leqslant k,$$

*where $\boldsymbol{H}_{k+1} = \boldsymbol{H}_{k+1}^T$. If $\boldsymbol{d}^{(i)} \neq \boldsymbol{0}$, $0 \leqslant i \leqslant k+1$, then $\boldsymbol{d}^{(0)}, \ldots, \boldsymbol{d}^{(k+1)}$ are $\boldsymbol{Q}$-conjugate.* □

*Proof.* We proceed by induction. For $k = 0$, the result holds trivially.

Assume the result is true for $k < n-1$. We now prove the result for $k+1$, that is, $\boldsymbol{d}^{(0)}, \ldots, \boldsymbol{d}^{(k+1)}$ are $\boldsymbol{Q}$-conjugate. It suffices to show that $\boldsymbol{d}^{(k+1)T} \boldsymbol{Q} \boldsymbol{d}^{(i)} = 0$, $0 \leqslant i \leqslant k$. To this end, note that since $\boldsymbol{d}^{(i)} \neq \boldsymbol{0}$, we have $\alpha_i > 0$, and $\boldsymbol{d}^{(i)} = \Delta \boldsymbol{x}^{(i)} / \alpha_i$. So, given $i, 0 \leqslant i \leqslant k$, we have

$$\begin{aligned} \boldsymbol{d}^{(k+1)T} \boldsymbol{Q} \boldsymbol{d}^{(i)} &= -\boldsymbol{g}^{(k+1)T} \boldsymbol{H}_{k+1} \boldsymbol{Q} \boldsymbol{d}^{(i)} \\ &= -\boldsymbol{g}^{(k+1)T} \boldsymbol{H}_{k+1} \frac{\boldsymbol{Q} \Delta \boldsymbol{x}^{(i)}}{\alpha_i} \\ &= -\boldsymbol{g}^{(k+1)T} \frac{\boldsymbol{H}_{k+1} \Delta \boldsymbol{g}^{(i)}}{\alpha_i} \\ &= -\boldsymbol{g}^{(k+1)T} \frac{\Delta \boldsymbol{x}^{(i)}}{\alpha_i} \\ &= -\boldsymbol{g}^{(k+1)T} \boldsymbol{d}^{(i)}. \end{aligned}$$

Since $\boldsymbol{d}^{(0)}, \ldots, \boldsymbol{d}^{(k)}$ are $\boldsymbol{Q}$-conjugate by assumption, we conclude from Lemma 10.2 that $\boldsymbol{g}^{(k+1)T} \boldsymbol{d}^{(i)} = 0$. Hence, $\boldsymbol{d}^{(k+1)T} \boldsymbol{Q} \boldsymbol{d}^{(i)} = 0$, which completes the proof. ■

By the above theorem, we conclude that a quasi-Newton algorithm solves a quadratic of $n$ variables in at most $n$ steps.

Note that the equations that the matrices $\boldsymbol{H}_k$ are required to satisfy do not determine those matrices uniquely. Thus, we have some freedom in the way we compute the $\boldsymbol{H}_k$. In the methods we will describe, we compute $\boldsymbol{H}_{k+1}$ by adding a correction to $\boldsymbol{H}_k$. In the following sections, we consider three specific updating formulas.

## 11.3. THE RANK ONE CORRECTION FORMULA

In the *rank one correction* formula, the correction term is symmetric and has the form $a_k \boldsymbol{z}^{(k)} \boldsymbol{z}^{(k)T}$, where $a_k \in \mathbb{R}$ and $\boldsymbol{z}^{(k)} \in \mathbb{R}^n$. Therefore, the update equation is

$$\boldsymbol{H}_{k+1} = \boldsymbol{H}_k + a_k \boldsymbol{z}^{(k)} \boldsymbol{z}^{(k)T}.$$

Note that

$$\operatorname{rank} \boldsymbol{z}^{(k)} \boldsymbol{z}^{(k)T} = \operatorname{rank} \left( \begin{bmatrix} z_1^{(k)} \\ \vdots \\ z_n^{(k)} \end{bmatrix} [z_1^{(k)} \cdots z_n^{(k)}] \right) = 1$$

and hence the name rank one correction. The product $\boldsymbol{z}^{(k)} \boldsymbol{z}^{(k)T}$ is sometimes referred to as the *dyadic product* or *outer product*. Observe that if $\boldsymbol{H}_k$ is symmetric, then so is $\boldsymbol{H}_{k+1}$.

Our goal now is to determine $a_k$ and $\boldsymbol{z}^{(k)}$, given $\boldsymbol{H}_k, \Delta \boldsymbol{g}^{(k)}, \Delta \boldsymbol{x}^{(k)}$, so that the required relationship discussed in the previous section is satisfied, namely $\boldsymbol{H}_{k+1} \Delta \boldsymbol{g}^{(i)} = \Delta \boldsymbol{x}^{(i)}$, $i = 1, \ldots, k$. To begin, let us first consider the condition $\boldsymbol{H}_{k+1} \Delta \boldsymbol{g}^{(k)} = \Delta \boldsymbol{x}^{(k)}$. In other words, given $\boldsymbol{H}_k, \Delta \boldsymbol{g}^{(k)}, \Delta \boldsymbol{x}^{(k)}$, we wish to find $a_k$ and $\boldsymbol{z}^{(k)}$ to ensure that

$$\boldsymbol{H}_{k+1} \Delta \boldsymbol{g}^{(k)} = \boldsymbol{H}_k + a_k \boldsymbol{z}^{(k)} \boldsymbol{z}^{(k)T}) \Delta \boldsymbol{g}^{(k)} = \Delta \boldsymbol{x}^{(k)}.$$

First note that $\boldsymbol{z}^{(k)T} \Delta \boldsymbol{g}^{(k)}$ is a scalar. Thus

$$\Delta \boldsymbol{x}^{(k)} - \boldsymbol{H}_k \Delta \boldsymbol{g}^{(k)} = (a_k \boldsymbol{z}^{(k)T} \Delta \boldsymbol{g}^{(k)}) \boldsymbol{z}^{(k)},$$

and hence

$$\boldsymbol{z}^{(k)} = \frac{\Delta \boldsymbol{x}^{(k)} - \boldsymbol{H}_k \Delta \boldsymbol{g}^{(k)}}{a_k (\boldsymbol{z}^{(k)T} \Delta \boldsymbol{g}^{(k)})}.$$

We can now determine

$$a_k \boldsymbol{z}^{(k)} \boldsymbol{z}^{(k)T} = \frac{(\Delta \boldsymbol{x}^{(k)} - \boldsymbol{H}_k \Delta \boldsymbol{g}^{(k)})(\Delta \boldsymbol{x}^{(k)} - \boldsymbol{H}_k \Delta \boldsymbol{g}^{(k)})^T}{a_k (\boldsymbol{z}^{(k)T} \Delta \boldsymbol{g}^{(k)})^2}.$$

Hence

$$\boldsymbol{H}_{k+1} = \boldsymbol{H}_k + \frac{(\Delta \boldsymbol{x}^{(k)} - \boldsymbol{H}_k \Delta \boldsymbol{g}^{(k)})(\Delta \boldsymbol{x}^{(k)} - \boldsymbol{H}_k \Delta \boldsymbol{g}^{(k)})^T}{a_k (\boldsymbol{z}^{(k)T} \Delta \boldsymbol{g}^{(k)})^2}.$$

The next step is to express the denominator of the second term on the right-hand side of the above equation as a function of the given quantities $\boldsymbol{H}_k$, $\Delta\boldsymbol{g}^{(k)}$, and $\Delta\boldsymbol{x}^{(k)}$. To accomplish this, premultiply $\Delta\boldsymbol{x}^{(k)} - \boldsymbol{H}_k\Delta\boldsymbol{g}^{(k)}$ by $\Delta\boldsymbol{g}^{(k)T}$ to obtain

$$\Delta\boldsymbol{g}^{(k)T}\Delta\boldsymbol{x}^{(k)} - \Delta\boldsymbol{g}^{(k)T}\boldsymbol{H}_k\Delta\boldsymbol{g}^{(k)} = \Delta\boldsymbol{g}^{(k)T}a_k\boldsymbol{z}^{(k)}\boldsymbol{z}^{(k)T}\Delta\boldsymbol{g}^{(k)}.$$

Observe that $a_k$ is a scalar and so is $\Delta\boldsymbol{g}^{(k)T}\boldsymbol{z}^{(k)} = \boldsymbol{z}^{(k)T}\Delta\boldsymbol{g}^{(k)}$. Thus

$$\Delta\boldsymbol{g}^{(k)T}\Delta\boldsymbol{x}^{(k)} - \Delta\boldsymbol{g}^{(k)T}\boldsymbol{H}_k\Delta\boldsymbol{g}^{(k)} = a_k(\boldsymbol{z}^{(k)T}\Delta\boldsymbol{g}^{(k)})^2.$$

Taking the above relation into account yields

$$\boldsymbol{H}_{k+1} = \boldsymbol{H}_k + \frac{(\Delta\boldsymbol{x}^{(k)} - \boldsymbol{H}_k\Delta\boldsymbol{g}^{(k)})(\Delta\boldsymbol{x}^{(k)} - \boldsymbol{H}_k\Delta\boldsymbol{g}^{(k)})^T}{\Delta\boldsymbol{g}^{(k)T}(\Delta\boldsymbol{x}^{(k)} - \boldsymbol{H}_k\Delta\boldsymbol{g}^{(k)})}.$$

We summarize the above development in the following algorithm.

**Rank One Algorithm**

**1.** Set $k := 0$; select $\boldsymbol{x}^{(0)}$, and a real symmetric positive definite $\boldsymbol{H}_0$.
**2.** If $\boldsymbol{g}^{(k)} = \boldsymbol{0}$, stop; else $\boldsymbol{d}^{(k)} = -\boldsymbol{H}_k\boldsymbol{g}^{(k)}$.
**3.** Compute

$$\alpha_k = \arg\min_{\alpha \geqslant 0} f(\boldsymbol{x}^{(k)} + \alpha\boldsymbol{d}^{(k)})$$
$$\boldsymbol{x}^{(k+1)} = \boldsymbol{x}^{(k)} + \alpha_k\boldsymbol{d}^{(k)}.$$

**4.** Compute

$$\Delta\boldsymbol{x}^{(k)} = \alpha_k\boldsymbol{d}^{(k)}$$
$$\Delta\boldsymbol{g}^{(k)} = \boldsymbol{g}^{(k+1)} - \boldsymbol{g}^{(k)}$$
$$\boldsymbol{H}_{k+1} = \boldsymbol{H}_k + \frac{(\Delta\boldsymbol{x}^{(k)} - \boldsymbol{H}_k\Delta\boldsymbol{g}^{(k)})(\Delta\boldsymbol{x}^{(k)} - \boldsymbol{H}_k\Delta\boldsymbol{g}^{(k)})^T}{\Delta\boldsymbol{g}^{(k)T}(\Delta\boldsymbol{x}^{(k)} - \boldsymbol{H}_k\Delta\boldsymbol{g}^{(k)})}.$$

**5.** Set $k := k + 1$; go to 2.

The rank one algorithm is based on satisfying the equation

$$\boldsymbol{H}_{k+1}\Delta\boldsymbol{g}^{(k)} = \Delta\boldsymbol{x}^{(k)}.$$

However, what we want is

$$\boldsymbol{H}_{k+1}\Delta\boldsymbol{g}^{(i)} = \Delta\boldsymbol{x}^{(i)}, \quad i = 0, 1, \ldots, k.$$

It turns out that the above is in fact automatically true, as stated in the following theorem.

**Theorem 11.2.** *For the rank one algorithm applied to the quadratic with Hessian $\boldsymbol{Q}=\boldsymbol{Q}^T$, we have $\boldsymbol{H}_{k+1}\Delta\boldsymbol{g}^{(i)}=\Delta\boldsymbol{x}^{(i)}$, $0\leqslant i\leqslant k$.* □

*Proof.* We prove the result by induction. From the discussion before the theorem, it is clear that the claim is true for $k=0$. Suppose now that the theorem is true for $k-1$, that is, $\boldsymbol{H}_k\Delta\boldsymbol{g}^{(i)}=\Delta\boldsymbol{x}^{(i)}$, $i<k$. We now show that the theorem is true for $k$. Our construction of the correction term ensures that

$$\boldsymbol{H}_{k+1}\Delta\boldsymbol{g}^{(k)}=\Delta\boldsymbol{x}^{(k)}.$$

So we only have to show

$$\boldsymbol{H}_{k+1}\Delta\boldsymbol{g}^{(i)}=\Delta\boldsymbol{x}^{(i)},\quad i<k.$$

To this end, fix $i<k$. We have

$$\boldsymbol{H}_{k+1}\Delta\boldsymbol{g}^{(i)}=\boldsymbol{H}_k\Delta\boldsymbol{g}^{(i)}+\frac{(\Delta\boldsymbol{x}^{(k)}-\boldsymbol{H}_k\Delta\boldsymbol{g}^{(k)})(\Delta\boldsymbol{x}^{(k)}-\boldsymbol{H}_k\Delta\boldsymbol{g}^{(k)})^T}{\Delta\boldsymbol{g}^{(k)T}(\Delta\boldsymbol{x}^{(k)}-\boldsymbol{H}_k\Delta\boldsymbol{g}^{(k)})}\Delta\boldsymbol{g}^{(i)}.$$

By the inductive hypothesis, $\boldsymbol{H}_k\Delta\boldsymbol{g}^{(i)}=\Delta\boldsymbol{x}^{(i)}$. To complete the proof, it is enough to show that the second term on the right-hand side of the above equation is equal to zero. For this to be true it is enough that

$$(\Delta\boldsymbol{x}^{(k)}-\boldsymbol{H}_k\Delta\boldsymbol{g}^{(k)})^T\Delta\boldsymbol{g}^{(i)}=\Delta\boldsymbol{x}^{(k)T}\Delta\boldsymbol{g}^{(i)}-\Delta\boldsymbol{g}^{(k)T}\boldsymbol{H}_k\Delta\boldsymbol{g}^{(i)}=0.$$

Indeed, since

$$\Delta\boldsymbol{g}^{(k)T}\boldsymbol{H}_k\Delta\boldsymbol{g}^{(i)}=\Delta\boldsymbol{g}^{(k)T}(\boldsymbol{H}_k\Delta\boldsymbol{g}^{(i)})=\Delta\boldsymbol{g}^{(k)T}\Delta\boldsymbol{x}^{(i)},$$

and since $\Delta\boldsymbol{g}^{(k)}=\boldsymbol{Q}\Delta\boldsymbol{x}^{(k)}$, we have

$$\Delta\boldsymbol{g}^{(k)T}\boldsymbol{H}_k\Delta\boldsymbol{g}^{(i)}=\Delta\boldsymbol{g}^{(k)T}\Delta\boldsymbol{x}^{(i)}=\Delta\boldsymbol{x}^{(k)T}\boldsymbol{Q}\Delta\boldsymbol{x}^{(i)}=\Delta\boldsymbol{x}^{(k)T}\Delta\boldsymbol{g}^{(i)}.$$

Hence

$$(\Delta\boldsymbol{x}^{(k)}-\boldsymbol{H}_k\Delta\boldsymbol{g}^{(k)})^T\Delta\boldsymbol{g}^{(i)}=\Delta\boldsymbol{x}^{(k)T}\Delta\boldsymbol{g}^{(i)}-\Delta\boldsymbol{x}^{(k)T}\Delta\boldsymbol{g}^{(i)}=0,$$

which completes the proof. ■

***Example 11.1.*** Let

$$f(x_1,x_2)=x_1^2+\tfrac{1}{2}x_2^2+3.$$

Apply the rank one correction algorithm to minimize $f$. Use $\boldsymbol{x}^{(0)}=[1,2]^T$ and $\boldsymbol{H}_0=\boldsymbol{I}_2$ ($2\times 2$ identity matrix).

We can represent $f$ as

$$f(\boldsymbol{x})=\tfrac{1}{2}\boldsymbol{x}^T\begin{bmatrix}2&0\\0&1\end{bmatrix}\boldsymbol{x}+3.$$

Thus,

$$g^{(k)} = \begin{bmatrix} 2 & 0 \\ 0 & 1 \end{bmatrix} x^{(k)}.$$

Since $H_0 = I_2$

$$d^{(0)} = -g^{(0)} = [-2, -2]^T.$$

The objective function is quadratic, and hence

$$\begin{aligned}\alpha_0 &= \arg\min_{\alpha \geqslant 0} f(x^{(0)} + \alpha d^{(0)}) = -\frac{g^{(0)T} d^{(0)}}{d^{(0)T} Q d^{(0)}} \\ &= \frac{[2,2]\begin{bmatrix} 2 \\ 2 \end{bmatrix}}{[2,2]\begin{bmatrix} 2 & 0 \\ 0 & 1 \end{bmatrix}\begin{bmatrix} 2 \\ 2 \end{bmatrix}} \\ &= \frac{2}{3},\end{aligned}$$

and thus

$$x^{(1)} = x^{(0)} + \alpha_0 d^{(0)} = [-\tfrac{1}{3}, \tfrac{2}{3}]^T.$$

We then compute

$$\begin{aligned}\Delta x^{(0)} &= \alpha_0 d^{(0)} = [-\tfrac{4}{3}, -\tfrac{4}{3}]^T, \\ g^{(1)} &= Q x^{(1)} = [-\tfrac{2}{3}, \tfrac{2}{3}]^T, \\ \Delta g^{(0)} &= g^{(1)} - g^{(0)} = [-\tfrac{8}{3}, -\tfrac{4}{3}]^T.\end{aligned}$$

Since

$$\Delta g^{(0)T}(\Delta x^{(0)} - H_0 \Delta g^{(0)}) = [-\tfrac{8}{3}, -\tfrac{4}{3}]\begin{bmatrix} \frac{4}{3} \\ 0 \end{bmatrix} = -\frac{32}{9},$$

then

$$H_1 = H_0 + \frac{(\Delta x^{(0)} - H_0 \Delta g^{(0)})(\Delta x^{(0)} - H_0 \Delta g^{(0)})^T}{\Delta g^{(0)T}(\Delta x^{(0)} - H_0 \Delta g^{(0)})} = \begin{bmatrix} \frac{1}{2} & 0 \\ 0 & 1 \end{bmatrix}.$$

Therefore

$$d^{(1)} = -H_1 g^{(1)} = [\tfrac{1}{3}, -\tfrac{2}{3}]^T,$$

and

$$\alpha_1 = -\frac{g^{(1)T} d^{(1)}}{d^{(1)T} Q d^{(1)}} = 1.$$

We now compute

$$x^{(2)} = x^{(1)} + \alpha_1 d^{(1)} = [0, 0]^T.$$

Note that $g^{(2)} = \mathbf{0}$, and therefore $x^{(2)} = x^*$. As expected, the algorithm solves the problem in two steps.

Note that the directions $\boldsymbol{d}^{(0)}$ and $\boldsymbol{d}^{(1)}$ are $\boldsymbol{Q}$-conjugate, in accordance with Theorem 11.1. ■

The rank one correction algorithm, also called the *single-rank symmetric* (SRS) algorithm, works well for the case of a constant Hessian matrix, that is, the quadratic case. Our analysis was in fact done for this case. Unfortunately, for the nonquadratic case, the rank one correction algorithm is not very satisfactory for several reasons. For a nonquadratic objective function, $\boldsymbol{H}_{k+1}$ may not be positive definite (see Exercise 11.2) and thus $\boldsymbol{d}^{(k+1)}$ may not be a descent direction. Furthermore, if

$$\Delta \boldsymbol{g}^{(k)T}(\Delta \boldsymbol{x}^{(k)} - \boldsymbol{H}_k \Delta \boldsymbol{g}^{(k)})$$

is close to zero, there may be numerical problems in evaluating $\boldsymbol{H}_{k+1}$.

Fortunately, alternative algorithms have been developed for updating $\boldsymbol{H}_k$. In particular, if we use a rank two update, then $\boldsymbol{H}_k$ is guaranteed to be positive definite for all $k$, provided the line search is exact. We discuss this in the next section.

## 11.4. THE DFP ALGORITHM

The rank two update was originally developed by Davidon in 1959 and was subsequently modified by Fletcher and Powell in 1963; hence the name DFP algorithm. The DFP algorithm is also known as the *variable metric algorithm*. We summarize the algorithm below.

### DFP Algorithm

1. Set $k := 0$; select $\boldsymbol{x}^{(0)}$, and a real symmetric positive definite $\boldsymbol{H}_0$.
2. If $\boldsymbol{g}^{(k)} = \boldsymbol{0}$, stop; else $\boldsymbol{d}^{(k)} = -\boldsymbol{H}_k \boldsymbol{g}^{(k)}$.
3. Compute

$$\alpha_k = \arg\min_{\alpha \geqslant 0} f(\boldsymbol{x}^{(k)} + \alpha \boldsymbol{d}^{(k)})$$
$$\boldsymbol{x}^{(k+1)} = \boldsymbol{x}^{(k)} + \alpha_k \boldsymbol{d}^{(k)}.$$

4. Compute

$$\Delta \boldsymbol{x}^{(k)} = \alpha_k \boldsymbol{d}^{(k)}$$
$$\Delta \boldsymbol{g}^{(k)} = \boldsymbol{g}^{(k+1)} - \boldsymbol{g}^{(k)}$$
$$\boldsymbol{H}_{k+1} = \boldsymbol{H}_k + \frac{\Delta \boldsymbol{x}^{(k)} \Delta \boldsymbol{x}^{(k)T}}{\Delta \boldsymbol{x}^{(k)T} \Delta \boldsymbol{g}^{(k)}} - \frac{[\boldsymbol{H}_k \Delta \boldsymbol{g}^{(k)}][\boldsymbol{H}_k \Delta \boldsymbol{g}^{(k)}]^T}{\Delta \boldsymbol{g}^{(k)T} \boldsymbol{H}_k \Delta \boldsymbol{g}^{(k)}}.$$

5. Set $k := k + 1$; go 2.

We now show that the DFP algorithm is a quasi-Newton method, in the sense that when applied to quadratic problems, we have $\boldsymbol{H}_{k+1}\Delta\boldsymbol{g}^{(i)} = \Delta\boldsymbol{x}^{(i)}$, $0 \leqslant i \leqslant k$.

**Theorem 11.3.** *In the DFP algorithm applied to the quadratic with Hessian* $\boldsymbol{Q} = \boldsymbol{Q}^T$, *we have* $\boldsymbol{H}_{k+1}\Delta\boldsymbol{g}^{(i)} = \Delta\boldsymbol{x}^{(i)}$, $0 \leqslant i \leqslant k$. □

*Proof.* We use induction. For $k = 0$, we have

$$\begin{aligned}
\boldsymbol{H}_1\Delta\boldsymbol{g}^{(0)} &= \boldsymbol{H}_0\Delta\boldsymbol{g}^{(0)} + \frac{\Delta\boldsymbol{x}^{(0)}\Delta\boldsymbol{x}^{(0)T}}{\Delta\boldsymbol{x}^{(0)T}\Delta\boldsymbol{g}^{(0)}}\Delta\boldsymbol{g}^{(0)} - \frac{\boldsymbol{H}_0\Delta\boldsymbol{g}^{(0)}\Delta\boldsymbol{g}^{(0)T}\boldsymbol{H}_0}{\Delta\boldsymbol{g}^{(0)T}\boldsymbol{H}_0\Delta\boldsymbol{g}^{(0)}}\Delta\boldsymbol{g}^{(0)} \\
&= \Delta\boldsymbol{x}^{(0)}.
\end{aligned}$$

Assume the result is true for $k-1$, that is, $\boldsymbol{H}_k\Delta\boldsymbol{g}^{(i)} = \Delta\boldsymbol{x}^{(i)}$, $0 \leqslant i \leqslant k-1$. We now show that $\boldsymbol{H}_{k+1}\Delta\boldsymbol{g}^{(i)} = \Delta\boldsymbol{x}^{(i)}$, $0 \leqslant i \leqslant k$. First, consider $i = k$. We have

$$\begin{aligned}
\boldsymbol{H}_{k+1}\Delta\boldsymbol{g}^{(k)} &= \boldsymbol{H}_k\Delta\boldsymbol{g}^{(k)} + \frac{\Delta\boldsymbol{x}^{(k)}\Delta\boldsymbol{x}^{(k)T}}{\Delta\boldsymbol{x}^{(k)T}\Delta\boldsymbol{g}^{(k)}}\Delta\boldsymbol{g}^{(k)} - \frac{\boldsymbol{H}_k\Delta\boldsymbol{g}^{(k)}\Delta\boldsymbol{g}^{(k)T}\boldsymbol{H}_k}{\Delta\boldsymbol{g}^{(k)T}\boldsymbol{H}_k\Delta\boldsymbol{g}^{(k)}}\Delta\boldsymbol{g}^{(k)} \\
&= \Delta\boldsymbol{x}^{(k)}.
\end{aligned}$$

It remains to consider the case $i < k$. To this end,

$$\begin{aligned}
\boldsymbol{H}_{k+1}\Delta\boldsymbol{g}^{(i)} &= \boldsymbol{H}_k\Delta\boldsymbol{g}^{(i)} + \frac{\Delta\boldsymbol{x}^{(k)}\Delta\boldsymbol{x}^{(k)T}}{\Delta\boldsymbol{x}^{(k)T}\Delta\boldsymbol{g}^{(k)}}\Delta\boldsymbol{g}^{(i)} - \frac{\boldsymbol{H}_k\Delta\boldsymbol{g}^{(k)}\Delta\boldsymbol{g}^{(k)T}\boldsymbol{H}_k}{\Delta\boldsymbol{g}^{(k)T}\boldsymbol{H}_k\Delta\boldsymbol{g}^{(k)}}\Delta\boldsymbol{g}^{(i)} \\
&= \Delta\boldsymbol{x}^{(i)} + \frac{\Delta\boldsymbol{x}^{(k)}}{\Delta\boldsymbol{x}^{(k)T}\Delta\boldsymbol{g}^{(k)}}(\Delta\boldsymbol{x}^{(k)T}\Delta\boldsymbol{g}^{(i)}) - \frac{\boldsymbol{H}_k\Delta\boldsymbol{g}^{(k)}}{\Delta\boldsymbol{g}^{(k)T}\boldsymbol{H}_k\Delta\boldsymbol{g}^{(k)}}(\Delta\boldsymbol{g}^{(k)T}\Delta\boldsymbol{x}^{(i)}).
\end{aligned}$$

Now,

$$\begin{aligned}
\Delta\boldsymbol{x}^{(k)T}\Delta\boldsymbol{g}^{(i)} &= \Delta\boldsymbol{x}^{(k)T}\boldsymbol{Q}\Delta\boldsymbol{x}^{(i)} \\
&= \alpha_k\alpha_i\boldsymbol{d}^{(k)T}\boldsymbol{Q}\boldsymbol{d}^{(i)} \\
&= 0
\end{aligned}$$

by the induction hypothesis and Theorem 11.1. The same arguments yield $\Delta\boldsymbol{g}^{(k)T}\Delta\boldsymbol{x}^{(i)} = 0$. Hence,

$$\boldsymbol{H}_{k+1}\Delta\boldsymbol{g}^{(i)} = \Delta\boldsymbol{x}^{(i)},$$

and the proof is completed. ■

By the above theorem and Theorem 11.1, we conclude that the DFP algorithm is a conjugate direction algorithm.

***Example 11.2.*** Locate the minimizer of

$$f(\boldsymbol{x})=\frac{1}{2}\boldsymbol{x}^T\begin{bmatrix}4 & 2\\ 2 & 2\end{bmatrix}\boldsymbol{x}-\boldsymbol{x}^T\begin{bmatrix}-1\\ 1\end{bmatrix}, \quad \boldsymbol{x}\in\mathbb{R}^2.$$

Use the initial point $\boldsymbol{x}^{(0)}=[0,0]^T$ and $\boldsymbol{H}_0=\boldsymbol{I}_2$.

Note that in this case,

$$\boldsymbol{g}^{(k)}=\begin{bmatrix}4 & 2\\ 2 & 2\end{bmatrix}\boldsymbol{x}^{(k)}-\begin{bmatrix}-1\\ 1\end{bmatrix}.$$

Hence

$$\boldsymbol{g}^{(0)}=[1,-1]^T,$$

$$\boldsymbol{d}^{(0)}=-\boldsymbol{H}_0\boldsymbol{g}^{(0)}=-\begin{bmatrix}1 & 0\\ 0 & 1\end{bmatrix}\begin{bmatrix}1\\ -1\end{bmatrix}=\begin{bmatrix}-1\\ 1\end{bmatrix}.$$

Since $f$ is a quadratic function

$$\alpha_0=\arg\min_{\alpha\geqslant 0} f(\boldsymbol{x}^{(0)}+\alpha\boldsymbol{d}^{(0)})=-\frac{\boldsymbol{g}^{(0)T}\boldsymbol{d}^{(0)}}{\boldsymbol{d}^{(0)T}\boldsymbol{Q}\boldsymbol{d}^{(0)}}$$

$$=\frac{[1,-1]\begin{bmatrix}-1\\ 1\end{bmatrix}}{[-1,1]\begin{bmatrix}4 & 2\\ 2 & 2\end{bmatrix}\begin{bmatrix}-1\\ 1\end{bmatrix}}=1.$$

Therefore

$$\boldsymbol{x}^{(1)}=\boldsymbol{x}^{(0)}+\alpha_0\boldsymbol{d}^{(0)}=[-1,1]^T.$$

We then compute

$$\Delta\boldsymbol{x}^{(0)}=\boldsymbol{x}^{(1)}-\boldsymbol{x}^{(0)}=[-1,1]^T.$$

$$\boldsymbol{g}^{(1)}=\begin{bmatrix}4 & 2\\ 2 & 2\end{bmatrix}\begin{bmatrix}-1\\ 1\end{bmatrix}-\begin{bmatrix}-1\\ 1\end{bmatrix}=\begin{bmatrix}-1\\ -1\end{bmatrix},$$

and

$$\Delta\boldsymbol{g}^{(0)}=\boldsymbol{g}^{(1)}-\boldsymbol{g}^{(0)}=[-2,0]^T.$$

Observe that

$$\Delta\boldsymbol{x}^{(0)}\Delta\boldsymbol{x}^{(0)T}=\begin{bmatrix}-1\\ 1\end{bmatrix}[-1,1]=\begin{bmatrix}1 & -1\\ -1 & 1\end{bmatrix},$$

$$\Delta\boldsymbol{x}^{(0)T}\Delta\boldsymbol{g}^{(0)}=[-1,1]\begin{bmatrix}-2\\ 0\end{bmatrix}=2,$$

$$\boldsymbol{H}_0\Delta\boldsymbol{g}^{(0)}=\begin{bmatrix}1 & 0\\ 0 & 1\end{bmatrix}\begin{bmatrix}-2\\ 0\end{bmatrix}=\begin{bmatrix}-2\\ 0\end{bmatrix}.$$

Thus

$$(\boldsymbol{H}_0\Delta\boldsymbol{g}^{(0)})(\boldsymbol{H}_0\Delta\boldsymbol{g}^{(0)})^T = \begin{bmatrix} -2 \\ 0 \end{bmatrix}[-2, 0] = \begin{bmatrix} 4 & 0 \\ 0 & 0 \end{bmatrix},$$

and

$$\Delta\boldsymbol{g}^{(0)T}\boldsymbol{H}_0\Delta\boldsymbol{g}^{(0)} = [-2, 0]\begin{bmatrix} 1 & 0 \\ 0 & 1 \end{bmatrix}\begin{bmatrix} -2 \\ 0 \end{bmatrix} = 4.$$

Using the above, we now compute $\boldsymbol{H}_1$,

$$\begin{aligned}\boldsymbol{H}_1 &= \boldsymbol{H}_0 + \frac{\Delta\boldsymbol{x}^{(0)}\Delta\boldsymbol{x}^{(0)T}}{\Delta\boldsymbol{x}^{(0)T}\Delta\boldsymbol{g}^{(0)}} - \frac{(\boldsymbol{H}_0\Delta\boldsymbol{g}^{(0)})(\boldsymbol{H}_0\Delta\boldsymbol{g}^{(0)})^T}{\Delta\boldsymbol{g}^{(0)T}\boldsymbol{H}_0\Delta\boldsymbol{g}^{(0)}} \\ &= \begin{bmatrix} 1 & 0 \\ 0 & 1 \end{bmatrix} + \tfrac{1}{2}\begin{bmatrix} 1 & -1 \\ -1 & 1 \end{bmatrix} - \tfrac{1}{4}\begin{bmatrix} 4 & 0 \\ 0 & 0 \end{bmatrix} \\ &= \begin{bmatrix} \frac{1}{2} & -\frac{1}{2} \\ -\frac{1}{2} & \frac{3}{2} \end{bmatrix}.\end{aligned}$$

We now compute $\boldsymbol{d}^{(1)} = -\boldsymbol{H}_1\boldsymbol{g}^{(1)} = [0, 1]^T$, and

$$\alpha_1 = \arg\min_{\alpha \geqslant 0} f(\boldsymbol{x}^{(1)} + \alpha\boldsymbol{d}^{(1)}) = -\frac{\boldsymbol{g}^{(1)T}\boldsymbol{d}^{(1)}}{\boldsymbol{d}^{(1)T}\boldsymbol{Q}\boldsymbol{d}^{(1)}} = \tfrac{1}{2}.$$

Hence

$$\boldsymbol{x}^{(2)} = \boldsymbol{x}^{(1)} + \alpha_1\boldsymbol{d}^{(1)} = [-1, \tfrac{3}{2}]^T = \boldsymbol{x}^*$$

since $f$ is a quadratic function of two variables.

Note that we have $\boldsymbol{d}^{(0)T}\boldsymbol{Q}\boldsymbol{d}^{(1)} = \boldsymbol{d}^{(1)T}\boldsymbol{Q}\boldsymbol{d}^{(0)} = 0$, that is, $\boldsymbol{d}^{(0)}$ and $\boldsymbol{d}^{(1)}$ are $\boldsymbol{Q}$-conjugate directions. ■

We now show that in the DFP algorithm, $\boldsymbol{H}_{k+1}$ inherits positive definiteness from $\boldsymbol{H}_k$.

**Theorem 11.4.** *Suppose* $\boldsymbol{g}^{(k)} \neq \boldsymbol{0}$. *In the DFP algorithm, if* $\boldsymbol{H}_k$ *is positive definite, then so is* $\boldsymbol{H}_{k+1}$. □

*Proof.* We first write the following quadratic form

$$\begin{aligned}\boldsymbol{x}^T\boldsymbol{H}_{k+1}\boldsymbol{x} &= \boldsymbol{x}^T\boldsymbol{H}_k\boldsymbol{x} + \frac{\boldsymbol{x}^T\Delta\boldsymbol{x}^{(k)}\Delta\boldsymbol{x}^{(k)T}\boldsymbol{x}}{\Delta\boldsymbol{x}^{(k)T}\Delta\boldsymbol{g}^{(k)}} - \frac{\boldsymbol{x}^T(\boldsymbol{H}_k\Delta\boldsymbol{g}^{(k)})(\boldsymbol{H}_k\Delta\boldsymbol{g}^{(k)})^T\boldsymbol{x}}{\Delta\boldsymbol{g}^{(k)T}\boldsymbol{H}_k\Delta\boldsymbol{g}^{(k)}} \\ &= \boldsymbol{x}^T\boldsymbol{H}_k\boldsymbol{x} + \frac{(\boldsymbol{x}^T\Delta\boldsymbol{x}^{(k)})^2}{\Delta\boldsymbol{x}^{(k)T}\Delta\boldsymbol{g}^{(k)}} - \frac{(\boldsymbol{x}^T\boldsymbol{H}_k\Delta\boldsymbol{g}^{(k)})^2}{\Delta\boldsymbol{g}^{(k)T}\boldsymbol{H}_k\Delta\boldsymbol{g}^{(k)}}.\end{aligned}$$

Define

$$\begin{aligned}\boldsymbol{a} &\triangleq \boldsymbol{H}_k^{1/2}\boldsymbol{x}, \\ \boldsymbol{b} &\triangleq \boldsymbol{H}_k^{1/2}\Delta\boldsymbol{g}^{(k)},\end{aligned}$$

where

$$H_k = H_k^{1/2} H_k^{1/2}.$$

Note that, since $H_k > 0$, its square root is well defined; see Part I for more information on this property of positive definite matrices. Using the definitions of $a$ and $b$, we obtain

$$x^T H_k x = x^T H_k^{1/2} H_k^{1/2} x = a^T a,$$

$$x^T H_k \Delta g^{(k)} = x^T H_k^{1/2} H_k^{1/2} \Delta g^{(k)} = a^T b,$$

and

$$\Delta g^{(k)T} H_k \Delta g^{(k)} = \Delta g^{(k)T} H_k^{1/2} H_k^{1/2} \Delta g^{(k)} = b^T b.$$

Hence

$$\begin{aligned} x^T H_{k+1} x &= a^T a + \frac{(x^T \Delta x^{(k)})^2}{\Delta x^{(k)T} \Delta g^{(k)}} - \frac{(a^T b)^2}{b^T b} \\ &= \frac{\|a\|^2 \|b\|^2 - (\langle a, b \rangle)^2}{\|b\|^2} + \frac{(x^T \Delta x^{(k)})^2}{\Delta x^{(k)T} \Delta g^{(k)}}. \end{aligned}$$

We also have

$$\Delta x^{(k)T} \Delta g^{(k)} = \Delta x^{(k)T} (g^{(k+1)} - g^{(k)}) = -\Delta x^{(k)T} g^{(k)}$$

since $\Delta x^{(k)T} g^{(k+1)} = \alpha_k d^{(k)T} g^{(k+1)} = 0$ by Lemma 10.2 (see also Exercise 11.1). Since

$$\Delta x^{(k)} = \alpha_k d^{(k)} = -\alpha_k H_k g^{(k)},$$

we have

$$\Delta x^{(k)T} \Delta g^{(k)} = -\Delta x^{(k)T} g^{(k)} = \alpha_k g^{(k)T} H_k g^{(k)}.$$

The above yields

$$x^T H_{k+1} x = \frac{\|a\|^2 \|b\|^2 - (\langle a, b \rangle)^2}{\|b\|^2} + \frac{(x^T \Delta x^{(k)})^2}{\alpha_k g^{(k)T} H_k g^{(k)}}.$$

Both terms on the right-hand side of the above equation are nonnegative—the first term is nonnegative because of the Cauchy–Schwarz Inequality, and the second term is nonnegative because $H_k > 0$ and $\alpha_k > 0$ (by Proposition 11.1). Therefore, to show that $x^T H_{k+1} x > 0$ for $x \neq 0$, we only need to demonstrate that these terms do not both vanish simultaneously.

The first term vanishes only if $a$ and $b$ are proportional, that is if $a = \beta b$ for some scalar $\beta$. Thus, to complete the proof, it is enough to show that if $a = \beta b$, then $(x^T \Delta x^{(k)})^2 / \alpha_k g^{(k)T} H_k g^{(k)} > 0$. Indeed, first observe that

$$H_k^{1/2} x = a = \beta b = \beta H_k^{1/2} \Delta g^{(k)} = H_k^{1/2} (\beta \Delta g^{(k)}).$$

Hence

$$x = \beta \Delta g^{(k)}.$$

Using the above expression for $\boldsymbol{x}$ and the expression $\Delta\boldsymbol{x}^{(k)T}\Delta\boldsymbol{g}^{(k)} = -\alpha_k\boldsymbol{g}^{(k)T}\boldsymbol{H}_k\boldsymbol{g}^{(k)}$, we obtain

$$\frac{(\boldsymbol{x}^T\Delta\boldsymbol{x}^{(k)})^2}{\alpha_k\boldsymbol{g}^{(k)T}\boldsymbol{H}_k\boldsymbol{g}^{(k)}} = \frac{\beta^2(\Delta\boldsymbol{g}^{(k)T}\Delta\boldsymbol{x}^{(k)})^2}{\alpha_k\boldsymbol{g}^{(k)T}\boldsymbol{H}_k\boldsymbol{g}^{(k)}} = \frac{\beta^2(\alpha_k\boldsymbol{g}^{(k)T}\boldsymbol{H}_k\boldsymbol{g}^{(k)})^2}{\alpha_k\boldsymbol{g}^{(k)T}\boldsymbol{H}_k\boldsymbol{g}^{(k)}} = \beta^2\alpha_k\boldsymbol{g}^{(k)T}\boldsymbol{H}_k\boldsymbol{g}^{(k)} > 0.$$

Thus, for all $\boldsymbol{x} \neq \mathbf{0}$,

$$\boldsymbol{x}^T\boldsymbol{H}_{k+1}\boldsymbol{x} > 0,$$

and the proof is completed. ■

The DFP algorithm is superior to the rank one algorithm in that it preserves the positive definiteness of $\boldsymbol{H}_k$. However, it turns out that in the case of larger nonquadratic problems the algorithm has the tendency of sometimes getting "stuck." This phenomenon is attributed to $\boldsymbol{H}_k$ becoming nearly singular [see Ref. 12]. In the next section, we discuss an algorithm that alleviates this problem.

## 11.5. THE BFGS ALGORITHM

In 1970, an alternative update formula was suggested independently by Broyden, Fletcher, Goldfarb, and Shanno. The method is now called the BFGS algorithm, which we discuss in this section.

To derive the BFGS update, we use the concept of *duality*, or *complementarity*, as presented in Refs. 22 and 51. To discuss this concept, recall that the updating formulas for the approximation of the inverse of the Hessian matrix were based on satisfying the equations

$$\boldsymbol{H}_{k+1}\Delta\boldsymbol{g}^{(i)} = \Delta\boldsymbol{x}^{(i)}, \quad 0 \leqslant i \leqslant k$$

which were derived from

$$\Delta\boldsymbol{g}^{(i)} = \boldsymbol{Q}\Delta\boldsymbol{x}^{(i)}, \quad 0 \leqslant i \leqslant k.$$

We then formulated update formulas for the approximations to the inverse of the Hessian matrix $\boldsymbol{Q}^{-1}$. An alternative to approximating $\boldsymbol{Q}^{-1}$ is to approximate $\boldsymbol{Q}$ itself. To do this, let $\boldsymbol{B}_k$ be our estimate of $\boldsymbol{Q}$ at the $k$th step. We will require $\boldsymbol{B}_{k+1}$ to satisfy

$$\Delta\boldsymbol{g}^{(i)} = \boldsymbol{B}_{k+1}\Delta\boldsymbol{x}^{(i)}, \quad 0 \leqslant i \leqslant k.$$

Notice that the above set of equations is similar to the previous set of equations for $\boldsymbol{H}_{k+1}$, the only difference being that the roles of $\Delta\boldsymbol{x}^{(i)}$ and $\Delta\boldsymbol{g}^{(i)}$ are interchanged. Thus, given any update formula for $\boldsymbol{H}_k$, a corresponding update formula for $\boldsymbol{B}_k$ can be found by interchanging the roles of $\boldsymbol{B}_k$ and $\boldsymbol{H}_k$, and of $\Delta \boldsymbol{g}^{(k)}$ and $\Delta \boldsymbol{x}^{(k)}$. In particular, the BFGS update for $\boldsymbol{B}_k$ corresponds to the DFP update for $\boldsymbol{H}_k$. Formulas related in this way are said to be dual or complementary [Ref. 22].

Recall that the DFP update for the approximation $\boldsymbol{H}_k$ of the inverse Hessian is

$$\boldsymbol{H}_{k+1}^{\mathrm{DFP}} = \boldsymbol{H}_k + \frac{\Delta \boldsymbol{x}^{(k)} \Delta \boldsymbol{x}^{(k)T}}{\Delta \boldsymbol{x}^{(k)T} \Delta \boldsymbol{g}^{(k)}} - \frac{\boldsymbol{H}_k \Delta \boldsymbol{g}^{(k)} \Delta \boldsymbol{g}^{(k)T} \boldsymbol{H}_k}{\Delta \boldsymbol{g}^{(k)T} \boldsymbol{H}_k \Delta \boldsymbol{g}^{(k)}}.$$

Using the complementarity concept, we can easily obtain an update equaton for the approximation $\boldsymbol{B}_k$ of the Hessian:

$$\boldsymbol{B}_{k+1} = \boldsymbol{B}_k + \frac{\Delta \boldsymbol{g}^{(k)} \Delta \boldsymbol{g}^{(k)T}}{\Delta \boldsymbol{g}^{(k)T} \Delta \boldsymbol{x}^{(k)}} - \frac{\boldsymbol{B}_k \Delta \boldsymbol{x}^{(k)} \Delta \boldsymbol{x}^{(k)T} \boldsymbol{B}_k}{\Delta \boldsymbol{x}^{(k)T} \boldsymbol{B}_k \Delta \boldsymbol{x}^{(k)}}.$$

This is the BFGS update of $\boldsymbol{B}_k$.

Now, to obtain the BFGS update for the approximation of the inverse Hessian, we take the inverse of $\boldsymbol{B}_{k+1}$ to obtain

$$\begin{aligned} \boldsymbol{H}_{k+1}^{\mathrm{BFGS}} &= (\boldsymbol{B}_{k+1})^{-1} \\ &= \left( \boldsymbol{B}_k + \frac{\Delta \boldsymbol{g}^{(k)} \Delta \boldsymbol{g}^{(k)T}}{\Delta \boldsymbol{g}^{(k)T} \Delta \boldsymbol{x}^{(k)}} - \frac{\boldsymbol{B}_k \Delta \boldsymbol{x}^{(k)} \Delta \boldsymbol{x}^{(k)T} \boldsymbol{B}_k}{\Delta \boldsymbol{x}^{(k)T} \boldsymbol{B}_k \Delta \boldsymbol{x}^{(k)}} \right)^{-1}. \end{aligned}$$

To compute $\boldsymbol{H}_{k+1}^{\mathrm{BFGS}}$ by inverting the right-hand side of the above equation, we apply the following formula for a matrix inverse, known as the Sherman–Morrison formula [see Ref. 36 (p. 123) or Ref. 29 (p. 3)].

**Lemma 11.1.** *Let $\boldsymbol{A}$ be a nonsingular matrix. Let $\boldsymbol{u}$ and $\boldsymbol{v}$ be column vectors and assume that $1 + \boldsymbol{v}^T \boldsymbol{A}^{-1} \boldsymbol{u} \neq 0$. Then, $\boldsymbol{A} + \boldsymbol{u}\boldsymbol{v}^T$ is nonsingular, and*

$$(\boldsymbol{A} + \boldsymbol{u}\boldsymbol{v}^T)^{-1} = \boldsymbol{A}^{-1} - \frac{(\boldsymbol{A}^{-1}\boldsymbol{u})(\boldsymbol{v}^T \boldsymbol{A}^{-1})}{1 + \boldsymbol{v}^T \boldsymbol{A}^{-1} \boldsymbol{u}}. \qquad \square$$

*Proof.* We can prove the result easily by verification. ∎

From the above lemma it follows that, if $\boldsymbol{A}^{-1}$ is known, then the inverse of the matrix $\boldsymbol{A}$ augmented by a rank one matrix can be obtained by a modification of the matrix $\boldsymbol{A}^{-1}$.

Applying the above lemma twice to $\boldsymbol{B}_{k+1}$ yields

$$\begin{aligned} \boldsymbol{H}_{k+1}^{\mathrm{BFGS}} = \boldsymbol{H}_k &+ \left( 1 + \frac{\Delta \boldsymbol{g}^{(k)T} \boldsymbol{H}_k \Delta \boldsymbol{g}^{(k)}}{\Delta \boldsymbol{g}^{(k)T} \Delta \boldsymbol{x}^{(k)}} \right) \frac{\Delta \boldsymbol{x}^{(k)} \Delta \boldsymbol{x}^{(k)T}}{\Delta \boldsymbol{x}^{(k)T} \Delta \boldsymbol{g}^{(k)}} \\ &- \frac{\boldsymbol{H}_k \Delta \boldsymbol{g}^{(k)} \Delta \boldsymbol{x}^{(k)T} + (\boldsymbol{H}_k \Delta \boldsymbol{g}^{(k)} \Delta \boldsymbol{x}^{(k)T})^T}{\Delta \boldsymbol{g}^{(k)T} \Delta \boldsymbol{x}^{(k)}}. \end{aligned}$$

The above represents the BFGS formula for updating $\boldsymbol{H}_k$.

Recall that for the quadratic case, the DFP algorithm satisfies $\boldsymbol{H}_{k+1}^{\mathrm{DFP}}\Delta\boldsymbol{g}^{(i)} = \Delta\boldsymbol{x}^{(i)}$, $0 \leqslant i \leqslant k$. Therefore, the BFGS update for $\boldsymbol{B}_k$ satisfies $\boldsymbol{B}_{k+1}\Delta\boldsymbol{x}^{(i)} = \Delta\boldsymbol{g}^{(i)}$, $0 \leqslant i \leqslant k$. By construction of the BFGS formula for $\boldsymbol{H}_{k+1}^{\mathrm{BFGS}}$, we conclude that $\boldsymbol{H}_{k+1}^{\mathrm{BFGS}}\Delta\boldsymbol{g}^{(i)} = \Delta\boldsymbol{x}^{(i)}$, $0 \leqslant i \leqslant k$. Hence, the BFGS algorithm enjoys all the properties of quasi-Newton methods, including the conjugate direction property.

The BFGS update is reasonably robust when the line searches are sloppy (see Ref. 12). This property allows us to save time in the line search part of the algorithm. The BFGS formula is often far more efficient than the DFP formula (see Ref. 60 for further discussion).

We conclude our discussion of the BFGS algorithm with the following numerical example.

***Example 11.3.*** Use the BFGS method to minimize

$$f(\boldsymbol{x}) = \tfrac{1}{2}\boldsymbol{x}^T\boldsymbol{Q}\boldsymbol{x} - \boldsymbol{x}^T\boldsymbol{b} + \log(\pi),$$

where

$$\boldsymbol{Q} = \begin{bmatrix} 5 & -3 \\ -3 & 2 \end{bmatrix}, \quad \boldsymbol{b} = \begin{bmatrix} 0 \\ 1 \end{bmatrix}.$$

Take $\boldsymbol{H}_0 = \boldsymbol{I}_2$ and $\boldsymbol{x}^{(0)} = [0, 0]^T$. Verify that $\boldsymbol{H}_2 = \boldsymbol{Q}^{-1}$.

We have

$$\boldsymbol{d}^{(0)} = -\boldsymbol{g}^{(0)} = -(\boldsymbol{Q}\boldsymbol{x}^{(0)} - \boldsymbol{b}) = \boldsymbol{b} = \begin{bmatrix} 0 \\ 1 \end{bmatrix}.$$

The objective function is a quadratic, and hence we can use the following formula to compute $\alpha_0$:

$$\alpha_0 = -\frac{\boldsymbol{g}^{(0)T}\boldsymbol{d}^{(0)}}{\boldsymbol{d}^{(0)T}\boldsymbol{Q}\boldsymbol{d}^{(0)}} = \tfrac{1}{2}.$$

Therefore,

$$\boldsymbol{x}^{(1)} = \boldsymbol{x}^{(0)} + \alpha_0\boldsymbol{d}^{(0)} = \begin{bmatrix} 0 \\ \frac{1}{2} \end{bmatrix}.$$

To compute $\boldsymbol{H}_1 = \boldsymbol{H}_1^{\mathrm{BFGS}}$, we need the following quantities:

$$\Delta\boldsymbol{x}^{(0)} = \boldsymbol{x}^{(1)} - \boldsymbol{x}^{(0)} = \begin{bmatrix} 0 \\ \frac{1}{2} \end{bmatrix},$$

$$\boldsymbol{g}^{(1)} = \boldsymbol{Q}\boldsymbol{x}^{(1)} - \boldsymbol{b} = \begin{bmatrix} -\frac{3}{2} \\ 0 \end{bmatrix},$$

$$\Delta\boldsymbol{g}^{(0)} = \boldsymbol{g}^{(1)} - \boldsymbol{g}^{(0)} = \begin{bmatrix} -\frac{3}{2} \\ 1 \end{bmatrix}.$$

Therefore,

$$H_1 = H_0 + \left(1 + \frac{\Delta g^{(0)T} H_0 \Delta g^{(0)}}{\Delta g^{(0)T} \Delta x^{(0)}}\right) \frac{\Delta x^{(0)} \Delta x^{(0)T}}{\Delta x^{(0)T} \Delta g^{(0)}} - \frac{\Delta x^{(0)} \Delta g^{(0)T} H_0 + H_0 \Delta g^{(0)} \Delta x^{(0)T}}{\Delta g^{(0)T} \Delta x^{(0)}}$$

$$= \begin{bmatrix} 1 & \frac{3}{2} \\ \frac{3}{2} & \frac{11}{4} \end{bmatrix}.$$

Hence, we have

$$d^{(1)} = -H_1 g^{(1)} = \begin{bmatrix} \frac{3}{2} \\ \frac{9}{4} \end{bmatrix},$$

$$\alpha_1 = -\frac{g^{(1)T} d^{(1)}}{d^{(1)T} Q d^{(1)}} = 2.$$

Therefore,

$$x^{(2)} = x^{(1)} + \alpha_1 d^{(1)} = \begin{bmatrix} 3 \\ 5 \end{bmatrix}.$$

Since our objective function is a quadratic on $\mathbb{R}^2$, then $x^{(2)}$ is the minimizer. Notice that the gradient at $x^{(2)}$ is $\mathbf{0}$ [that is, $g^{(2)} = \mathbf{0}$].

To verify that $H_2 = Q^{-1}$, we compute:

$$\Delta x^{(1)} = x^{(2)} - x^{(1)} = \begin{bmatrix} 3 \\ \frac{9}{2} \end{bmatrix},$$

$$\Delta g^{(1)} = g^{(2)} - g^{(1)} = \begin{bmatrix} \frac{3}{2} \\ 0 \end{bmatrix}.$$

Hence,

$$H_2 = H_1 + \left(1 + \frac{\Delta g^{(1)T} H_1 \Delta g^{(1)}}{\Delta g^{(1)T} \Delta x^{(1)}}\right) \frac{\Delta x^{(1)} \Delta x^{(1)T}}{\Delta x^{(1)T} \Delta g^{(1)}} - \frac{\Delta x^{(1)} \Delta g^{(1)T} H_1 + H_1 \Delta g^{(1)} \Delta x^{(1)T}}{\Delta g^{(1)T} \Delta x^{(1)}}$$

$$= \begin{bmatrix} 2 & 3 \\ 3 & 5 \end{bmatrix}.$$

Note that indeed $H_2 Q = Q H_2 = I_2$, and hence $H_2 = Q^{-1}$. ■

For nonquadratic problems, the quasi-Newton algorithm will not usually converge in $n$ steps. As in the case of the conjugate gradient methods, here too some modifications may be necessary to deal with nonquadratic problems. For example, we may reinitialize the direction vector to the negative gradient after every few iterations (for example, $n$ or $n+1$), and continue until the algorithm satisfies the stopping criterion.

## EXERCISES

**11.1** Given $f:\mathbb{R}^n \to \mathbb{R}$, $f \in \mathscr{C}^1$, consider the algorithm

$$x^{(k+1)} = x^{(k)} + \alpha_k d^{(k)}$$

where $d^{(1)}, d^{(2)}, \ldots$ are vectors in $\mathbb{R}^n$, and $\alpha_k \geqslant 0$ is chosen to minimize $f(x^{(k)} + \alpha d^{(k)})$; that is,

$$\alpha_k = \arg\min_{\alpha \geqslant 0} f(x^{(k)} + \alpha d^{(k)}).$$

Note that the above general algorithm encompasses almost all algorithms that we discussed in this part, including the steepest descent, Newton, conjugate gradient, and quasi-Newton algorithms.

Let $g^{(k)} = \nabla f(x^{(k)})$, and assume that $d^{(k)T} g^{(k)} < 0$.

**a.** Show that $d^{(k)}$ is a descent direction for $f$, in the sense that there exists $\bar{\alpha} > 0$ such that for all $\alpha \in (0, \bar{\alpha}]$,

$$f(x^{(k)} + \alpha d^{(k)}) < f(x^{(k)}).$$

**b.** Show that $\alpha_k > 0$.

**c.** Show that $d^{(k)T} g^{(k+1)} = 0$.

**d.** Show that the following algorithms all satisfy the condition $d^{(k)T} g^{(k)} < 0$, if $g^{(k)} \neq 0$:

i. Steepest descent algorithm
ii. Newton's method, assuming the Hessian is positive definite
iii. Conjugate gradient algorithm
iv. Quasi-Newton algorithm, assuming $H_k > 0$

**e.** For the case where $f(x) = \frac{1}{2} x^T Q x - x^T b$, with $Q = Q^T > 0$, derive an expression for $\alpha_k$ in terms of $Q$, $d^{(k)}$, and $g^{(k)}$.

**11.2** Consider the rank one algorithm. Assume that $H_k > 0$. Show that if $\Delta g^{(k)T}(\Delta x^{(k)} - H_k \Delta g^{(k)}) > 0$, then $H_{k+1} > 0$. On the other hand, show that, if $\Delta g^{(k)T}(\Delta x^{(k)} - H_k \Delta g^{(k)}) < 0$, then $H_{k+1}$ may not be positive definite.

**11.3** Based on the rank one update equation, derive an update formula using complementarity and the matrix inverse formula.

**11.4** Consider the DFP algorithm applied to the quadratic function

$$f(\boldsymbol{x}) = \tfrac{1}{2}\boldsymbol{x}^T\boldsymbol{Q}\boldsymbol{x} - \boldsymbol{x}^T\boldsymbol{b},$$

where $\boldsymbol{Q} = \boldsymbol{Q}^T > 0$.

**a.** Write down a formula for $\alpha_k$ in terms of $\boldsymbol{Q}, \boldsymbol{g}^{(k)}$, and $\boldsymbol{d}^{(k)}$.
**b.** Show that if $\boldsymbol{g}^{(k)} \neq \boldsymbol{0}$, then $\alpha_k > 0$.

**11.5** Assuming exact line search, show that if $\boldsymbol{H}_0 = \boldsymbol{I}_n$ ($n \times n$ identity matrix), then the first two steps of the BFGS algorithm yield the same points $\boldsymbol{x}^{(1)}$ and $\boldsymbol{x}^{(2)}$ as conjugate gradient algorithms with the Hestenes–Stiefel, the Polak–Ribiere, as well as the Fletcher–Reeves formulas.

**11.6** Write a MATLAB routine to implement the quasi-Newton algorithm for general functions. Use the secant method for the line search (for example, the MATLAB function of Exercise 7.8). Test the different update formulas for $\boldsymbol{H}_k$ on Rosenbrock's function (see Exercise 9.3), with an initial condition $\boldsymbol{x}^{(0)} = [-2, 2]^T$. For this exercise, reinitialize the update direction to the negative gradient every six iterations.

**11.7** Consider the function

$$f(\boldsymbol{x}) = \frac{x_1^4}{4} + \frac{x_2^2}{2} - x_1x_2 + x_1 - x_2.$$

**a.** Use MATLAB to plot the level sets of $f$ at levels $-0.72, -0.6, -0.2, 0.5, 2$. Locate the minimizers of $f$ from the plots of the level sets.
**b.** Apply the DFP algorithm to minimize the above function with the following starting initial conditions: (i) $[0, 0]^T$; (ii) $[1.5, 1]^T$. Use $\boldsymbol{H}_0 = \boldsymbol{I}_2$. Does the algorithm converge to the same point for the two initial conditions? If not, explain.

# 12

# Solving $Ax = b$

## 12.1. LEAST-SQUARES ANALYSIS

Consider a system of linear equations

$$Ax = b,$$

where $A \in \mathbb{R}^{m \times n}$, $b \in \mathbb{R}^m$, $m \geqslant n$, and rank $A = n$. Note that the number of unknowns, $n$, is no larger than the number of equations, $m$. If $b$ does not belong to the range of $A$, that is, if $b \notin \mathcal{R}(A)$, then this system of equations is said to be inconsistent or overdetermined. In this case, there is no solution to the above set of equations. Our goal, then, is to find the vector (or vectors) $x$ minimizing $\|Ax - b\|^2$.

Let $x^*$ be a vector that minimizes $\|Ax - b\|^2$; that is, for all $x \in \mathbb{R}^n$,

$$\|Ax - b\|^2 \geqslant \|Ax^* - b\|^2.$$

We refer to the vector $x^*$ as a *least-squares solution* to $Ax = b$. In the case where $Ax = b$ has a solution, then the solution is a least-squares solution. Otherwise, a least-squares solution minimizes the norm of the difference between the left- and right-hand sides of the equation $Ax = b$. To characterize least-squares solutions, we need the following lemma.

**Lemma 12.1.** *Let* $A \in \mathbb{R}^{m \times n}$, $m \geqslant n$. *Then,* rank $A = n$ *if, and only if,* rank $A^T A = n$ *(i.e., the square matrix* $A^T A$ *is nonsingular).* □

*Proof.* ⇒: Suppose rank $A = n$. To show rank $A^T A = n$, it is equivalent to show $\mathcal{N}(A^T A) = \{0\}$. To proceed, let $x \in \mathcal{N}(A^T A)$, i.e., $A^T A x = 0$. Therefore,

$$\|Ax\|^2 = x^T A^T A x = 0,$$

which implies that $Ax = 0$. Since rank $A = n$, we have $x = 0$.
⇐: Suppose rank $A^T A = n$, that is, $\mathcal{N}(A^T A) = \{0\}$. To show rank $A = n$, it is equivalent to show that $\mathcal{N}(A) = \{0\}$. To proceed, let $x \in \mathcal{N}(A)$, i.e., $Ax = 0$. Then, $A^T A x = 0$, and hence $x = 0$. ■

Recall that we assume throughout that rank $A = n$. By the above lemma, we conclude that $(A^TA)^{-1}$ exists. The following theorem characterizes the least-squares solution.

**Theorem 12.1.** *There exists a unique vector $x^*$ that minimizes $\|Ax - b\|^2$, given by the solution to the equation $A^TAx = A^Tb$, that is, $x^* = (A^TA)^{-1}A^Tb$.* □

*Proof.* Let $x^* = (A^TA)^{-1}A^Tb$. First, observe that

$$\begin{aligned}\|Ax - b\|^2 &= \|A(x - x^*) + (Ax^* - b)\|^2 \\ &= (A(x - x^*) + (Ax^* - b))^T(A(x - x^*) + (Ax^* - b)) \\ &= \|A(x - x^*)\|^2 + \|Ax^* - b\|^2 + 2[A(x - x^*)]^T(Ax^* - b).\end{aligned}$$

We now show that the last term in the above equation is zero. Indeed, substituting the above expression for $x^*$,

$$\begin{aligned}[A(x - x^*)]^T(Ax^* - b) &= (x - x^*)^TA^T[A(A^TA)^{-1}A^T - I_n]b \\ &= (x - x^*)^T[(A^TA)(A^TA)^{-1}A^T - A^T]b \\ &= (x - x^*)^T(A^T - A^T)b \\ &= 0.\end{aligned}$$

Hence

$$\|Ax - b\|^2 = \|A(x - x^*)\|^2 + \|Ax^* - b\|^2.$$

If $x \neq x^*$, then $\|A(x - x^*)\|^2 > 0$, since rank $A = n$. Thus, if $x \neq x^*$, we have

$$\|Ax - b\|^2 > \|Ax^* - b\|^2.$$

Thus, $x^* = (A^TA)^{-1}A^Tb$ is the unique minimizer of $\|Ax - b\|^2$. ■

We now give a geometric interpretation of the above theorem. First note that the columns of $A$ span the range $\mathscr{R}(A)$ of $A$, which is a $n$-dimensional subspace of $\mathbb{R}^m$. The equation $Ax = b$ has a solution if, and only if, $b$ lies in this $n$-dimensional subspace $\mathscr{R}(A)$. If $m = n$, then $b \in \mathscr{R}(A)$ always, and the solution is $x^* = A^{-1}b$. Suppose now that $m > n$. Intuitively, we would expect the "likelihood" of $b \in \mathscr{R}(A)$ to be small, since the subspace spanned by the columns of $A$ is very "thin". Therefore, let us suppose that $b$ does not belong to $\mathscr{R}(A)$. We wish to find a point $h \in \mathscr{R}(A)$ that is "closest" to $b$. Geometrically, the point $h$ should be such that the vector $e = h - b$ is orthogonal to the subspace $\mathscr{R}(A)$ (see Figure 12.1). Recall that a vector $e \in \mathbb{R}^m$ is said to be orthogonal to the subspace $\mathscr{R}(A)$ if it is orthogonal to every vector in this subspace. We call $h$ the *orthogonal projection* of $b$ onto the subspace $\mathscr{R}(A)$. It turns out that $h = Ax^* = A(A^TA)^{-1}A^Tb$. Hence, the vector $h \in \mathscr{R}(A)$ minimizing $\|b - h\|$ is exactly the orthogonal projection of $b$ onto $\mathscr{R}(A)$.

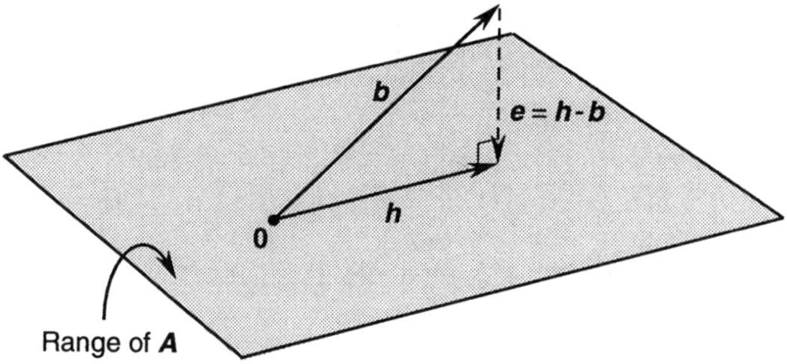


**Figure 12.1.** Orthogonal projection of $\boldsymbol{b}$ on the subspace $\mathcal{R}(A)$

In other words, the vector $\boldsymbol{x}^*$ minimizing $\|\boldsymbol{Ax} - \boldsymbol{b}\|$ is exactly the vector that makes $\boldsymbol{Ax} - \boldsymbol{b}$ orthogonal to $\mathcal{R}(\boldsymbol{A})$.

To proceed further, we write $\boldsymbol{A} = [\boldsymbol{a}_1, \ldots, \boldsymbol{a}_n]$, where $\boldsymbol{a}_1, \ldots, \boldsymbol{a}_n$ are the columns of $\boldsymbol{A}$. The vector $\boldsymbol{e}$ is orthogonal to $\mathcal{R}(\boldsymbol{A})$ if, and only if, it is orthogonal to each of the columns $\boldsymbol{a}_1, \ldots, \boldsymbol{a}_n$ of $\boldsymbol{A}$. To see that, note that

$$\langle \boldsymbol{e}, \boldsymbol{a}_i \rangle = 0, \quad i = 1, \ldots, n$$

if, and only if, for any set of scalars $\{x_1, x_2, \ldots, x_n\}$, we also have

$$\langle \boldsymbol{e}, x_1\boldsymbol{a}_1 + \cdots + x_n\boldsymbol{a}_n \rangle = 0.$$

Any vector in $\mathcal{R}(\boldsymbol{A})$ has the form $x_1\boldsymbol{a}_1 + \cdots + x_n\boldsymbol{a}_n$.

**Proposition 12.1.** *Let $\boldsymbol{h} \in \mathcal{R}(\boldsymbol{A})$ be such that $\boldsymbol{h} - \boldsymbol{b}$ is orthogonal to $\mathcal{R}(\boldsymbol{A})$. Then, $\boldsymbol{h} = \boldsymbol{Ax}^* = \boldsymbol{A}(\boldsymbol{A}^T\boldsymbol{A})^{-1}\boldsymbol{A}^T\boldsymbol{b}$.* □

*Proof.* Since $\boldsymbol{h} \in \mathcal{R}(\boldsymbol{A}) = \operatorname{span}[\boldsymbol{a}_1, \ldots, \boldsymbol{a}_n]$, it has the form $\boldsymbol{h} = x_1\boldsymbol{a}_1 + \cdots + x_n\boldsymbol{a}_n$, where $x_1, \ldots, x_n \in \mathbb{R}$. To find $x_1, \ldots, x_n$, we use the assumption that $\boldsymbol{e} = \boldsymbol{h} - \boldsymbol{b}$ is orthogonal to $\operatorname{span}[\boldsymbol{a}_1, \ldots, \boldsymbol{a}_n]$, that is, for all $i = 1, \ldots, n$, we have

$$\langle \boldsymbol{h} - \boldsymbol{b}, \boldsymbol{a}_i \rangle = 0,$$

or, equivalently,

$$\langle \boldsymbol{h}, \boldsymbol{a}_i \rangle = \langle \boldsymbol{b}, \boldsymbol{a}_i \rangle.$$

Substituting $\boldsymbol{h}$ into the above equations, we obtain a set of $n$ linear equations of the form

$$\langle \boldsymbol{a}_1, \boldsymbol{a}_i \rangle x_1 + \cdots + \langle \boldsymbol{a}_n, \boldsymbol{a}_i \rangle x_n = \langle \boldsymbol{b}, \boldsymbol{a}_i \rangle, \quad i = 1, \ldots, n.$$

In matrix notation this system of $n$ equations can be represented as

$$\begin{bmatrix} \langle \boldsymbol{a}_1, \boldsymbol{a}_1 \rangle & \cdots & \langle \boldsymbol{a}_n, \boldsymbol{a}_1 \rangle \\ \vdots & & \vdots \\ \langle \boldsymbol{a}_1, \boldsymbol{a}_n \rangle & \cdots & \langle \boldsymbol{a}_n, \boldsymbol{a}_n \rangle \end{bmatrix} \begin{bmatrix} x_1 \\ \vdots \\ x_n \end{bmatrix} = \begin{bmatrix} \langle \boldsymbol{b}, \boldsymbol{a}_1 \rangle \\ \vdots \\ \langle \boldsymbol{b}, \boldsymbol{a}_n \rangle \end{bmatrix}.$$

Note that we can write

$$\begin{bmatrix} \langle \boldsymbol{a}_1, \boldsymbol{a}_1 \rangle & \cdots & \langle \boldsymbol{a}_n, \boldsymbol{a}_1 \rangle \\ \vdots & & \vdots \\ \langle \boldsymbol{a}_1, \boldsymbol{a}_n \rangle & \cdots & \langle \boldsymbol{a}_n, \boldsymbol{a}_n \rangle \end{bmatrix} = A^T A = \begin{bmatrix} \boldsymbol{a}_1^T \\ \vdots \\ \boldsymbol{a}_n^T \end{bmatrix} [\boldsymbol{a}_1 \cdots \boldsymbol{a}_n].$$

We also note that

$$\begin{bmatrix} \langle \boldsymbol{b}, \boldsymbol{a}_1 \rangle \\ \vdots \\ \langle \boldsymbol{b}, \boldsymbol{a}_n \rangle \end{bmatrix} = A^T \boldsymbol{b} = \begin{bmatrix} \boldsymbol{a}_1^T \\ \vdots \\ \boldsymbol{a}_n^T \end{bmatrix} \boldsymbol{b}.$$

Since rank $A = n$, then $A^T A$ is nonsingular, and thus we conclude that

$$\boldsymbol{x} = \begin{bmatrix} x_1 \\ \vdots \\ x_n \end{bmatrix} = (A^T A)^{-1} A^T \boldsymbol{b} = \boldsymbol{x}^*.$$

■

In the above proof, we noticed that the matrix

$$A^T A = \begin{bmatrix} \langle \boldsymbol{a}_1, \boldsymbol{a}_1 \rangle & \cdots & \langle \boldsymbol{a}_n, \boldsymbol{a}_1 \rangle \\ \vdots & & \vdots \\ \langle \boldsymbol{a}_1, \boldsymbol{a}_n \rangle & \cdots & \langle \boldsymbol{a}_n, \boldsymbol{a}_n \rangle \end{bmatrix}$$

plays an important role. This matrix is often called the *Gram matrix*.

An alternative method of arriving at the least-squares solution is to proceed as follows. First, we write

$$\begin{aligned} f(\boldsymbol{x}) &= \|A\boldsymbol{x} - \boldsymbol{b}\|^2 \\ &= (A\boldsymbol{x} - \boldsymbol{b})^T (A\boldsymbol{x} - \boldsymbol{b}) \\ &= \tfrac{1}{2}\boldsymbol{x}^T (2A^T A)\boldsymbol{x} - \boldsymbol{x}^T (2A^T \boldsymbol{b}) + \boldsymbol{b}^T \boldsymbol{b}. \end{aligned}$$

Therefore, $f$ is a quadratic function. The quadratic term is positive definite because rank $A = n$. Thus, the unique minimizer of $f$ is obtained by solving the FONC (see Exercise 6.14); that is,

$$\nabla f(\boldsymbol{x}) = 2A^T A\boldsymbol{x} - 2A^T \boldsymbol{b} = \boldsymbol{0}.$$

The only solution to the equaion $\nabla f(\boldsymbol{x}) = \boldsymbol{0}$ is $\boldsymbol{x}^* = (A^T A)^{-1} A^T \boldsymbol{b}$.

**Table 12.1 Experimental Data for Example 12.1**

| $i$ | 0 | 1 | 2 |
|---|---|---|---|
| $t_i$ | 2 | 3 | 4 |
| $y_i$ | 3 | 4 | 15 |

We now give an example in which least-squares analysis is used to fit measurements by a straight line.

***Example 12.1.*** Suppose that a process has a single input $t \in \mathbb{R}$ and a single output $y \in \mathbb{R}$. Suppose we perform an experiment on the process, resulting in a number of measurements, as displayed in Table 12.1. The $i$th measurement results in the input labeled $t_i$ and the output labeled $y_i$. We would like to find a straight line given by

$$y = mt + c$$

that fits the experimental data. In other words, we wish to find two numbers, $m$ and $c$, such that $y_i = mt_i + c, i = 0, 1, 2$. However, it is apparent that there is no choice of $m$ and $c$ that results in the above requirement; that is, there is no straight line that passes through all three points simultaneously. Therefore, we would like to find the values of $m$ and $c$ that "best fit" the data. A graphical illustration of our problem is shown in Figure 12.2.

We can represent our problem as a system of three linear equations of the form:

$$\begin{aligned} 2m + c &= 3 \\ 3m + c &= 4 \\ 4m + c &= 15. \end{aligned}$$

We can write the above system of equation as

$$\boldsymbol{Ax} = \boldsymbol{b},$$

where

$$\boldsymbol{A} = \begin{bmatrix} 2 & 1 \\ 3 & 1 \\ 4 & 1 \end{bmatrix}, \quad \boldsymbol{b} = \begin{bmatrix} 3 \\ 4 \\ 15 \end{bmatrix}, \quad \boldsymbol{x} = \begin{bmatrix} m \\ c \end{bmatrix}.$$

Note that since

$$\operatorname{rank} \boldsymbol{A} < \operatorname{rank}[\boldsymbol{A}, \boldsymbol{b}],$$

the vector $\boldsymbol{b}$ does not belong to the range of $\boldsymbol{A}$. Thus, as we have noted before, the above system of equations is inconsistent.

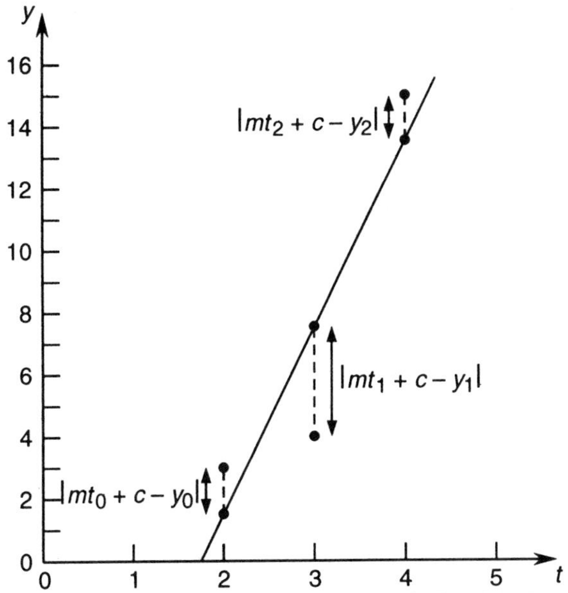


**Figure 12.2.** Fitting a straight line to experimental data

The straight line of best fit is the one that minimizes

$$\|\boldsymbol{Ax}-\boldsymbol{b}\|^2=\sum_{i=0}^{2}(mt_i+c-y_i)^2.$$

Therefore, our problem lies in the class of least-squares problems. Note that the above function of $m$ and $c$ is simply the total squared vertical distance (squared error) between the straight line defined by $m$ and $c$ and the experimental points. The solution to our least-squares problem is

$$\boldsymbol{x}^*=\begin{bmatrix} m^* \\ c^* \end{bmatrix}=(\boldsymbol{A}^T\boldsymbol{A})^{-1}\boldsymbol{A}^T\boldsymbol{b}=\begin{bmatrix} 6 \\ -32/3 \end{bmatrix}.$$

Note that the error vector $\boldsymbol{e}=\boldsymbol{Ax}^*-\boldsymbol{b}$ is orthogonal to each column of $\boldsymbol{A}$. ■

We now give a simple example where the least-squares method is used in digital signal processing.

***Example 12.2.*** *Discrete Fourier Series.* Suppose we are given a "discrete-time signal," represented by the vector

$$\boldsymbol{b}=[b_1,b_2,\dots,b_m]^T.$$

We wish to approximate the above signal by a sum of sinusoids. Specifically, we approximate $\boldsymbol{b}$ by the vector

$$y_0\boldsymbol{c}^{(0)} + \sum_{k=1}^{n} (y_k\boldsymbol{c}^{(k)} + z_k\boldsymbol{s}^{(k)}),$$

where $y_0, y_1, \ldots, y_n, z_1, \ldots, z_n \in \mathbb{R}$, and the vectors $\boldsymbol{c}^{(k)}$ and $\boldsymbol{s}^{(k)}$ are given by

$$\boldsymbol{c}^{(0)} = \left[\frac{1}{\sqrt{2}}, \frac{1}{\sqrt{2}}, \ldots, \frac{1}{\sqrt{2}}\right]^T,$$

$$\boldsymbol{c}^{(k)} = \left[\cos\left(1\frac{2k\pi}{m}\right), \cos\left(2\frac{2k\pi}{m}\right), \ldots, \cos\left(m\frac{2k\pi}{m}\right)\right]^T, \quad k = 1, \ldots, n,$$

$$\boldsymbol{s}^{(k)} = \left[\sin\left(1\frac{2k\pi}{m}\right), \sin\left(2\frac{2k\pi}{m}\right), \ldots, \sin\left(m\frac{2k\pi}{m}\right)\right]^T, \quad k = 1, \ldots, n.$$

We call the above sum of sinusoids a *discrete Fourier series* (although, strictly speaking, it is not a series but a finite sum). We wish to find $y_0, y_1, \ldots, y_n, z_1, \ldots, z_n$ such that

$$\left\| \left( y_0\boldsymbol{c}^{(0)} + \sum_{k=1}^{n} y_k\boldsymbol{c}^{(k)} + z_k\boldsymbol{s}^{(k)} \right) - \boldsymbol{b} \right\|^2$$

is minimized.

To proceed, we define

$$\boldsymbol{A} = [\boldsymbol{c}^{(0)}, \boldsymbol{c}^{(1)}, \ldots, \boldsymbol{c}^{(n)}, \boldsymbol{s}^{(1)}, \ldots, \boldsymbol{s}^{(n)}]$$

$$\boldsymbol{x} = [y_0, y_1, \ldots, y_n, z_1, \ldots, z_n]^T.$$

Our problem can be reformulated as minimizing

$$\|\boldsymbol{A}\boldsymbol{x} - \boldsymbol{b}\|^2.$$

We assume that $m \geqslant 2n + 1$. To find the solution, we first compute $\boldsymbol{A}^T\boldsymbol{A}$. We make use of the following trigonometric identities: for any nonzero integer $k$ that is not an integral multiple of $m$, we have

$$\sum_{i=1}^{m} \cos\left(i\frac{2k\pi}{m}\right) = 0$$

$$\sum_{i=1}^{m} \sin\left(i\frac{2k\pi}{m}\right) = 0.$$

With the aid of the above identities, we can verify that

$$c^{(k)T}c^{(j)} = \begin{cases} m/2, & \text{if } k=j \\ 0, & \text{otherwise} \end{cases}$$

$$s^{(k)T}s^{(j)} = \begin{cases} m/2, & \text{if } k=j \\ 0, & \text{otherwise} \end{cases}$$

$$c^{(k)T}s^{(j)} = 0, \quad \text{for any } k, j.$$

Hence,

$$A^T A = \frac{m}{2} I_{2n+1}$$

which is clearly nonsingular, with inverse

$$(A^T A)^{-1} = \frac{2}{m} I_{2n+1}.$$

Therefore, the solution to our problem is

$$\begin{aligned} x^* &= [y_0^*, y_1^*, \ldots, y_n^*, z_1^*, \ldots, z_n^*]^T \\ &= (A^T A)^{-1} A^T b \\ &= \frac{2}{m} A^T b. \end{aligned}$$

We represent the above solution as

$$y_0^* = \frac{\sqrt{2}}{m} \sum_{i=1}^{m} b_i$$

$$y_k^* = \frac{2}{m} \sum_{i=1}^{m} b_i \cos\left( i \frac{2k\pi}{m} \right), \quad k = 1, \ldots, n$$

$$z_k^* = \frac{2}{m} \sum_{i=1}^{m} b_i \sin\left( i \frac{2k\pi}{m} \right), \quad k = 1, \ldots, n.$$

We call the above *discrete Fourier coefficients.* ■

## 12.2. RECURSIVE LEAST-SQUARES ALGORITHM

Consider again the example in the last section. We are given experimental points $(t_0, y_0)$, $(t_1, y_1)$, and $(t_2, y_2)$, and we find the parameters $m^*$ and $c^*$ of the straight line that best fits these data in the least-squares sense. Suppose we are now given an extra measurement point $(t_3, y_3)$, so that we now have a set of four experimental data points, $(t_0, y_0), (t_1, y_1), (t_2, y_2)$, and $(t_3, y_3)$. We can similarly go through the procedure for finding the parameters of the line of best fit for this set of four points. However, as we shall see, there is a more efficient way: we can use previous calculations of $m^*$ and $c^*$ for the three data points to calculate the parameters for the four data points. In effect, we simply "update" our values of $m^*$ and $c^*$ to accommodate the new data point. This procedure is called the *recursive least-squares* (RLS) algorithm, which is the topic of this section.

To derive the RLS algorithm, first consider the problem of minimizing $\|\boldsymbol{A}_0\boldsymbol{x} - \boldsymbol{b}^{(0)}\|^2$. We know that the solution to this is given by $\boldsymbol{x}^{(0)} = \boldsymbol{G}_0^{-1}\boldsymbol{A}_0^T\boldsymbol{b}^{(0)}$, where $\boldsymbol{G}_0 = \boldsymbol{A}_0^T\boldsymbol{A}_0$. Suppose now that we are given more data, in the form of a matrix $\boldsymbol{A}_1$ and a vector $\boldsymbol{b}^{(1)}$. Consider now the problem of minimizing

$$\left\|\begin{bmatrix}\boldsymbol{A}_0\\ \boldsymbol{A}_1\end{bmatrix}\boldsymbol{x} - \begin{bmatrix}\boldsymbol{b}^{(0)}\\ \boldsymbol{b}^{(1)}\end{bmatrix}\right\|^2.$$

The solution is given by

$$\boldsymbol{x}^{(1)} = \boldsymbol{G}_1^{-1}\begin{bmatrix}\boldsymbol{A}_0\\ \boldsymbol{A}_1\end{bmatrix}^T\begin{bmatrix}\boldsymbol{b}^{(0)}\\ \boldsymbol{b}^{(1)}\end{bmatrix},$$

where

$$\boldsymbol{G}_1 = \begin{bmatrix}\boldsymbol{A}_0\\ \boldsymbol{A}_1\end{bmatrix}^T\begin{bmatrix}\boldsymbol{A}_0\\ \boldsymbol{A}_1\end{bmatrix}.$$

Our goal is to write $\boldsymbol{x}^{(1)}$ as a function of $\boldsymbol{x}^{(0)}$, $\boldsymbol{G}_0$, and the new data $\boldsymbol{A}_1$ and $\boldsymbol{b}^{(1)}$. To this end, we first write $\boldsymbol{G}_1$ as

$$\begin{aligned}\boldsymbol{G}_1 &= [\boldsymbol{A}_0^T\ \boldsymbol{A}_1^T]\begin{bmatrix}\boldsymbol{A}_0\\ \boldsymbol{A}_1\end{bmatrix}\\ &= \boldsymbol{A}_0^T\boldsymbol{A}_0 + \boldsymbol{A}_1^T\boldsymbol{A}_1\\ &= \boldsymbol{G}_0 + \boldsymbol{A}_1^T\boldsymbol{A}_1.\end{aligned}$$

Next, we write

$$\begin{aligned}\begin{bmatrix}\boldsymbol{A}_0\\ \boldsymbol{A}_1\end{bmatrix}^T\begin{bmatrix}\boldsymbol{b}^{(0)}\\ \boldsymbol{b}^{(1)}\end{bmatrix} &= [\boldsymbol{A}_0^T\ \boldsymbol{A}_1^T]\begin{bmatrix}\boldsymbol{b}^{(0)}\\ \boldsymbol{b}^{(1)}\end{bmatrix}\\ &= \boldsymbol{A}_0^T\boldsymbol{b}^{(0)} + \boldsymbol{A}_1^T\boldsymbol{b}^{(1)}.\end{aligned}$$

To proceed further, we write $\boldsymbol{A}_0^T\boldsymbol{b}^{(0)}$ as

$$\begin{aligned}\boldsymbol{A}_0^T\boldsymbol{b}^{(0)} &= \boldsymbol{G}_0\boldsymbol{G}_0^{-1}\boldsymbol{A}_0^T\boldsymbol{b}^{(0)}\\ &= \boldsymbol{G}_0\boldsymbol{x}^{(0)}\\ &= (\boldsymbol{G}_1 - \boldsymbol{A}_1^T\boldsymbol{A}_1)\boldsymbol{x}^{(0)}\\ &= \boldsymbol{G}_1\boldsymbol{x}^{(0)} - \boldsymbol{A}_1^T\boldsymbol{A}_1\boldsymbol{x}^{(0)}.\end{aligned}$$

Combining the above formulas, we see that we can write $x^{(1)}$ as

$$\begin{aligned} x^{(1)} &= G_1^{-1}\begin{bmatrix} A_0 \\ A_1 \end{bmatrix}^T \begin{bmatrix} b^{(0)} \\ b^{(1)} \end{bmatrix} \\ &= G_1^{-1}(G_1 x^{(0)} - A_1^T A_1 x^{(0)} + A_1^T b^{(1)}) \\ &= x^{(0)} + G_1^{-1} A_1^T (b^{(1)} - A_1 x^{(0)}), \end{aligned}$$

where $G_1$ can be calculated using

$$G_1 = G_0 + A_1^T A_1.$$

We note that with the above formula, $x^{(1)}$ can be computed using only $x^{(0)}$, $A_1$, $b^{(1)}$, and $G_0$. Hence, we have a way of using our previous efforts in calculating $x^{(0)}$ to compute $x^{(1)}$, without having to directly compute $x^{(1)}$ from scratch. The solution $x^{(1)}$ is obtained from $x^{(0)}$ by a simple update equation that adds to $x^{(0)}$ a correction term $G_1^{-1} A_1^T (b^{(1)} - A_1 x^{(0)})$. Observe that, if the new data are consistent with the old data [that is, $A_1 x^{(0)} = b^{(1)}$], then the correction term is $\mathbf{0}$, and the updated solution $x^{(1)}$ is equal to the previous solution $x^{(0)}$.

We can generalize the above argument to write a recursive algorithm for updating the least-squares solution as new data arrive. At the $(k+1)$st iteration, we have

$$\begin{aligned} G_{k+1} &= G_k + A_{k+1}^T A_{k+1} \\ x^{(k+1)} &= x^{(k)} + G_{k+1}^{-1} A_{k+1}^T (b^{(k+1)} - A_{k+1} x^{(k)}). \end{aligned}$$

The vector $b^{(k+1)} - A_{k+1} x^{(k)}$ is often called the *innovation*. As before, observe that if the innovation is zero, then the updated solution $x^{(k+1)}$ is equal to the previous solution $x^{(k)}$.

We can see from the above that, to compute $x^{(k+1)}$ from $x^{(k)}$, we need $G_{k+1}^{-1}$ rather than $G_{k+1}$. It turns out that we can derive an update formula for $G_{k+1}^{-1}$ itself. To do so, we need the following technical lemma, which is a generalization of the Sherman–Morrison formula (Lemma 11.1) [see Ref. 36, p. 124, or Ref. 29, p. 3].

**Lemma 12.2.** *Let $A$ be a nonsingular matrix. Let $U$ and $V$ be matrices such that $I + VA^{-1}U$ is nonsingular. Then $A + UV$ is nonsingular, and*

$$(A + UV)^{-1} = A^{-1} - (A^{-1}U)(I + VA^{-1}U)^{-1}(VA^{-1}).$$

□

*Proof.* We can prove the result easily by verification. ■

Using the above lemma, we get

$$\begin{aligned} G_{k+1}^{-1} &= (G_k + A_{k+1}^T A_{k+1})^{-1} \\ &= G_k^{-1} - G_k^{-1} A_{k+1}^T (I + A_{k+1} G_k^{-1} A_{k+1}^T)^{-1} A_{k+1} G_k^{-1}. \end{aligned}$$

For simplicity of notation, we rewrite $G_k^{-1}$ as $P_k$.

We summarize by writing the RLS algorithm using $P_k$.

$$\begin{aligned} P_{k+1} &= P_k - P_k A_{k+1}^T (I + A_{k+1} P_k A_{k+1}^T)^{-1} A_{k+1} P_k \\ x^{(k+1)} &= x^{(k)} + P_{k+1} A_{k+1}^T (b^{(k+1)} - A_{k+1} x^{(k)}). \end{aligned}$$

In the special case where the new data at each step are such that $A_{k+1}$ is a matrix consisting of a single row, $A_{k+1} = a_{k+1}^T$, and $b^{(k+1)}$ is a scalar, $b^{(k+1)} = b_{k+1}$, we get

$$\begin{aligned} P_{k+1} &= P_k - \frac{P_k a_{k+1} a_{k+1}^T P_k}{1 + a_{k+1}^T P_k a_{k+1}} \\ x^{(k+1)} &= x^{(k)} + P_{k+1} a_{k+1} (b_{k+1} - a_{k+1}^T x^{(k)}). \end{aligned}$$

*Example 12.3.* Let

$$A_0 = \begin{bmatrix} 1 & 0 \\ 0 & 1 \\ 1 & 1 \end{bmatrix}, \qquad b^{(0)} = \begin{bmatrix} 1 \\ 1 \\ 1 \end{bmatrix}$$

$$\begin{aligned} A_1 &= a_1^T = [2 \quad 1], \quad b^{(1)} = b_1 = [3] \\ A_2 &= a_2^T = [3 \quad 1], \quad b^{(2)} = b_2 = [4]. \end{aligned}$$

First compute the vector $x^{(0)}$ minimizing $\|A_0 x - b^{(0)}\|^2$. Then, use the RLS algorithm to find $x^{(2)}$ minimizing

$$\left\| \begin{bmatrix} A_0 \\ A_1 \\ A_2 \end{bmatrix} x - \begin{bmatrix} b^{(0)} \\ b^{(1)} \\ b^{(2)} \end{bmatrix} \right\|^2.$$

We have

$$P_0 = (A_0^T A_0)^{-1} = \begin{bmatrix} \frac{2}{3} & -\frac{1}{3} \\ -\frac{1}{3} & \frac{2}{3} \end{bmatrix}$$

$$x^{(0)} = P_0 A_0^T b^{(0)} = \begin{bmatrix} \frac{2}{3} \\ \frac{2}{3} \end{bmatrix}.$$

Applying the RLS algorithm twice, we get

$$P_1 = P_0 - \frac{P_0 a_1 a_1^T P_0}{1 + a_1^T P_0 a_1} = \begin{bmatrix} \frac{1}{3} & -\frac{1}{3} \\ -\frac{1}{3} & \frac{2}{3} \end{bmatrix}$$

$$x^{(1)} = x^{(0)} + P_1 a_1 (b_1 - a_1^T x^{(0)}) = \begin{bmatrix} 1 \\ \frac{2}{3} \end{bmatrix}$$

$$P_2 = P_1 - \frac{P_1 a_2 a_2^T P_1}{1 + a_2^T P_1 a_2} = \begin{bmatrix} \frac{1}{6} & -\frac{1}{4} \\ -\frac{1}{4} & \frac{5}{8} \end{bmatrix}$$

$$x^{(2)} = x^{(1)} + P_2 a_2 (b_2 - a_2^T x^{(1)}) = \begin{bmatrix} \frac{13}{12} \\ \frac{5}{8} \end{bmatrix}.$$

We can easily check our solution by directly computing $x^{(2)}$ using the formula $x^{(2)} = (A^T A)^{-1} A^T b$, where

$$A = \begin{bmatrix} A_0 \\ A_1 \\ A_2 \end{bmatrix}, \quad b = \begin{bmatrix} b^{(0)} \\ b^{(1)} \\ b^{(2)} \end{bmatrix}.$$

■

## 12.3. SOLUTION TO $Ax = b$ MINIMIZING $\|x\|$

Consider now a system of linear equations

$$Ax = b,$$

where $A \in \mathbb{R}^{m \times n}$, $b \in \mathbb{R}^m$, $m \leqslant n$, and rank $A = m$. Note that the number of equations is no larger than the number of unknowns. There may exist an infinite number of solutions to this system of equations. However, as we shall see, there is only one solution that is closest to the origin: the solution to $Ax = b$ whose norm $\|x\|$ is minimal. Let $x^*$ be this solution; that is, $Ax^* = b$ and $\|x^*\| \leqslant \|x\|$ for any $x$ such that $Ax = b$. In other words, $x^*$ is the solution to the problem

$$\begin{aligned} &\text{minimize} \quad \|x\| \\ &\text{subject to} \quad Ax = b. \end{aligned}$$

In Part IV, we study problems of the above type in more detail.

**Theorem 12.2.** *The unique solution $x^*$ to $Ax = b$ that minimizes the norm $\|x\|$ is given by*

$$x^* = A^T (AA^T)^{-1} b.$$

□

*Proof.* Let $x^* = A^T(AA^T)^{-1}b$. Note that

$$\begin{aligned}\|x\|^2 &= \|(x - x^*) + x^*\|^2 \\ &= ((x - x^*) + x^*)^T((x - x^*) + x^*) \\ &= \|x - x^*\|^2 + \|x^*\|^2 + 2x^{*T}(x - x^*).\end{aligned}$$

We now show that

$$x^{*T}(x - x^*) = 0.$$

Indeed,

$$\begin{aligned}x^{*T}(x - x^*) &= [A^T(AA^T)^{-1}b]^T[x - A^T(AA^T)^{-1}b] \\ &= b^T(AA^T)^{-1}[Ax - (AA^T)(AA^T)^{-1}b] \\ &= b^T(AA^T)^{-1}[b - b] = 0.\end{aligned}$$

Therefore

$$\|x\|^2 = \|x^*\|^2 + \|x - x^*\|^2.$$

Since, for all $x \neq x^*$, we have

$$\|x - x^*\|^2 > 0,$$

it follows that for all $x \neq x^*$,

$$\|x\|^2 > \|x^*\|^2,$$

which implies

$$\|x\| > \|x^*\|.$$

■

***Example 12.4.*** Find the point closest to the origin of $\mathbb{R}^3$ on the line of intersection of the two planes defined by the following two equations:

$$\begin{aligned}x_1 + 2x_2 - x_3 &= 1 \\ 4x_1 + x_2 + 3x_3 &= 0.\end{aligned}$$

Note that the above problem is equivalent to the problem

$$\begin{aligned}&\text{minimize} \quad \|x\| \\ &\text{subject to} \quad Ax = b,\end{aligned}$$

where

$$A = \begin{bmatrix} 1 & 2 & -1 \\ 4 & 1 & 3 \end{bmatrix}, \quad b = \begin{bmatrix} 1 \\ 0 \end{bmatrix}.$$

Thus, the solution to the problem is

$$x^* = A^T(AA^T)^{-1}b = \begin{bmatrix} 0.0952 \\ 0.3333 \\ -0.2381 \end{bmatrix}.$$

■

In the next section, we discuss an iterative algorithm for solving $Ax = b$.

## 12.4. KACZMARZ'S ALGORITHM

As in the previous section, let $A \in \mathbb{R}^{m \times n}$, $b \in \mathbb{R}^m$, $m \leqslant n$, and rank $A = m$. We now discuss an iterative algorithm for solving $Ax = b$, originally analyzed by Kaczmarz in 1937 [see Ref. 41]. The algorithm converges to the vector $x^* = A^T(AA^T)^{-1}b$ without having to explicitly invert the matrix $AA^T$. This is important from a practical point of view, especially when $A$ has many rows.

Let $a_j^T$ denote the $j$th row of $A$, $b_j$ the $j$th component of $b$, and $\mu$ a positive scalar, $0 < \mu < 2$. With this notation, Kaczmarz's algorithm is

1. Set $i := 0$, initial condition $x^{(0)}$.
2. For $j = 1, \ldots, m$, set $x^{(im+j)} = x^{(im+j-1)} + \mu(b_j - a_j^T x^{(im+j-1)}) a_j / a_j^T a_j$.
3. Set $i := i + 1$; go to step 2.

In words, Kaczmarz's algorithm works as follows. For the first $m$ iterations ($k = 0, \ldots, m-1$), we have

$$x^{(k+1)} = x^{(k)} + \mu(b_{k+1} - a_{k+1}^T x^{(k)}) \frac{a_{k+1}}{a_{k+1}^T a_{k+1}},$$

where, in each iteration, we use rows of $A$ and corresponding components of $b$ successively. For the $(m+1)$st iteration, we revert back to the first row of $A$ and the first component of $b$; that is,

$$x^{(m+1)} = x^{(m)} + \mu(b_1 - a_1^T x^{(m)}) \frac{a_1}{a_1^T a_1}.$$

We continue with the $(m+2)$nd iteration using the second row of $A$ and second component of $b$, and so on, repeating the cycle every $m$ iteration. We can view the scalar $\mu$ as the step size of the algorithm. The reason for requiring that $\mu$ be between 0 and 2 will become apparent from the convergence analysis.

We now prove the convergence of Kaczmarz's algorithm, using ideas from Kaczmarz's original paper [Ref. 41] and subsequent exposition by Parks [see Ref. 58].

**Theorem 12.3.** *In Kaczmarz's algorithm, if* $x^{(0)} = 0$ *then* $x^{(k)} \to x^* = A^T(AA^T)^{-1}b$ as $k \to \infty$. □

*Proof.* We may assume without loss of generality that $\|a_i\| = 1$, $i = 1, \dots, m$. For if not, we simply replace each $a_i$ by $a_i/\|a\|$ and each $b_i$ by $b_i/\|a_i\|$.

We first introduce the following notation. For each $j = 0, 1, 2, \dots$, let $R(j)$ denote the unique integer in $\{0, \dots, m-1\}$ satisfying $j = lm + R(j)$ for some integer $l$; that is, $R(j)$ is the remainder that results if we divide $j$ by $m$.

Using the above notation, we can write Kaczmarz's algorithm as

$$x^{(k+1)} = x^{(k)} + \mu(b_{R(k)+1} - a^T_{R(k)+1}x^{(k)})a_{R(k)+1}.$$

Using the identity $\|x + y\|^2 = \|x\|^2 + \|y\|^2 + 2\langle x, y\rangle$, we obtain

$$\begin{aligned}\|x^{(k+1)} - x^*\|^2 &= \|x^{(k)} - x^* + \mu(b_{R(k)+1} - a^T_{R(k)+1}x^{(k)})a_{R(k)+1}\|^2 \\ &= \|x^{(k)} - x^*\|^2 + \mu^2(b_{R(k)+1} - a^T_{R(k)+1}x^{(k)})^2 \\ &\quad + 2\mu(b_{R(k)+1} - a^T_{R(k)+1}x^{(k)})a^T_{R(k)+1}(x^{(k)} - x^*).\end{aligned}$$

Substituting $a^T_{R(k)+1}x^* = b_{R(k)+1}$ into the above equation, we get

$$\begin{aligned}\|x^{(k+1)} - x^*\|^2 &= \|x^{(k)} - x^*\|^2 - \mu(2 - \mu)(b_{R(k)+1} - a^T_{R(k)+1}x^{(k)})^2 \\ &= \|x^{(k)} - x^*\|^2 - \mu(2 - \mu)(a^T_{R(k)+1}(x^{(k)} - x^*))^2.\end{aligned}$$

Since $0 < \mu < 2$, the second term on the right-hand side is nonnegative, and hence

$$\|x^{(k+1)} - x^*\|^2 \leqslant \|x^{(k)} - x^*\|^2.$$

Therefore, $\{\|x^{(k)} - x^*\|^2\}$ is a nonincreasing sequence that is bounded below, since $\|x^{(k)} - x^*\|^2 \geqslant 0$ for all $k$. Hence, $\{\|x^{(k)} - x^*\|^2\}$ converges [see Ref. 3, p. 105]. Furthermore, we may write

$$\|x^{(k)} - x^*\|^2 = \|x^{(0)} - x^*\|^2 - \mu(2 - \mu)\sum_{i=0}^{k-1}(a^T_{R(i+1)}(x^{(i)} - x^*))^2.$$

Since $\{\|x^{(k)} - x^*\|^2\}$ converges, we conclude that

$$\sum_{i=0}^{\infty}(a^T_{R(i+1)}(x^{(i)} - x^*))^2 < \infty,$$

which implies that

$$a^T_{R(k)+1}(x^{(k)} - x^*) \to 0.$$

Observe that

$$\|x^{(k+1)} - x^{(k)}\|^2 = \mu^2(b_{R(k)+1} - a^T_{R(k)+1}x^{(k)})^2 = \mu^2(a^T_{R(k)+1}(x^{(k)} - x^*))^2$$

and therefore $\|x^{(k+1)} - x^{(k)}\|^2 \to 0$. Note also that, since $\{\|x^{(k)} - x^*\|^2\}$ converges, $\{x^{(k)}\}$ is a bounded sequence.

Following Kaczmarz [Ref. 41], we introduce the notation $x^{(r,s)} \triangleq x^{(rm+s)}$, $r = 0, 1, 2, \ldots, s = 0, \ldots, m-1$. With this notation, we have, for each $s = 0, \ldots, m-1$,

$$a_{s+1}^T(x^{(r,s)} - x^*) \to 0$$

as $r \to \infty$. Consider now the sequence $\{x^{(r,0)}: r \geqslant 0\}$. Since this sequence is bounded, we conclude that it has a convergent subsequence (this follows from the Bolzano–Weierstrass theorem [see Ref. 3, p. 70]). Denote this convergent subsequence by $\{x^{(r,0)}: r \in \mathscr{E}\}$, where $\mathscr{E}$ is a subset of $\{0, 1, \ldots\}$. Let $z^*$ be the limit of $\{x^{(r,0)}: r \in \mathscr{E}\}$. Hence,

$$a_1^T(z^* - x^*) = 0.$$

Next, note that since $\|x^{(k+1)} - x^{(k)}\|^2 \to 0$ as $k \to \infty$, we also have $\|x^{(r,1)} - x^{(r,0)}\|^2 \to 0$ as $r \to \infty$. Therefore, the subsequence $\{x^{(r,1)}: r \in \mathscr{E}\}$ also converges to $z^*$. Hence,

$$a_2^T(z^* - x^*) = 0.$$

Repeating the argument, we conclude that for each $i = 1, \ldots, m$,

$$a_i^T(z^* - x^*) = 0.$$

In matrix notation, the above equations take the form

$$A(z^* - x^*) = \mathbf{0}.$$

Now, $x^{(k)} \in \mathscr{R}(A^T)$ for all $k$ because $x^{(0)} = \mathbf{0}$ (see Exercise 12.9). Therefore, $z^* \in \mathscr{R}(A^T)$, because $\mathscr{R}(A^T)$ is closed. Hence, there exists $y^*$ such that $z^* = A^T y^*$. Thus

$$\begin{aligned} A(z^* - x^*) &= A(A^T y^* - A^T(AA^T)^{-1} b) \\ &= (AA^T) y^* - b \\ &= \mathbf{0}. \end{aligned}$$

Since rank $A = m$, $y^* = (AA^T)^{-1} b$ and hence $z^* = x^*$. Therefore, the subsequence $\{\|x^{r,0} - x^*\|^2: r \in \mathscr{E}\}$ converges to 0. Since $\{\|x^{r,0} - x^*\|^2: r \in \mathscr{E}\}$ is a subsequence of the convergent sequence $\{\|x^{(k)} - x^*\|^2\}$, we conclude that the sequence $\{\|x^{(k)} - x^*\|^2\}$ converges to 0; that is,

$$x^{(k)} \to x^*.$$

■

For the case when $x^{(0)} \neq 0$, Kaczmarz's algorithm converges to the unique point on $\{x : Ax = b\}$ minimizing the distance $\|x - x^{(0)}\|$ (see Exercise 12.10).

If we set $\mu = 1$, Kaczmarz's algorithm has the property that at each iteration $k$, the error $b_{R(k)} - a_{R(k)}^T x^{(k)}$ satisfies

$$b_{R(k)} - a_{R(k)}^T x^{(k)} = 0$$

(see Exercise 12.12). Substituting $b_{R(k)} = a_{R(k)}^T x^*$, we may write

$$a_{R(k)}^T (x^{(k)} - x^*) = 0.$$

Hence, the difference between $x^{(k)}$ and the solution $x^*$ is orthogonal to $a_{R(k)}$. This property is illustrated in following example.

***Example 12.5.*** Let

$$A = \begin{bmatrix} 1 & -1 \\ 0 & 1 \end{bmatrix}, \quad b = \begin{bmatrix} 2 \\ 3 \end{bmatrix}.$$

In this case, $x^* = [5, 3]^T$. Figure 12.3 shows a few iterations of Kaczmarz's algorithm with $\mu = 1$ and $x^{(0)} = 0$. We have $a_1^T = [1, -1]$, $a_2^T = [0, 1]$, $b_1 = 2$, $b_2 = 3$. In Figure 12.3, the diagonal line passing through the point $[2, 0]^T$ corresponds to the set $\{x : a_1^T x = b_1\}$, and the horizontal line passing through the point $[0, 3]^T$ corresponds to the set $\{x : a_2^T x = b_2\}$. To illustrate the algorithm, we perform three iterations:

$$x^{(1)} = \begin{bmatrix} 0 \\ 0 \end{bmatrix} + (2 - 0)\frac{1}{2}\begin{bmatrix} 1 \\ -1 \end{bmatrix} = \begin{bmatrix} 1 \\ -1 \end{bmatrix},$$

$$x^{(2)} = \begin{bmatrix} 1 \\ -1 \end{bmatrix} + (3 - (-1))\begin{bmatrix} 0 \\ 1 \end{bmatrix} = \begin{bmatrix} 1 \\ 3 \end{bmatrix}$$

$$x^{(3)} = \begin{bmatrix} 1 \\ 3 \end{bmatrix} + (2 - (-2))\frac{1}{2}\begin{bmatrix} 1 \\ -1 \end{bmatrix} = \begin{bmatrix} 3 \\ 1 \end{bmatrix}.$$

$$\vdots$$

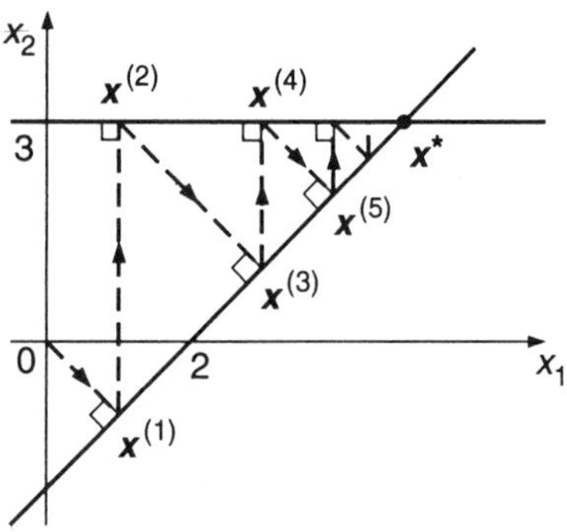


**Figure 12.3.** Iterations of Kaczmarz's algorithm in Example 12.5

As Figure 12.3 illustrates, the property

$$a_{R(k)}^T(x^{(k)} - x^*) = 0$$

holds at every iteration. Note the convergence of the iterations of the algorithm to the solution. ■

## 12.5. SOLVING $Ax = b$ IN GENERAL

Consider the general problem of solving a system of linear equations

$$Ax = b,$$

where $A \in \mathbb{R}^{m \times n}$, and rank $A = r$. Note that we always have $r \leqslant \min(m, n)$. In the case when $A \in \mathbb{R}^{n \times n}$ and rank $A = n$, the unique solution to the above equation has the form $x^* = A^{-1}b$. Thus, to solve the problem in this case it is enough to know the inverse $A^{-1}$. In this section, we analyze a general approach to solving $Ax = b$. The approach involves defining a *pseudoinverse* or *generalized inverse* of a given matrix $A \in \mathbb{R}^{m \times n}$, which plays the role of $A^{-1}$ when $A$ does not have an inverse (for example, when $A$ is not a square matrix). In particular, we discuss the Moore–Penrose inverse of a given matrix $A$, denoted by $A^{\dagger}$.

In our discussion of the Moore–Penrose inverse, we use the fact that a nonzero matrix of rank $r$ can be expressed as the product of a matrix of full column rank $r$ and a matrix of full row rank $r$. Such a factorization is referred to as the *full-rank factorization.* The term full-rank factorization in this context was proposed by Gantmacher [Ref. 23] and Ben-Israel and Greville [Ref. 5]. We state and prove the above result in the following lemma.

**Lemma 12.3. Full-rank Factorization Lemma.** *Let* $A \in \mathbb{R}^{m \times n}$, rank $A = r \leqslant \min(m, n)$. *Then, there exist matrices* $B \in \mathbb{R}^{m \times r}$ *and* $C \in \mathbb{R}^{r \times n}$ *such that*

$$A = BC,$$

*where*

$$\text{rank}\, A = \text{rank}\, B = \text{rank}\, C = r.$$

□

*Proof.* Since rank $A = r$, we can find $r$ linearly independent columns of $A$. Without loss of generality, let $a_1, a_2, \ldots, a_r$ be such columns, where $a_i$ is the $i$th column of $A$. The remaining columns of $A$ can be expressed as linear combinations of $a_1, a_2, \ldots, a_r$. Thus, a possible choice for the full-rank matrices $B$ and $C$ are

$$B = [a_1, \ldots, a_r] \in \mathbb{R}^{m \times r},$$

$$C = \begin{bmatrix} 1 & \cdots & 0 & c_{1,r+1} & \cdots & c_{1,n} \\ \vdots & \ddots & \vdots & \vdots & \ddots & \vdots \\ 0 & \cdots & 1 & c_{r,r+1} & \cdots & c_{r,n} \end{bmatrix} \in \mathbb{R}^{r \times n},$$

where the entries $c_{i,j}$ are such that, for each $j = r+1, \ldots, n$, we have $\boldsymbol{a}_j = c_{1,j}\boldsymbol{a}_1 + \cdots + c_{r,j}\boldsymbol{a}_r$. Thus, $A = BC$. ■

Note that if $m < n$ and rank $A = m$, then we can take

$$B = I_m, \quad C = A,$$

where $I_m$ is the $m \times m$ identity matrix. If on the other hand, $m > n$ and rank $A = n$, then we can take

$$B = A, \quad C = I_n.$$

***Example 12.6.*** Let $A$ be given by

$$A = \begin{bmatrix} 2 & 1 & -2 & 5 \\ 1 & 0 & -3 & 2 \\ 3 & -1 & -13 & 5 \end{bmatrix}.$$

Note that rank $A = 2$. We can write a full-rank factorization of $A$ based on the proof of Lemma 12.3:

$$A = \begin{bmatrix} 2 & 1 \\ 1 & 0 \\ 3 & -1 \end{bmatrix} \begin{bmatrix} 1 & 0 & -3 & 2 \\ 0 & 1 & 4 & 1 \end{bmatrix} = BC.$$ ■

We now define the Moore–Penrose inverse and discuss its existence and uniqueness. For this, we first consider the matrix equation

$$AXA = A,$$

where $A \in \mathbb{R}^{m \times n}$ is a given matrix, and $X \in \mathbb{R}^{n \times m}$ is a matrix we wish to determine. Observe that, if $A$ is a nonsingular square matrix, then the above equation has the unique solution $X = A^{-1}$. We now define the Moore–Penrose inverse, also called the pseudoinverse or generalized inverse.

***Definition 12.1.*** Given $A \in \mathbb{R}^{m \times n}$, a matrix $A^\dagger \in \mathbb{R}^{n \times m}$ is called a *pseudoinverse* of the matrix $A$ if

$$AA^\dagger A = A,$$

and there exist matrices $U \in \mathbb{R}^{n \times n}$, $V \in \mathbb{R}^{m \times m}$ such that

$$A^\dagger = UA^T,$$

$$A^\dagger = A^T V.$$ ■

The requirement $A^\dagger = UA^T = A^T V$ can be interpreted as follows. Each row of the pseudoinverse matrix $A^\dagger$ of $A$ is a linear combination of the rows of $A^T$, and each column of $A^\dagger$ is a linear combination of the columns of $A^T$.

For the case in which a matrix $A \in \mathbb{R}^{m \times n}$ with $m \geqslant n$ and rank $A = n$, we can easily check that the following is a pseudoinverse of $A$:

$$A^\dagger = (A^T A)^{-1} A^T.$$

Indeed, $A(A^T A)^{-1} A^T A = A$, and if we define $U = (A^T A)^{-1}$ and $V = A(A^T A)^{-1}(A^T A)^{-1} A^T$, then $A^\dagger = UA^T = A^T V$. Note that we in fact have $A^\dagger A = I_n$. For this reason, $(A^T A)^{-1} A^T$ is often called the *left pseudoinverse* of $A$. This formula also appears in least-squares analysis (see Section 12.1).

For the case in which a matrix $A \in \mathbb{R}^{m \times n}$ with $m \leqslant n$ and rank $A = m$, we can easily check, as we did in the previous case, that the following is a pseudoinverse of $A$:

$$A^\dagger = A^T (A A^T)^{-1}.$$

Note that in this case, we have $AA^\dagger = I_m$. For this reason, $A^T(AA^T)^{-1}$ is often called the *right pseudoinverse* of $A$. This formula also appears in the problem of minimizing $\|x\|$ subject to $Ax = b$ (see Section 12.3).

**Theorem 12.4.** *Let $A \in \mathbb{R}^{m \times n}$. If a pseudoinverse $A^\dagger$ of $A$ exists, then it is unique.* □

*Proof.* Let $A_1^\dagger$ and $A_2^\dagger$ be pseudoinverses of $A$. We show that $A_1^\dagger = A_2^\dagger$. By definition,

$$AA_1^\dagger A = AA_2^\dagger A = A,$$

and there are matrices $U_1, U_2 \in \mathbb{R}^{n \times n}$, $V_1, V_2 \in \mathbb{R}^{m \times m}$ such that

$$A_1^\dagger = U_1 A^T = A^T V_1,$$
$$A_2^\dagger = U_2 A^T = A^T V_2.$$

Let

$$D = A_2^\dagger - A_1^\dagger, \quad U = U_2 - U_1, \quad V = V_2 - V_1.$$

Then, we have

$$O = ADA, \quad D = UA^T = A^T V.$$

Therefore, using the above two equations, we have

$$(DA)^T DA = A^T D^T DA = A^T V^T ADA = O,$$

which implies that

$$DA = O.$$

On the other hand, since $DA = O$, we have

$$DD^T = DAU^T = O,$$

which implies that

$$D = A_2^\dagger - A_1^\dagger = O$$

and hence

$$A_2^\dagger = A_1^\dagger.$$

■

From the above theorem, we know that, if a pseudoinverse matrix exists, then it is unique. Our goal now is to show that the pseudoinverse always exists. In fact, we show that the pseudoinverse of any given matrix $A$ is given by the formula

$$A^\dagger = C^\dagger B^\dagger$$

where $B^\dagger$ and $C^\dagger$ are the pseudoinverses of the matrices $B$ and $C$ that form a full-rank factorization of $A$; that is, $A = BC$ and $B$ and $C$ are full rank (see Lemma 12.3). Note that we already know how to compute $B^\dagger$ and $C^\dagger$, namely,

$$B^\dagger = (B^T B)^{-1} B^T,$$

and

$$C^\dagger = C^T (CC^T)^{-1}.$$

**Theorem 12.5.** *Let a matrix $A \in \mathbb{R}^{m \times n}$ have a full-rank factorization $A = BC$, with* rank $A$ = rank $B$ = rank $C = r$, $B \in \mathbb{R}^{m \times r}$, $C \in \mathbb{R}^{r \times n}$. *Then*

$$A^\dagger = C^\dagger B^\dagger.$$

□

*Proof.* We show that

$$A^\dagger = C^\dagger B^\dagger = C^T (CC^T)^{-1} (B^T B)^{-1} B^T$$

satisfies the conditions of Definition 12.1 for a pseudoinverse. Indeed, first observe that

$$AC^\dagger B^\dagger A = BCC^T (CC^T)^{-1} (B^T B)^{-1} B^T BC = BC = A.$$

Next, define

$$U = C^T (CC^T)^{-1} (B^T B)^{-1} (CC^T)^{-1} C$$

and

$$V = B (B^T B)^{-1} (CC^T)^{-1} (B^T B)^{-1} B^T.$$

It is easy to verify that the matrices $U$ and $V$ above satisfy

$$A^\dagger = C^\dagger B^\dagger = UA^T = A^T V.$$

Hence

$$A^{\dagger} = C^{\dagger} B^{\dagger}$$

is the pseudoinverse of $A$. ■

***Example 12.7.*** Continued from Example 12.6. Recall that

$$A = \begin{bmatrix} 2 & 1 & -2 & 5 \\ 1 & 0 & -3 & 2 \\ 3 & -1 & -13 & 5 \end{bmatrix} = \begin{bmatrix} 2 & 1 \\ 1 & 0 \\ 3 & -1 \end{bmatrix} \begin{bmatrix} 1 & 0 & -3 & 2 \\ 0 & 1 & 4 & 1 \end{bmatrix} = BC.$$

We compute

$$B^{\dagger} = (B^T B)^{-1} B^T = \frac{1}{27} \begin{bmatrix} 5 & 2 & 5 \\ 16 & 1 & -11 \end{bmatrix},$$

and

$$C^{\dagger} = C^T (CC^T)^{-1} = \frac{1}{76} \begin{bmatrix} 9 & 5 \\ 5 & 7 \\ -7 & 13 \\ 23 & 17 \end{bmatrix}.$$

Thus,

$$A^{\dagger} = C^{\dagger} B^{\dagger} = \frac{1}{2052} \begin{bmatrix} 125 & 23 & -10 \\ 137 & 17 & -52 \\ 173 & -1 & -178 \\ 387 & 63 & -73 \end{bmatrix}.$$

■

We emphasize that the formula $A^{\dagger} = C^{\dagger} B^{\dagger}$ does not necessarily hold if $A = BC$ is not a full-rank factorization. The following example, which is taken from Ref. 23, illustrates this point.

***Example 12.8.*** Let

$$A = [1].$$

Obviously, $A^{\dagger} = A^{-1} = A = [1]$. Observe that $A$ can be represented as

$$A = [0, 1] \begin{bmatrix} 1 \\ 1 \end{bmatrix} = BC.$$

The above is not a full-rank factorization of $A$. Let us now compute $B^{\dagger}$ and $C^{\dagger}$. We have

$$B^{\dagger} = (B^T B)^{-1} B^T = \begin{bmatrix} 0 \\ 1 \end{bmatrix},$$

$$C^{\dagger} = C^T (CC^T)^{-1} = [\tfrac{1}{2} \;\; \tfrac{1}{2}].$$

Thus,

$$C^{\dagger}B^{\dagger} = [\tfrac{1}{2}],$$

which is not equal to $A^{\dagger}$. ■

We can simplify the expression

$$A^{\dagger} = C^{\dagger}B^{\dagger} = C^T(CC^T)^{-1}(B^TB)^{-1}B^T$$

to

$$A^{\dagger} = C^T(B^TAC^T)^{-1}B^T.$$

The above expression is easily verified simply by substituting $A = BC$. This explicit formula for $A^{\dagger}$ is attributed to C. C. MacDuffee by Ben-Israel and Greville [Ref. 5]. Ben-Israel and Greville report that, around 1959, MacDuffee was the first to point out that a full-rank factorization of $A$ leads to the above explicit formula. However, they mention that MacDuffee did it in a private communication, so there is no published work by MacDuffee that contains the result.

We now prove two important properties of $A^{\dagger}$ in the context of solving a system of linear equations $Ax = b$.

**Theorem 12.6.** *Consider a system of linear equations $Ax = b$, $A \in \mathbb{R}^{m \times n}$, rank $A = r$. The vector $x^* = A^{\dagger}b$ minimizes $\|Ax - b\|^2$ on $\mathbb{R}^n$. Furthermore, among all vectors in $\mathbb{R}^n$ that minimize $\|Ax - b\|^2$, the vector $x^* = A^{\dagger}b$ is the unique vector with minimum norm.* □

*Proof.* We first show that $x^* = A^{\dagger}b$ minimizes $\|Ax - b\|^2$ over $\mathbb{R}^n$. To this end, observe that for any $x \in \mathbb{R}^n$,

$$\begin{aligned}\|Ax - b\|^2 &= \|A(x - x^*) + Ax^* - b\|^2 \\ &= \|A(x - x^*)\|^2 + \|Ax^* - b\|^2 + 2[A(x - x^*)]^T(Ax^* - b).\end{aligned}$$

We now show that

$$[A(x - x^*)]^T(Ax^* - b) = 0.$$

Indeed

$$\begin{aligned}[A(x - x^*)]^T(Ax^* - b) &= (x - x^*)^T(A^TAx^* - A^Tb) \\ &= (x - x^*)^T(A^TAA^{\dagger}b - A^Tb).\end{aligned}$$

However,

$$A^TAA^{\dagger} = C^TB^TBCC^T(CC^T)^{-1}(B^TB)^{-1}B^T = A^T.$$

Hence

$$[A(x - x^*)]^T(Ax^* - b) = (x - x^*)^T(A^Tb - A^Tb) = 0.$$

Thus, we have

$$\|Ax - b\|^2 = \|A(x - x^*)\|^2 + \|Ax^* - b\|^2.$$

Since

$$\|A(x - x^*)\|^2 \geqslant 0,$$

we obtain

$$\|Ax - b\|^2 \geqslant \|Ax^* - b\|^2$$

and thus $x^*$ minimizes $\|Ax - b\|^2$.

We now show that among all $x$ that minimize $\|Ax - b\|^2$, the vector $x^* = A^\dagger b$ is the unique vector with minimum norm. So let $\tilde{x}$ be a vector minimizing $\|Ax - b\|^2$. We have

$$\begin{aligned}\|\tilde{x}\|^2 &= \|(\tilde{x} - x^*) + x^*\|^2 \\ &= \|\tilde{x} - x^*\|^2 + \|x^*\|^2 + 2x^{*T}(\tilde{x} - x^*).\end{aligned}$$

Observe that

$$x^{*T}(\tilde{x} - x^*) = 0.$$

To see this, note that

$$\begin{aligned}x^{*T}(\tilde{x} - x^*) &= (A^\dagger b)^T(\tilde{x} - A^\dagger b) \\ &= b^T B(B^T B)^{-T}(CC^T)^{-T} C(\tilde{x} - C^T(CC^T)^{-1}(B^T B)^{-1} B^T b) \\ &= b^T B(B^T B)^{-T}(CC^T)^{-T}[C\tilde{x} - (B^T B)^{-1} B^T b],\end{aligned}$$

where the superscript $-T$ denotes the transpose of the inverse. Now, $\|Ax - b\|^2 = \|B(Cx) - b\|^2$. Since $\tilde{x}$ minimizes $\|Ax - b\|^2$ and $C$ is of full rank, then $y^* = C\tilde{x}$ minimizes $\|By - b\|^2$ over $\mathbb{R}^r$ (see Exercise 12.13). Since $B$ is of full rank, then by Theorem 12.1, we have $C\tilde{x} = y^* = (B^T B)^{-1} B^T b$. Substituting this into the above equation, we get $x^{*T}(\tilde{x} - x^*) = 0$.

Therefore, we have

$$\|\tilde{x}\|^2 = \|x^*\|^2 + \|\tilde{x} - x^*\|^2.$$

For all $\tilde{x} \neq x^*$, we have

$$\|\tilde{x} - x^*\|^2 > 0,$$

and hence

$$\|\tilde{x}\|^2 > \|x^*\|^2$$

or equivalently

$$\|\tilde{x}\| > \|x^*\|.$$

Hence, among all vectors minimizing $\|Ax - b\|^2$, the vector $x^* = A^\dagger b$ is the unique vector with minimum norm. ■

The generalized inverse has the following useful properties (see Exercise 12.4):

1. $(A^T)^\dagger = (A^\dagger)^T$
2. $(A^\dagger)^\dagger = A$

The above two properties are similar to those that are satisfied by the usual matrix inverse. However, we point out that the property $(A_1 A_2)^\dagger = A_2^\dagger A_1^\dagger$ does not hold in general (see Exercise 12.16).

Finally, it is important to note that we can define the generalized inverse in an alternative way, following the definition of Penrose. Specifically, the Penrose definition of the generalized inverse of a matrix $A \in \mathbb{R}^{m \times n}$ is the unique matrix $A^\dagger \in \mathbb{R}^{n \times m}$ satisfying the following properties:

1. $AA^\dagger A = A$
2. $A^\dagger A A^\dagger = A^\dagger$
3. $(AA^\dagger)^T = AA^\dagger$
4. $(A^\dagger A)^T = A^\dagger A$

The Penrose definition above is equivalent to Definition 12.1 (see Exercise 12.15). For more information on generalized inverses and their applications, we refer the reader to the books by Ben-Israel and Greville [Ref. 5] and Campbell and Meyer [Ref. 14].

## EXERCISES

**12.1** Suppose that we perform an experiment to calculate the gravitational constant $g$ as follows. We drop a ball from a certain height and measure its distance from the original point at certain time instants. The results of the experiment are shown in the following table:

| | | | |
|---|---|---|---|
| Time (sec) | 1.00 | 2.00 | 3.00 |
| Distance (m) | 5.00 | 19.5 | 44.0 |

The equation relating the distance $s$ and the time $t$ at which $s$ is measured is given by

$$s = \tfrac{1}{2} g t^2.$$

**a.** Find a least-squares estimate of $g$ using the experimental results from the above table.

**b.** Suppose that we take an additional measurement at time 4.00, and obtain a distance of 78.5. Use the recursive least-squares algorithm to calculate an updated least-squares estimate of $g$.

**12.2** *Linear Regression.* Let $[x_1, y_1]^T, \ldots, [x_p, y_p]^T$, $p \geqslant 2$, be points in $\mathbb{R}^2$. We wish to find the straight line of best fit through these points ("best" in the sense that the total squared error is minimized). That is, we wish to find $a^*, b^* \in \mathbb{R}$ to minimize

$$f(a,b) = \sum_{i=1}^{p} (ax_i + b - y_i)^2.$$

Assume that the $x_i$, $i = 1, \ldots, p$, are not all equal. Show that there exist unique parameters $a^*$ and $b^*$ for the line of best fit, and find the parameters in terms of the following quantities:

$$\bar{X} = \frac{1}{p} \sum_{i=1}^{p} x_i$$

$$\bar{Y} = \frac{1}{p} \sum_{i=1}^{p} y_i$$

$$\overline{X^2} = \frac{1}{p} \sum_{i=1}^{p} x_i^2$$

$$\overline{Y^2} = \frac{1}{p} \sum_{i=1}^{p} y_i^2$$

$$\overline{XY} = \frac{1}{p} \sum_{i=1}^{p} x_i y_i.$$

**12.3** Suppose we take measurements of a sinusoidal signal $y(t) = \sin(\omega t + \theta)$ at times $t_1, \ldots, t_p$, and obtain values $y_1, \ldots, y_p$, where $-\pi/2 \leqslant \omega t_i + \theta \leqslant \pi/2$, $i = 1, \ldots, p$, and the $t_i$ are not all equal. We wish to determine the values of the frequency $\omega$ and phase $\theta$.

**a.** Express the problem as a system of linear equations.

**b.** Find the least-squares estimate of $\omega$ and $\theta$ based on part a. Use the following notation:

$$\bar{T} = \frac{1}{p} \sum_{i=1}^{p} t_i,$$

$$\overline{T^2} = \frac{1}{p} \sum_{i=1}^{p} t_i^2,$$

$$\overline{TY} = \frac{1}{p} \sum_{i=1}^{p} t_i \arcsin y_i,$$

$$\bar{Y} = \frac{1}{p} \sum_{i=1}^{p} \arcsin y_i.$$

**12.4** Consider the affine function $f: \mathbb{R}^n \to \mathbb{R}$ of the form $f(x) = a^T x + c$, where $a \in \mathbb{R}^n$ and $c \in \mathbb{R}$.

**a.** We are given a set of $p$ pairs $(x_1, y_1), \ldots, (x_p, y_p)$, where $x_i \in \mathbb{R}^n$, $y_i \in \mathbb{R}$, $i = 1, \ldots, p$. We wish to find the affine function of best fit to these points, where "best" is in the sense of minimizing the total square error

$$\sum_{i=1}^{p} (f(x_i) - y_i)^2.$$

Formulate the above as an optimization problem of the form: minimize $\|Az - b\|^2$ with respect to $z$. Specify the dimensions of $A$, $z$, and $b$.

**b.** Suppose the points satisfy

$$x_1 + \cdots + x_p = 0$$

and

$$y_1 x_1 + \cdots + y_p x_p = 0.$$

Find the affine function of best fit in this case, assuming it exists and is unique.

**12.5** Consider a discrete time linear system $x_{k+1} = a x_k + b u_k$, where $u_k$ is the input at time $k$, $x_k$ is the output at time $k$, and $a, b \in \mathbb{R}$ are system parameters. Suppose that we apply a constant input $u_k = 1$ for all $k \geqslant 0$, and measure the first 4 values of the output to be $x_0 = 0$, $x_1 = 1$, $x_2 = 2$, $x_3 = 8$. Find the least-squares estimate of $a$ and $b$ based on the above data.

**12.6** Consider a discrete time linear system $x_{k+1} = a x_k + b u_k$, where $u_k$ is the input at time $k$, $x_k$ is the output at time $k$, and $a, b \in \mathbb{R}$ are system parameters. Given the first $n + 1$ values of the impulse response $h_0, \ldots, h_n$, find the least squares estimate of $a$ and $b$. You may assume that at least one $h_k$ is nonzero. *Note*: The *impulse response* is the output sequence resulting from an input of $u_0 = 1, u_k = 0$ for $k \neq 0$, and zero initial condition $x_0 = 0$.

**12.7** Let $A \in \mathbb{R}^{m \times n}$, $b \in \mathbb{R}^m$, $m \leqslant n$, rank $A = m$, and $x_0 \in \mathbb{R}^n$. Consider the problem

$$\begin{aligned} &\text{minimize} && \|x - x_0\| \\ &\text{subject to} && Ax = b. \end{aligned}$$

Show that the above problem has a unique solution given by

$$x^* = A^T(AA^T)^{-1} b + (I_n - A^T(AA^T)^{-1} A) x_0.$$

**12.8** Let $A \in \mathbb{R}^{m \times n}$, $b \in \mathbb{R}^m$, $m \leqslant n$, and rank $A = m$. Show that $x^* = A^T(AA^T)^{-1}b$ is the only vector in $\mathcal{R}(A^T)$ satisfying $Ax^* = b$.

**12.9** Show that in Kaczmarz's algorithm, if $x^{(0)} = 0$, then $x^{(k)} \in \mathcal{R}(A^T)$ for all $k$.

**12.10** Consider Kaczmarz's algorithm with $x^{(0)} \neq 0$.

**a.** Show that there exists a unique point minimizing $\|x - x^{(0)}\|$ subject to $\{x : Ax = b\}$.
**b.** Show that Kaczmarz's algorithm converges to the point in part a.

**12.11** Consider Kaczmarz's algorithm with $x^{(0)} = 0$, where $m = 1$; that is, $A = [a^T] \in \mathbb{R}^{1 \times n}$ and $a \neq 0$, and $0 < \mu < 2$. Show that there exists $0 \leqslant \gamma < 1$ such that $\|x^{(k+1)} - x^*\| \leqslant \gamma \|x^{(k)} - x^*\|$ for all $k \geqslant 0$.

**12.12** Show that in Kaczmarz's algorithm, if $\mu = 1$, then $a_{R(k)}^T(x^{(k)} - x^*) = 0$ for each $k$.

**12.13** Consider the problem of minimizing $\|Ax - b\|^2$ over $\mathbb{R}^n$, where $A \in \mathbb{R}^{m \times n}$, $b \in \mathbb{R}^m$. Let $x^*$ be a solution. Suppose $A = BC$ is a full-rank factorization of $A$, that is, rank $A =$ rank $B =$ rank $C = r$, and $B \in \mathbb{R}^{m \times r}$, $C \in \mathbb{R}^{r \times n}$. Show that the minimizer of $\|By - b\|$ over $\mathbb{R}^r$ is $Cx^*$.

**12.14** Prove the following properties of generalized inverses:

**a.** $(A^T)^\dagger = (A^\dagger)^T$
**b.** $(A^\dagger)^\dagger = A$

**12.15** Show that the Penrose definition of the generalized inverse is equivalent to Definition 12.1.

**12.16** Construct matrices $A_1$ and $A_2$ such that $(A_1A_2)^\dagger \neq A_2^\dagger A_1^\dagger$.

# 13

# Unconstrained Optimization and Feedforward Neural Networks

## 13.1. INTRODUCTION

In this chapter, we apply the techniques of the previous chapters to the training of feedforward neural networks. Neural networks have found numerous practical applications, ranging from telephone echo cancellation to aiding in the interpretation of EEG data [see, for example, Refs. 61 and 43]. The essence of neural networks lies in the connection weights between neurons. The selection of these weights is referred to as training or learning. For this reason, we often refer to the weights as the learning parameters. A popular method for training a neural network is called the backpropagation algorithm, based on an unconstrained optimization problem, and an associated gradient algorithm applied to the problem. This chapter is devoted to a description of neural networks and the use of techniques we have developed in preceding chapters for the training of neural networks.

An *artificial neural network* is a circuit composed of interconnected simple circuit elements called *neurons*. Each neuron represents a map, typically with multiple inputs and a single output. Specifically, the output of the neuron is a function of the sum of the inputs, as illustrated in Figure 13.1. The function at the output of the neuron is called the *activation function*. We use the symbol given in Figure 13.2 to pictorially represent a single neuron. Note that the single output of the neuron may be applied as inputs to several other neurons, and, therefore, the symbol for a single neuron shows multiple arrows emanating from it. A neural network may be implemented using an analog circuit. In this case, inputs and outputs may be represented by currents and voltages.

A neural network consists of interconnected neurons, where the inputs to each neuron consist of weighted outputs of other neurons. The interconnections allow exchange of data or information between neurons. In a *feedforward* neural network, the neurons are interconnected in layers, so that the data flows only in one direction. Thus, each neuron receives information only from neurons in the previous layer: the inputs to each neuron are weighted outputs of neurons in the

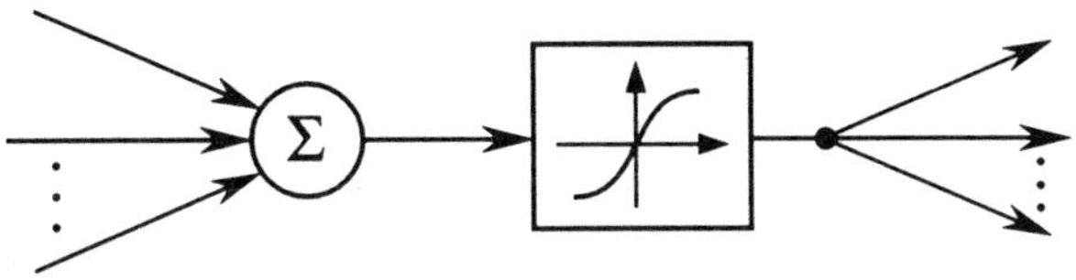


**Figure 13.1.** A single neuron

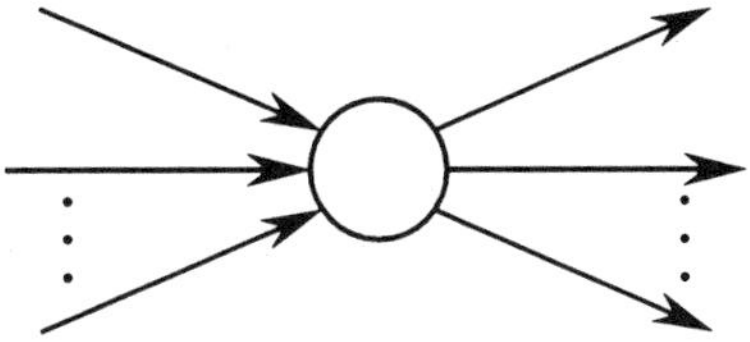

**Figure 13.2.** Symbol for a single neuron

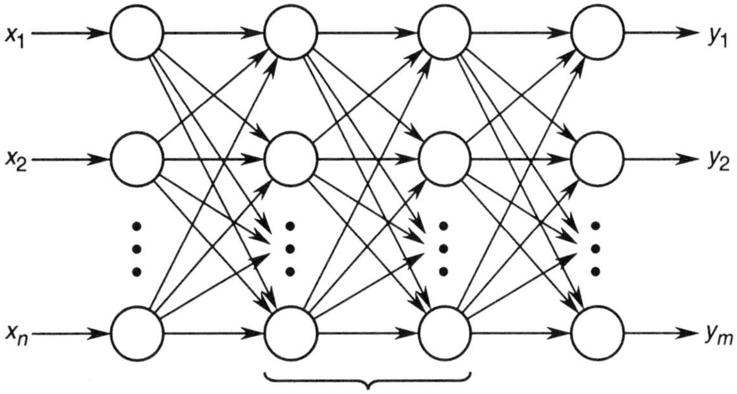


**Figure 13.3.** Structure of a feedforward neural network

previous layer. Figure 13.3 illustrates the structure of feedforward neural networks. The first layer in the network is called the *input layer*, and the last layer is called the *output layer*. The layers in between the input and output layers are called *hidden layers*.

We can view a neural network as simply a particular implementation of a map from $\mathbb{R}^n$ to $\mathbb{R}^m$, where $n$ is the number of inputs $x_1, \ldots, x_n$, and $m$ is the number of outputs $y_1, \ldots, y_m$. The map that is implemented by a neural network depends on the weights of the interconnections in the network. Therefore, we can change the mapping that is implemented by the network by adjusting the values of the weights in the network. The information about the mapping is stored in the weights over all the neurons, and thus the neural network is a *distributed* representation of the mapping. Moreover, for any given input, the computation of the corresponding output is achieved through the collective effect of individual input–output characteristics of each neuron; therefore, the neural network can be considered as a *parallel* computation device. We point out that the ability to implement or approximate a map encompasses many important practical applications. For example, pattern recognition and classification problems can be viewed as function implementation or approximation problems.

Suppose we are given a map $\boldsymbol{F}: \mathbb{R}^n \to \mathbb{R}^m$ that we wish to implement using a given neural network. Our task boils down to appropriately selecting the interconnection weights in the network. As mentioned earlier, we refer to this task as training of the neural network, or learning by the neural network. We use input–output examples of the given map to train the neural network. Specifically, let $(\boldsymbol{x}_{d,1}, \boldsymbol{y}_{d,1}), \ldots, (\boldsymbol{x}_{d,p}, \boldsymbol{y}_{d,p}) \in \mathbb{R}^n \times \mathbb{R}^m$, where each $\boldsymbol{y}_{d,i}$ is the output of the map $\boldsymbol{F}$ corresponding to the input $\boldsymbol{x}_{d,i}$; that is, $\boldsymbol{y}_{d,i} = \boldsymbol{F}(\boldsymbol{x}_{d,i})$. We refer to the set $\{(\boldsymbol{x}_{d,1}, \boldsymbol{y}_{d,1}), \ldots, (\boldsymbol{x}_{d,p}, \boldsymbol{y}_{d,p})\}$ as the *training set*. We train the neural network by adjusting the weights such that the map that is implemented by the network is close to the desired map $\boldsymbol{F}$. For this reason, we can think of neural networks as function approximators.

The form of learning described above can be thought of as learning with a teacher. The teacher supplies questions to the network in the form of $\boldsymbol{x}_{d,1}, \ldots, \boldsymbol{x}_{d,p}$, and also tells the network the correct answers $\boldsymbol{y}_{d,1}, \ldots, \boldsymbol{y}_{d,p}$. Training of the network then comprises applying a training algorithm that adjusts weights based on the error between the computed output and the desired output, or the difference between $\boldsymbol{y}_{d,i} = \boldsymbol{F}(\boldsymbol{x}_{d,i})$ and the output of the neural network corresponding to $\boldsymbol{x}_{d,i}$. Having trained the neural network, our hope is that the network correctly generalizes the examples used in the training set. By this, we mean that the network should correctly implement the mapping $\boldsymbol{F}$ and produce the correct output corresponding to any input, including those that were not a part of the training set.

As we shall see in the remainder of this chapter, the training problem can be formulated as an optimization problem. We can then use optimization techniques (e.g., steepest descent, conjugate gradients [Ref. 40], and quasi-Newton) for selection of the weights. The training algorithms are based on such optimization algorithms.

In the literature, the form of learning described above is referred to as *supervised learning*, for obvious reasons. The term supervised learning suggests that there is also a form of learning called unsupervised learning. Indeed, this is the case. However, unsupervised learning does not fit into the framework described above. Therefore, we shall not discuss the idea of unsupervised learning any further [we refer the reader interested in unsupervised learning to Ref. 33].

## 13.2. SINGLE-NEURON TRAINING

Consider a single neuron, as shown in Figure 13.4. For this particular neuron, the activation function is simply the identity (linear function with unit slope). The

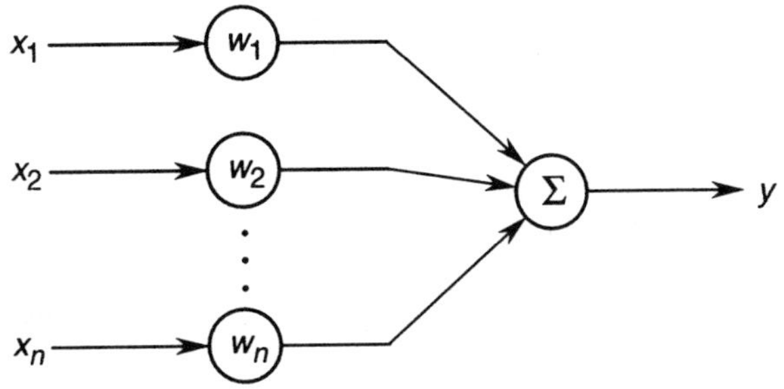


**Figure 13.4.** A single linear neuron

neuron implements the following (linear) map from $\mathbb{R}^n$ to $\mathbb{R}$:

$$y = \sum_{i=1}^{n} w_i x_i = \boldsymbol{x}^T \boldsymbol{w}$$

where $\boldsymbol{x} = [x_1, \ldots, x_n]^T \in \mathbb{R}^n$ is the vector of inputs, $y \in \mathbb{R}$ is the output, and $\boldsymbol{w} = [w_1, \ldots, w_n]^T \in \mathbb{R}^n$ is the vector of weights. Suppose we are given a map $F: \mathbb{R}^n \to \mathbb{R}$. We wish to find the value of the weights $w_1, \ldots, w_n$ such that the neuron approximates the map $F$ as closely as possible. To do this, we use a training set consisting of $p$ pairs $\{(\boldsymbol{x}_{d,1}, y_{d,1}), \ldots, (\boldsymbol{x}_{d,p}, y_{d,p})\}$, where $\boldsymbol{x}_{d,i} \in \mathbb{R}^n$ and $y_{d,i} \in \mathbb{R}$, $i = 1, \ldots, p$. For each $i$, $y_{d,i} = F(\boldsymbol{x}_{d,i})$ is the "desired" output corresponding to the given input $\boldsymbol{x}_{d,i}$. The training problem can then be formulated as the following optimization problem:

$$\text{minimize } \frac{1}{2} \sum_{i=1}^{p} (y_{d,i} - \boldsymbol{x}_{d,i}^T \boldsymbol{w})^2$$

where the minimization is taken over all $\boldsymbol{w} = [w_1, \ldots, w_n]^T \in \mathbb{R}^n$. Note that the objective function represents the sum of the squared errors between the desired outputs $y_{d,i}$ and the corresponding outputs of the neuron $\boldsymbol{x}_{d,i}^T \boldsymbol{w}$. The factor of $\frac{1}{2}$ is added for notational convenience and does not change the minimizer.

The above objective function can be written in matrix form as follows. First define the matrix $\boldsymbol{X}_d \in \mathbb{R}^{n \times p}$ and vector $\boldsymbol{y}_d \in \mathbb{R}^p$ by

$$\boldsymbol{X}_d = [\boldsymbol{x}_{d,1} \cdots \boldsymbol{x}_{d,p}]$$

$$\boldsymbol{y}_d = \begin{bmatrix} y_{d,1} \\ \vdots \\ y_{d,p} \end{bmatrix}.$$

Then, the optimization problem becomes

$$\text{minimize } \frac{1}{2} \| \boldsymbol{y}_d - \boldsymbol{X}_d^T \boldsymbol{w} \|^2.$$

There are two cases to consider in the above optimization problem: $p \leqslant n$ and $p > n$. We first consider the case where $p \leqslant n$, that is, where we have at most as many training pairs as the number of weights. For convenience, we assume that rank $X_d^T = p$. In this case, there are infinitely many points satisfying $y_d = X_d^T w$. Hence, there are infinitely many solutions to the above optimization problem, with the optimal objective function value of 0. Therefore, we have a choice of which optimal solution to select. A possible criterion for this selection is that of minimizing the solution norm. This is exactly the problem considered in Section 12.3. Recall that the minimum norm solution is $w^* = X_d(X_d^T X_d)^{-1} y_d$. An efficient iterative algorithm for finding this solution is Kaczmarz's algorithm (discussed in Section 12.4). Kaczmarz's algorithm in this setting takes the form

$$w^{(k+1)} = w^{(k)} + \mu \frac{e_k x_{d,R(k)}}{\| x_{d,R(k)} \|^2},$$

where $w^{(0)} = \mathbf{0}$, and

$$e_k = y_{d,R(k)} - x_{d,R(k)}^T w^{(k)}.$$

Recall that $R(k)$ is the unique integer in $\{0, \ldots, p-1\}$ satisfying $k = lp + R(k)$ for some integer $l$. That is, $R(k)$ is the remainder that results if we divide $k$ by $p$ (see Section 12.4 for further details on the algorithm).

The above algorithm was applied to the training of linear neurons by Widrow and Hoff [see Ref. 77 for some historical remarks]. The single neuron together with the above training algorithm is illustrated in Figure 13.5, and is often called *Adaline*, an acronym for "adaptive linear element."

We now consider the case where $p > n$. Here, we have more training points than the number of weights. We assume that rank $X_d^T = n$. In this case, the objective function $\frac{1}{2} \| y_d - X_d^T w \|^2$ is simply a strictly convex quadratic function of

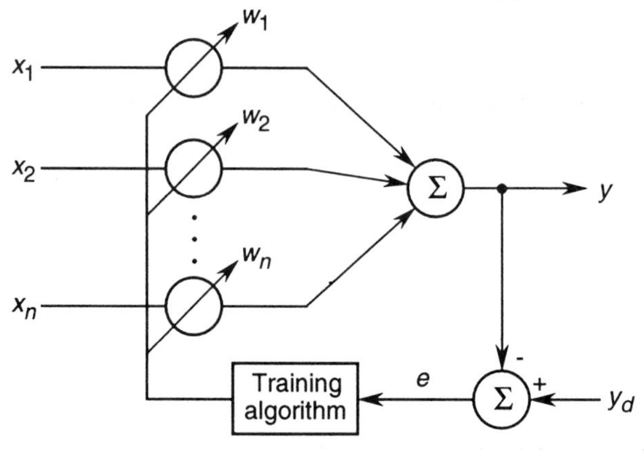


**Figure 13.5.** Adaline

$w$, since the matrix $X_d X_d^T$ is a positive definite matrix. To solve this optimization problem, we have at our disposal the whole slew of unconstrained optimization algorithms considered in the previous chapters. For example, we can use a gradient algorithm, which in this case takes the form

$$w^{(k+1)} = w^{(k)} + \alpha_k X_d e^{(k)},$$

where $e^{(k)} = y_d - X_d^T w^{(k)}$.

## 13.3. THE BACKPROPAGATION ALGORITHM

In the previous section, we considered the problem of training a single neuron. In this section, we consider a neural network consisting of many layers. For simplicity of presentation, we restrict our attention to networks with three layers, as depicted in Figure 13.6. The three layers are referred to as the input, hidden, and output layers. There are $n$ inputs $x_i$, where $i = 1, \ldots, n$. We have $m$ outputs $y_s$, $s = 1, \ldots, m$. There are $l$ neurons in the hidden layer. The outputs of the neurons in the hidden layer are $z_j$, where $j = 1, \ldots, l$. The inputs $x_1, \ldots, x_n$ are distributed to the neurons in the hidden layer. We may think of the neurons in the input layer as single-input–single-output linear elements, with each activation function being the identity map. In Figure 13.6, we do not explicitly depict the neurons in the input layer; instead, we illustrate the neurons as signal spliters. We denote the activation functions of the neurons in the hidden layer by $f_j^h$, where $j = 1, \ldots, l$, and the activation functions of the neurons in the output layer by $f_s^o$, where $s = 1, \ldots, m$. Note that each activation function is a function from $\mathbb{R}$ to $\mathbb{R}$.

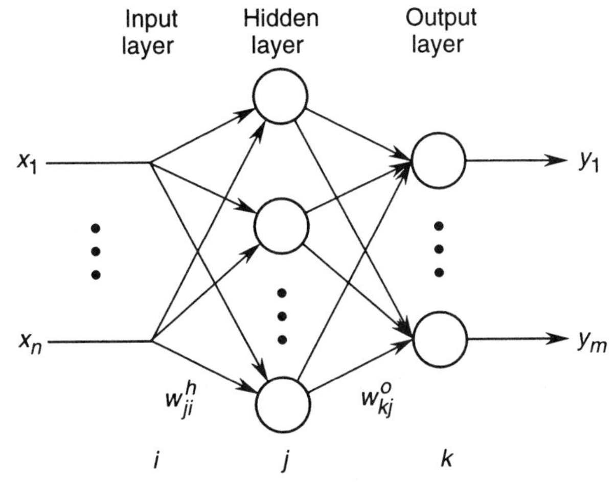


**Figure 13.6.** A three-layered neural network

We denote the weights for inputs into the hidden layer by $w_{ji}^h$, $i = 1, \ldots, n$, $j = 1, \ldots, l$. We denote the weights for inputs from the hidden layer into the output layer by $w_{sj}^o$, $j = 1, \ldots, l$, $s = 1, \ldots, m$. Given the weights $w_{ji}^h$ and $w_{sj}^o$, the neural network implements a map from $\mathbb{R}^n$ to $\mathbb{R}^m$. To find an explicit formula for this map, let us denote the input to the $j$th neuron in the hidden layer by $v_j$, and the output of the $j$th neuron in the hidden layer by $z_j$. Then, we have

$$v_j = \sum_{i=1}^{n} w_{ji}^h x_i,$$

$$z_j = f_j^h\left(\sum_{i=1}^{n} w_{ji}^h x_i\right).$$

The output from the $s$th neuron of the output layer is

$$y_s = f_s^o\left(\sum_{j=1}^{l} w_{sj}^o z_j\right).$$

Therefore, the relationship between the inputs $x_i$, $i = 1, \ldots, n$, and the $s$th output $y_s$ is given by

$$\begin{aligned} y_s &= f_s^o\left(\sum_{j=1}^{l} w_{sj}^o f_j^h(v_j)\right) \\ &= f_s^o\left(\sum_{j=1}^{l} w_{sj}^o f_j^h\left(\sum_{i=1}^{n} w_{ji}^h x_i\right)\right) \\ &= F_s(x_1, \ldots, x_n). \end{aligned}$$

The overall mapping that the neural network implements is therefore given by

$$\begin{bmatrix} y_1 \\ \vdots \\ y_m \end{bmatrix} = \begin{bmatrix} F_1(x_1, \ldots, x_n) \\ \vdots \\ F_m(x_1, \ldots, x_n) \end{bmatrix}.$$

We now consider the problem of training the neural network. As for the single neuron considered in the last section, we analyze the case where the training set consists of a single pair $(\boldsymbol{x}_d, \boldsymbol{y}_d)$, where $\boldsymbol{x}_d \in \mathbb{R}^n$ and $\boldsymbol{y}_d \in \mathbb{R}^m$. In practice, the training set consists of many such pairs, and training is typically performed with each pair at a time [see, for example, Refs. 37 and 66]. Our analysis is therefore also relevant to the general training problem with multiple training pairs.

The training of the neural network involves adjusting the weights of the network such that the output generated by the network for the given input $\boldsymbol{x}_d = [x_{d1}, \ldots, x_{dn}]^T$ is as "close" to $\boldsymbol{y}_d$ as possible. Formally, this can be formulated as the following optimization problem:

$$\text{minimize} \ \frac{1}{2} \sum_{s=1}^{m} (y_{ds} - y_s)^2,$$

where $y_s$, $s = 1, \ldots, m$, are the actual outputs of the neural network in response to the inputs $x_{d1}, \ldots, x_{dn}$, as given by

$$y_s = f_s^o\left(\sum_{j=1}^{l} w_{sj}^o f_j^h\left(\sum_{i=1}^{n} w_{ji}^h x_i\right)\right).$$

The above minimization is taken over all $w_{ji}^h, w_{sj}^o, i = 1, \ldots, n, j = 1, \ldots, l, s = 1, \ldots, m$. For simplicity of notation, we use the symbol $\boldsymbol{w}$ for the vector

$$\boldsymbol{w} = \{w_{ji}^h, w_{sj}^o : i = 1, \ldots, n, j = 1, \ldots, l, s = 1, \ldots, m\},$$

and the symbol $E$ for the objective function to be minimized, that is,

$$\begin{aligned} E(\boldsymbol{w}) &= \frac{1}{2}\sum_{s=1}^{m}(y_{ds} - y_s)^2 \\ &= \frac{1}{2}\sum_{s=1}^{m}\left(y_{ds} - f_s^o\left(\sum_{j=1}^{l} w_{sj}^o f_j^h\left(\sum_{i=1}^{n} w_{ji}^h x_{di}\right)\right)\right)^2. \end{aligned}$$

To solve the above optimization problem, we use a gradient algorithm with fixed step size. To formulate the algorithm, we will need to compute the partial derivatives of $E$ with respect to each component of $\boldsymbol{w}$. For this, let us first fix the indices $i$, $j$, and $s$. We first compute the partial derivative of $E$ with respect to $w_{sj}^o$. For this, we write

$$E(\boldsymbol{w}) = \frac{1}{2}\sum_{p=1}^{m}\left(y_{dp} - f_p^o\left(\sum_{q=1}^{l} w_{pq}^o z_q\right)\right)^2,$$

where, for each $q = 1, \ldots, l$,

$$z_q = f_q^h\left(\sum_{i=1}^{n} w_{qi}^h x_{di}\right).$$

Using the chain rule, we obtain

$$\frac{\partial E}{\partial w_{sj}^o}(\boldsymbol{w}) = -(y_{ds} - y_s) f_s^{o\prime}\left(\sum_{q=1}^{l} w_{sq}^o z_q\right) z_j,$$

where $f_s^{o\prime} : \mathbb{R} \to \mathbb{R}$ is the derivative of $f_s^o$. For simplicity of notation, we write

$$\delta_s = (y_{ds} - y_s) f_s^{o\prime}\left(\sum_{q=1}^{l} w_{sq}^o z_q\right).$$

We can think of each $\delta_s$ as a scaled output error, since it is the difference between the actual output $y_s$ of the neural network and the desired output $y_{ds}$, scaled by

$f_s^{o'}(\sum_{q=1}^{l} w_{sq}^o z_q)$. Using the $\delta_s$ notation, we have

$$\frac{\partial E}{\partial w_{sj}^o}(\boldsymbol{w}) = -\delta_s z_j.$$

We next compute the partial derivative of $E$ with respect to $w_{ji}^h$. We start with the equation

$$E(\boldsymbol{w}) = \frac{1}{2}\sum_{p=1}^{m}\left(y_{dp} - f_p^o\left(\sum_{q=1}^{l} w_{pq}^o f_q^h\left(\sum_{r=1}^{n} w_{qr}^h x_{dr}\right)\right)\right)^2.$$

Using the chain rule once again, we get

$$\frac{\partial E}{\partial w_{ji}^h}(\boldsymbol{w}) = -\sum_{p=1}^{m}(y_{dp} - y_p) f_p^{o'}\left(\sum_{q=1}^{l} w_{pq}^o z_q\right) w_{pj}^o f_j^{h'}\left(\sum_{r=1}^{n} w_{jr}^h x_{dr}\right) x_{di},$$

where $f_j^{h'}: \mathbb{R} \to \mathbb{R}$ is the derivative of $f_j^h$. Simplifying the above yields

$$\frac{\partial E}{\partial w_{ji}^h}(\boldsymbol{w}) = -\left(\sum_{p=1}^{m} \delta_p w_{pj}^o\right) f_j^{h'}(v_j) x_{di}.$$

We are now ready to formulate the gradient algorithm for updating the weights of the neural network. We will write the update equations for the two sets of weights $w_{sj}^o$ and $w_{ji}^h$ separately. We have

$$w_{sj}^{o(k+1)} = w_{sj}^{o(k)} + \eta \delta_s^{(k)} z_j^{(k)}$$

$$w_{ji}^{h(k+1)} = w_{ji}^{h(k)} + \eta \left(\sum_{p=1}^{m} \delta_p^{(k)} w_{pj}^{o(k)}\right) f_j^{h'}(v_j^{(k)}) x_{di},$$

where $\eta$ is the (fixed) step size and

$$v_j^{(k)} = \sum_{i=1}^{n} w_{ji}^{h(k)} x_{di}$$

$$z_j^{(k)} = f_j^h(v_j^{(k)})$$

$$y_s^{(k)} = f_s^o\left(\sum_{q=1}^{l} w_{sq}^{o(k)} z_q^{(k)}\right)$$

$$\delta_s^{(k)} = (y_{ds} - y_s^{(k)}) f_s^{o'}\left(\sum_{q=1}^{l} w_{sq}^{o(k)} z_q^{(k)}\right).$$

The update equation for the weights $w_{sj}^o$ of the output layer neurons is illustrated in Figure 13.7, whereas the update equation for the weights $w_{ji}^h$ of the hidden layer neurons is illustrated in Figure 13.8.

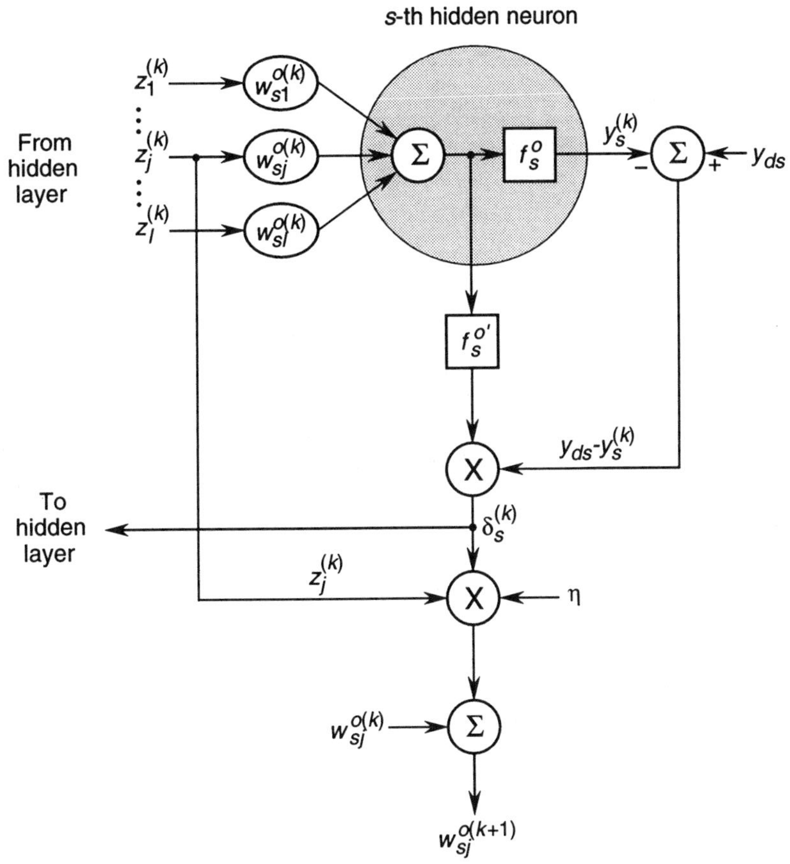


**Figure 13.7.** Illustration of the update equation for the output layer weights

The above update equations are referred to in the literature as the *backpropagation algorithm*. The reason for the name "backpropagation" is that the output errors $\delta_1^{(k)}, \ldots, \delta_m^{(k)}$ are "propagated back" from the output layer to the hidden layer, and are used in the update equation for the hidden layer weights, as illustrated in Figure 13.8. In the above discussion, we assumed only a single hidden layer. In general, we may have multiple hidden layers—in this case, the update equations for the weights will resemble the equations derived above. In the general case, the output errors are propagated backward from layer to layer and are used to update the weights at each layer.

We summarize the backpropagation algorithm qualitatively as follows. Using the inputs $x_{di}$ and the current set of weights, we first compute the quantities $v_j^{(k)}$, $z_j^{(k)}$, $y_s^{(k)}$, and $\delta_s^{(k)}$, in turn. This is called the *forward pass* of the algorithm, because it involves propagating the input forward from the input layer to the output layer. Next, we compute the updated weights using the quantities computed in the

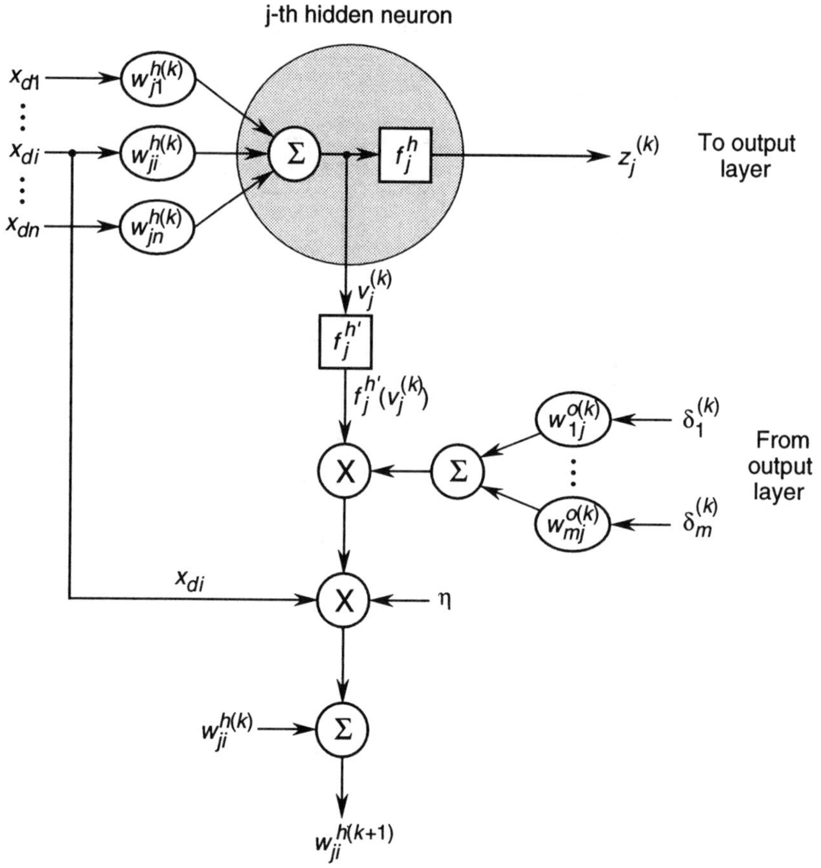


**Figure 13.8.** Illustration of the update equation for the hidden layer weights

forward pass. This is called the *reverse pass* of the algorithm, because it involves propagating the computed output errors $\delta_s^{(k)}$ backward through the network. We illustrate the backpropagation procedure numerically in the following example.

***Example 13.1.*** Consider the simple feedforward neural network shown in Figure 13.9. The activation functions for all the neurons are given by $f(v) = 1/(1 + e^{-v})$. This particular activation function has the convenient property that $f'(v) = f(v)(1 - f(v))$. Therefore, using this property, we can write

$$\begin{aligned}
\delta_1 &= (y_d - y_1) f'\left( \sum_{q=1}^{2} w_{1q}^o z_q \right) \\
&= (y_d - y_1) f\left( \sum_{q=1}^{2} w_{1q}^o z_q \right)\left( 1 - f\left( \sum_{q=1}^{2} w_{1q}^o z_q \right)\right) \\
&= (y_d - y_1) y_1 (1 - y_1).
\end{aligned}$$

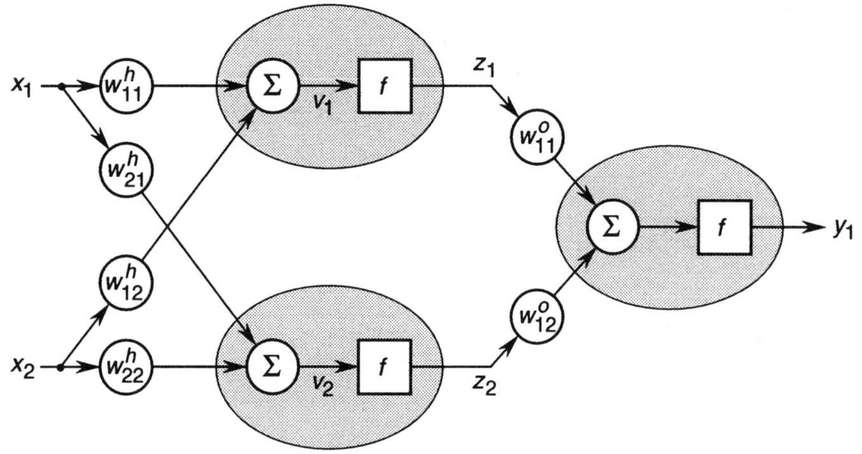


**Figure 13.9.** Neural network for Example 13.1

Suppose the initial weights are $w_{11}^{h(0)} = 0.1$, $w_{12}^{h(0)} = 0.3$, $w_{21}^{h(0)} = 0.3$, $w_{22}^{h(0)} = 0.4$, $w_{11}^{o(0)} = 0.4$, and $w_{12}^{o(0)} = 0.6$. Let $\boldsymbol{x}_d = [0.2, 0.6]^T$ and $y_d = 0.7$. Perform one iteration of the backpropagation algorithm to update the weights of the network. Use a step size of $\eta = 10$.

To proceed, we first compute

$$v_1^{(0)} = w_{11}^{h(0)} x_{d1} + w_{12}^{h(0)} x_{d2} = 0.2$$
$$v_2^{(0)} = w_{21}^{h(0)} x_{d1} + w_{22}^{h(0)} x_{d2} = 0.3.$$

Next, we compute

$$z_1^{(0)} = f(v_1^{(0)}) = \frac{1}{1 + e^{-0.2}} = 0.5498$$

$$z_2^{(0)} = f(v_2^{(0)}) = \frac{1}{1 + e^{-0.3}} = 0.5744.$$

We then compute

$$y_1^{(0)} = f(w_{11}^{o(0)} z_1^{(0)} + w_{12}^{o(0)} z_2^{(0)}) = f(0.5646) = 0.6375,$$

which gives an output error of

$$\delta_1^{(0)} = (y_d - y_1^{(0)}) y_1^{(0)} (1 - y_1^{(0)}) = 0.01444.$$

This completes the forward pass.

To update the weights, we use

$$w_{11}^{o(1)} = w_{11}^{o(0)} + \eta \delta_1^{(0)} z_1^{(0)} = 0.4794$$
$$w_{12}^{o(1)} = w_{12}^{o(0)} + \eta \delta_1^{(0)} z_2^{(0)} = 0.6830,$$

and, using the fact that $f'(v_j^{(0)}) = f(v_j^{(0)})(1 - f(v_j^{(0)})) = z_j^{(0)}(1 - z_j^{(0)})$, we get

$$w_{11}^{h(1)} = w_{11}^{h(0)} + \eta \delta_1^{(0)} w_{11}^{o(0)} z_1^{(0)}(1 - z_1^{(0)}) x_{d1} = 0.1029$$
$$w_{12}^{h(1)} = w_{12}^{h(0)} + \eta \delta_1^{(0)} w_{11}^{o(0)} z_1^{(0)}(1 - z_1^{(0)}) x_{d2} = 0.3086$$
$$w_{21}^{h(1)} = w_{21}^{h(0)} + \eta \delta_1^{(0)} w_{12}^{o(0)} z_2^{(0)}(1 - z_2^{(0)}) x_{d1} = 0.3042$$
$$w_{22}^{h(1)} = w_{22}^{h(0)} + \eta \delta_1^{(0)} w_{12}^{o(0)} z_2^{(0)}(1 - z_2^{(0)}) x_{d2} = 0.4127.$$

Thus, we have completed one iteration of the backpropagation algorithm. We can easily check that $y_1^{(1)} = 0.6588$, and hence $|y_d - y_1^{(1)}| < |y_d - y_1^{(0)}|$; that is, the actual output of the neural network has become closer to the desired output as a result of updating the weights.

After 15 iterations of the backpropagation algorithm, we get

$$w_{11}^{o(15)} = 0.6365$$
$$w_{12}^{o(15)} = 0.8474$$
$$w_{11}^{h(15)} = 0.1105$$
$$w_{12}^{h(15)} = 0.3315$$
$$w_{21}^{h(15)} = 0.3146$$
$$w_{22}^{h(15)} = 0.4439.$$

The resulting value of the output corresponding to the input $\boldsymbol{x}_d = [0.2, 0.6]^T$ is $y_1^{(15)} = 0.6997$.

In the above example, we considered an activation function of the form

$$f(v) = \frac{1}{1 + e^{-v}}.$$

The above function is called a *sigmoid*, and is a popular activation function used in practice. The sigmoid function is illustrated in Figure 13.10. It is possible to use a more general version of the sigmoid function, of the form

$$g(v) = \frac{\beta}{1 + e^{-(v-\theta)}}.$$

The parameters $\beta$ and $\theta$ represent scale and shift parameters, respectively. The

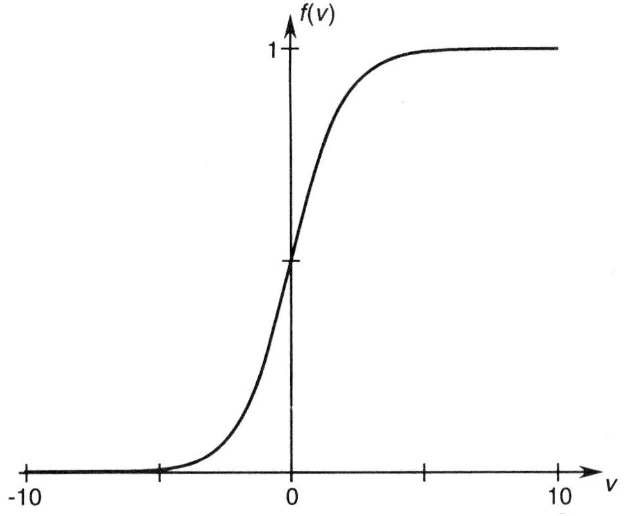


**Figure 13.10.** The sigmoid function

parameter $\theta$ is often interpreted as a threshold. If such an activation function is used in a neural network, we would also want to adjust the values of the parameters $\beta$ and $\theta$, which also affect the value of the objective function to be minimized. However, it turns out that these parameters can be incorporated into the backpropagation algorithm simply by treating them as additional weights in the network. Specifically, we can represent a neuron with activation function $g$ as one with activation function $f$ with the addition of two extra weights, as shown in Figure 13.11.

***Example 13.2.*** Consider the same neural network as in Example 13.1. We introduce shift parameters $\theta_1$, $\theta_2$, and $\theta_3$ to the activation functions in the neurons. Using the the configuration illustrated in Figure 13.11, we can incorporate the shift parameters into the backpropagation algorithm. We have

$$\begin{aligned}
v_1 &= w_{11}^h x_{d1} + w_{12}^h x_{d2} - \theta_1 \\
v_2 &= w_{21}^h x_{d1} + w_{22}^h x_{d2} - \theta_2 \\
z_1 &= f(v_1) \\
z_2 &= f(v_2) \\
y_1 &= f(w_{11}^o z_1 + w_{12}^o z_2 - \theta_3) \\
\delta_1 &= (y_d - y_1) y_1 (1 - y_1),
\end{aligned}$$

where $f$ is the sigmoid function

$$f(v) = \frac{1}{1 + e^{-v}}.$$

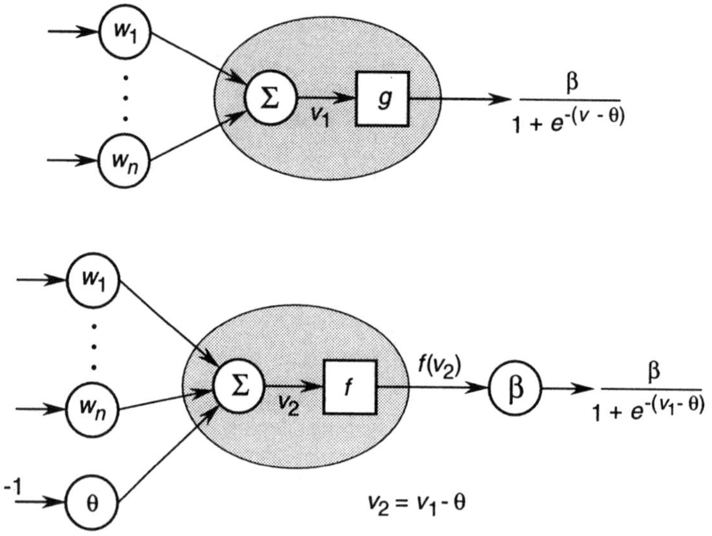


**Figure 13.11.** These two configurations are equivalent

The components of the gradient of the objective function $E$ with respect to the shift parameters are

$$\frac{\partial E}{\partial \theta_1}(\boldsymbol{w}) = \delta_1 w_{11}^o z_1(1 - z_1)$$

$$\frac{\partial E}{\partial \theta_2}(\boldsymbol{w}) = \delta_1 w_{12}^o z_2(1 - z_2)$$

$$\frac{\partial E}{\partial \theta_3}(\boldsymbol{w}) = \delta_1.$$

■

In the next example, we apply the network discussed in Example 13.2 to solve the celebrated Exclusive OR (XOR) problem [see Ref. 66].

***Example 13.3.*** Consider the neural network of Example 13.2. We wish to train the neural network to approximate the *Exclusive OR* (XOR) function, defined in Table 13.1. Note that the XOR function has two inputs and one output.

To train the neural network, we use the following training pairs:

$$\boldsymbol{x}_{d,0} = [0, 0]^T, \quad y_{d,0} = 0$$

$$\boldsymbol{x}_{d,1} = [0, 1]^T, \quad y_{d,1} = 1$$

$$\boldsymbol{x}_{d,2} = [1, 0]^T, \quad y_{d,2} = 1$$

$$\boldsymbol{x}_{d,3} = [1, 1]^T, \quad y_{d,3} = 0.$$

**Table 13.1 Truth Table for XOR Function**

| $x_1$ | $x_2$ | $F(x_1, x_2)$ |
|---|---|---|
| 0 | 0 | 0 |
| 0 | 1 | 1 |
| 1 | 0 | 1 |
| 1 | 1 | 0 |

We now apply the backpropagation algorithm to train the network using the above training pairs. To do this, we apply the above pairs one per iteration, in a cyclic fashion. In other words, in the $k$th iteration of the algorithm, we apply the pair $(\boldsymbol{x}_{d,R(k)}, y_{d,R(k)})$, where, as in Kaczmarz's algorithm, $R(k)$ is the unique integer in $\{0, 1, 2, 3\}$ satisfying $k = 4l + R(k)$ for some integer $l$, that is, $R(k)$ is the remainder that results if we divide $k$ by 4 (see Section 12.4).

The experiment yields the following weights (see Exercise 13.5):

$$
\begin{aligned}
w_{11}^o &= -11.01 \\
w_{12}^o &= 10.92 \\
w_{11}^h &= -7.777 \\
w_{12}^h &= -8.403 \\
w_{21}^h &= -5.593 \\
w_{22}^h &= -5.638 \\
\theta_1 &= -3.277 \\
\theta_2 &= -8.357 \\
\theta_3 &= 5.261.
\end{aligned}
$$

Table 13.2 shows the output of the network with the above weights corresponding to the training input data. ■

For a more comprehensive treatment of neural networks, see Refs. 31, 32, and 82.

**Table 13.2 Response of the Trained Network for Example 13.3**

| $x_1$ | $x_2$ | $y_1$ |
|---|---|---|
| 0 | 0 | 0.007 |
| 0 | 1 | 0.99 |
| 1 | 0 | 0.99 |
| 1 | 1 | 0.009 |

## EXERCISES

**13.1** Consider a single linear neuron with $n$ inputs (see Figure 13.4). Suppose we are given $X_d \in \mathbb{R}^{n \times p}$ and $y_d \in \mathbb{R}^p$ representing $p$ training pairs, where $p > n$. The objective function to be minimized in the training of the neuron is

$$f(w) = \frac{1}{2} \| y_d - X_d^T w \|^2.$$

**a.** Find the gradient of the objective function.

**b.** Write the conjugate gradient algorithm for training the neuron.

**c.** Suppose we wish to approximate the function $F: \mathbb{R}^2 \to \mathbb{R}$ given by

$$F(x) = (\sin x_1)(\cos x_2).$$

Use the conjugate gradient algorithm from part b to train the linear neuron, using the following training points:

$$\{x : x_1, x_2 = -0.5, 0, 0.5\}.$$

It may be helpful to use the MATLAB routine from Exercise 10.6.

**d.** Plot the level sets of the objective function for the problem in part c, at levels 0.01, 0.1, 0.2, 0.4. Check if the solution in part c agrees with the level sets.

**e.** Plot the error function $e(x) = F(x) - w^{*T} x$ versus $x_1$ and $x_2$, where $w^*$ is the optimal weight vector obtained in part c.

**13.2** Consider the Adaline, depicted in Figure 13.5. Assume we have a single training pair $(x_d, y_d)$, where $x_d \neq 0$. Suppose we use the Widrow–Hoff algorithm to adjust the weights:

$$w^{(k+1)} = w^{(k)} + \mu \frac{e_k x_d}{x_d^T x_d},$$

where $e_k = y_d - x_d^T w^{(k)}$.

**a.** Write an expression for $e_{k+1}$ as a function of $e_k$ and $\mu$.

**b.** Find the largest range of values for $\mu$ for which $e_k \to 0$ (for any initial condition $w^{(0)}$).

**13.3** As in Exercise 13.2, consider the Adaline. Consider the case in which there are multiple pairs in the training set $\{(x_{d,1}, y_{d,1}), \ldots, (x_{d,p}, y_{d,p})\}$, where $p \leq n$, and rank $X_d = p$ (the matrix $X_d$ has $x_{d,i}$ as its $i$th column). Suppose we use the following training algorithm:

$$w^{(k+1)} = w^{(k)} + X_d (X_d^T X_d)^{-1} \mu e^{(k)},$$

where $e^{(k)} = y_d - X_d^T w^{(k)}$, and $\mu$ is a given constant $p \times p$ matrix.

**a.** Find an expression for $\boldsymbol{e}^{(k+1)}$ as a function of $\boldsymbol{e}^{(k)}$ and $\boldsymbol{\mu}$.

**b.** Find a necessary and sufficient condition on $\boldsymbol{\mu}$ for which $\boldsymbol{e}^{(k)} \to \boldsymbol{0}$ (for any initial condition $\boldsymbol{w}^{(0)}$).

**13.4** Consider the three-layered neural network described in Example 13.1 (see Figure 13.9). Implement the backpropagation algorithm for this network in MATLAB. Test the algorithm for the training pair $\boldsymbol{x}_d = [0, 1]^T$ and $y_d = 1$. Use a step size of $\eta = 50$ and initial weights as in Example 13.1.

**13.5** Consider the neural network of Example 13.3, with the training pairs for the XOR problem. Use MATLAB to implement the training algorithm described in Example 13.3, with a step size of $\eta = 10.0$. Tabulate the outputs of the trained network corresponding to the training input data.

# 14

# Genetic Algorithms

## 14.1. BASIC DESCRIPTION

In this chapter, we discuss genetic algorithms and their application to solving optimization problems. Genetic algorithms are radically different from the optimization algorithms discussed in previous chapters. For example, genetic algorithms do not use gradients or Hessians. Consequently, they are applicable to a much wider class of optimization problems.

A genetic algorithm is a probabilistic search technique that has its roots in the principles of genetics. The beginnings of genetic algorithms is credited to John Holland, who developed the basic ideas in the late 1960s and early 1970s. Since its conception, genetic algorithms have been widely used as a tool in computer programming and artificial intelligence [e.g., Refs. 34, 47], optimization [e.g., Refs. 17, 75], neural network training [e.g., Refs. 48], and many other areas.

Suppose we wish to solve an optimization problem of the form

$$\begin{aligned} &\text{maximize} \quad f(\boldsymbol{x}) \\ &\text{subject to} \quad \boldsymbol{x} \in \Omega. \end{aligned}$$

The underlying idea of genetic algorithms applied to the above problem is as follows. We start with an initial set of points in $\Omega$, denoted by $P(0)$. We call $P(0)$ the *initial population.* We then evaluate the objective function at points in $P(0)$. Based on this evaluation, we create a new set of points $P(1)$. The creation of $P(1)$ involves certain operations on points in $P(0)$, called *crossover* and *mutation*, to be discussed later. We repeat the above procedure iteratively, generating populations $P(2)$, $P(3), \ldots,$ until an appropriate stopping criterion is reached. The purpose of the crossover and mutation operations is to create a new population with an average objective function value that is higher than the previous population. To summarize, the genetic algorithm iteratively performs the operations of crossover and mutation on each population to produce a new population until a chosen termination criterion is met.

The terminology used in describing genetic algorithms is adopted from genetics. To proceed with describing the details of the algorithm, we need the additional ideas and terms described below.

### 14.1.1. Chromosomes and Representation Schemes

First, we point out that genetic algorithms in fact do not work directly with points in the set $\Omega$, but rather with an encoding of the points in $\Omega$. Specifically, we need to first map $\Omega$ onto a set consisting of strings of symbols, all of equal length. The strings are called *chromosomes*. Each chromosome consists of elements from a chosen set of symbols, called the *alphabet*. For example, a common alphabet is the set $\{0, 1\}$, in which case the chromosomes are simply binary strings. We denote by $L$ the length of chromosomes (that is, the number of symbols in the strings). To each chromosome there corresponds a value of the objective function, referred to as the *fitness* of the chromosome. For each chromosome $\boldsymbol{x}$, we write $f(\boldsymbol{x})$ for its fitness. Note that, for convenience, we use $f$ to denote both the original objective function as well as the fitness measure on the set of chromosomes.

The choice of chromosome length, alphabet, and encoding (that is, the mapping from $\Omega$ onto the set of chromosomes), is called the *representation scheme* for the problem. Identification of an appropriate representation scheme is the first step in using genetic algorithms to solve a given optimization problem.

Once a suitable representation scheme has been chosen, the next phase is to initialize the first population $P(0)$ of chromosomes. This is usually done by a random selection of a set of chromosomes. After we form the initial population of chromosomes, we then apply the operations of crossover and mutation on the population. During each iteration $k$ of the process, we evaluate the fitness $f(\boldsymbol{x}^{(k)})$ of each member $\boldsymbol{x}^{(k)}$ of the population $P(k)$. After the fitness of the whole population has been evaluated, we then form a new population $P(k+1)$ in two stages.

### 14.1.2. Selection and Evolution

In the first stage, we apply an operation called *selection*, where we form a set $M(k)$ with the same number of elements as $P(k)$. The set $M(k)$, called the *mating pool*, is formed from $P(k)$ using a random procedure as follows: each point $\boldsymbol{m}^{(k)}$ in $M(k)$ is equal to $\boldsymbol{x}^{(k)}$ in $P^{(k)}$ with probability

$$\frac{f(\boldsymbol{x}^{(k)})}{F(k)},$$

where

$$F(k) = \sum f(\boldsymbol{x}_i^{(k)})$$

and the sum is taken over the whole of $P(k)$. In other words, we select chromosomes into the mating pool with probabilities proportional to their fitness.

The second stage is called *evolution*: in this stage, we apply the crossover and mutation operations. The *crossover* operation takes a pair of chromosomes, called the *parents*, and gives a pair of *offspring* chromosomes. The operation involves exchanging substrings of the two parent chromosomes, described below. Pairs of parents for crossover are chosen from the mating pool randomly, such that the probability that a chromosome is chosen for crossover is $p_c$. We assume that whether a given chromosome is chosen or not is independent of whether or not any other chromosome is chosen for crossover.

We can pick parents for crossover in several ways. For example, we may randomly choose two chromosomes from the mating pool as parents. In this case, if $N$ is the number of chromosomes in the mating pool, then $p_c = 2/N$. Similarly, if we randomly pick $2k$ chromosomes from the mating pool (where $k < N/2$), forming $k$ pairs of parents, we have $p_c = 2k/N$.

Once the parents for crossover have been determined, we apply the crossover operation to the parents. There are many types of possible crossover operations. The simplest crossover operation is the *one-point crossover*. In this operation, we first choose a number randomly between 1 and $L-1$ according to a uniform distribution, where $L$ is the length of chromosomes. We refer to this number as the *crossing site*. Crossover then involves exchanging substrings of the parents to the left of the crossing site, as illustrated in the following example.

***Example 14.1.*** Suppose we have chromosomes of length $L=6$ over the binary alphabet $\{0,1\}$. Consider the pair of parents 000000 and 111111. Suppose the crossing site is 4. Then, the crossover operation applied to the above parent chromosomes yields the two offsprings 000011 and 111100. ■

We can also have crossover operations with multiple crossing sites, as illustrated in this example .

***Example 14.2.*** Consider two chromosomes of length $L=9$: 000000000 and 111111111. Suppose we have two crossing sites, at 3 and 7. Then, the crossover operation applied to the above parent chromosomes yields the two offsprings 000111100 and 111000011. ■

After the crossover operation, we replace the parents in the mating pool by their offsprings. The mating pool has therefore been modified, but still maintains the same number of elements.

Next, we apply the mutation operation. The mutation operation takes each chromosome from the mating pool and randomly changes each symbol of the chromosome with a given probability $p_m$. In the case of the binary alphabet, this change corresponds to complementing the corresponding bits; that is, we replace each bit with probability $p_m$ from 0 to 1, or vice versa. If the alphabet contains more than two symbols, then the change involves randomly substituting the symbol with another symbol from the alphabet. Typically, the value of $p_m$ is very small, so that only a few chromosomes will undergo a change due to mutation, and, of those that are affected, only a few of the symbols are modified. Therefore, the mutation operation plays only a minor role in the genetic algorithm relative to the crossover operation.

After applying the crossover and mutation operations to the mating pool $M(k)$, we obtain the new population $P(k+1)$. We then repeat the procedure of evaluation, selection, and evolution, iteratively. We summarize the genetic algorithm as follows.

### Genetic Algorithm

1. Set $k := 0$; form initial population $P(0)$
2. Evaluate $P(k)$
3. If stopping criterion satisfied, then stop
4. Select $M(k)$ from $P(k)$
5. Evolve $M(k)$ to form $P(k+1)$
6. Set $k := k+1$, go to step 2

A flow chart illustrating the above algorithm is shown in Figure 14.1.

During the execution of the genetic algorithm, we keep track of the best-so-far chromosome—the chromosome with the highest fitness of all the chromosomes

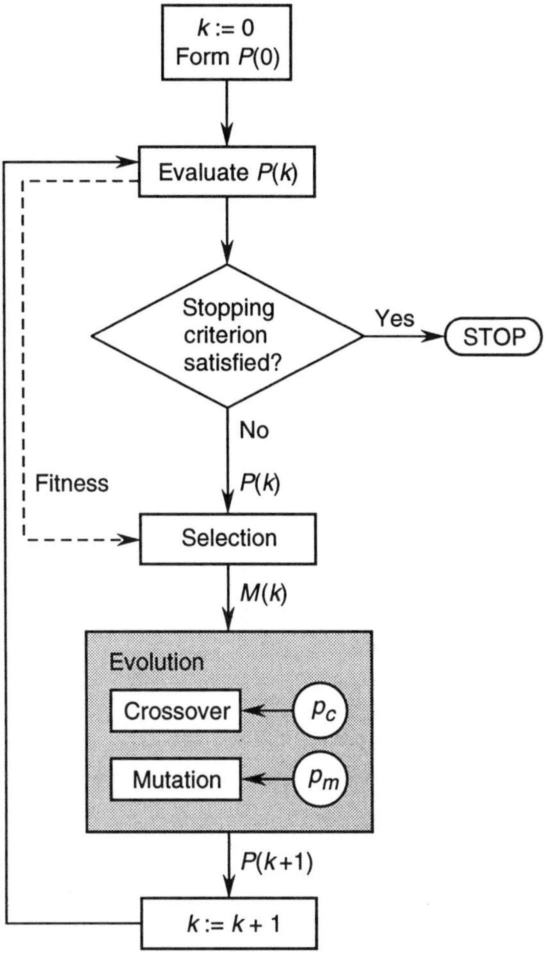


**Figure 14.1.** Flow chart for the genetic algorithm

evaluated. After each iteration, the best-so-far chromosome serves as the candidate for the solution to the original problem.

The stopping criterion can be implemented in a number of ways. For example, a simple stopping criterion is to stop after a prespecified number of iterations. Another possible criterion is to stop when the fitness for the best-so-far chromosome does not change significantly from iteration to iteration.

The genetic algorithm differs from the algorithms discussed in previous chapters in several respects:

1. It works with an encoding of the feasible set rather than the set itself
2. It searches from a set of points rather than a single point at each iteration
3. It does not use derivatives of the objective function
4. It uses operations that are random within each iteration

Application of the genetic algorithm to an optimization problem is illustrated in the following example.

***Example 14.3.*** Consider the function $f: \mathbb{R}^2 \to \mathbb{R}$ given by

$$f(x, y) = 3(1-x)^2 e^{-x^2-(y+1)^2} - 10\left(\frac{x}{5} - x^3 - y^5\right)e^{-x^2-y^2} - \frac{e^{-(x+1)^2-y^2}}{3}.$$

The above function comes from Ref. 1 (p. 2-61). We wish to maximize $f$ over the set $\Omega = \{[x, y]^T \in \mathbb{R}^2 : -3 \leqslant x, y \leqslant 3\}$. A plot of the objective function $f$ over the feasible set $\Omega$ is shown in Figure 14.2. Using MATLAB, we found the optimal point to be $[-0.0303, 1.5455]^T$, with objective function value 8.0926.

To apply the genetic algorithm to solve the above optimization problem, we use a simple binary representation scheme with length $L = 32$, where the first 16 bits of each chromosome encode the $x$ component, while the remaining 16 bits encode the $y$ component. Recall that $x$ and $y$ take values in the interval $[-3, 3]$. We first map the interval $[-3, 3]$ onto the interval $[0, 2^{16} - 1]$, via a simple translation and scaling. The integers in the interval $[0, 2^{16} - 1]$ are then expressed as binary 16 bit strings. This defines the encoding of each component $x$ and $y$. The chromosome is obtained by juxtaposing the two 16 bit strings. For example, the point $[x, y]^T = [-1, 3]^T$ is encoded as (see Exercise 14.1 for a simple algorithm for converting from decimal into binary)

$$\underbrace{0101010101010101}_{\text{encoded } x = -1}\ \underbrace{1111111111111111}_{\text{encoded } y = 3}.$$

Using a population size of 20, we apply 20 iterations of the genetic algorithm on the above problem. Figure 14.3 shows plots of the objective function values for a single run of the algorithm, and averaged objective function values over 10 runs. The candidate solution from the single run is $[-0.004944, 1.645203]^T$, with

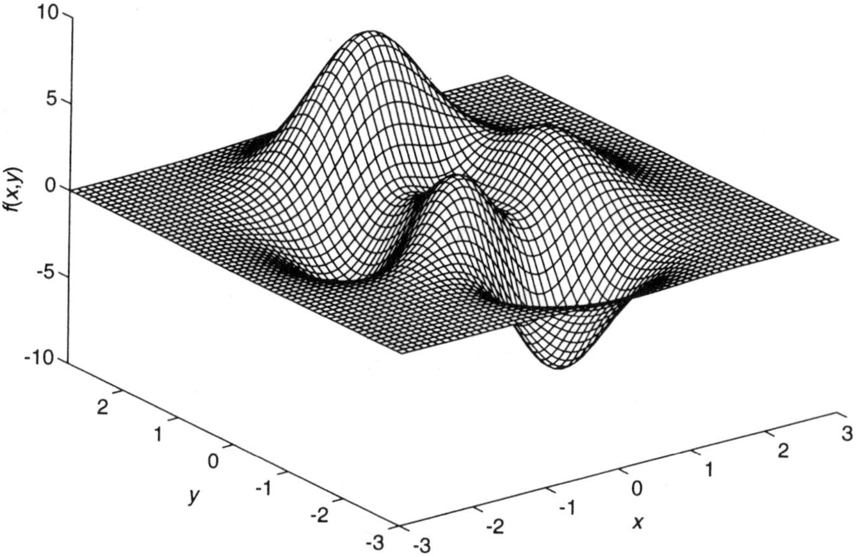


**Figure 14.2.** Plot of $f$ for Example 14.3

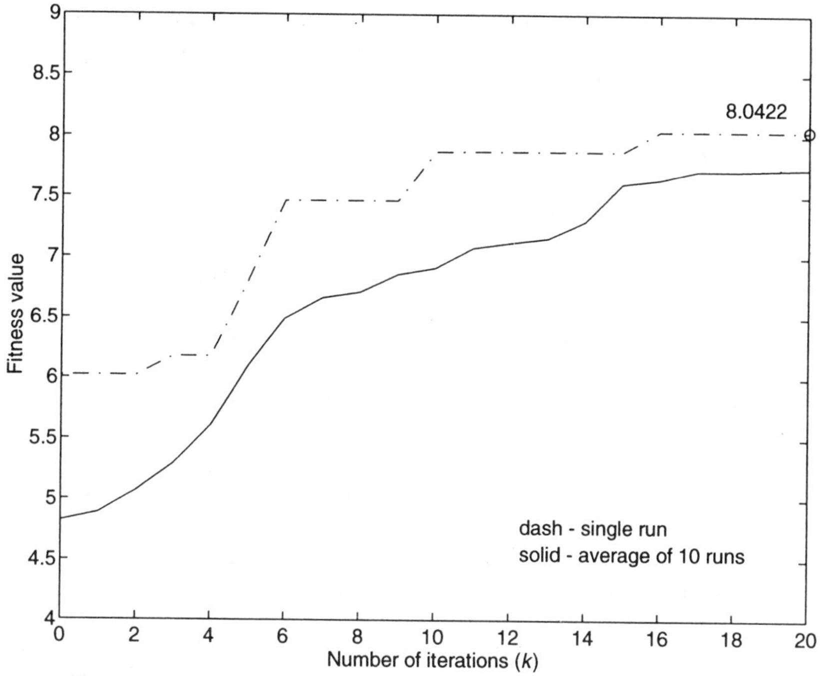


**Figure 14.3.** Values of the objective function for a single run and for an average of ten runs for Example 14.3

objective function value 8.042244. Note that the objective function value obtained for the candidate solution agrees with the one obtained using MATLAB to within 0.05. The purpose for showing the averaged objective function values over 10 runs is to illustrate the variation between individual runs of the algorithm. ■

## 14.2. ANALYSIS OF GENETIC ALGORITHMS

In this section, we use heuristic arguments to describe why genetic algorithms work. As pointed out before, the genetic algorithm was motivated by ideas from natural genetics [Ref. 34]. Specifically, the notion of "survival of the fittest" plays a central role. The mechanisms used in the genetic algorithm mimic this principle. We start with a population of chromosomes, and selectively pick the fittest ones for reproduction. From these selected chromosomes, we form the new generation by combining information encoded in them. In this way, the goal is to ensure that the fittest members of the population survive, and their information content is preserved and combined to produce even better offsprings.

To further analyze the genetic algorithm in a more quantitative fashion, we need to define a few terms. For convenience, we only consider chromosomes over the binary alphabet. We introduce the notion of a *schema* (plural: *schemata*) as a set of chromosomes with certain common features. Specifically, a schema is a set of chromosomes that contains 1s and 0s in particular locations. We represent a schema using a string notation over an extended alphabet $\{0, 1, *\}$. For example, the notation $1*01$ represents the schema

$$1*01 = \{1001, 1101\},$$

and the notation $0*101*$ represents the schema

$$0*101* = \{001010, 001011, 011010, 011011\}.$$

In the schema notation, the numbers 0 and 1 denote the fixed binary values in the chromosomes that belong to the schema. The symbol $*$, meaning "don't care," matches either 0 or 1 at the positions it occupies. Thus, a schema describes a set of chromosomes that have certain specified similarities. A chromosome belongs to a particular schema if for all positions $j = 1, \ldots, L$ the symbol found in the $j$th position of the chromosome matches the symbol found in the $j$th position of the schema, with the understanding that any symbol matches $*$. Note that if a schema has $r$ "don't care" symbols, then it contains $2^r$ chromosomes. Moreover, any chromosome of length $L$ belongs to $2^L$ schemata.

Given a schema that represents good solutions to our optimization problem, we would like the number of matching chromosomes in the population $P(k)$ to grow as $k$ increases. This growth is affected by several factors, which we analyze in the following discussion.

The first key idea in explaining why the genetic algorithm works is the observation that if a scheme has chromosomes with better-than-averge fitness, then the expected (mean) number of chromosomes matching this schema in the mating pool $M(k)$ is larger than the number of chromosomes matching this schema in the population $P(k)$. To quantify this assertion, we need some additional notation. Let $H$ be a given schema, and let $e(H,k)$ be the number of chromosomes in $P(k)$ that match $H$, that is, $e(H,k)$ is the number of elements in the set $P(k) \cap H$. Let $f(H,k)$ be the average fitness of chromosomes in $P(k)$ that match schema $H$. This means that if $H \cap P(k) = \{\boldsymbol{x}_1, \ldots, \boldsymbol{x}_{e(H,k)}\}$, then

$$f(H,k) = \frac{f(\boldsymbol{x}_1) + \cdots + f(\boldsymbol{x}_{e(H,k)})}{e(H,k)}.$$

Let $N$ be the number of chromosomes in the population, and $F(k)$ be the sum of the fitness values of chromosomes in $P(k)$, as before. Denote by $\bar{F}(k)$ the average fitness of chromosomes in the population; that is,

$$\bar{F}(k) = \frac{F(k)}{N} = \frac{1}{N}\sum f(\boldsymbol{x}_i^{(k)}).$$

Finally, let $m(H,k)$ be the number of chromosomes in $M(k)$ that match $H$, in other words, the number of elements in the set $M(k) \cap H$.

**Lemma 14.1.** *Let $H$ be a given schema, and $\mathscr{M}(H,k)$ the expected value of $m(H,k)$ given $P(k)$. Then,*

$$\mathscr{M}(H,k) = \frac{f(H,k)}{\bar{F}(k)} e(H,k). \qquad \square$$

*Proof.* Let $P(k) \cap H = \{\boldsymbol{x}_1, \ldots, \boldsymbol{x}_{e(H,k)}\}$. In the remainder of the proof, the term "expected" should be taken to mean "expected, given $P(k)$." For each element $\boldsymbol{m}^{(k)} \in M(k)$ and each $i = 1, \ldots, e(H,k)$, the probability that $\boldsymbol{m}^{(k)} = \boldsymbol{x}_i$ is given by $f(\boldsymbol{x}_i)/F(k)$. Therefore, the expected number of chromosomes in $M(k)$ equal to $\boldsymbol{x}_i$ is

$$N\frac{f(\boldsymbol{x}_i)}{F(k)} = \frac{f(\boldsymbol{x}_i)}{\bar{F}(k)}.$$

Hence, the expected number of chromosomes in $P(k) \cap H$ that are selected into $M(k)$ is

$$\begin{aligned}\sum_{i=1}^{e(H,k)} \frac{f(\boldsymbol{x}_i)}{\bar{F}(k)} &= e(H,k)\frac{\sum_{i=1}^{e(H,k)} f(\boldsymbol{x}_i)}{e(H,k)} \frac{1}{\bar{F}(k)} \\ &= \frac{f(H,k)}{\bar{F}(k)} e(H,k).\end{aligned}$$

Since any chromosome in $M(k)$ is also a chromosome in $P(k)$, the chromosomes in $M(k) \cap H$ are simply those in $P(k) \cap H$ that are selected into $M(k)$. Hence,

$$\mathcal{M}(H,k) = \frac{f(H,k)}{\bar{F}(k)} e(H,k). \qquad \blacksquare$$

The above lemma quantifies our assertion that if a schema $H$ has chromosomes with better than average fitness [that is, $f(H,k)/\bar{F}(k) > 1$], then the expected number of chromosomes matching $H$ in the mating pool $M(k)$ is larger than the number of chromosomes matching $H$ in the population $P(k)$.

We now analyze the effect of the evolution operations on the chromosomes in the mating pool. For this, we need to introduce two parameters that are useful in the characterization of a schema, namely its *order* and *length*. The order $o(S)$ of a schema $S$ is the number of fixed symbols (non-$*$ symbols) in its representation. If the length of chromosomes in $S$ is $L$, then $o(S)$ is $L$ minus the number of $*$ symbols in $S$. For example,

$$o(1*01) = 4 - 1 = 3,$$

whereas

$$o(0*1*01) = 6 - 2 = 4.$$

The length $l(S)$ of a schema $S$ is the distance between the first and last fixed symbols (i.e., the difference between the positions of the rightmost fixed symbol and the leftmost fixed symbol). For example,

$$l(1*01) = 4 - 1 = 3,$$

whereas

$$l(0*101*) = 5 - 1 = 4.$$

Note that for a schema $S$ with chromosomes of length $L$, the order $o(S)$ is a number between 0 and $L$, and the length $l(S)$ is a number between 0 and $L-1$. The order of a schema with all $*$ symbols is 0; its length is also 0. The order of a schema containing only a single element (i.e., its representation has no $*$ symbols) is $L$ [for example, $o(1011) = 4 - 0 = 4$]. The length of a schema with fixed symbols in its first and last positions is $L-1$ [for example, $l(0{*}{*}1) = 4 - 1 = 3$].

We first consider the effect of the crossover operation on the mating pool. The basic observation in the following lemma is that given a chromosome in $M(k) \cap H$, the probability that it leaves $H$ after crossover is bounded above by a quantity that is proportional to $p_c$ and $l(H)$.

**Lemma 14.2.** *Given a chromosome in $M(k) \cap H$, the probability that it is chosen for crossover and neither of its offsprings is in $H$ is bounded above by*

$$p_c \frac{l(H)}{L-1}. \qquad \square$$

*Proof.* Consider a given chromosome in $M(k) \cap H$. The probability that it is chosen for crossover is $p_c$. If neither of its offsprings is in $H$, then the crossover point must be between the corresponding first and last fixed symbols of $H$. The probability of this is $l(H)/(L-1)$. Hence, the probability that the given chromosome is chosen for crossover and neither of its offsprngs is in $H$ is bounded above by

$$p_c \frac{l(H)}{L-1}. \qquad \blacksquare$$

From the above lemma, we conclude that given a chromosome in $M(k) \cap H$, the probability that either it is not selected for crossover, or at least one of its offsprings is in $H$ after the crossover operation, is bounded below by

$$1 - p_c \frac{l(H)}{L-1}.$$

Note that if a chromosome in $H$ is chosen for crossover, and the other parent chromosome is also in $H$, then both offsprings are automatically in $H$ (see Exercise 14.2). Hence, for each chromosome in $M(k) \cap H$, there is a certain probability that it will result in an associated chromosome in $H$ (either itself or one of its offsprings) after going through crossover (including selection for crossover), and that probability is bounded below by the above expression.

We next consider the effect of the mutation operation on the mating pool $M(k)$.

**Lemma 14.3.** *Given a chromosome in $M(k) \cap H$, the probability that it remains in $H$ after the mutation operation is given by*

$$(1 - p_m)^{o(H)}. \qquad \square$$

*Proof.* Given a chromosome in $M(k) \cap H$, it remains in $H$ after the mutation operation if, and only if, none of the symbols in this chromosome that correspond to fixed symbols in $H$ is changed by the mutation operation. The probability of this event is $(1 - p_m)^{o(H)}$. $\blacksquare$

Note that if $p_m$ is small, the expression $(1 - p_m)^{o(H)}$ above is approximately equal to

$$1 - p_m o(H).$$

The following theorem combines the results of the preceding lemmas.

**Theorem 14.1.** *Let $H$ be a given schema, and $\mathscr{E}(H, k+1)$ the expected value of $e(H, k+1)$ given $P(k)$. Then,*

$$\mathscr{E}(H, k+1) \geqslant \left(1 - p_c \frac{l(H)}{L-1}\right)(1 - p_m)^{o(H)} \left(\frac{f(H,k)}{\bar{F}(k)}\right) e(H,k). \qquad \square$$

*Proof.* Consider a given chromosome in $M(k) \cap H$. If, after the evolution operations, it has a resulting chromosome that is in $H$, then that chromosome is in $P(k+1) \cap H$. By Lemmas 14.2 and 14.3, the probability of this event is bounded below by

$$\left(1 - p_c \frac{l(H)}{L-1}\right)(1 - p_m)^{o(H)}.$$

Therefore, since each chromosome in $M(k) \cap H$ results in a chromosome in $P(k+1) \cap H$ with a probability bounded below by the above expression, the expected value of $e(H, k+1)$ given $M(k)$ is bounded below by

$$\left(1 - p_c \frac{l(H)}{L-1}\right)(1 - p_m)^{o(H)} m(H, k).$$

Taking the expectation given $P(k)$, we get

$$\mathscr{E}(H, k+1) \geqslant \left(1 - p_c \frac{l(H)}{L-1}\right)(1 - p_m)^{o(H)} \mathscr{M}(H, k).$$

Finally, using Lemma 14.1, we arrive at the desired result. ∎

The above theorem indicates how the number of chromosomes in a given schema changes from one population to the next. Three factors influence this change, reflected by the three terms on the right-hand side of the inequality in the above theorem, namely, $1 - p_c l(H)/(L-1)$, $(1 - p_m)^{o(H)}$, and $f(H, k)/\bar{F}(k)$. Note that the larger the values of these terms, the higher the expected number of matches of the schema $H$ in the next population. The effect of each term is summarized as follows:

The term $f(H, k)/\bar{F}(k)$ reflects the role of average fitness of the given schema $H$—the higher the average fitness, the higher the expected number of matches in the next population.

The term $1 - p_c l(H)/(L-1)$ reflects the effect of crossover—the smaller the term $p_c l(H)/(L-1)$, the higher the expected number of matches in the next population.

The term $(1 - p_m)^{o(H)}$ reflects the effect of mutation—the larger the term, the higher the expected number of matches in the next population.

In summary, we see that a schema that is short, low order, and has above average fitness will have on average an increasing number of its representatives in the population from iteration to iteration. Observe that the encoding is relevant to the performance of the algorithm. Specifically, a good encoding is one that results in high-fitness schemata having small lengths and orders.

## EXERCISES

**14.1**

**a.** Let $(I)_{10}$ be the decimal representation for a given integer, and let $a_m a_{m-1} \cdots a_0$ be its binary representation; that is, each $a_i$ is either 0 or 1, and

$$(I)_{10} = a_m 2^m + a_{m-1} 2^{m-1} + \cdots + a_1 2^1 + a_0 2^0.$$

Verify that the following is true:

$$(I)_{10} = (((\cdots(((a_m 2 + a_{m-1})2 + a_{m-2})2 + a_{m-3})\cdots)2 + a_1)2 + a_0).$$

**b.** The second expression in part a suggests a simple algorithm for converting from decimal representation to equivalent binary representation, as follows. Dividing both sides of the expression in part a by 2, the remainder is $a_0$. Subsequent divisions by two yield the remaining bits $a_1, a_2, \ldots, a_m$ as remainders. Use this algorithm to find the binary representation of the integer $(I)_{10} = 1995$.

**c.** Let $(F)_{10}$ be the decimal representation for a given number in $[0, 1]$, and let $0.a_{-1}a_{-2}\cdots$ be its binary representation; that is,

$$(F)_{10} = a_{-1}2^{-1} + a_{-2}2^{-2} + \cdots.$$

If the above expression is multiplied by 2, the integer part of the product is $a_{-1}$. Subsequent multiplications yield the remaining bits $a_{-2}, a_{-3}, \ldots$. As in part b, the above gives a simple algorithm for converting from a decimal fraction to its binary representations. Use this algorithm to find the binary representation of $(F)_{10} = 0.7265625$. Note that we can combine the algorithms from parts b and c to convert an arbitrary positive decimal representation into its equivalent binary representation. Specifically, we apply the algorithms in parts b and c separately to the integer and fraction parts of the given decimal number, respectively.

**d.** The procedure in part c may yield an infinitely long binary representation. If this is the case, we need to determine the number of bits required to keep at least the same accuracy as the given decimal number. If we have a $d$-digit decimal fraction, then the number of bits $b$ in the binary representation must satisfy $2^{-b} \leqslant 10^{-d}$, which yields $b \geqslant 3.32d$. Convert 19.95 to its equivalent binary representation with at least the same degree of accuracy (i.e., to two decimal places).

**14.2** Given two chromosomes in a schema $H$, suppose we swap some (or all) of the symbols between them at corresponding positions. Show that the resulting two chromosomes are also in $H$. From this fact, we conclude that given two chromosomes in $H$, both offsprings after the crossover operation are also in $H$. In other words, the crossover operation preserves membership in $H$.

**14.3** Consider a two-point crossover scheme (see Example 14.2), described as follows. Given a pair of binary chromosomes of length $L$, we independently choose two random numbers, uniform over $1, \ldots, L-1$. We call the two numbers $c_1$ and $c_2$, where $c_1 \leqslant c_2$. If $c_1 = c_2$, we do not swap any symbols (i.e., leave the two given parent chromosomes unchanged). If $c_1 < c_2$, we interchange the $(c_1 + 1)$st through $c_2$th bits in the given parent chromosomes. Prove the analog of Lemma 14.2 for this case, given below.

**Lemma.** *Given a chromosome in $M(k) \cap H$, the probability that it is chosen for crossover and neither of its offsprings is in $H$ is bounded above by*

$$p_c\left(1-\left(1-\frac{l(H)}{L-1}\right)^2\right). \qquad \square$$

*Hint*: Note that the two-point crossover operation is equivalent to a composition of two one-point crossover operations (that is, doing two one-point crossover operations in succession).

**14.4** State and prove the analog of Lemma 14.2 for an $n$-point crossover operation. *Hint*: See Exercise 14.3.

# Part III

# LINEAR PROGRAMMING

# 15

# Introduction to Linear Programming

## 15.1. A BRIEF HISTORY OF LINEAR PROGRAMMING

The goal of linear programming is to determine the values of decision variables that maximize or minimize a linear objective function, where the decision variables are subject to linear constraints. A linear programming problem is a special case of a general constrained optimization problem. In the general setting, the goal is to find a point that minimizes the objective function and at the same time satisfies the constraints. We refer to any point that satisfies the constraints as a *feasible* point. In a linear programming problem, the objective function is linear, and the set of feasible points is determined by a set of linear equations and/or inequalities.

In this part, we study methods for solving linear programming problems. Linear programming methods provide a way of choosing the best feasible point among the many possible feasible points. In general, the number of feasible points is infinitely large. However, as we shall see, the solution to a linear programming problem can be found by searching through a particular finite number of feasible points, known as basic feasible solutions. Therefore, in principle, we can solve a linear programming problem simply by comparing the finite number of basic feasible solutions and finding one that minimizes or maximizes the objective function—we refer to this approach as the "brute-force approach." For most practical decision problems, even this finite number of basic feasible solutions is so large that the method of choosing the best solution by comparing them to each other is impractical. To get a feel for the amount of computation needed in a brute-force approach, consider the following example. Suppose we have a small factory with 20 different machines producing 20 different parts. Assume that any of the machines can produce any part. We also assume that the time for producing each part on each machine is known. The problem then is to assign a part to each machine so that the overall production time is minimized. We see that there are 20! (20 factorial) possible assignments. The brute-force approach to solving this assignment problem would involve writing down all the possible assignments and then choosing the best one by comparing them. Suppose that we have at our disposal a computer that takes 1 $\mu$sec ($10^{-6}$ seconds) to determine each assignment. Then, to find the best (optimal) assignment, this computer would need 77,147 years (working 24 hours a day, 365 days a year) to find the best solution. An alternative approach to solving this problem is to use experts to optimize this assignment problem. Such an approach relies on heuristics. Heuristics usually give suboptimal solutions. Heuristics that do reasonably well, with an error of, say, 10%, may still not be good enough. For example, in a business that operates on a small profit margin, a 10% error could mean the difference between loss and profit.

Efficient methods for solving linear programming problems became available in the late 1930s. In 1939, Kantorovich presented a number of solutions to some problems related to production and transportation planning. During World War II, Koopmans contributed significantly to the solution of transportation problems. Kantorovich and Koopmans were awarded a Nobel Prize in economics in 1975 for their work on the theory of optimal allocation of resources. In 1947, Dantzig developed a new method for solving linear programs, known today as the *simplex* method [see Ref. 16 for Dantzig's own treatment of the algorithm]. In the following chapters, we discuss the simplex method in detail. The simplex method is efficient and elegant. However, it has the undesirable property that, in the worst case, the number of steps (and hence total time) required to find a solution grows exponentially with the number of variables. Thus, the simplex method is said to have exponential worst-case complexity. This led to an interest in devising algorithms for solving linear programs that have polynomial complexity; that is, algorithms that find a solution in an amount of time that is bounded by a polynomial in the number of variables. Khachiyan, in 1979, was the first to devise such an algorithm. However, his algorithm gained more theoretical than practical interest. Then, in 1984, Karmarkar of Bell Laboratories proposed a new linear programming algorithm that has polynomial complexity, and appears to solve some complicated, real-world problems of scheduling, routing, and planning more efficiently than the simplex method. Karmarkar's work led to the development of many other non-simplex methods, commonly referred to as *interior point* methods. This approach is currently still an active research area. [For more details on Karmarkar's and related algorithms, see Refs. 30, 42, 69, 72.] Some basic ideas illustrating Khachiyan's and Karmarkar's algorithms will be presented at the end of this part.

## 15.2. SIMPLE EXAMPLES OF LINEAR PROGRAMS

Formally, a linear program is an optimization problem of the form:

$$\begin{aligned} &\text{minimize} \quad \boldsymbol{c}^T\boldsymbol{x}, \qquad \boldsymbol{x}\in\mathbb{R}^n \\ &\text{subject to} \quad \boldsymbol{A}\boldsymbol{x}=\boldsymbol{b} \\ &\qquad\qquad\quad \boldsymbol{x}\geqslant\boldsymbol{0} \end{aligned}$$

where $c \in \mathbb{R}^n$, $b \in \mathbb{R}^m$, and $A \in \mathbb{R}^{m \times n}$. The vector inequality $x \geqslant 0$ means that each component of $x$ is nonnegative. Several variations to the above problem are possible; for example, instead of minimizing, we can maximize, or the constraints may be in the form of inequalities, such as $Ax \geqslant b$, or $Ax \leqslant b$. We also refer to these variations as linear programs. In fact, as we shall see later, these variations can all be rewritten into the standard form shown above.

The purpose of this section is to give some simple examples of linear programming problems illustrating the importance and the various applications of linear programming methods.

***Example 15.1.*** This example is adapted from Ref. 71. A manufacturer produces four different products $X_1$, $X_2$, $X_3$, and $X_4$. There are three inputs to this production process: labor in man weeks, kilograms of raw material A, and boxes of raw material B. Each product has different input requirements. In determining each week's production schedule, the manufacturer cannot use more than the available amounts of manpower and the two raw materials. The relevant information is presented in Table 15.1. Every production decision must satisfy the restrictions on the availability of inputs. These constraints can be written using the data in Table 15.1. In particular, we have

$$\begin{aligned} x_1 + 2x_2 + x_3 + 2x_4 &\leqslant 20 \\ 6x_1 + 5x_2 + 3x_3 + 2x_4 &\leqslant 100 \\ 3x_1 + 4x_2 + 9x_3 + 12x_4 &\leqslant 75. \end{aligned}$$

Since negative production levels are not meaningful, we must impose the following nonnegativity constraints on the production levels:

$$x_i \geqslant 0, \quad i = 1, 2, 3, 4.$$

Now, suppose that one unit of product $X_1$ sells for \$6, and $X_2$, $X_3$, and $X_4$ sell for \$4, \$7, and \$5, respectively. Then, the total revenue for any production decision $(x_1, x_2, x_3, x_4)$ is

$$f(x_1, x_2, x_3, x_4) = 6x_1 + 4x_2 + 7x_3 + 5x_4.$$

**Table 15.1 Data for Example 15.1**

| | Product | | | | |
|---|---|---|---|---|---|
| Inputs | $X_1$ | $X_2$ | $X_3$ | $X_4$ | Input Availability |
| Man weeks | 1 | 2 | 1 | 2 | 20 |
| Kilograms of material A | 6 | 5 | 3 | 2 | 100 |
| Boxes of material B | 3 | 4 | 9 | 12 | 75 |
| Production levels | $x_1$ | $x_2$ | $x_3$ | $x_4$ | |

The problem is then to maximize $f$, subject to the given constraints (the three inequalities and four nonnegativity constraints). Note that the above problem can be written in the compact form:

$$\begin{aligned} &\text{maximize} \quad f(\boldsymbol{x}) = \text{maximize } \boldsymbol{c}^T\boldsymbol{x} \\ &\text{subject to} \quad \boldsymbol{A}\boldsymbol{x} \leqslant \boldsymbol{b}, \\ &\qquad\qquad\quad \boldsymbol{x} \geqslant \boldsymbol{0}, \end{aligned}$$

where

$$\boldsymbol{c}^T = [6, 4, 7, 5],$$

$$\boldsymbol{x} = [x_1, x_2, x_3, x_4]^T,$$

$$\boldsymbol{A} = \begin{bmatrix} 1 & 2 & 1 & 2 \\ 6 & 5 & 3 & 2 \\ 3 & 4 & 9 & 12 \end{bmatrix},$$

$$\boldsymbol{b} = \begin{bmatrix} 20 \\ 100 \\ 75 \end{bmatrix}.$$

■

Another example that illustrates linear programming involves determining the most economical diet that satisfies the basic minimum requirements for good health.

***Example 15.2.*** This example is adapted from Ref. 51. Assume there are $n$ different food types available. The $j$th food sells at a price $c_j$ per unit. In addition there are $m$ basic nutrients. To achieve a balanced diet, you must receive at least $b_i$ units of the $i$th nutrient per day. Assume that each unit of food $j$ contains $a_{ij}$ units of the $i$th nutrient. Denote by $x_j$ the number of units of food $j$ in the diet. The objective is to select the $x_j$'s to minimize the total cost of the diet, that is,

$$\text{minimize } c_1x_1 + c_2x_2 + \cdots + c_nx_n$$

subject to the nutritional constraints

$$\begin{aligned} a_{11}x_1 + a_{12}x_2 + \cdots + a_{1n}x_n &\geqslant b_1 \\ a_{21}x_1 + a_{22}x_2 + \cdots + a_{2n}x_n &\geqslant b_2 \\ &\vdots \\ a_{m1}x_1 + a_{m2}x_2 + \cdots + a_{mn}x_n &\geqslant b_m, \end{aligned}$$

and the nonnegativity constraints

$$x_1 \geqslant 0, x_2 \geqslant 0, \ldots, x_n \geqslant 0.$$

In the more compact vector notation, this problem becomes

$$\begin{aligned} &\text{minimize} \quad \boldsymbol{c}^T\boldsymbol{x} \\ &\text{subject to} \quad \boldsymbol{A}\boldsymbol{x} \geqslant \boldsymbol{b}, \\ &\qquad\qquad\quad \boldsymbol{x} \geqslant \boldsymbol{0}, \end{aligned}$$

where $\boldsymbol{x}$ is an $n$-dimensional column vector (that is, $\boldsymbol{x} = [x_1, x_2, \ldots, x_n]^T$), $\boldsymbol{c}^T$ is an $n$-dimensional row vector, $\boldsymbol{A}$ is an $m \times n$ matrix, and $\boldsymbol{b}$ is an $m$-dimensional column vector. We call the above problem the "diet problem," and will return to it in Chapter 17. ■

In the next example, we consider a linear programming problem that arises in manufacturing.

***Example 15.3.*** A manufacturer produces two different products $X_1$ and $X_2$ using three machines $M_1$, $M_2$, and $M_3$. Each machine can be used only for a limited amount of time. Production times of each product on each machine are given in Table 15.2. The objective is to maximize the combined time of utilization of all three machines.

Every production decision must satisfy the constraints on the available time. These restrictions can be written down using data from Table 15.2. In particular, we have

$$\begin{aligned} x_1 + x_2 &\leqslant 8 \\ x_1 + 3x_2 &\leqslant 18 \\ 2x_1 + x_2 &\leqslant 14, \end{aligned}$$

where $x_1$ and $x_2$ denote the production levels. The combined production time of all three machines is

$$f(x_1, x_2) = 4x_1 + 5x_2.$$

**Table 15.2 Data for Example 15.3**

| | Production Time (hours/unit) | | Available Time |
|---|---|---|---|
| Machine | $X_1$ | $X_2$ | (hours) |
| $M_1$ | 1 | 1 | 8 |
| $M_2$ | 1 | 3 | 18 |
| $M_3$ | 2 | 1 | 14 |
| Total | 4 | 5 | |

Thus, the problem in compact notation has the form

$$\begin{aligned}\text{maximize}\quad & \boldsymbol{c}^T\boldsymbol{x}\\ \text{subject to}\quad & \boldsymbol{A}\boldsymbol{x}\leqslant \boldsymbol{b},\\ & \boldsymbol{x}\geqslant \boldsymbol{0},\end{aligned}$$

where

$$\boldsymbol{c}^T=[4,5],$$

$$\boldsymbol{x}=[x_1,x_2]^T,$$

$$\boldsymbol{A}=\begin{bmatrix}1 & 1\\ 1 & 3\\ 2 & 1\end{bmatrix},$$

$$\boldsymbol{b}=[8,18,14]^T.$$

■

In the following example, we discuss an application of linear programming in transportation.

***Example 15.4.*** A manufacturing company has plants in cities A, B, and C. The company produces and distributes its product to dealers in various cities. On a particular day, the company has 60 units of its product in A, 80 in B, and 60 in C. The company plans to ship 40 units to D, 40 to E, 50 to F, and 70 to G, following orders received from dealers. The transportation costs per unit of each product between the cities are given in Table 15.3. In the table, the quantities supplied and demanded appear at the right and along the bottom of the table. The quantities to be transported from the plants to different destinations are represented by the decision variables.

This problem can be stated in the form:

$$\begin{aligned}\text{minimize}\quad & 7x_{11}+10x_{12}+14x_{13}+8x_{14}+7x_{21}+11x_{22}+12x_{23}\\ & +6x_{24}+5x_{31}+8x_{32}+15x_{33}+9x_{34}\end{aligned}$$

subject to

$$\begin{aligned}x_{11}+x_{12}+x_{13}+x_{14}&=60\\ x_{21}+x_{22}+x_{23}+x_{24}&=80\\ x_{31}+x_{32}+x_{33}+x_{34}&=60\\ x_{11}+x_{21}+x_{31}&=40\\ x_{12}+x_{22}+x_{32}&=40\\ x_{13}+x_{23}+x_{33}&=50\\ x_{14}+x_{24}+x_{34}&=70\end{aligned}$$

and

$$x_{11},x_{12},\ldots,x_{34}\geqslant 0.$$

**Table 15.3 Data for Example 15.4**

| From | To: D | E | F | G | Supply |
|---|---|---|---|---|---|
| A | \$7 | \$10 | \$14 | \$8 | 60 |
| B | \$7 | \$11 | \$12 | \$6 | 80 |
| C | \$5 | \$8 | \$15 | \$9 | 60 |
| Demand | 40 | 40 | 50 | 70 | 200 |

In this problem, one of the constraint equations is redundant, because it can be derived from the rest of the constraint equations. The mathematical formulation of the transportation problem is then in a linear programming form with twelve ($3 \times 4$) decision variables and six ($3+4-1$) linearly independent constraint equations. Obviously, we also require nonnegativity of the decision variables, since a negative shipment is impossible and does not have any valid interpretation. ■

Finally, we give an example of a linear programming problem arising in electrical engineering.

***Example 15.5.*** This example is adapted from Ref. 56. Figure 15.1 shows an electric circuit that is designed to use a 30-V source to charge 10-V, 6-V, and 20-V batteries connected in parallel. Physical constraints limit the currents $I_1, I_2, I_3, I_4$, and $I_5$ to a maximum of 4 A, 3 A, 3 A, 2 A, and 2 A, respectively. In addition, the batteries must not be discharged (that is, the currents $I_1, I_2, I_3, I_4$, and $I_5$ must not

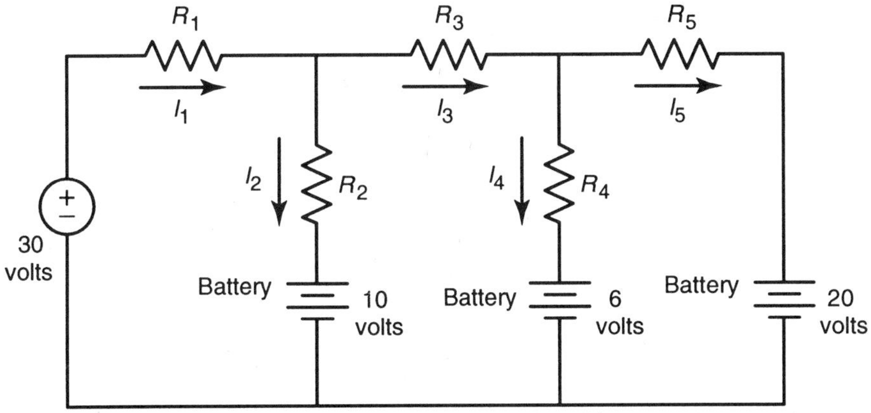


**Figure 15.1.** Battery charger circuit for Example 15.5

be negative). We wish to find the values of the currents $I_1, \ldots, I_5$ such that the total power transferred to the batteries is maximized.

The total power transferred to the batteries is the sum of the powers transferred to each battery, and is given by $10I_2 + 6I_4 + 20I_5$ W. From the circuit in Figure 15.1, we observe that the currents satisfy the constraints $I_1 = I_2 + I_3$, and $I_3 = I_4 + I_5$. Therefore, the problem can be posed as the following linear program:

$$\begin{aligned}
\text{maximize} \quad & 10I_2 + 6I_4 + 20I_5 \\
\text{subject to} \quad & I_1 = I_2 + I_3 \\
& I_3 = I_4 + I_5 \\
& I_1 \leqslant 4 \\
& I_2 \leqslant 3 \\
& I_3 \leqslant 3 \\
& I_4 \leqslant 2 \\
& I_5 \leqslant 2, \\
& I_1, I_2, I_3, I_4, I_5 \geqslant 0.
\end{aligned}$$

■

For more examples of linear programming and their applications in a variety of engineering problems, we refer the reader to Refs. 2, 16, 24, and 62. Linear programming also provides the basis for theoretical applications, as, for example, in matrix game theory [discussed in Ref. 11].

## 15.3. TWO-DIMENSIONAL LINEAR PROGRAMS

Many fundamental concepts of linear programming are easily illustrated in two-dimensional space. Therefore, we first consider linear problems in $\mathbb{R}^2$ before discussing general linear programming problems.

Consider the following linear program [adapted from Ref. 71]:

$$\begin{aligned}
\text{maximize} \quad & \boldsymbol{c}^T \boldsymbol{x} \\
\text{subject to} \quad & \boldsymbol{A}\boldsymbol{x} \leqslant \boldsymbol{b}, \\
& \boldsymbol{x} \geqslant \boldsymbol{0},
\end{aligned}$$

where

$$\boldsymbol{c}^T = [1, 5],$$

$$\boldsymbol{x} = [x_1, x_2]^T,$$

$$\boldsymbol{A} = \begin{bmatrix} 5 & 6 \\ 3 & 2 \end{bmatrix},$$

$$\boldsymbol{b} = [30, 12]^T.$$

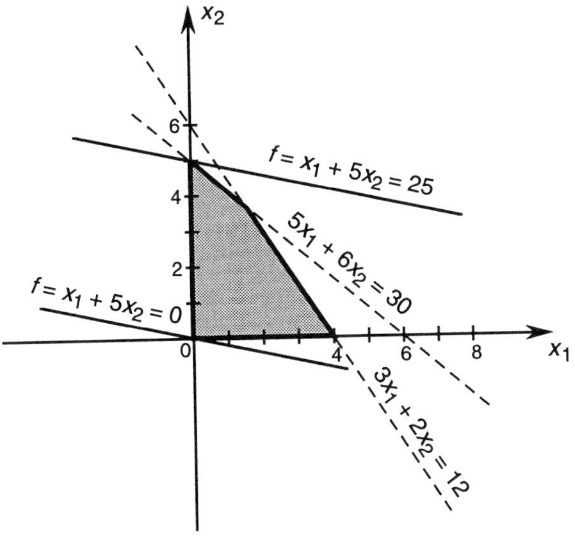


**Figure 15.2.** Geometric solution of a linear program in $\mathbb{R}^2$

First, we note that the set of equations $\{c^T x = x_1 + 5x_2 = f, f \in \mathbb{R}\}$ specifies a family of straight lines in $\mathbb{R}^2$. Each member of this family can be obtained by setting $f$ equal to some real number. Thus, for example, $x_1 + 5x_2 = -5$, $x_1 + 5x_2 = 0$, and $x_1 + 5x_2 = 3$ are three parallel lines belonging to the family. Now, suppose we try to choose several values for $x_1$ and $x_2$ and observe how large we can make $f$, while still satisfying the constraints on $x_1$ and $x_2$. We first try $x_1 = 1$ and $x_2 = 3$. This point satisfies the constraints. For this point, $f = 16$. If we now select $x_1 = 0$ and $x_2 = 5$ then $f = 25$, and this point yields a larger value for $f$ than does $x = [1, 3]^T$. There are infinitely many points $[x_1, x_2]^T$ satisfying the constraints. Therefore, we need a better method than "trial-and-error" to solve the problem. In the following sections, we develop a systematic approach that considerably simplifies the process of solving linear programming problems.

In the case of the above example, we can easily solve the problem using geometric arguments. First let us sketch the constraints in $\mathbb{R}^2$. The region of feasible points, that is, the set of points $x$ satisfying the constraints $Ax \leqslant b$, $x \geqslant 0$, is depicted by the shaded region in Figure 15.2.

Geometrically, maximizing $c^T x = x_1 + 5x_2$ subject to the constraints can be thought of as finding the straight line $f = x_1 + 5x_2$ that intersects the shaded region and has the largest $f$. The coordinates of the point of intersection will then yield a maximum value of $c^T x$. In our example, the point $[0, 5]^T$ is the solution (see Figure 15.2). in some cases, there may be more than one point of intersection; all of them will yield the same value for the objective function $c^T x$, and therefore any one of them is a solution.

## 15.4. CONVEX POLYHEDRA AND LINEAR PROGRAMMING

The goal of linear programming is to minimize (or maximize) a linear objective function

$$c^T x = c_1 x_1 + c_2 x_2 + \cdots + c_n x_n$$

subject to constraints that are represented by linear equalities and/or inequalities. For the time being, let us only consider constraints of the form $Ax \leqslant b$, $x \geqslant 0$. In this section, we discuss linear programs from a geometric point of view (for a review of geometric concepts used in the section, see Chapter 4). The set of points satisfying these constraints can be represented as the intersection of a finite number of closed half-spaces. Thus, the constraints define a convex polytope. We assume, for simplicity, that this polytope is nonempty and bounded. In other words the equations of constraints define a polyhedron $M$ in $\mathbb{R}^n$. Let $H$ be a hyperplane of support of this polyhedron. If the dimension of $M$ is less than $n$, then the set of all points common to the hyperplane $H$ and the polyhedron $M$ coincides with $M$. If the dimension of $M$ is equal to $n$ then the set of all points common to the hyperplane $H$ and the polyhedron $M$ is a face of the polyhedron. If this face is $(n-1)$-dimensional, then there exists only one hyperplane of support, namely, the carrier of this face. If the dimension of the face is less than $n-1$, then there exists an infinite number of hyperplanes of support whose intersection with this polyhedron yields this face—see Figure 15.3.

The goal of our linear programming problem is to maximize a linear objective function $f(x) = c^T x = c_1 x_1 + \cdots + c_n x_n$ on the convex polyhedron $M$. Next, let $H$ be the hyperplane defined by the equation

$$c^T x = 0.$$

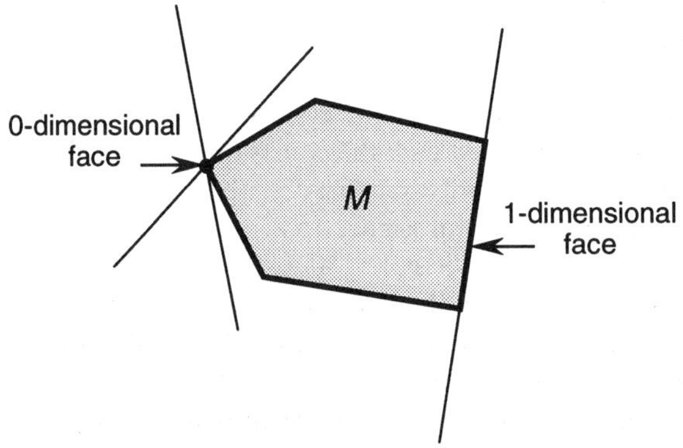


**Figure 15.3.** Hyperplanes of support at different boundary points of the polyhedron $M$

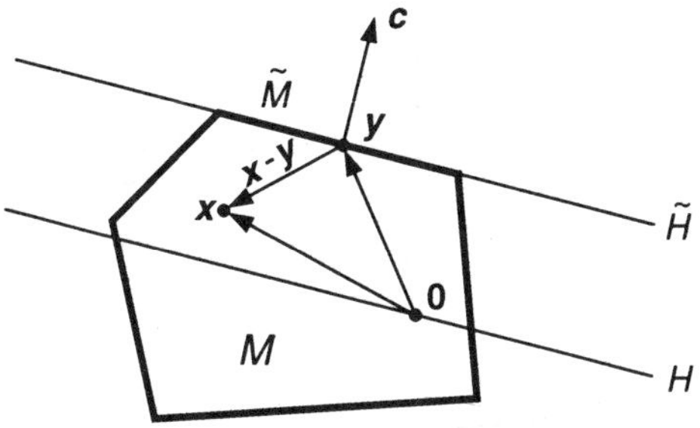


**Figure 15.4.** Maximization of a linear function on the polyhedron $M$

Draw a hyperplane of support $\tilde{H}$ to the polyhedron $M$, which is parallel to $H$ and positioned in such a way that the vector $\boldsymbol{c}$ points in the direction of the half-space that does not contain $M$ (see Figure 15.4). The equation of the hyperplane $\tilde{H}$ has the form

$$\boldsymbol{c}^T\boldsymbol{x} = \beta,$$

and, for all $\boldsymbol{x} \in M$, we have $\boldsymbol{c}^T\boldsymbol{x} \leqslant \beta$. Denote by $\tilde{M}$ the convex polyhedron that is the intersection of the hyperplane of support $\tilde{H}$ with the polyhedron $M$. We now show that $f$ is constant on $\tilde{M}$ and that $\tilde{M}$ is the set of all points in $M$ for which $f$ attains its maximum value. To this end, let $\boldsymbol{y}$ and $\boldsymbol{z}$ be two arbitrary points in $\tilde{M}$. This implies that both $\boldsymbol{y}$ and $\boldsymbol{z}$ belong to $\tilde{H}$. Hence

$$f(\boldsymbol{y}) = \boldsymbol{c}^T\boldsymbol{y} = \beta = \boldsymbol{c}^T\boldsymbol{z} = f(\boldsymbol{z}),$$

which means that $f$ is constant on $\tilde{M}$.

Let $\boldsymbol{y}$ be a point of $\tilde{M}$, and let $\boldsymbol{x}$ be a point of $M \backslash \tilde{M}$ (that is, $x$ is a point of $M$ that does not belong to $\tilde{M}$—see Figure 15.4). Then

$$\boldsymbol{c}^T\boldsymbol{x} < \beta = \boldsymbol{c}^T\boldsymbol{y},$$

which implies that

$$f(\boldsymbol{x}) < f(\boldsymbol{y}).$$

Thus, the values of $f$ at the points of $M$ that do not belong to $\tilde{M}$ are smaller than the values at points of $\tilde{M}$. Hence, $f$ achieves its maximum on $M$ at points in $\tilde{M}$.

It may happen that $\tilde{M}$ contains only a single point, in which case $f$ achieves its maximum at a unique point. This occurs when the hyperplane of support passes through an extreme point of $M$ (see Figure 15.5).

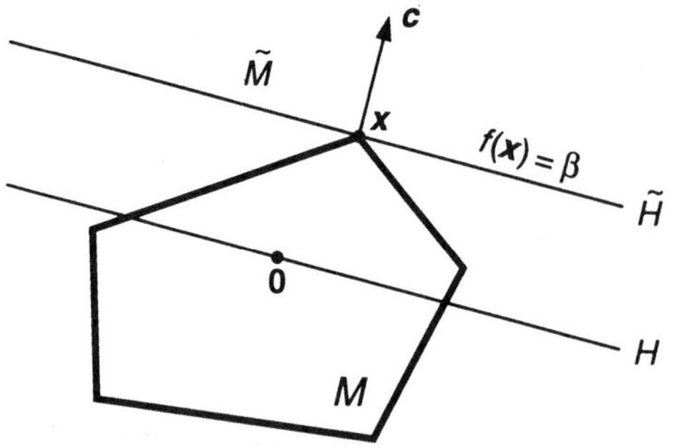


**Figure 15.5.** Unique maximum point of $f$ on the polyhedron $M$

## 15.5. STANDARD-FORM LINEAR PROGRAMS

We refer to a linear program of the form

$$\begin{aligned} &\text{minimize} && \boldsymbol{c}^T\boldsymbol{x} \\ &\text{subject to} && \boldsymbol{A}\boldsymbol{x} = \boldsymbol{b}, \\ & && \boldsymbol{x} \geqslant \boldsymbol{0}, \end{aligned}$$

as a linear program in *standard form*. Here $\boldsymbol{A}$ is an $m \times n$ matrix composed of real entries, $m < n$, rank $\boldsymbol{A} = m$. Without loss of generality, we assume $\boldsymbol{b} \geqslant \boldsymbol{0}$. If a component of $\boldsymbol{b}$ is negative, say the $i$th component, we multiply the $i$th constraint by $-1$ to obtain a positive right-hand side.

Theorems and solution techniques for linear programs are usually stated for problems in standard form. Other forms of linear programs can be converted to the standard form, as we shall now show. If a linear program is in the form

$$\begin{aligned} &\text{minimize} && \boldsymbol{c}^T\boldsymbol{x} \\ &\text{subject to} && \boldsymbol{A}\boldsymbol{x} \geqslant \boldsymbol{b}, \\ & && \boldsymbol{x} \geqslant \boldsymbol{0}, \end{aligned}$$

then by introducing so-called *surplus variables* $y_i$, we can convert the original problem into the standard form

$$\begin{aligned} &\text{minimize} && \boldsymbol{c}^T\boldsymbol{x} \\ &\text{subject to} && a_{i1}x_1 + a_{i2}x_2 + \cdots + a_{in}x_n - y_i = b_i, \quad i = 1, \ldots, m \end{aligned}$$

with

$$x_1 \geqslant 0, x_2 \geqslant 0, \ldots, x_n \geqslant 0,$$

and

$$y_1 \geqslant 0, y_2 \geqslant 0, \ldots, y_m \geqslant 0.$$

In more compact notation the above formulation can be represented as

$$\begin{aligned} &\text{minimize} \quad \boldsymbol{c}^T\boldsymbol{x} \\ &\text{subject to} \quad \boldsymbol{A}\boldsymbol{x} - \boldsymbol{I}_m\boldsymbol{y} = [\boldsymbol{A}, -\boldsymbol{I}_m]\begin{bmatrix}\boldsymbol{x}\\ \boldsymbol{y}\end{bmatrix} = \boldsymbol{b}, \\ &\qquad\qquad \boldsymbol{x} \geqslant \boldsymbol{0}, \quad \boldsymbol{y} \geqslant \boldsymbol{0}, \end{aligned}$$

where $\boldsymbol{I}_m$ is the $m \times m$ identity matrix.

If, on the other hand, the constraints have the form

$$\begin{aligned} \boldsymbol{A}\boldsymbol{x} &\leqslant \boldsymbol{b}, \\ \boldsymbol{x} &\geqslant \boldsymbol{0}, \end{aligned}$$

then we introduce *slack variables* $y_i$ to convert the constraints into the form

$$\begin{aligned} &\boldsymbol{A}\boldsymbol{x} + \boldsymbol{I}_m\boldsymbol{y} = [\boldsymbol{A}, \boldsymbol{I}_m]\begin{bmatrix}\boldsymbol{x}\\ \boldsymbol{y}\end{bmatrix} = \boldsymbol{b}, \\ &\boldsymbol{x} \geqslant \boldsymbol{0}, \quad \boldsymbol{y} \geqslant \boldsymbol{0}, \end{aligned}$$

where $\boldsymbol{y}$ is the vector of slack variables. Note that neither surplus nor slack variables contribute to the objective function $\boldsymbol{c}^T\boldsymbol{x}$.

At first glance, it may appear that the two problems

$$\begin{aligned} &\text{minimize} \quad \boldsymbol{c}^T\boldsymbol{x} \\ &\text{subject to} \quad \boldsymbol{A}\boldsymbol{x} = \boldsymbol{b}, \\ &\qquad\qquad \boldsymbol{x} \geqslant \boldsymbol{0}, \end{aligned}$$

and

$$\begin{aligned} &\text{minimize} \quad \boldsymbol{c}^T\boldsymbol{x} \\ &\text{subject to} \quad \boldsymbol{A}\boldsymbol{x} - \boldsymbol{I}_m\boldsymbol{y} = \boldsymbol{b}, \\ &\qquad\qquad \boldsymbol{x} \geqslant \boldsymbol{0}, \quad \boldsymbol{y} \geqslant \boldsymbol{0}, \end{aligned}$$

are different, in that the first problem refers to intersection of half-spaces in the $n$-dimensional space, whereas the second problem refers to an intersection of half-spaces and hyperplanes in the $(n+m)$-dimensional space. It turns out that both formulations are algebraically equivalent in the sense that a solution to one of the problems gives rise to a solution to the other. To illustrate this equivalence, we consider the following examples.

***Example 15.6.*** Suppose we are given the inequality constraint

$$x_1 \leqslant 7.$$

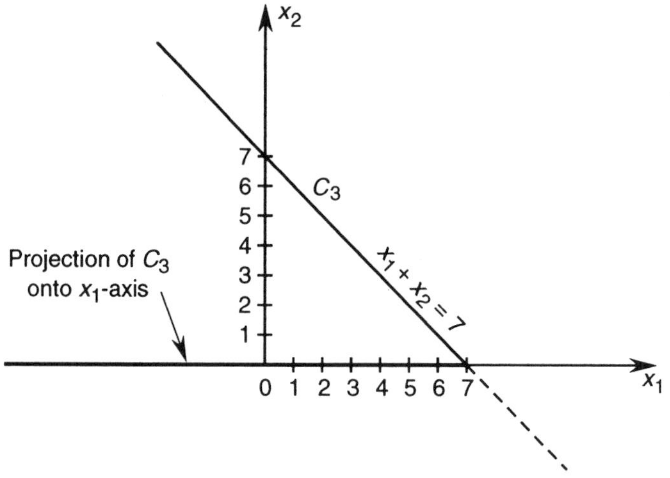


**Figure 15.6.** Projection of the set $C_3$ onto the $x_1$ axis

We convert this to an equality constraint by introducing a slack variable $x_2 \geqslant 0$ to obtain

$$x_1 + x_2 = 7,$$
$$x_2 \geqslant 0.$$

Consider the sets $C_1 = \{x_1 : x_1 \leqslant 7\}$ and $C_2 = \{x_1 : x_1 + x_2 = 7, x_2 \geqslant 0\}$. Are the sets $C_1$ and $C_2$ equal? It is clear that indeed they are; in this example, we give a geometric interpretation for their equality. Consider a third set $C_3 = \{[x_1, x_2]^T : x_1 + x_2 = 7, x_2 \geqslant 0\}$. From Figure 15.6, we can see that the set $C_3$ consists of all points on the line to the left and above the point of intersection of the line with the $x_1$-axis. This set, being a subset of $\mathbb{R}^2$, is of course not the same set as the set $C_1$ (a subset of $\mathbb{R}$). However, we can project the set $C_3$ onto the $x_1$ axis (see Figure 15.6). We can associate, with each point $x_1 \in C_1$, a point $[x_1, 0]^T$ on the projection of $C_3$ onto the $x_1$ axis, and vice versa. Note that $C_2 = \{x_1 : [x_1, x_2]^T \in C_3\} = C_1$. ∎

***Example 15.7.*** Consider the inequality constraints [as in Ref. 71]

$$a_1 x_1 + a_2 x_2 \leqslant b,$$
$$x_1, x_2 \geqslant 0,$$

where $a_1, a_2$, and $b$ are positive numbers. Again, we introduce a slack variable $x_3 \geqslant 0$ to get

$$a_1 x_1 + a_2 x_2 + x_3 = b,$$
$$x_1, x_2, x_3 \geqslant 0.$$

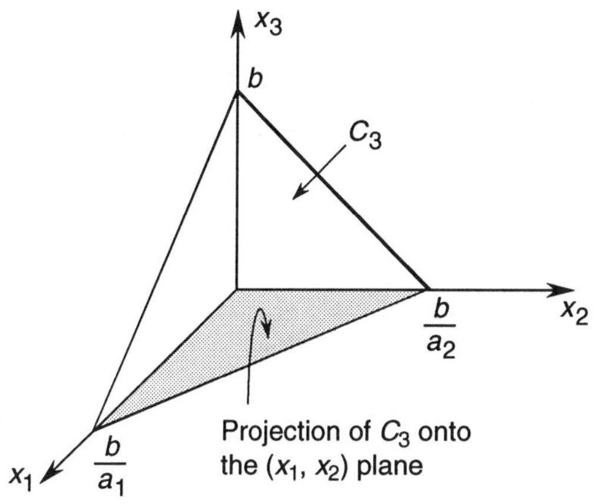


**Figure 15.7.** Projection of the set $C_3$ onto the $(x_1, x_2)$ plane

Define the sets

$$C_1 = \{[x_1, x_2]^T : a_1x_1 + a_2x_2 \leqslant b, x_1, x_2 \geqslant 0\}$$
$$C_2 = \{[x_1, x_2]^T : a_1x_1 + a_2x_2 + x_3 = b, x_1, x_2, x_3 \geqslant 0\}$$
$$C_3 = \{[x_1, x_2, x_3]^T : a_1x_1 + a_2x_2 + x_3 = b, x_1, x_2, x_3 \geqslant 0\}.$$

We again see that $C_3$ is not the same as $C_1$. However, the projection of $C_3$ onto the $(x_1, x_2)$ plane allows us to associate the resulting set with the set $C_1$. We associate the points $[x_1, x_2, 0]^T$ resulting from the projection of $C_3$ onto the $(x_1, x_2)$ plane with the points in $C_1$ (see Figure 15.7). Note that $C_2 = \{[x_1, x_2]^T : [x_1, x_2, x_3]^T \in C_3\} = C_1$. ■

***Example 15.8.*** Suppose we wish to maximize [as in Ref. 71]

$$f(x_1, x_2) = c_1x_1 + c_2x_2$$

subject to the constraints

$$a_{11}x_1 + a_{12}x_2 \leqslant b_1$$
$$a_{21}x_1 + a_{22}x_2 = b_2$$
$$x_1, x_2, \geqslant 0,$$

where, for simplicity, we assume that each $a_{ij} > 0$ and $b_1, b_2 \geqslant 0$. The set of feasible points is depicted in Figure 15.8. Let $C_1 \subset \mathbb{R}^2$ be the set of points satisfying the above constraints.

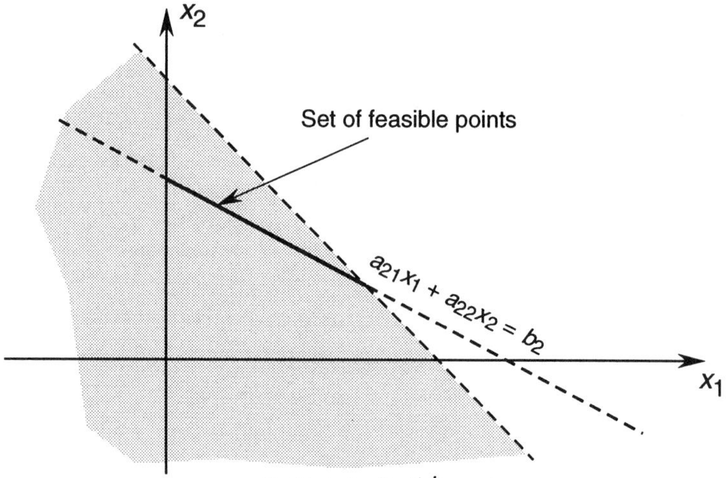


**Figure 15.8.** The feasible set for Example 15.8

Introducing a slack variable, we convert the above constraints into standard form:

$$\begin{aligned} a_{11}x_1 + a_{12}x_2 + x_3 &= b_1, \\ a_{21}x_1 + a_{22}x_2 &= b_2, \\ x_i \geqslant 0, \quad i &= 1, 2, 3. \end{aligned}$$

Let $C_2 \subset \mathbb{R}^3$ be the set of points satisfying the above constraints. As illustrated in Figure 15.9, this set is a line segment (in $\mathbb{R}^3$). We now project $C_2$ onto the $(x_1, x_2)$ plane. The projected set consists of the points $[x_1, x_2, 0]^T$, with $[x_1, x_2, x_3]^T \in C_2$ for some $x_3 \geqslant 0$. In Figure 15.9 this set is marked by a heavy line in the $(x_1, x_2)$ plane. We can associate the points on the projection with the corresponding points in the set $C_1$. ■

## 15.6. BASIC SOLUTIONS

In Section 15.5, we have seen that any linear programming problem involving inequalities can be converted to *standard form*, that is, a problem involving linear equations with nonnegative variables:

$$\begin{aligned} &\text{minimize} \quad \boldsymbol{c}^T\boldsymbol{x}, \quad \boldsymbol{x} \in \mathbb{R}^n \\ &\text{subject to} \quad \boldsymbol{A}\boldsymbol{x} = \boldsymbol{b}, \\ &\qquad\qquad\quad \boldsymbol{x} \geqslant \boldsymbol{0}, \end{aligned}$$

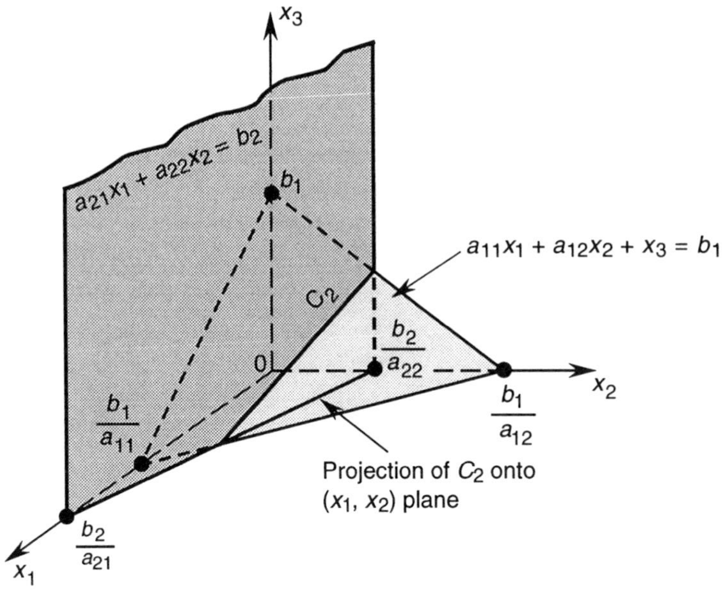


**Figure 15.9.** Projection of $C_2$ onto the $(x_1, x_2)$ plane

where $\boldsymbol{c} \in \mathbb{R}^n$, $\boldsymbol{A} \in \mathbb{R}^{m \times n}$, $\boldsymbol{b} \in \mathbb{R}^m$, $m < n$, rank $\boldsymbol{A} = m$, and $\boldsymbol{b} \geqslant \boldsymbol{0}$. In the following discussion, we only consider linear programming problems in standard form.

Consider the system of equalities

$$\boldsymbol{A}\boldsymbol{x} = \boldsymbol{b},$$

where rank $\boldsymbol{A} = m$. In dealing with this system of equations, we frequently need to consider a subset of columns of the matrix $\boldsymbol{A}$. For convenience, we often reorder the columns of $\boldsymbol{A}$ so that the columns we are interested in appear first. Specifically, let $\boldsymbol{B}$ be a square matrix whose columns are $m$ linearly independent columns of $\boldsymbol{A}$. If necessary, we reorder the columns of $\boldsymbol{A}$ so that the columns in $\boldsymbol{B}$ appear first: $\boldsymbol{A}$ has the form $\boldsymbol{A} = [\boldsymbol{B}, \boldsymbol{D}]$, where $\boldsymbol{D}$ is an $m \times (n - m)$ matrix whose columns are the remaining columns of $\boldsymbol{A}$. The matrix $\boldsymbol{B}$ is nonsingular, and thus we can solve the equation

$$\boldsymbol{B}\boldsymbol{x}_B = \boldsymbol{b}$$

for the $m$-vector $\boldsymbol{x}_B$. The solution is $\boldsymbol{x}_B = \boldsymbol{B}^{-1}\boldsymbol{b}$. Let $\boldsymbol{x}$ be the $n$-vector whose first $m$ components are equal to $\boldsymbol{x}_B$, and the remaining components are equal to zero, that is, $\boldsymbol{x} = [\boldsymbol{x}_B^T, \boldsymbol{0}^T]^T$. Then, $\boldsymbol{x}$ is a solution to $\boldsymbol{A}\boldsymbol{x} = \boldsymbol{b}$.

***Definition 15.1***

We call $[\boldsymbol{x}_B^T, \boldsymbol{0}^T]^T$ a *basic solution* to $\boldsymbol{Ax} = \boldsymbol{b}$ with respect to the *basis* $\boldsymbol{B}$. We refer to the components of the vector $\boldsymbol{x}_B$ as *basic variables*, and the columns of $\boldsymbol{B}$ as *basic columns*.

If some of the basic variables of a basic solution are zero, then the basic solution is said to be a *degenerate basic solution*.

A vector $\boldsymbol{x}$ satisfying $\boldsymbol{Ax} = \boldsymbol{b}$, $\boldsymbol{x} \geqslant \boldsymbol{0}$, is said to be a *feasible solution*.

A feasible solution that is also basic is called a *basic feasible solution*.

If the basic feasible solution is a degenerate basic solution, then it is called a *degenerate basic feasible solution*. ■

Note that in any basic feasible solution, $\boldsymbol{x}_B \geqslant \boldsymbol{0}$.

***Example 15.9.*** Consider the equation $\boldsymbol{Ax} = \boldsymbol{b}$ with

$$\boldsymbol{A} = [\boldsymbol{a}_1, \boldsymbol{a}_2, \boldsymbol{a}_3, \boldsymbol{a}_4] = \begin{bmatrix} 1 & 1 & -1 & 4 \\ 1 & -2 & -1 & 1 \end{bmatrix}, \quad \boldsymbol{b} = \begin{bmatrix} 8 \\ 2 \end{bmatrix},$$

where $\boldsymbol{a}_i$ denotes the $i$th column of the matrix $\boldsymbol{A}$.

Then, $\boldsymbol{x} = [6, 2, 0, 0]^T$ is a basic feasible solution with respect to the basis $\boldsymbol{B} = [\boldsymbol{a}_1, \boldsymbol{a}_2]$, $\boldsymbol{x} = [0, 0, 0, 2]^T$ is a degenerate basic feasible solution with respect to the basis $\boldsymbol{B} = [\boldsymbol{a}_3, \boldsymbol{a}_4]$ (as well as $[\boldsymbol{a}_1, \boldsymbol{a}_4]$ and $[\boldsymbol{a}_2, \boldsymbol{a}_4]$), $\boldsymbol{x} = [3, 1, 0, 1]^T$ is a feasible solution that is not basic, and $\boldsymbol{x} = [0, 2, -6, 0]^T$ is a basic solution with respect to the basis $\boldsymbol{B} = [\boldsymbol{a}_2, \boldsymbol{a}_3]$, but is not feasible. ■

***Example 15.10.*** As another example, consider the system of linear equations $\boldsymbol{Ax} = \boldsymbol{b}$ (adapted from Ref. 38, p. 162), where

$$\boldsymbol{A} = \begin{bmatrix} 2 & 3 & -1 & -1 \\ 4 & 1 & 1 & -2 \end{bmatrix}, \quad \boldsymbol{b} = \begin{bmatrix} -1 \\ 9 \end{bmatrix}.$$

We will now find all solutions of this system. Note that every solution $\boldsymbol{x}$ of $\boldsymbol{Ax} = \boldsymbol{b}$ has the form $\boldsymbol{x} = \boldsymbol{v} + \boldsymbol{h}$, where $\boldsymbol{v}$ is a particular solution of $\boldsymbol{Ax} = \boldsymbol{b}$ and $\boldsymbol{h}$ is a solution to $\boldsymbol{Ax} = \boldsymbol{0}$.

We form the augmented matrix $[\boldsymbol{A}, \boldsymbol{b}]$ of the system:

$$[\boldsymbol{A}, \boldsymbol{b}] = \begin{bmatrix} 2 & 3 & -1 & -1 & -1 \\ 4 & 1 & 1 & -2 & 9 \end{bmatrix}.$$

Using elementary row operations, we transform the above matrix into the form (see the next chapter) given by

$$\begin{bmatrix} 1 & 0 & \frac{2}{5} & -\frac{1}{2} & \frac{14}{5} \\ 0 & 1 & -\frac{3}{5} & 0 & -\frac{11}{5} \end{bmatrix}.$$

The corresponding system of equations is given by

$$\begin{aligned} x_1 + \frac{2}{5}x_3 - \frac{1}{2}x_4 &= \frac{14}{5} \\ x_2 - \frac{3}{5}x_3 &= -\frac{11}{5}. \end{aligned}$$

Solving for the leading unknowns $x_1$ and $x_2$, we obtain

$$\begin{aligned} x_1 &= \frac{14}{5} - \frac{2}{5}x_3 + \frac{1}{2}x_4 \\ x_2 &= -\frac{11}{5} + \frac{3}{5}x_3 \end{aligned}$$

where $x_3$ and $x_4$ are arbitrary real numbers. If $[x_1, x_2, x_3, x_4]^T$ is a solution, then we have

$$\begin{aligned} x_1 &= \frac{14}{5} - \frac{2}{5}s + \frac{1}{2}t \\ x_2 &= -\frac{11}{5} + \frac{3}{5}s \\ x_3 &= s \\ x_4 &= t, \end{aligned}$$

where we have substituted $s$ and $t$ for $x_3$ and $x_4$, respectively, to indicate that they are arbitrary real numbers.

Using vector notation, we may write the above system of equations as

$$\begin{bmatrix} x_1 \\ x_2 \\ x_3 \\ x_4 \end{bmatrix} = \begin{bmatrix} \frac{14}{5} \\ -\frac{11}{5} \\ 0 \\ 0 \end{bmatrix} + s \begin{bmatrix} -\frac{2}{5} \\ \frac{3}{5} \\ 1 \\ 0 \end{bmatrix} + t \begin{bmatrix} \frac{1}{2} \\ 0 \\ 0 \\ 1 \end{bmatrix}.$$

Note that we have infinitely many solutions, parameterized by $s, t \in \mathbb{R}$. For the choice $s = t = 0$ we obtain a particular solution to $\boldsymbol{Ax} = \boldsymbol{b}$, given by

$$\boldsymbol{v} = \begin{bmatrix} \frac{14}{5} \\ -\frac{11}{5} \\ 0 \\ 0 \end{bmatrix}.$$

Any other solution has the form $\boldsymbol{v} + \boldsymbol{h}$, where

$$\boldsymbol{h} = s \begin{bmatrix} -\frac{2}{5} \\ \frac{3}{5} \\ 1 \\ 0 \end{bmatrix} + t \begin{bmatrix} \frac{1}{2} \\ 0 \\ 0 \\ 1 \end{bmatrix}.$$

The total number of possible basic solutions is at most

$$\binom{n}{m} = \frac{n!}{m!(n-m)!} = \frac{4!}{2!(4-2)!} = 6.$$

To find basic solutions that are feasible, we check each of the basic solutions for feasibility.

Our first candidate for a basic feasible solution is obtained by setting $x_3 = x_4 = 0$, which corresponds to the basis $\boldsymbol{B} = [\boldsymbol{a}_1, \boldsymbol{a}_2]$. Solving $\boldsymbol{Bx}_B = \boldsymbol{b}$, we obtain $\boldsymbol{x}_B = [14/5, -11/5]^T$, and hence $\boldsymbol{x} = [14/5, -11/5, 0, 0]^T$ is a basic solution that is not feasible.

For our second candidate basic feasible solution, we set $x_2 = x_4 = 0$. We have the basis $\boldsymbol{B} = [\boldsymbol{a}_1, \boldsymbol{a}_3]$. Solving $\boldsymbol{Bx}_B = \boldsymbol{b}$ yields $\boldsymbol{x}_B = [4/3, 11/3]^T$. Hence, $\boldsymbol{x} = [4/3, 0, 11/3, 0]^T$ is a basic feasible solution.

A third candidate basic feasible solution is obtained by setting $x_2 = x_3 = 0$. However, the matrix

$$\boldsymbol{B} = [\boldsymbol{a}_1, \boldsymbol{a}_4] = \begin{bmatrix} 2 & -1 \\ 4 & -2 \end{bmatrix}$$

is singular. Therefore, $\boldsymbol{B}$ cannot be a basis, and we do not have a basic solution corresponding to $\boldsymbol{B} = [\boldsymbol{a}_1, \boldsymbol{a}_4]$.

We get our fourth candidate for a basic feasible solution by setting $x_1 = x_4 = 0$. We have a basis $\boldsymbol{B} = [\boldsymbol{a}_2, \boldsymbol{a}_3]$, resulting in $\boldsymbol{x} = [0, 2, 7, 0]^T$, which is a basic feasible solution.

Our fifth candidate for a basic feasible solution corresponds to setting $x_1 = x_3 = 0$, with the basis $\boldsymbol{B} = [\boldsymbol{a}_2, \boldsymbol{a}_4]$. This results in $\boldsymbol{x} = [0, -11/5, 0, -28/5]^T$, which is a basic solution that is not feasible.

Finally, the sixth candidate for a basic feasible solution is obtained by setting $x_1 = x_2 = 0$. This results in the basis $\boldsymbol{B} = [\boldsymbol{a}_3, \boldsymbol{a}_4]$, and $\boldsymbol{x} = [0, 0, 11/3, -8/3]^T$, which is a basic solution but is not feasible. ■

## 15.7. PROPERTIES OF BASIC SOLUTIONS

In this section, we discuss the importance of basic feasible solutions in solving linear programming (LP) problems. We first prove the fundamental theorem of LP, which states that when solving an LP problem, we need only consider basic feasible solutions. This is because the optimal value (if it exists) is always achieved at a basic feasible solution. We will need the following definitions.

***Definition 15.2***

Any vector $\boldsymbol{x}$ that yields the minimum value of the objective function $\boldsymbol{c}^T\boldsymbol{x}$ over the set of vectors satisfying the constraints $\boldsymbol{A}\boldsymbol{x} = \boldsymbol{b}$, $\boldsymbol{x} \geqslant \boldsymbol{0}$, is said to be an *optimal feasible solution.*

An optimal feasible solution that is basic is said to be an *optimal basic feasible solution.* ■

**Theorem 15.1. Fundamental Theorem of LP.** *Consider a linear program in standard form.*

1. *If there exists a feasible solution, then there exists a basic feasible solution.*
2. *If there exists an optimal feasible solution, then there exists an optimal basic feasible solution.* □

*Proof.* We first prove part 1. Suppose $\boldsymbol{x} = [x_1, \ldots, x_n]^T$ is a feasible solution, and it has $p$ positive components. Without loss of generality, we can assume that the first $p$ components are positive, whereas the remaining components are zero. Then, in terms of the columns of $\boldsymbol{A} = [\boldsymbol{a}_1, \ldots, \boldsymbol{a}_p, \ldots, \boldsymbol{a}_n]$ this solution satisfies

$$x_1\boldsymbol{a}_1 + x_2\boldsymbol{a}_2 + \cdots + x_p\boldsymbol{a}_p = \boldsymbol{b}.$$

There are now two cases to consider.

*Case 1.* If $\boldsymbol{a}_1, \boldsymbol{a}_2, \ldots, \boldsymbol{a}_p$ are linearly independent, then $p \leqslant m$. If $p = m$, then the solution $\boldsymbol{x}$ is basic and the proof is completed. If, on the other hand, $p < m$, then, since rank $\boldsymbol{A} = m$, we can find $m - p$ columns of $\boldsymbol{A}$ from the remaining $n - p$ columns so that the resulting set of $m$ columns forms a basis. Hence, the solution $\boldsymbol{x}$ is a (degenerate) basic feasible solution corresponding to the above basis.

*Case 2.* Assume that $\boldsymbol{a}_1, \boldsymbol{a}_2, \ldots, \boldsymbol{a}_p$ are linearly dependent. Then, there exist numbers $y_i$, $i = 1, \ldots, p$, not all zero, such that

$$y_1\boldsymbol{a}_1 + y_2\boldsymbol{a}_2 + \cdots + y_p\boldsymbol{a}_p = \boldsymbol{0}.$$

We can assume that there exists at least one $y_i$ that is positive, for if all the $y_i$ are nonpositive, we can multiply the above equation by $-1$. Multiply the above equation by a scalar $\varepsilon$ and subtract the resulting equation from $x_1\boldsymbol{a}_1 + x_2\boldsymbol{a}_2 + \cdots + x_p\boldsymbol{a}_p = \boldsymbol{b}$ to obtain

$$(x_1 - \varepsilon y_1)\boldsymbol{a}_1 + (x_2 - \varepsilon y_2)\boldsymbol{a}_2 + \cdots + (x_p - \varepsilon y_p)\boldsymbol{a}_p = \boldsymbol{b}.$$

Let

$$\boldsymbol{y} = [y_1, \ldots, y_p, 0, \ldots, 0]^T.$$

Then, for any $\varepsilon$, we can write

$$\boldsymbol{A}[\boldsymbol{x} - \varepsilon\boldsymbol{y}] = \boldsymbol{b}.$$

Let $\varepsilon = \min\{x_i/y_i : i = 1, \ldots, p,\ y_i > 0\}$. Then, the first $p$ components of $\boldsymbol{x} - \varepsilon\boldsymbol{y}$ are nonnegative, and at least one of these components is zero. We then have a feasible solution with at most $p - 1$ positive components. We can repeat this process until we get linearly independent columns of $\boldsymbol{A}$, after which we are back to Case 1. Therefore, part 1 is proved.

We now prove part 2. Suppose $\boldsymbol{x} = [x_1, \ldots, x_n]^T$ is an optimal feasible solution, and only the first $p$ variables are nonzero. Then, we have two cases to consider. The first case (Case 1) is exactly the same as in part 1. The second case (Case 2) follows the same arguments as in part 1, but in addition we must show that $\boldsymbol{x} - \varepsilon\boldsymbol{y}$ is optimal for any $\varepsilon$. We do this by showing that $\boldsymbol{c}^T\boldsymbol{y} = 0$. To this end, assume $\boldsymbol{c}^T\boldsymbol{y} \neq 0$. Note that for $\varepsilon$ of sufficiently small magnitude ($|\varepsilon| \leqslant \min\{|x_i/y_i| : i = 1, \ldots, p,\ y_i \neq 0\}$), the vector $\boldsymbol{x} - \varepsilon\boldsymbol{y}$ is feasible. We can choose $\varepsilon$ such that $\boldsymbol{c}^T\boldsymbol{x} > \boldsymbol{c}^T\boldsymbol{x} - \varepsilon\boldsymbol{c}^T\boldsymbol{y} = \boldsymbol{c}^T(\boldsymbol{x} - \varepsilon\boldsymbol{y})$. This contradicts the optimality of $\boldsymbol{x}$. We can now use the procedure from part 1 to obtain an optimal basic feasible solution from a given optimal feasible solution. ■

***Example 15.11.*** Consider the system of equations given in the previous example. Find a nonbasic feasible solution to this system, and then use the method in the proof of the fundamental theorem of LP to find a basic feasible solution.

Recall that solutions for the system given in the previous example have the form

$$\boldsymbol{x} = \begin{bmatrix} \frac{14}{5} \\ -\frac{11}{5} \\ 0 \\ 0 \end{bmatrix} + s\begin{bmatrix} -\frac{2}{5} \\ \frac{3}{5} \\ 1 \\ 0 \end{bmatrix} + t\begin{bmatrix} \frac{1}{2} \\ 0 \\ 0 \\ 1 \end{bmatrix},$$

where $s, t \in \mathbb{R}$. Note that if $s = 4$ and $t = 0$ then

$$x_0 = \begin{bmatrix} \frac{6}{5} \\ \frac{1}{5} \\ 4 \\ 0 \end{bmatrix}$$

is a nonbasic feasible solution.

There are constants $y_i$, $i = 1, 2, 3$, such that

$$y_1 \boldsymbol{a}_1 + y_2 \boldsymbol{a}_2 + y_3 \boldsymbol{a}_3 = \boldsymbol{0}.$$

For example, let

$$y_1 = -\tfrac{2}{5},$$
$$y_2 = \tfrac{3}{5},$$
$$y_3 = 1.$$

Note that

$$\boldsymbol{A}(\boldsymbol{x}_0 - \varepsilon \boldsymbol{y}) = \boldsymbol{b},$$

where

$$\boldsymbol{y} = \begin{bmatrix} -\frac{2}{5} \\ \frac{3}{5} \\ 1 \\ 0 \end{bmatrix}.$$

If $\varepsilon = \frac{1}{3}$ then

$$\boldsymbol{x}_1 = \boldsymbol{x}_0 - \varepsilon \boldsymbol{y} = \begin{bmatrix} \frac{4}{3} \\ 0 \\ \frac{11}{3} \\ 0 \end{bmatrix}$$

is a basic feasible solution. ■

Observe that the fundamental theorem of LP reduces the task of solving an LP problem to that of searching over a finite number of basic feasible solutions. That is, we need only check basic feasible solutions for optimality. As mentioned before, the total number of basic solutions is at most

$$\binom{n}{m} = \frac{n!}{m!(n-m)!}.$$

Although this number is finite, it may be quite large. For example, if $m = 5$ and $n = 50$, then

$$\binom{n}{m} = \binom{50}{4} = 2{,}118{,}760.$$

This is potentially the number of basic feasible solutions to be checked for optimality. Therefore, a more efficient method of solving linear programs is needed. To this end, in the next section, we analyze a geometric interpretation of the fundamental theorem of LP. This will lead us to the simplex method for solving linear programs, which we discuss in the following chapter.

## 15.8. A GEOMETRIC VIEW OF LINEAR PROGRAMS

Recall that a set $\Theta \subset \mathbb{R}^n$ is said to be *convex* if, for every $\boldsymbol{x}, \boldsymbol{y} \in \Theta$ and every real number $\alpha, 0 < \alpha < 1$, the point $\alpha\boldsymbol{x} + (1-\alpha)\boldsymbol{y} \in \Theta$. In other words, a set is convex if, given two points in the set, every point on the line segment joining these two points is also a member of the set.

Note that the set of points satisfying the constraints

$$\boldsymbol{Ax} = \boldsymbol{b}, \quad \boldsymbol{x} \geqslant \boldsymbol{0},$$

is convex. To see this, let $\boldsymbol{x}_1$ and $\boldsymbol{x}_2$ satisfy the constraints [that is, $\boldsymbol{Ax}_i = \boldsymbol{b}, \boldsymbol{x}_i \geqslant \boldsymbol{0}$, $i = 1, 2$]. Then, for all $\alpha \in (0, 1)$, $\boldsymbol{A}[\alpha\boldsymbol{x}_1 + (1-\alpha)\boldsymbol{x}_2] = \alpha\boldsymbol{Ax}_1 + (1-\alpha)\boldsymbol{Ax}_2 = \boldsymbol{b}$. Also, for $\alpha \in (0, 1)$, we have $\alpha\boldsymbol{x}_1 + (1-\alpha)\boldsymbol{x}_2 \geqslant \boldsymbol{0}$.

Recall that a point $\boldsymbol{x}$ in a convex set $\Theta$ is said to be an *extreme point* of $\Theta$ if there are no two distinct points $\boldsymbol{x}_1$ and $\boldsymbol{x}_2$ in $\Theta$ such that $\boldsymbol{x} = \alpha\boldsymbol{x}_1 + (1-\alpha)\boldsymbol{x}_2$ for some $\alpha \in (0, 1)$. In other words, an extreme point is a point that does not lie strictly within the line segment connecting two other points of the set. Therefore, if $\boldsymbol{x}$ is an extreme point, and $\boldsymbol{x} = \alpha\boldsymbol{x}_1 + (1-\alpha)\boldsymbol{x}_2$ for some $\boldsymbol{x}_1, \boldsymbol{x}_2 \in \Theta$ and $\alpha \in (0, 1)$, then $\boldsymbol{x}_1 = \boldsymbol{x}_2$. In the following theorem, we show that all extreme points of the constraint set are basic feasible solutions.

**Theorem 15.2.** *Let $\Omega$ be the convex set consisting of all feasible solutions, that is, all n-vectors $\boldsymbol{x}$ satisfying*

$$\boldsymbol{Ax} = \boldsymbol{b}, \quad \boldsymbol{x} \geqslant \boldsymbol{0},$$

*where $\boldsymbol{A} \in \mathbb{R}^{m \times n}, m < n$. If $\boldsymbol{x}$ is an extreme point of $\Omega$, then $\boldsymbol{x}$ is a basic feasible solution to $\boldsymbol{Ax} = \boldsymbol{b}, \boldsymbol{x} \geqslant \boldsymbol{0}$.* □

*Proof.* Suppose $\boldsymbol{x}$ satisfies $\boldsymbol{Ax} = \boldsymbol{b}, \boldsymbol{x} \geqslant \boldsymbol{0}$, and it has $p$ positive components. As before, without loss of generality, we can assume that the first $p$ components are positive, and the remaining components are zero. We have

$$x_1\boldsymbol{a}_1 + x_2\boldsymbol{a}_2 + \cdots + x_p\boldsymbol{a}_p = \boldsymbol{b}.$$

Let $y_i$, $i = 1, \ldots, p$, be numbers such that

$$y_1\boldsymbol{a}_1 + y_2\boldsymbol{a}_2 + \cdots + y_p\boldsymbol{a}_p = \boldsymbol{0}.$$

We will show that each $y_i = 0$. To begin, multiply this equation by $\varepsilon > 0$; then add and subtract the result from the equation $x_1\boldsymbol{a}_1 + x_2\boldsymbol{a}_2 + \cdots + x_p\boldsymbol{a}_p = \boldsymbol{b}$ to get

$$(x_1 + \varepsilon y_1)\boldsymbol{a}_1 + (x_2 + \varepsilon y_2)\boldsymbol{a}_2 + \cdots + (x_p + \varepsilon y_p)\boldsymbol{a}_p = \boldsymbol{b}$$
$$(x_1 - \varepsilon y_1)\boldsymbol{a}_1 + (x_2 - \varepsilon y_2)\boldsymbol{a}_2 + \cdots + (x_p - \varepsilon y_p)\boldsymbol{a}_p = \boldsymbol{b}.$$

Since each $x_i > 0$, $\varepsilon > 0$ can be chosen such that each $x_i + \varepsilon y_i, x_i - \varepsilon y_i \geqslant 0$ (for example, $\varepsilon = \min\{|x_i/y_i| : i = 1, \ldots, p,\ y_i \neq 0\}$). For such a choice of $\varepsilon$, the vectors

$$\boldsymbol{z}_1 = [x_1 + \varepsilon y_1, x_2 + \varepsilon y_2, \ldots, x_p + \varepsilon y_p, 0, \ldots, 0]^T$$
$$\boldsymbol{z}_2 = [x_1 - \varepsilon y_1, x_2 - \varepsilon y_2, \ldots, x_p - \varepsilon y_p, 0, \ldots, 0]^T$$

belong to $\Omega$. Observe that $\boldsymbol{x} = \frac{1}{2}\boldsymbol{z}_1 + \frac{1}{2}\boldsymbol{z}_2$. Since $\boldsymbol{x}$ is an extreme point, then $\boldsymbol{z}_1 = \boldsymbol{z}_2$. Hence, each $y_i = 0$, which implies that the $\boldsymbol{a}_i$ are linearly independent. ■

The above theorem states that if the vector $\boldsymbol{x} \in \Omega$ is an extreme point of $\Omega$, then it is a basic feasible solution of the constraints $\boldsymbol{A}\boldsymbol{x} = \boldsymbol{b}$, $\boldsymbol{x} \geqslant \boldsymbol{0}$. The next theorem states that the converse is also true.

**Theorem 15.3.** *If $\boldsymbol{x}$ is a basic feasible solution to $\boldsymbol{A}\boldsymbol{x} = \boldsymbol{b}$, $\boldsymbol{x} \geqslant \boldsymbol{0}$, then $\boldsymbol{x}$ is an extreme point of $\Omega$.* □

*Proof.* Let $\boldsymbol{x} \in \Omega$ be a basic feasible solution. Let $\boldsymbol{y}, \boldsymbol{z} \in \Omega$ be such that

$$\boldsymbol{x} = \alpha\boldsymbol{y} + (1 - \alpha)\boldsymbol{z}$$

for some $\alpha \in (0, 1)$. We shall show that $\boldsymbol{y} = \boldsymbol{z}$, and conclude that $\boldsymbol{x}$ is an extreme point. Since $\boldsymbol{y}, \boldsymbol{z} \geqslant \boldsymbol{0}$, and the last $n - m$ components of $\boldsymbol{x}$ are zero, the last $n - m$ components of $\boldsymbol{y}$ and $\boldsymbol{z}$ are zero as well. Furthermore, since $\boldsymbol{A}\boldsymbol{y} = \boldsymbol{A}\boldsymbol{z} = \boldsymbol{b}$,

$$y_1\boldsymbol{a}_1 + \cdots + y_m\boldsymbol{a}_m = \boldsymbol{b}$$

and

$$z_1\boldsymbol{a}_1 + \cdots + z_m\boldsymbol{a}_m = \boldsymbol{b}.$$

Combining the above two equations yields

$$(y_1 - z_1)\boldsymbol{a}_1 + \cdots + (y_m - z_m)\boldsymbol{a}_m = \boldsymbol{0}.$$

Since the columns $\boldsymbol{a}_1, \ldots, \boldsymbol{a}_m$ are linearly independent, we have $y_i = z_i, i = 1, \ldots, m$. Therefore, $\boldsymbol{y} = \boldsymbol{z}$, and hence $\boldsymbol{x}$ is an extreme point of $\Omega$. ■

From the above two theorems, it follows that the set of extreme points of $\Omega = \{\boldsymbol{x}: \boldsymbol{A}\boldsymbol{x} = \boldsymbol{b}, \boldsymbol{x} \geqslant \boldsymbol{0}\}$ is equal to the set of basic feasible solutions to $\boldsymbol{A}\boldsymbol{x} = \boldsymbol{b}$, $\boldsymbol{x} \geqslant \boldsymbol{0}$. Combining the above observation with the fundamental theorem of LP, we can see that in solving linear programming problems we need only examine the extreme points of the constraint set.

***Example 15.12.*** Consider the following LP problem:

$$\begin{aligned} \text{maximize} \quad & 3x_1 + 5x_2 \\ \text{subject to} \quad & x_1 + 5x_2 \leqslant 40 \\ & 2x_1 + x_2 \leqslant 20 \\ & x_1 + x_2 \leqslant 12 \\ & x_1, x_2 \geqslant 0. \end{aligned}$$

We introduce slack variables $x_3, x_4, x_5$ to convert the above LP problem into standard form:

$$\begin{aligned} \text{minimize} \quad & -3x_1 - 5x_2 \\ \text{subject to} \quad & x_1 + 5x_2 + x_3 = 40 \\ & 2x_1 + x_2 + x_4 = 20 \\ & x_1 + x_2 + x_5 = 12 \\ & x_1, \ldots, x_5 \geqslant 0. \end{aligned}$$

In the remainder of the example, we consider only the problem in standard form. We can represent the above constraints as

$$x_1 \begin{bmatrix} 1 \\ 2 \\ 1 \end{bmatrix} + x_2 \begin{bmatrix} 5 \\ 1 \\ 1 \end{bmatrix} + x_3 \begin{bmatrix} 1 \\ 0 \\ 0 \end{bmatrix} + x_4 \begin{bmatrix} 0 \\ 1 \\ 0 \end{bmatrix} + x_5 \begin{bmatrix} 0 \\ 0 \\ 1 \end{bmatrix} = \begin{bmatrix} 40 \\ 20 \\ 12 \end{bmatrix}, \quad x_1, \ldots, x_5 \geqslant 0;$$

that is, $x_1\boldsymbol{a}_1 + x_2\boldsymbol{a}_2 + x_3\boldsymbol{a}_3 + x_4\boldsymbol{a}_4 + x_5\boldsymbol{a}_5 = \boldsymbol{b}, \boldsymbol{x} \geqslant \boldsymbol{0}$. Note that

$$\boldsymbol{x} = [0, 0, 40, 20, 12]^T$$

is a feasible solution. But, for this $\boldsymbol{x}$, the value of the objective function is 0. We already know that the minimum of the objective function (if it exists) is achieved at an extreme point of the constraint set $\Omega$ defined by the above constraints. The point $[0, 0, 40, 20, 12]^T$ is an extreme point of the set of feasible solutions, but it turns out that it does not minimize the objective function. Therefore, we need to seek the solution among the other extreme points. To do this, we move from one extreme point to an adjacent extreme point such that the value of the objective function decreases. Here, we define two extreme points to be adjacent if the corresponding basic columns differ by only one vector. We begin with $\boldsymbol{x} = [0, 0, 40, 20, 12]^T$. We have

$$0\boldsymbol{a}_1 + 0\boldsymbol{a}_2 + 40\boldsymbol{a}_3 + 20\boldsymbol{a}_4 + 12\boldsymbol{a}_5 = \boldsymbol{b}.$$

To select an adjacent extreme point, let us choose to include $\boldsymbol{a}_1$ as a basic column in the new basis. We will need to remove either $\boldsymbol{a}_3$, $\boldsymbol{a}_4$, or $\boldsymbol{a}_5$ from the old basis. We proceed as follows. We first express $\boldsymbol{a}_1$ as a linear combination of the old basic columns:

$$\boldsymbol{a}_1 = 1\boldsymbol{a}_3 + 2\boldsymbol{a}_4 + 1\boldsymbol{a}_5.$$

Multiplying both sides of this equation by $\varepsilon_1 > 0$, we get

$$\varepsilon_1\boldsymbol{a}_1 = \varepsilon_1\boldsymbol{a}_3 + 2\varepsilon_1\boldsymbol{a}_4 + \varepsilon_1\boldsymbol{a}_5.$$

We now add the above equation to the equation $0\boldsymbol{a}_1 + 0\boldsymbol{a}_2 + 40\boldsymbol{a}_3 + 20\boldsymbol{a}_4 + 12\boldsymbol{a}_5 = \boldsymbol{b}$. Collecting terms yields

$$\varepsilon_1\boldsymbol{a}_1 + 0\boldsymbol{a}_2 + (40 - \varepsilon_1)\boldsymbol{a}_3 + (20 - 2\varepsilon_1)\boldsymbol{a}_4 + (12 - \varepsilon_1)\boldsymbol{a}_5 = \boldsymbol{b}.$$

We want to choose $\varepsilon_1$ in such a way that each of the above coefficients is nonnegative, and at the same time one of the coefficients of either $\boldsymbol{a}_3$, $\boldsymbol{a}_4$, or $\boldsymbol{a}_5$ becomes zero. Clearly $\varepsilon_1 = 10$ does the job. The result is

$$10\boldsymbol{a}_1 + 30\boldsymbol{a}_3 + 2\boldsymbol{a}_5 = \boldsymbol{b}.$$

The corresponding basic feasible solution (extreme point) is

$$[10, 0, 30, 0, 2]^T.$$

For this solution, the objective function value is $-30$, which is an improvement relative to the objective function value at the old extreme point.

We now apply the same procedure as above to move to another adjacent extreme point, which hopefully further decreases the value of the objective function. This time, we choose $\boldsymbol{a}_2$ to enter the new basis. We have

$$\boldsymbol{a}_2 = \tfrac{1}{2}\boldsymbol{a}_1 + \tfrac{9}{2}\boldsymbol{a}_3 + \tfrac{1}{2}\boldsymbol{a}_5,$$

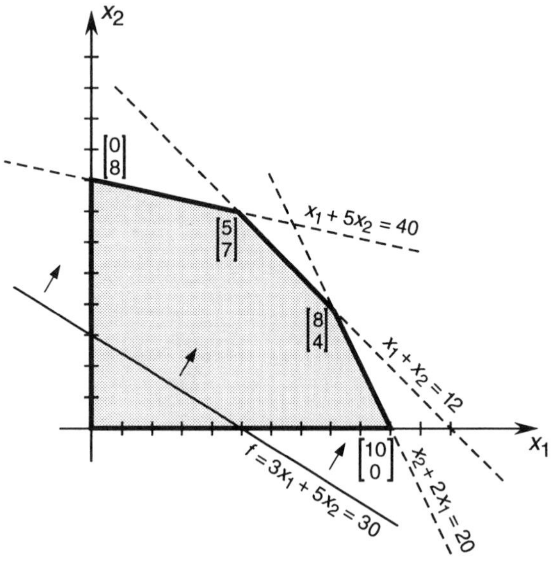


**Figure 15.10.** A graphical solution to the linear programming problem in Example 15.12

and

$$(10 - \tfrac{1}{2}\varepsilon_2)\boldsymbol{a}_1 + \varepsilon_2\boldsymbol{a}_2 + (30 - \tfrac{9}{2}\varepsilon_2)\boldsymbol{a}_3 + (2 - \tfrac{1}{2}\varepsilon_2)\boldsymbol{a}_5 = \boldsymbol{b}.$$

Substituting $\varepsilon_2 = 4$, we obtain

$$8\boldsymbol{a}_1 + 4\boldsymbol{a}_2 + 12\boldsymbol{a}_3 = \boldsymbol{b}.$$

The solution is $[8, 4, 12, 0, 0]^T$ and the corresponding value of the objective function is $-44$, which is smaller than the value at the previous extreme point. To complete the example, we repeat the procedure once more. This time, we select $\boldsymbol{a}_4$ and express it as a combination of the vectors in the previous basis, $\boldsymbol{a}_1, \boldsymbol{a}_2$, and $\boldsymbol{a}_3$:

$$\boldsymbol{a}_4 = \boldsymbol{a}_1 - \boldsymbol{a}_2 + 4\boldsymbol{a}_3,$$

and hence

$$(8 - \varepsilon_3)\boldsymbol{a}_1 + (4 + \varepsilon_3)\boldsymbol{a}_2 + (12 - 4\varepsilon_3)\boldsymbol{a}_3 + \varepsilon_3\boldsymbol{a}_4 = \boldsymbol{b}.$$

The largest permissible value for $\varepsilon_3$ is 3. The corresponding basic feasible solution is $[5, 7, 0, 3, 0]^T$, with an objective function value of $-50$. The solution $[5, 7, 0, 3, 0]^T$ turns out to be an optimal solution to our problem in standard form. Hence, the solution to the original problem is $[5, 7]^T$, which we can easily obtain graphically (see Figure 15.10). ■

The technique used in the above example for moving from one extreme point to an adjacent extreme point is also used in the simplex method for solving LP problems. The simplex method is essentially a refined method of performing these manipulations.

## EXERCISES

**15.1** Convert the following linear programming problem to standard form:

$$\begin{array}{ll} \text{maximize} & 2x_1 + x_2 \\ \text{subject to} & 0 \leqslant x_1 \leqslant 2 \\ & x_1 + x_2 \leqslant 3 \\ & x_1 + 2x_2 \leqslant 5 \\ & x_2 \geqslant 0 \end{array}$$

**15.2** A cereal manufacturer wishes to produce 1000 pounds of a cereal that contains exactly 10% fiber, 2% fat, and 5% sugar (by weight). The cereal is to be produced by combining four items of raw food material in appropriate proportions. These four items have certain combinations of fiber, fat, and sugar content, and are available at various prices per pound, as shown below:

| Item | 1 | 2 | 3 | 4 |
|---|---|---|---|---|
| % fiber | 3 | 8 | 16 | 4 |
| % fat | 6 | 46 | 9 | 9 |
| % sugar | 20 | 5 | 4 | 0 |
| Price/lb | 2 | 4 | 1 | 2 |

The manufacturer wishes to find the amounts of each of the above items to be used to produce the cereal in the least expensive way. Formulate the problem as a linear programming problem. What can you say about the existence of a solution to this problem?

**15.3** Consider the system of equations:

$$\begin{bmatrix} 2 & -1 & 2 & -1 & 3 \\ 1 & 2 & 3 & 1 & 0 \\ 1 & 0 & -2 & 0 & -5 \end{bmatrix} \begin{bmatrix} x_1 \\ x_2 \\ x_3 \\ x_4 \\ x_5 \end{bmatrix} = \begin{bmatrix} 14 \\ 5 \\ -10 \end{bmatrix}.$$

Check if the system has basic solutions. If yes, find all basic solutions.

**15.4** Solve the following linear program graphically:

$$\begin{aligned} \text{maximize} \quad & 2x_1 + 5x_2 \\ \text{subject to} \quad & 0 \leqslant x_1 \leqslant 4 \\ & 0 \leqslant x_2 \leqslant 6 \\ & x_1 + x_2 \leqslant 8 \end{aligned}$$

**15.5** The optimization toolbox in MATLAB provides a function, lp, for solving linear programming problems. Use the function lp to solve the problem in Example 15.5. Use the initial condition **0**.

# 16

# The Simplex Method

## 16.1. SOLVING LINEAR EQUATIONS USING ROW OPERATIONS

The examples in the previous chapters illustrate that solving linear programs involves the solution of systems of linear simultaneous algebraic equations. In this section, we present a method for solving a system of $n$ linear equations in $n$ unknowns, which we will use in subsequent sections. The method uses elementary row operations and corresponding elementary matrices. For a discussion of numerical issues involved in solving a system of simultaneous linear algebraic equations, we refer the reader to Refs. 20 and 29.

***Definition 16.1.*** We call $\boldsymbol{E}$ an *elementary matrix of the first kind* if $\boldsymbol{E}$ is obtained from the identity matrix $\boldsymbol{I}$ by interchanging any two of its rows. ■

An elementary matrix of the first kind formed from $\boldsymbol{I}$ by interchanging the $i$th and the $j$th rows has the form

$$\boldsymbol{E}=\begin{bmatrix} 1 & & & & & & & & & \\ & \ddots & & & & & & & & \\ & & 1 & & & & & & & \\ & & & 0 & \cdots & & 1 & & & \\ & & & & 1 & & & & & \\ & & & \vdots & & \ddots & \vdots & & & \\ & & & & & 1 & & & & \\ & & & 1 & \cdots & & 0 & & & \\ & & & & & & & 1 & & \\ & & & & & & & & \ddots & \\ & & & & & & & & & 1 \end{bmatrix} \begin{matrix} \\ \\ \\ \leftarrow i\text{th row} \\ \\ \\ \\ \leftarrow j\text{th row} \\ \\ \\ \\ \end{matrix}$$

Note that $\boldsymbol{E}$ is invertible and $\boldsymbol{E}=\boldsymbol{E}^{-1}$.

***Definition 16.2.*** We call $\boldsymbol{E}$ an *elementary matrix of the second kind* if $\boldsymbol{E}$ is obtained from the identity matrix $\boldsymbol{I}$ by multiplying one of its rows by a real number $\alpha \neq 0$. ■

The elementary matrix of the second kind formed from $\boldsymbol{I}$ by multiplying the $i$th row by $\alpha \neq 0$ has the form

$$\boldsymbol{E} = \begin{bmatrix} 1 & & & & & & \\ & \ddots & & & & 0 & \\ & & 1 & & & & \\ & & & \alpha & & & \\ & & & & 1 & & \\ & 0 & & & & \ddots & \\ & & & & & & 1 \end{bmatrix} \leftarrow i\text{th row}$$

Note that $\boldsymbol{E}$ is invertible and

$$\boldsymbol{E}^{-1} = \begin{bmatrix} 1 & & & & & & \\ & \ddots & & & & 0 & \\ & & 1 & & & & \\ & & & 1/\alpha & & & \\ & & & & 1 & & \\ & 0 & & & & \ddots & \\ & & & & & & 1 \end{bmatrix} \leftarrow i\text{th row}$$

***Definition 16.3.*** We call $\boldsymbol{E}$ an *elementary matrix of the third kind* if $\boldsymbol{E}$ is obtained from the identity matrix $\boldsymbol{I}$ by adding $\beta$ times one row to another row of $\boldsymbol{I}$. ■

An elementary matrix of the third kind obtained from $\boldsymbol{I}$ by adding $\beta$ times the $j$th row to the $i$th row has the form

$$\boldsymbol{E} = \begin{bmatrix} 1 & & & & & & \\ & \ddots & & & & 0 & \\ & & 1 & \cdots & \beta & & \\ & & & \ddots & \vdots & & \\ & & & & 1 & & \\ & 0 & & & & \ddots & \\ & & & & & & 1 \end{bmatrix} \begin{matrix} \\ \\ \leftarrow i\text{th row} \\ \\ \leftarrow j\text{th row} \\ \\ \\ \end{matrix}$$

Observe that $\boldsymbol{E}$ is the identity matrix with an extra $\beta$ in the $(i, j)$th location. Note that $\boldsymbol{E}$ is invertible and

$$\boldsymbol{E}^{-1} = \begin{bmatrix} 1 & & & & & & \\ & \ddots & & & & 0 & \\ & & 1 & \cdots & -\beta & & \\ & & & \ddots & \vdots & & \\ & & & & 1 & & \\ & 0 & & & & \ddots & \\ & & & & & & 1 \end{bmatrix} \begin{matrix} \\ \\ \leftarrow i\text{th row} \\ \\ \leftarrow j\text{th row} \\ \\ \\ \end{matrix}$$

***Definition 16.4.*** An *elementary row operation* (of first, second or third kind) on a given matrix is a premultiplication of the given matrix by a corresponding elementary matrix of the respective kind. ■

Since elementary matrices are invertible, the corresponding elementary row operations are also invertible. Consider a system of $n$ linear equations in $n$ unknowns $x_1, x_2, \ldots, x_n$ with right-hand sides $b_1, b_2, b_n$. In matrix form, this system may be written as

$$\boldsymbol{Ax} = \boldsymbol{b}$$

where

$$\begin{aligned} \boldsymbol{x} &= [x_1, \ldots, x_n]^T, \\ \boldsymbol{b} &= [b_1, \ldots, b_n]^T, \\ \boldsymbol{A} &\in \mathbb{R}^{n \times n}. \end{aligned}$$

If $\boldsymbol{A}$ is invertible then

$$\boldsymbol{x} = \boldsymbol{A}^{-1}\boldsymbol{b}.$$

Thus, the problem of solving the system of equations $\boldsymbol{Ax} = \boldsymbol{b}$, with $\boldsymbol{A} \in \mathbb{R}^{n \times n}$ invertible is related to the problem of computing $\boldsymbol{A}^{-1}$. We now show that $\boldsymbol{A}^{-1}$ can be effectively computed using elementary row operations. In particular, we prove the following theorem.

**Theorem 16.1.** *Let $\boldsymbol{A} \in \mathbb{R}^{n \times n}$ be a given matrix. Then, $\boldsymbol{A}$ is nonsingular (invertible) if, and only if, there exist elementary matrices $\boldsymbol{E}_i, i = 1, \ldots, t$, such that*

$$\boldsymbol{E}_t \cdots \boldsymbol{E}_2 \boldsymbol{E}_1 \boldsymbol{A} = \boldsymbol{I}. \qquad \square$$

*Proof.* ⇒: If $\boldsymbol{A}$ is nonsingular then its first column must have at least one nonzero element, say $a_{j1} \neq 0$. Premultiplying $\boldsymbol{A}$ by an elementary matrix of the first kind of the form

$$\boldsymbol{E}_1 = \begin{bmatrix} 0 & & & 1 & & & \\ & 1 & & & & & \\ & & \ddots & \vdots & & & \\ & & & 1 & & & \\ 1 & & \cdots & 0 & & & \\ & & & & 1 & & \\ & & & & & \ddots & \\ & & & & & & 1 \end{bmatrix} \leftarrow j\text{th row}$$

brings the nonzero element $a_{j1}$ to the location $(1, 1)$. Hence, in the matrix $\boldsymbol{E}_1\boldsymbol{A}$, the element $a_{11} \neq 0$. Note that since $\boldsymbol{E}_1$ is nonsingular, $\boldsymbol{E}_1\boldsymbol{A}$ is also nonsingular.

Next, we premultiply $\boldsymbol{E}_1\boldsymbol{A}$ by an elementary matrix of the second kind of the form

$$\boldsymbol{E}_2=\begin{bmatrix}1/a_{11} & & & \\ & 1 & & \\ & & \ddots & \\ & & & 1\end{bmatrix}.$$

The result of this operation is the matrix $\boldsymbol{E}_2\boldsymbol{E}_1\boldsymbol{A}$ with unity in the location $(1,1)$. We next apply a sequence of elementary row operations of the third kind on the matrix $\boldsymbol{E}_2\boldsymbol{E}_1\boldsymbol{A}$. Specifically, we premultiply $\boldsymbol{E}_2\boldsymbol{E}_1\boldsymbol{A}$ by $n-1$ elementary matrices of the form

$$\boldsymbol{E}_3=\begin{bmatrix}1 & & & & \\ -a_{21} & 1 & & & \\ & & 1 & & \\ & & & \ddots & \\ & & & & 1\end{bmatrix},\ldots,\quad \boldsymbol{E}_r=\begin{bmatrix}1 & & & & \\ & \ddots & & & \\ \vdots & & 1 & & \\ & & & \ddots & \\ -a_{n1} & & & & 1\end{bmatrix},$$

where $r=2+n-1=n+1$. The result of these operations is the nonsingular matrix

$$\boldsymbol{E}_r\boldsymbol{E}_{r-1}\cdots\boldsymbol{E}_2\boldsymbol{E}_1\boldsymbol{A}=\begin{bmatrix}1 & \bar{a}_{12} & \cdots & \bar{a}_{1n}\\ 0 & \bar{a}_{22} & \cdots & \bar{a}_{2n}\\ \vdots & \vdots & & \vdots\\ 0 & \bar{a}_{n2} & \cdots & \bar{a}_{nn}\end{bmatrix}.$$

Since the matrix $\boldsymbol{E}_r\cdots\boldsymbol{E}_1\boldsymbol{A}$ is nonsingular, its submatrix

$$\begin{bmatrix}\bar{a}_{22} & \cdots & \bar{a}_{2n}\\ \vdots & & \vdots\\ \bar{a}_{n2} & \cdots & \bar{a}_{nn}\end{bmatrix}$$

must be nonsingular too. The above implies that there is a nonzero element $\bar{a}_{j2}$, where $2\leqslant j\leqslant n$. Using an elementary operation of the first kind, we bring this element to the location $(2,2)$. Thus, in the matrix

$$\boldsymbol{E}_{r+1}\boldsymbol{E}_r\cdots\boldsymbol{E}_1\boldsymbol{A}$$

the $(2,2)$th element is nonzero. Premultiplying the above matrix by an elementary matrix of the second kind yields the matrix

$$\boldsymbol{E}_{r+2}\boldsymbol{E}_{r+1}\cdots\boldsymbol{E}_1\boldsymbol{A}$$

in which the element in the location $(2,2)$ is unity. As before, we premultiply this matrix by $n-1$ elementary row operations of the third kind, to get a matrix of the form

$$E_s \cdots E_r \cdots E_1 A = \begin{bmatrix} 1 & 0 & \tilde{a}_{13} & \cdots & \tilde{a}_{1n} \\ 0 & 1 & \tilde{a}_{23} & \cdots & \tilde{a}_{2n} \\ 0 & 0 & \tilde{a}_{33} & \cdots & \tilde{a}_{3n} \\ \vdots & \vdots & \vdots & & \vdots \\ 0 & 0 & \tilde{a}_{n3} & \cdots & \tilde{a}_{nn} \end{bmatrix},$$

where $s = r + 2 + n - 1$. The above matrix is nonsingular. Hence there is a nonzero element $\tilde{a}_{j3}$, $3 \leqslant j \leqslant n$. Proceeding in a similar fashion as before, we obtain

$$E_t \cdots E_s \cdots E_r \cdots E_1 A = I.$$

where $t = n(n+1)$.

$\Leftarrow$: If there exist elementary matrices $E_1, \ldots, E_t$ such that

$$E_t \cdots E_1 A = I,$$

then clearly $A$ is invertible, with

$$A^{-1} = E_t \cdots E_1. \qquad \blacksquare$$

The above theorem suggests the following procedure for finding $A^{-1}$, if it exists. We first form an augmented matrix

$$[A, I].$$

We then apply elementary row operations to $[A, I]$ so that $A$ is transformed into $I$, that is, we obtain

$$E_t \cdots E_1 [A, I] = [I, B].$$

It then follows that

$$B = E_t \cdots E_1 = A^{-1}.$$

***Example 16.1.*** Let

$$\mathbf{A} = \begin{bmatrix} 2 & 5 & 10 & 0 \\ 1 & 1 & 1 & 0 \\ -2 & -10 & -30 & 1 \\ -1 & -2 & -3 & 0 \end{bmatrix}.$$

Find $A^{-1}$.

We form an augmented matrix

$$[A,I]=\begin{bmatrix} 2 & 5 & 10 & 0 & 1 & 0 & 0 & 0 \\ 1 & 1 & 1 & 0 & 0 & 1 & 0 & 0 \\ -2 & -10 & -30 & 1 & 0 & 0 & 1 & 0 \\ -1 & -2 & -3 & 0 & 0 & 0 & 0 & 1 \end{bmatrix},$$

and perform row operations on this matrix. Applying row operations of the first and third kinds yields

$$\begin{bmatrix} 1 & 1 & 1 & 0 & 0 & 1 & 0 & 0 \\ 0 & 3 & 8 & 0 & 1 & -2 & 0 & 0 \\ 0 & -8 & -28 & 1 & 0 & 2 & 1 & 0 \\ 0 & -1 & -2 & 0 & 0 & 1 & 0 & 1 \end{bmatrix}.$$

We then interchange the second and fourth rows and apply elementary row operations of the second and third kinds to get

$$\begin{bmatrix} 1 & 0 & -1 & 0 & 0 & 2 & 0 & 1 \\ 0 & 1 & 2 & 0 & 0 & -1 & 0 & -1 \\ 0 & 0 & 2 & 0 & 1 & 1 & 0 & 3 \\ 0 & 0 & -12 & 1 & 0 & -6 & 1 & -8 \end{bmatrix}.$$

Now multiply the third row by $\frac{1}{2}$ and then perform a sequence of the elementary operations of the third kind to obtain

$$\begin{bmatrix} 1 & 0 & 0 & 0 & \frac{1}{2} & \frac{5}{2} & 0 & \frac{5}{2} \\ 0 & 1 & 0 & 0 & -1 & -2 & 0 & -4 \\ 0 & 0 & 1 & 0 & \frac{1}{2} & \frac{1}{2} & 0 & \frac{3}{2} \\ 0 & 0 & 0 & 1 & 6 & 0 & 1 & 10 \end{bmatrix}.$$

Hence

$$A^{-1}=\begin{bmatrix} \frac{1}{2} & \frac{5}{2} & 0 & \frac{5}{2} \\ -1 & -2 & 0 & -4 \\ \frac{1}{2} & \frac{1}{2} & 0 & \frac{3}{2} \\ 6 & 0 & 1 & 10 \end{bmatrix}.$$

■

We now return to the general problem of solving the system of equations $\boldsymbol{Ax}=\boldsymbol{b}, \boldsymbol{A}\in\mathbb{R}^{n\times n}$. If $\boldsymbol{A}^{-1}$ exists, then the solution is $\boldsymbol{x}=\boldsymbol{A}^{-1}\boldsymbol{b}$. However, we do not need an explicit expression for $\boldsymbol{A}^{-1}$ to find the solution. Indeed, let $\boldsymbol{A}^{-1}$ be expressed as a product of elementary matrices

$$\boldsymbol{A}^{-1}=\boldsymbol{E}_t\boldsymbol{E}_{t-1}\cdots\boldsymbol{E}_1.$$

Thus

$$\boldsymbol{E}_t\cdots\boldsymbol{E}_1\boldsymbol{Ax}=\boldsymbol{E}_t\cdots\boldsymbol{E}_1\boldsymbol{b}$$

and hence

$$x = E_t \cdots E_1 b.$$

The above discussion leads to the following procedure for solving the system $Ax = b$. Form an augmented matrix

$$[A, b].$$

Then, perform a sequence of elementary row operations on this augmented matrix until we obtain

$$[I, \tilde{b}].$$

From the above, we have that if $x$ is a solution to $Ax = b$, then it is also a solution to $EAx = Eb$, where $E = E_t \cdots E_1$ represents a sequence of elementary row operations. Since $EA = I$, and $Eb = \tilde{b}$, then it follows that $x = \tilde{b}$ is the solution to $Ax = b$, $A \in \mathbb{R}^{n \times n}$ invertible.

Suppose now that $A \in \mathbb{R}^{m \times n}$ where $m < n$, and rank $A = m$. Then, $A$ is not a square matrix. Clearly in this case the system of equations $Ax = b$ has infinitely many solutions. Without loss of generality, we can assume that the first $m$ columns of $A$ are linearly independent. Then, if we perform a sequence of elementary row operations on the augmented matrix $[A, b]$ as before, we obtain

$$[I, D, \tilde{b}],$$

where $D$ is an $m \times (n - m)$ matrix. Let $x \in \mathbb{R}^n$ be a solution to $Ax = b$, and write $x = [x_B^T, x_D^T]^T$, where $x_B \in \mathbb{R}^m$, $x_D \in \mathbb{R}^{(n-m)}$. Then, $[I, D]x = \tilde{b}$, which we can rewrite as $x_B + Dx_D = \tilde{b}$, or $x_B = \tilde{b} - Dx_D$. Note that for an arbitrary $x_D \in \mathbb{R}^{(n-m)}$, if $x_B = \tilde{b} - Dx_D$, then the resulting vector $x = [x_B^T, x_D^T]^T$ is a solution to $Ax = b$. In particular, $[\tilde{b}^T, 0^T]^T$ is a solution to $Ax = b$. We often refer to the basic solution $[\tilde{b}^T, 0^T]^T$ as a *particular solution* to $Ax = b$. Note that $[-(Dx_D)^T, x_D^T]^T$ is a solution to $Ax = 0$. Any solution to $Ax = b$ has the form

$$x = \begin{bmatrix} \tilde{b} \\ 0 \end{bmatrix} + \begin{bmatrix} -Dx_D \\ x_D \end{bmatrix}$$

for some $x_D \in \mathbb{R}^{(n-m)}$.

## 16.2. THE CANONICAL AUGMENTED MATRIX

Consider the system of simultaneous linear equations $Ax = b$, rank $A = m$. Using a sequence of elementary row operations and reordering the variables if necessary, we transform the system $\boldsymbol{Ax}=\boldsymbol{b}$ into the following canonical form:

$$\begin{aligned}
x_1 \qquad\quad &+ y_{1m+1}x_{m+1} + \cdots + y_{1n}x_n = y_{10}\\
x_2 \qquad &+ y_{2m+1}x_{m+1} + \cdots + y_{2n}x_n = y_{20}\\
\ddots \qquad &\qquad\qquad\qquad\qquad\qquad \vdots\\
x_m &+ y_{mm+1}x_{m+1} + \cdots + y_{mn}x_n = y_{m0}.
\end{aligned}$$

The above can be represented in matrix notation as

$$[\boldsymbol{I}_m, \boldsymbol{Y}_{m,n-m}]\boldsymbol{x} = \boldsymbol{y}_0.$$

Formally, we define the canonical form as follows.

***Definition 16.5.*** A system $\boldsymbol{Ax}=\boldsymbol{b}$ is said to be in canonical form if, among the $n$ variables, there are $m$ variables with the property that each appears in only one equation, and its coefficient in that equation is unity. ■

A system is in canonical form if, by some reordering of the equations and the variables, it takes the form $[\boldsymbol{I}_m, \boldsymbol{Y}_{m,n-m}]\boldsymbol{x}=\boldsymbol{y}_0$. If a system of equations $\boldsymbol{Ax}=\boldsymbol{b}$ is not in canonical form, we can transform the system into canonical form by a sequence of elementary row operations. The system in canonical form has the same solution as the original system $\boldsymbol{Ax}=\boldsymbol{b}$, and is called the *canonical representation* of the system with respect to the basis $\boldsymbol{a}_1,\ldots,\boldsymbol{a}_m$. There are, in general, many canonical representations of a given system, depending on which columns of $\boldsymbol{A}$ we transform into the columns of $\boldsymbol{I}_m$ (i.e., basic columns). We will call the augmented matrix $[\boldsymbol{I}_m, \boldsymbol{Y}_{m,n-m}, \boldsymbol{y}_0]$ of the canonical representation of a given system the *canonical augmented matrix* of the system with respect to the basis $\boldsymbol{a}_1,\ldots,\boldsymbol{a}_m$. Of course, there may be many canonical augmented matrices of a given system, depending on which columns of $\boldsymbol{A}$ are chosen as basic columns.

The variables corresponding to basic columns in a canonical representation of a given system are the basic variables, whereas the other variables are the nonbasic variables. In particular, in the canonical representation $[\boldsymbol{I}_m, \boldsymbol{Y}_{m,n-m}]\boldsymbol{x}=\boldsymbol{y}_0$ of a given system, the variables $x_1,\ldots,x_m$ are the basic variables, and the other variables are the nonbasic variables. Note that in general the basic variables need not be the first $m$ variables. However, in the following discussion we assume, for convenience and without loss of generality, that the basic variables are indeed the first $m$ variables in the system. Having done so, the corresponding basic solution is

$$\begin{aligned}
x_1 &= y_{10},\\
&\ \vdots\\
x_m &= y_{m0},\\
x_{m+1} &= 0,\\
&\ \vdots\\
x_n &= 0,
\end{aligned}$$

that is,

$$x = \begin{bmatrix} y_0 \\ 0 \end{bmatrix}.$$

Given a system of equations $Ax = b$, consider the associated canonical augmented matrix

$$[I_m, Y_{m,n-m}, y_0] = \begin{bmatrix} 1 & 0 & \cdots & 0 & y_{1m+1} & \cdots & y_{1n} & y_{10} \\ 0 & 1 & \cdots & 0 & y_{2m+1} & \cdots & y_{2n} & y_{20} \\ \vdots & \vdots & \ddots & \vdots & \vdots & & \vdots & \vdots \\ 0 & 0 & \cdots & 1 & y_{mm+1} & \cdots & y_{mn} & y_{m0} \end{bmatrix}.$$

From the arguments above, we conclude that

$$b = y_{10}a_1 + y_{20}a_2 + \cdots + y_{m0}a_m.$$

In other words, the entries in the last column of the canonical augmented matrix are the coordinates of the vector $b$ with respect to the basis $\{a_1, \ldots, a_m\}$. The entries of all the other columns of the canonical augmented matrix have a similar interpretation. Specifically, the entries of the $j$th column of the canonical augmented matrix, $j = 1, \ldots, n$, are the coordinates of $a_j$ with respect to the basis $\{a_1, \ldots, a_m\}$. To see this, note that the first $m$ columns of the augmented matrix form a basis (the standard basis). Every other vector in the augmented matrix can be expressed as a linear combination of these basis vectors by reading the coefficients down the corresponding column. Specifically, let $a'_i$, $i = 1, \ldots, n+1$, be the $i$th column in the above augmented matrix. Clearly, since $a'_1, \ldots, a'_m$ form the standard basis, then for $m < j \leqslant n$,

$$a'_j = y_{1j}a'_1 + y_{2j}a'_2 + \cdots + y_{mj}a'_m.$$

Let $a_i$, $i = 1, \ldots, n$ be the $i$th column of $A$, and $a_{n+1} = b$. Now, $a'_i = Ea_i$, $i = 1, \ldots, n+1$, where $E$ is a nonsingular matrix that represents the elementary row operations needed to transform $[A, b]$ into $[I_m, Y_{m,n-m}, y_0]$. Therefore, for $m < j \leqslant n$, we also have

$$a_j = y_{1j}a_1 + y_{2j}a_2 + \cdots + y_{mj}a_m.$$

## 16.3. UPDATING THE AUGMENTED MATRIX

To summarize the previous section, the canonical augmented matrix of a given system $Ax = b$ specifies the representations of the columns $a_j$, $m < j \leqslant n$, in terms of the basic columns $a_1, \ldots, a_m$. Thus, the elements of the $j$th column of the canonical augmented matrix are the coordinates of the vector $a_j$ with respect to the basis $a_1, \ldots, a_m$. The coordinates of $b$ are given in the last column.

Suppose we are given the canonical representation of a system $\boldsymbol{Ax} = \boldsymbol{b}$. We now consider the following question: If we replace a basic variable by a nonbasic variable, what is the new canonical representation corresponding to the new set of basic variables? Specifically, suppose we wish to replace the basis vector $\boldsymbol{a}_p$, $1 \leqslant p \leqslant m$, by the vector $\boldsymbol{a}_q, m < q \leqslant n$. Provided the first $m$ vectors with $\boldsymbol{a}_p$ replaced by $\boldsymbol{a}_q$ are linearly independent, these vectors constitute a basis and every vector can be expressed as a linear combination of the new basic columns.

Let us now find the coordinates of the vectors $\boldsymbol{a}_1, \ldots, \boldsymbol{a}_n$ with respect to the new basis. These coordinates form the entries of the canonical augmented matrix of the system with respect to the new basis. In terms of the old basis, we can express $\boldsymbol{a}_q$ as

$$\boldsymbol{a}_q = \sum_{i=1}^{m} y_{iq}\boldsymbol{a}_i = \sum_{\substack{i=1 \\ i \neq p}}^{m} y_{iq}\boldsymbol{a}_i + y_{pq}\boldsymbol{a}_p.$$

Note that the set of vectors $\{\boldsymbol{a}_1, \ldots, \boldsymbol{a}_{p-1}, \boldsymbol{a}_q, \boldsymbol{a}_{p+1}, \ldots, \boldsymbol{a}_m\}$ is linearly independent if, and only if, $y_{pq} \neq 0$. Solving the above equation for $\boldsymbol{a}_p$, we get

$$\boldsymbol{a}_p = \frac{1}{y_{pq}}\boldsymbol{a}_q - \sum_{\substack{i=1 \\ i \neq p}}^{m} \frac{y_{iq}}{y_{pq}}\boldsymbol{a}_i.$$

Recall that in terms of the old augmented matrix, any vector $\boldsymbol{a}_j$, $m < j \leqslant n$, can be expressed as

$$\boldsymbol{a}_j = y_{1j}\boldsymbol{a}_1 + y_{2j}\boldsymbol{a}_2 + \cdots + y_{mj}\boldsymbol{a}_m.$$

Combining the last two equations yields

$$\boldsymbol{a}_j = \sum_{\substack{i=1 \\ i \neq p}}^{m} \left( y_{ij} - \frac{y_{pj}}{y_{pq}} y_{iq} \right)\boldsymbol{a}_i + \frac{y_{pj}}{y_{pq}}\boldsymbol{a}_q.$$

Denoting the entries of the new augmented matrix by $y'_{ij}$, we obtain

$$y'_{ij} = y_{ij} - \frac{y_{pj}}{y_{pq}} y_{iq}, \quad i \neq p,$$

$$y'_{pj} = \frac{y_{pj}}{y_{pq}}.$$

Therefore, the entries of the new canonical augmented matrix can be obtained from the entries of the old canonical augmented matrix via the above formulas. The above equations are often called the *pivot equations*, and $y_{pq}$ the *pivot element*. We refer to the operation on a given matrix by the above formulas as *pivoting about the* $(p, q)$*th element*. Note that pivoting about the $(p, q)$th element results in a matrix whose $q$th column has all zero entries, except the $(p, q)$th entry, which is unity. The pivoting operation can be accomplished via a sequence of elementary row operations, as was done in the proof of Theorem 16.1.

## 16.4. THE SIMPLEX ALGORITHM

The essence of the simplex algorithm is to move from one basic feasible solution to another until an optimal basic feasible solution is found. The canonical augmented matrix discussed in the previous section plays a central role in the simplex algorithm.

Suppose we are given the basic feasible solution

$$\boldsymbol{x} = [x_1, \ldots, x_m, 0, \ldots, 0]^T, \quad x_i \geqslant 0, \quad i = 1, \ldots, m$$

or equivalently

$$x_1\boldsymbol{a}_1 + \cdots + x_m\boldsymbol{a}_m = \boldsymbol{b}.$$

In the previous section, we saw how to update the canonical augmented matrix if we wish to replace a basic column by a nonbasic column, that is, if we wish to change from one basis to another by replacing a single basic column. The values of the basic variables in a basic solution corresponding to a given basis are given in the last column of the canonical augmented matrix with respect to that basis, i.e., $x_i = y_{i0}$, $i = 1, \ldots, m$. Basic solutions are not necessarily feasible, that is, the values of the basic variables may be negative. In the simplex method, we want to move from one basic feasible solution to another. This means that we want to change basic columns in such a way that the last column of the canonical augmented matrix remains nonnegative. In this section, we discuss a systematic method for doing this.

In the remainder of this chapter, we assume that every basic feasible solution of

$$\begin{aligned} \boldsymbol{A}\boldsymbol{x} &= \boldsymbol{b}, \\ \boldsymbol{x} &\geqslant \boldsymbol{0}, \end{aligned}$$

is a nondegenerate basic feasible solution. We make this assumption primarily for convenience—all arguments can be extended to include degeneracy.

Let us start with the basic columns $\boldsymbol{a}_1, \ldots, \boldsymbol{a}_m$, and assume that the corresponding basic solution $\boldsymbol{x} = [y_{10}, \ldots, y_{m0}, 0, \ldots, 0]^T$ is feasible; that is, the entries $y_{i0}$, $i = 1, \ldots, m$, in the last column of the canonical augmented matrix are positive. Suppose we now decide to make the vector $\boldsymbol{a}_q$, $q > m$, a basic column. We first represent $\boldsymbol{a}_q$ in terms of the current basis as

$$\boldsymbol{a}_q = y_{1q}\boldsymbol{a}_1 + y_{2q}\boldsymbol{a}_2 + \cdots + y_{mq}\boldsymbol{a}_m.$$

Multiplying the above by $\varepsilon > 0$ yields

$$\varepsilon \boldsymbol{a}_q = \varepsilon y_{1q}\boldsymbol{a}_1 + \varepsilon y_{2q}\boldsymbol{a}_2 + \cdots + \varepsilon y_{mq}\boldsymbol{a}_m.$$

We combine the above equation with

$$y_{10}\boldsymbol{a}_1 + \cdots + y_{m0}\boldsymbol{a}_m = \boldsymbol{b}$$

to get

$$(y_{10} - \varepsilon y_{1q})\boldsymbol{a}_1 + (y_{20} - \varepsilon y_{2q})\boldsymbol{a}_2 + \cdots + (y_{m0} - \varepsilon y_{mq})\boldsymbol{a}_m + \varepsilon \boldsymbol{a}_q = \boldsymbol{b}.$$

Note that the vector

$$\begin{bmatrix} y_{10} - \varepsilon y_{1q} \\ \vdots \\ y_{m0} - \varepsilon y_{mq} \\ 0 \\ \vdots \\ \varepsilon \\ \vdots \\ 0 \end{bmatrix}$$

where $\varepsilon$ appears in the $q$th position, is a solution to $\boldsymbol{Ax} = \boldsymbol{b}$. If $\varepsilon = 0$, then we obtain the old basic feasible solution. As $\varepsilon$ is increased from zero, the $q$th component of the above vector increases. All other entries of this vector will increase or decrease linearly as $\varepsilon$ is increased, depending on whether the corresponding $y_{iq}$ is negative or positive. For small enough $\varepsilon$, we have a feasible but nonbasic solution. If any of the components decreases as $\varepsilon$ increases, we choose $\varepsilon$ to be the smallest value where one (or more) of the components vanishes. That is

$$\varepsilon = \min_i \{y_{i0}/y_{iq} : y_{iq} > 0\}.$$

With the above choice of $\varepsilon$, we have a new basic feasible solution, with the vector $\boldsymbol{a}_q$ replacing $\boldsymbol{a}_p$, where $p$ corresponds to the minimizing index $p = \arg\min_i \{y_{i0}/y_{iq} : y_{iq} > 0\}$. So, we now have a new basis $\boldsymbol{a}_1, \ldots, \boldsymbol{a}_{p-1}, \boldsymbol{a}_{p+1}, \ldots, \boldsymbol{a}_m, \boldsymbol{a}_q$. As we can see, $\boldsymbol{a}_p$ was replaced by $\boldsymbol{a}_q$ in the new basis. We say that $\boldsymbol{a}_q$ enters the basis, and $\boldsymbol{a}_p$ leaves the basis. If the minimum in $\min_i \{y_{i0}/y_{iq} : y_{iq} > 0\}$ is achieved by more than a single index, then the new solution is degenerate and any of the zero components can be regarded as the component corresponding to the basic column that leaves the basis. If none of the $y_{iq}$ are positive, then all components in the vector $[y_{10} - \varepsilon y_{1q}, \ldots, y_{m0} - \varepsilon y_{mq}, 0, \ldots, \varepsilon, \ldots, 0]^T$ increase (or remain constant) as $\varepsilon$ is increased, and no new basic feasible solution is obtained, no matter how large we make $\varepsilon$. In this case, there are feasible solutions having arbitrarily large components, that is, the set $\Omega$ of feasible solutions is unbounded.

So far, we have discussed how to change from one basis to another, while preserving feasibility of the corresponding basic solution, assuming that we have already chosen a nonbasic column to enter the basis. To complete our development of the simplex method, we need to consider two more issues. The first issue concerns the choice of which nonbasic column should enter the basis. The second issue is to find a stopping criterion, that is, a way to determine if a basic feasible solution is optimal or not. To this end, suppose that we have found a basic feasible solution. The main idea of the simplex method is to move from one basic feasible solution (extreme point of the set $\Omega$) to another basic feasible solution at which the value of the objective function is smaller. Since there is only a finite number of extreme points of the feasible set, the optimal point will be reached after a finite number of steps.

We already know how to move from one extreme point of the set $\Omega$ to a neighboring one, by updating the canonical augmented matrix. To see which neighboring solution we should move to and when to stop moving, consider the following basic feasible solution

$$[\boldsymbol{x}_B^T, \boldsymbol{0}^T]^T = [y_{10}, \ldots, y_{m0}, 0, \ldots, 0]^T$$

together with the corresponding canonical augmented matrix, having an identity matrix appearing in the first $m$ columns. The value of the objective function for any solution $\boldsymbol{x}$ is

$$z = c_1 x_1 + c_2 x_2 + \cdots + c_n x_n.$$

For our basic solution, the value of the objective function is

$$z = z_0 = \boldsymbol{c}_B^T \boldsymbol{x}_B = c_1 y_{10} + \cdots + c_m y_{m0},$$

where

$$\boldsymbol{c}_B^T = [c_1, c_2, \ldots, c_m].$$

To see how the value of the objective function changes when we move from one basic feasible solution to another, suppose we choose the $q$th column, $m < q \leqslant n$, to enter the basis. To update the canonical augmented matrix, let $p = \arg\min_i \{y_{i0}/y_{iq} : y_{iq} > 0\}$, and $\varepsilon = y_{p0}/y_{pq}$. The new basic feasible solution is

$$\begin{bmatrix} y_{10} - \varepsilon y_{1q} \\ \vdots \\ y_{m0} - \varepsilon y_{mq} \\ 0 \\ \vdots \\ \varepsilon \\ \vdots \\ 0 \end{bmatrix}.$$

Note that the single $\varepsilon$ appears in the $q$th component, whereas the $p$th component is 0. Observe that we could have arrived at the above basic feasible solution by simply updating the canonical augmented matrix using the pivot equations from the previous section:

$$y'_{ij} = y_{ij} - \frac{y_{pj}}{y_{pq}} y_{iq}, \quad i \neq p,$$

$$y'_{pj} = \frac{y_{pj}}{y_{pq}},$$

where the $q$th column enters the basis, and the $p$th column leaves [that is, we pivot about the $(p, q)$th element]. The values of the basic variables are entries in the last column of the updated canonical augmented matrix.

The cost for this new basic feasible solution is

$$\begin{aligned} z &= c_1(y_{10} - y_{1q}\varepsilon) + \cdots + c_m(y_{m0} - y_{mq}\varepsilon) + c_q\varepsilon \\ &= z_0 + [c_q - (c_1 y_{1q} + \cdots + c_m y_{mq})]\varepsilon, \end{aligned}$$

where $z_0 = c_1 y_{10} + \cdots + c_m y_{m0}$. Let

$$z_q = c_1 y_{1q} + \cdots + c_m y_{mq}.$$

Then,

$$z = z_0 + (c_q - z_q)\varepsilon.$$

Thus, if

$$z - z_0 = (c_q - z_q)\varepsilon < 0,$$

then the objective function value at the new basic feasible solution above is smaller than the objective function value at the original solution (i.e., $z < z_0$). Therefore, if $c_q - z_q < 0$, then the new basic feasible solution with $\boldsymbol{a}_q$ entering the basis has a lower objective function value.

On the other hand, if the given basic feasible solution is such that for all $q = m + 1, \ldots, n$,

$$c_q - z_q \geqslant 0,$$

then, we can show that this solution is in fact an optimal solution. To show this, recall from Section 16.1 that any solution to $\boldsymbol{A}\boldsymbol{x} = \boldsymbol{b}$ can be represented as

$$\boldsymbol{x} = \begin{bmatrix} \boldsymbol{y}_0 \\ \boldsymbol{0} \end{bmatrix} + \begin{bmatrix} -\boldsymbol{Y}_{m,n-m}\boldsymbol{x}_D \\ \boldsymbol{x}_D \end{bmatrix}$$

for some $\boldsymbol{x}_D = [x_{m+1}, \ldots, x_n]^T \in \mathbb{R}^{(n-m)}$. Using similar manipulations as the above, we obtain

$$\boldsymbol{c}^T\boldsymbol{x} = z_0 + \sum_{i=m+1}^{n} (c_i - z_i)x_i,$$

where $z_i = c_1 y_{1i} + \cdots + c_m y_{mi}$, $i = m+1, \ldots, n$. For a feasible solution, we have $x_i \geqslant 0$, $i = 1, \ldots, n$. Therefore, if $c_i - z_i \geqslant 0$ for all $i = m+1, \ldots, n$, then any feasible solution $\boldsymbol{x}$ will have objective function value $\boldsymbol{c}^T \boldsymbol{x}$ no smaller than $z_0$.

Let $r_i = 0$ for $i = 1, \ldots, m$, and $r_i = c_i - z_i$ for $i = m+1, \ldots, n$. We call $r_i$ the $i$th *relative cost coefficient* or *reduced cost coefficient*. Note that the relative cost coefficients corresponding to basic variables are zero.

We summarize the above the discussion with the following result.

**Theorem 16.2.** *A basic feasible solution is optimal if, and only if, the corresponding reduced cost coefficients are all nonnegative.* □

At this point, we have all the necessary steps for the simplex algorithm:

**The Simplex Algorithm**

1. Form a canonical augmented matrix corresponding to an initial basic feasible solution.
2. Calculate the relative cost coefficients corresponding to the nonbasic variables.
3. If $r_j \geqslant 0$ for all $j$, stop—the current basic feasible solution is optimal.
4. Select a $q$ such that $r_q < 0$.
5. If no $y_{iq} > 0$, stop—the problem is unbounded; else, calculate $p = \arg\min_i \{y_{i0}/y_{iq} : y_{iq} > 0\}$.
6. Update the canonical augmented matrix by pivoting about the $(p, q)$th element.
7. Go to step 2.

We state the following result for the simplex algorithm, which we have already proved in the foregoing discussion.

**Theorem 16.3.** *Suppose we have an LP problem in standard form that has an optimal feasible solution. If the simplex method applied to this problem terminates, and the reduced cost coefficients in the last step are all nonnegative, then the resulting basic feasible solution is optimal.* □

***Example 16.2.*** Consider the following linear program (see also Exercise 15.4):

$$\begin{aligned} \text{maximize} \quad & [2 \ \ 5] \begin{bmatrix} x_1 \\ x_2 \end{bmatrix} \\ \text{subject to} \quad & x_1 \leqslant 4 \\ & x_2 \leqslant 6 \\ & x_1 + x_2 \leqslant 8 \\ & x_1, x_2 \geqslant 0. \end{aligned}$$

We solve this problem using the simplex method.

Introducing slack variables, we transform the problem into standard form

$$\begin{array}{ll} \text{minimize} & -2x_1 - 5x_2 - 0x_3 - 0x_4 - 0x_5 \\ \text{subject to} & x_1 \quad\quad + x_3 \quad\quad\quad\quad = 4 \\ & \quad\quad x_2 \quad\quad + x_4 \quad\quad = 6 \\ & x_1 + x_2 \quad\quad\quad\quad + x_5 = 8 \\ & \quad\quad\quad\quad x_1, \ldots, x_5 \geqslant 0. \end{array}$$

The starting canonical augmented matrix for this problem is

| $a_1$ | $a_2$ | $a_3$ | $a_4$ | $a_5$ | $b$ |
|---|---|---|---|---|---|
| 1 | 0 | 1 | 0 | 0 | 4 |
| 0 | 1 | 0 | 1 | 0 | 6 |
| 1 | 1 | 0 | 0 | 1 | 8 |

Observe that the columns forming the identity matrix in the above canonical augmented matrix do not appear at the beginning. We could rearrange the augmented matrix so that the identity matrix would appear first. However, this is not essential from the computational point of view.

The starting basic feasible solution to the problem in standard form is

$$\boldsymbol{x} = [0, 0, 4, 6, 8]^T.$$

The columns $\boldsymbol{a}_3, \boldsymbol{a}_4$, and $\boldsymbol{a}_5$ corresponding to $x_3, x_4$, and $x_5$ are basic, and they form the identity matrix. The basis matrix is $\boldsymbol{B} = [\boldsymbol{a}_3, \boldsymbol{a}_4, \boldsymbol{a}_5] = \boldsymbol{I}_3$.

The value of the objective function corresponding to this basic feasible solution is $z = 0$. We next compute the reduced cost coefficients corresponding to the nonbasic variables $x_1$ and $x_2$. They are

$$r_1 = c_1 - z_1 = c_1 - (c_3 y_{11} + c_4 y_{21} + c_5 y_{31}) = -2,$$
$$r_2 = c_2 - z_2 = c_2 - (c_3 y_{12} + c_4 y_{22} + c_5 y_{32}) = -5.$$

We would like now to move to an adjacent basic feasible solution for which the objective function value is lower. Naturally, if there is more than one such solution, it is desirable to move to the adjacent basic feasible solution with the lowest objective value. A common practice is to select the most negative value of $r_j$ and then to bring the corresponding column into the basis (see Exercise 16.7 for an alternative rule for choosing the column to bring into the basis). In our example, we bring $\boldsymbol{a}_2$ into the basis, that is, we choose $\boldsymbol{a}_2$ as the new basic column. We then compute $p = \arg\min\{y_{i0}/y_{i2} : y_{i2} > 0\} = 2$. We now update the canonical augmented matrix by pivoting about the $(2, 2)$th entry using the pivot equations:

$$y'_{ij} = y_{ij} - \frac{y_{2j}}{y_{22}} y_{i2}, \quad i \neq 2,$$

$$y'_{2j} = \frac{y_{2j}}{y_{22}}.$$

The resulting updated canonical augmented matrix is:

| $a_1$ | $a_2$ | $a_3$ | $a_4$ | $a_5$ | $b$ |
|---|---|---|---|---|---|
| 1 | 0 | 1 | 0 | 0 | 4 |
| 0 | 1 | 0 | 1 | 0 | 6 |
| 1 | 0 | 0 | −1 | 1 | 2 |

Note that $a_2$ entered the basis, and $a_4$ left the basis. The corresponding basic feasible solution is $x = [0, 6, 4, 0, 2]^T$. We now compute the reduced cost coefficients for the nonbasic columns:

$$r_1 = c_1 - z_1 = -2$$
$$r_4 = c_4 - z_4 = 5.$$

Since $r_1 = -2 < 0$ this means that the current solution is not optimal, and a lower objective function value can be obtained by bringing $a_1$ into the basis. Proceeding to update the canonical augmented matrix by pivoting about the (3, 1)th element, we obtain:

| $a_1$ | $a_2$ | $a_3$ | $a_4$ | $a_5$ | $b$ |
|---|---|---|---|---|---|
| 0 | 0 | 1 | 1 | −1 | 2 |
| 0 | 1 | 0 | 1 | 0 | 6 |
| 1 | 0 | 0 | −1 | 1 | 2 |

The corresponding basic feasible solution is $x = [2, 6, 2, 0, 0]^T$. The relative cost coefficients are

$$r_4 = c_4 - z_4 = 3$$
$$r_5 = c_5 - z_5 = 2.$$

Since no reduced cost coefficient is negative the current basic feasible solution $x = [2, 6, 2, 0, 0]^T$ is optimal. The solution to the original problem is therefore $x_1 = 2$, $x_2 = 6$, and the objective function value is 34. ■

We can see from the above example that we can solve a linear programming problem of any size using the simplex algorithm. However, to make the calculations in the algorithm more efficient, we discuss the matrix form of the simplex method in the next section.

## 16.5. MATRIX FORM OF THE SIMPLEX METHOD

Consider a linear programming problem in standard form

$$\begin{array}{ll} \text{minimize} & c^T x \\ \text{subject to} & Ax = b \\ & x \geqslant 0. \end{array}$$

Let the first $m$ columns of $\boldsymbol{A}$ be the basic columns. The columns form a square $m \times m$ nonsingular matrix $\boldsymbol{B}$. The nonbasic columns of $\boldsymbol{A}$ form a $m \times (n-m)$ matrix $\boldsymbol{D}$. We correspondingly partition the cost vector as $\boldsymbol{c}^T = [\boldsymbol{c}_B^T, \boldsymbol{c}_D^T]$. Then, the original linear program can be represented as follows:

$$
\begin{aligned}
&\text{minimize} \quad (\boldsymbol{c}_B^T \boldsymbol{x}_B + \boldsymbol{c}_D^T \boldsymbol{x}_D)\\
&\text{subject to} \quad [\boldsymbol{B}, \boldsymbol{D}] \begin{bmatrix} \boldsymbol{x}_B \\ \boldsymbol{x}_D \end{bmatrix} = \boldsymbol{B}\boldsymbol{x}_B + \boldsymbol{D}\boldsymbol{x}_D = \boldsymbol{b},\\
&\qquad\qquad\quad \boldsymbol{x}_B \geqslant \boldsymbol{0}, \quad \boldsymbol{x}_D \geqslant \boldsymbol{0}.
\end{aligned}
$$

If $\boldsymbol{x}_D = \boldsymbol{0}$, then the solution $\boldsymbol{x} = [\boldsymbol{x}_B^T, \boldsymbol{x}_D^T]^T = [\boldsymbol{x}_B^T, \boldsymbol{0}^T]^T$ is the basic feasible solution corresponding to the basis $\boldsymbol{B}$. It is clear that for this to be a solution, we need $\boldsymbol{x}_B = \boldsymbol{B}^{-1}\boldsymbol{b}$, i.e., the basic feasible solution is

$$
\boldsymbol{x} = \begin{bmatrix} \boldsymbol{B}^{-1}\boldsymbol{b} \\ \boldsymbol{0} \end{bmatrix}.
$$

The corresponding objective function value is

$$
z_0 = \boldsymbol{c}_B^T \boldsymbol{B}^{-1}\boldsymbol{b}.
$$

If, on the other hand, $\boldsymbol{x}_D \neq \boldsymbol{0}$, then the solution $\boldsymbol{x} = [\boldsymbol{x}_B^T, \boldsymbol{x}_D^T]^T$ is not basic with respect to $\boldsymbol{B}$. In this case, $\boldsymbol{x}_B$ is given by

$$
\boldsymbol{x}_B = \boldsymbol{B}^{-1}\boldsymbol{b} - \boldsymbol{B}^{-1}\boldsymbol{D}\boldsymbol{x}_D,
$$

and the corresponding objective function value is

$$
\begin{aligned}
z &= \boldsymbol{c}_B^T \boldsymbol{x}_B + \boldsymbol{c}_D^T \boldsymbol{x}_D\\
&= \boldsymbol{c}_B^T(\boldsymbol{B}^{-1}\boldsymbol{b} - \boldsymbol{B}^{-1}\boldsymbol{D}\boldsymbol{x}_D) + \boldsymbol{c}_D^T \boldsymbol{x}_D\\
&= \boldsymbol{c}_B^T \boldsymbol{B}^{-1}\boldsymbol{b} + (\boldsymbol{c}_D^T - \boldsymbol{c}_B^T \boldsymbol{B}^{-1}\boldsymbol{D})\boldsymbol{x}_D.
\end{aligned}
$$

Defining

$$
\boldsymbol{r}_D^T = \boldsymbol{c}_D^T - \boldsymbol{c}_B^T \boldsymbol{B}^{-1}\boldsymbol{D},
$$

we obtain

$$
z = z_0 + \boldsymbol{r}_D^T \boldsymbol{x}_D.
$$

The elements of the vector $\boldsymbol{r}_D$ are the relative cost coefficients corresponding to the nonbasic variables.

If $\boldsymbol{r}_D \geqslant \boldsymbol{0}$, then the basic feasible solution corresponding to the basis $\boldsymbol{B}$ is optimal. If, on the other hand, a component of $\boldsymbol{r}_D$ is negative, then the value of the objective can be reduced by increasing a corresponding component of $\boldsymbol{x}_D$ (that is, by changing the basis).

We now use the above observations to develop a matrix form of the simplex method. To this end, we first add the cost coefficient vector $\boldsymbol{c}^T$ to the bottom of the augmented matrix $[\boldsymbol{A}, \boldsymbol{b}]$ as follows:

$$\begin{bmatrix} \boldsymbol{A} & \boldsymbol{b} \\ \boldsymbol{c}^T & 0 \end{bmatrix} = \begin{bmatrix} \boldsymbol{B} & \boldsymbol{D} & \boldsymbol{b} \\ \boldsymbol{c}_B^T & \boldsymbol{c}_D^T & 0 \end{bmatrix}.$$

We refer to the above matrix as the *tableau* of the given LP problem. The tableau contains all relevant information about the linear program.

Suppose we now apply elementary row operations to the tableau such that the top part of the tableau corresponding to the augmented matrix $[\boldsymbol{A}, \boldsymbol{b}]$ is transformed into canonical form. This corresponds to premultiplying the tableau by the matrix

$$\begin{bmatrix} \boldsymbol{B}^{-1} & \boldsymbol{0} \\ \boldsymbol{0}^T & 1 \end{bmatrix}.$$

The result of this operation is

$$\begin{bmatrix} \boldsymbol{B}^{-1} & \boldsymbol{0} \\ \boldsymbol{0}^T & 1 \end{bmatrix} \begin{bmatrix} \boldsymbol{B} & \boldsymbol{D} & \boldsymbol{b} \\ \boldsymbol{c}_B^T & \boldsymbol{c}_D^T & 0 \end{bmatrix} = \begin{bmatrix} \boldsymbol{I}_m & \boldsymbol{B}^{-1}\boldsymbol{D} & \boldsymbol{B}^{-1}\boldsymbol{b} \\ \boldsymbol{c}_B^T & \boldsymbol{c}_D^T & 0 \end{bmatrix}.$$

We now apply elementary row operations to the above tableau so that the entries of the last row corresponding to the basic columns become zero. Specifically, this corresponds to premultiplication of the above tableau by the matrix

$$\begin{bmatrix} \boldsymbol{I}_m & \boldsymbol{0} \\ -\boldsymbol{c}_B^T & 1 \end{bmatrix}.$$

The result is

$$\begin{bmatrix} \boldsymbol{I}_m & \boldsymbol{0} \\ -\boldsymbol{c}_B^T & 1 \end{bmatrix} \begin{bmatrix} \boldsymbol{I}_m & \boldsymbol{B}^{-1}\boldsymbol{D} & \boldsymbol{B}^{-1}\boldsymbol{b} \\ \boldsymbol{c}_B^T & \boldsymbol{c}_D^T & 0 \end{bmatrix} = \begin{bmatrix} \boldsymbol{I}_m & \boldsymbol{B}^{-1}\boldsymbol{D} & \boldsymbol{B}^{-1}\boldsymbol{b} \\ \boldsymbol{0}^T & \boldsymbol{c}_D^T - \boldsymbol{c}_B^T\boldsymbol{B}^{-1}\boldsymbol{D} & -\boldsymbol{c}_B^T\boldsymbol{B}^{-1}\boldsymbol{b} \end{bmatrix}.$$

We refer to the resulting tableau above as the *canonical tableau corresponding to the basis* $\boldsymbol{B}$. Note that the first $m$ entries of the last column of the canonical tableau, $\boldsymbol{B}^{-1}\boldsymbol{b}$, are the values of the basic variables corresponding to the basis $\boldsymbol{B}$. The entries of $\boldsymbol{c}_D^T - \boldsymbol{c}_B^T\boldsymbol{B}^{-1}\boldsymbol{D}$ in the last row are the relative cost coefficients. The last element in the last row of the tabuleau, $-\boldsymbol{c}_B^T\boldsymbol{B}^{-1}\boldsymbol{b}$, is the negative of the value of the objective corresponding to the basic feasible solution.

Given an LP problem, we can in general construct many different canonical tableaus, depending on which columns are basic. Suppose we have a canonical tableau corresponding to a particular basis. Consider the task of computing the tableau corresponding to another basis that differs from the previous basis by a single vector. This can be accomplished by applying elementary row operations to the tableau in a similar fashion as discussed above. We refer to this operation as *updating* the canonical tableau. Note that updating of the tableau involves using exactly the same update equations as we used before in updating the canonical augmented matrix, namely, for $i = 1, \ldots, m+1$.

$$y'_{ij} = y_{ij} - \frac{y_{pj}}{y_{pq}} y_{iq}, \quad i \neq p,$$

$$y'_{pj} = \frac{y_{pj}}{y_{pq}},$$

where $y_{ij}$ and $y'_{ij}$ are the $(i, j)$th entries of the original and updated canonical tableaus, respectively.

Working with the tableau is a convenient way of implementing the simplex algorithm, since updating the tableau immediately gives us both the values of the basic variables and the reduced cost coefficients. In addition, the (negative of the) value of the objective function can be found in the lower right-hand corner of the tableau. We illustrate the use of the tableau in the following example.

***Example 16.3.*** Consider the following linear programming problem:

$$\begin{aligned} \text{maximize} \quad & 7x_1 + 6x_2 \\ \text{subject to} \quad & 2x_1 + x_2 \leqslant 3, \\ & x_1 + 4x_2 \leqslant 4, \\ & x_1, x_2 \geqslant 0. \end{aligned}$$

We first transform the problem into standard form so that the simplex method can be applied. To do this, we change the maximization to minimization by multiplying the objective function by $-1$. We then introduce two nonnegative slack variables, $x_3$ and $x_4$, and construct the tableau for the problem:

| | $\boldsymbol{a}_1$ | $\boldsymbol{a}_2$ | $\boldsymbol{a}_3$ | $\boldsymbol{a}_4$ | $\boldsymbol{b}$ |
|---|---|---|---|---|---|
| | 2 | 1 | 1 | 0 | 3 |
| | 1 | 4 | 0 | 1 | 4 |
| $\boldsymbol{c}^T$ | $-7$ | $-6$ | 0 | 0 | 0 |

Notice that the above tabuleau is already in canonical form with respect to the basis $[\boldsymbol{a}_3, \boldsymbol{a}_4]$. Hence, the last row contains the reduced cost coefficients, and the rightmost column contains the values of the basic variables. Since $r_1 = -7$ is the most negative reduced cost coefficient, we bring $\boldsymbol{a}_1$ into the basis. We then compute the ratios $y_{10}/y_{11} = 3/2$ and $y_{20}/y_{21} = 4$. Since $y_{10}/y_{11} < y_{20}/y_{21}$, then $p = \arg\min_i \{y_{i0}/y_{i1} : y_{i1} > 0\} = 1$. We pivot about the $(1, 1)$th element of the tableau to obtain

$$\begin{matrix} 1 & \frac{1}{2} & \frac{1}{2} & 0 & \frac{3}{2} \\ 0 & \frac{7}{2} & -\frac{1}{2} & 1 & \frac{5}{2} \\ 0 & -\frac{5}{2} & \frac{7}{2} & 0 & \frac{21}{2} \end{matrix}$$

In the second tableau above, only $r_2$ is negative. Therefore, $q = 2$ (i.e., we bring $\boldsymbol{a}_2$ into the basis). Since

$$\frac{y_{10}}{y_{12}} = 3, \quad \frac{y_{20}}{y_{22}} = \frac{5}{7}$$

we have $p = 2$. We thus pivot about the (2, 2)th element of the second tableau to obtain the third tabuleau below:

$$\begin{matrix} 1 & 0 & \frac{4}{7} & -\frac{1}{7} & \frac{8}{7} \\ 0 & 1 & -\frac{1}{7} & \frac{2}{7} & \frac{5}{7} \\ 0 & 0 & \frac{22}{7} & \frac{5}{7} & \frac{86}{7} \end{matrix}$$

Since the last row of the third tableau above has no negative elements, we conclude that the basic feasible solution corresponding to the third tableau is optimal. Thus $x_1 = 8/7$, $x_2 = 5/7$, $x_3 = 0$, $x_4 = 0$ is the solution to our LP in standard form, and the corresponding objective value is $-86/7$. The solution to the original problem is simply $x_1 = 8/7, x_2 = 5/7$, and the corresponding objective value is 86/7. ■

Degenerate basic feasible solutions may arise in the course of applying the simplex algorithm. In such a situation, the minimum ratio $y_{i0}/y_{iq}$ is 0. Therefore, even though the basis changes after we pivot about the $(p, q)$th element, the basic feasible solution does not (and remains degenerate). It is possible that if we start with a basis corresponding to a degenerate solution, several iterations of the simplex algorithm will involve the same degenerate solution, and eventually the original basis will occur. The whole process will then repeat indefinitely, leading to what is called *cycling*. Such a scenario, although rare in practice, is clearly undesirable. Fortunately, there is a simple rule for choosing $q$ and $p$, due to Bland, that eliminates the cycling problem (see Exercise 16.7):

$$q = \min\{i : r_i < 0\}$$
$$p = \min\{j : y_{j0}/y_{jq} = \min_i \{y_{i0}/y_{iq} : y_{iq} > 0\}\}.$$

## 16.6. THE TWO-PHASE SIMPLEX METHOD

The simplex method requires starting with a tableau for the problem in canonical form; that is, we need an initial basic feasible solution. A brute force approach to finding a starting basic feasible solution is to arbitrarily choose $m$ basic columns and transform the tableau for the problem into canonical form. If the rightmost column is positive, then we have a legitimate (initial) basic feasible solution. Otherwise, we would have to pick another candidate basis. Potentially, this brute force procedure requires $\binom{n}{m}$ tries, and is therefore not practical.

Certain LP problems have obvious initial basic feasible solutions. For example, if we have constraints of the form $\boldsymbol{Ax} \leqslant \boldsymbol{b}$ and we add $m$ slack variables $z_1, \ldots, z_m$, then the constraints in standard form become

$$[\boldsymbol{A}, \boldsymbol{I}_m]\begin{bmatrix} \boldsymbol{x} \\ \boldsymbol{z} \end{bmatrix} = \boldsymbol{b}, \quad \begin{bmatrix} \boldsymbol{x} \\ \boldsymbol{z} \end{bmatrix} \geqslant \boldsymbol{0},$$

where $\boldsymbol{z} = [z_1, \ldots, z_m]^T$. The obvious initial basic feasible solution is

$$\begin{bmatrix} \boldsymbol{0} \\ \boldsymbol{b} \end{bmatrix},$$

and the basic variables are the slack variables. This was the case in the example in the previous section.

Suppose we are given a linear program in standard form:

$$\begin{aligned} \text{minimize} \quad & \boldsymbol{c}^T\boldsymbol{x} \\ \text{subject to} \quad & \boldsymbol{Ax} = \boldsymbol{b} \\ & \boldsymbol{x} \geqslant \boldsymbol{0}. \end{aligned}$$

In general, an initial basic feasible solution is not always apparent. We therefore need a systematic method for finding an initial basic feasible solution for general LP problems, so that the simplex method can be initialized. Consider the following associated *artificial problem*:

$$\begin{aligned} \text{minimize} \quad & y_1 + y_2 + \cdots + y_m \\ \text{subject to} \quad & [\boldsymbol{A}, \boldsymbol{I}_m]\begin{bmatrix} \boldsymbol{x} \\ \boldsymbol{y} \end{bmatrix} = \boldsymbol{b} \\ & \begin{bmatrix} \boldsymbol{x} \\ \boldsymbol{y} \end{bmatrix} \geqslant 0, \end{aligned}$$

where $\boldsymbol{y} = [y_1, \ldots, y_m]^T$. We call $\boldsymbol{y}$ the *vector of artificial variables*. Note that the artificial problem has an obvious initial basic feasible solution:

$$\begin{bmatrix} \boldsymbol{0} \\ \boldsymbol{b} \end{bmatrix}.$$

We can therefore solve this problem by the simplex method.

**Proposition 16.1.** *The original LP problem has a basic feasible solution if, and only if, the associated artificial problem has an optimal feasible solution with objective function value zero.* □

*Proof.* ⇒: If the original problem has a basic feasible solution $\boldsymbol{x}$, then the vector $[\boldsymbol{x}^T, \boldsymbol{0}^T]^T$ is a basic feasible solution to the artificial problem. Clearly, this solution has an objective function value of zero. This solution is therefore optimal for the artificial problem, since there can be no feasible solution with negative objective function value.

⇐: Suppose the artificial problem has an optimal feasible solution with objective function value zero. Then, this solution must have the form $[\boldsymbol{x}^T, \boldsymbol{0}^T]^T$, where $\boldsymbol{x} \geqslant \boldsymbol{0}$. Hence, we have $\boldsymbol{Ax} = \boldsymbol{b}$, and $\boldsymbol{x}$ is a feasible solution to the original problem. By the fundamental theorem of LP, there also exists a basic feasible solution. ■

Assume that the original LP problem has a basic feasible solution. Suppose that the simplex method applied to the associated artificial problem has terminated. Then, as suggested by the proof above, the solution to the artificial problem will have all $y_i = 0$, $i = 1, \ldots, m$. Hence, assuming nondegeneracy, the basic variables are in the first $n$ components; that is, none of the artificial variables are basic. Therefore, the first $n$ components form a basic feasible solution to the original problem. We can then use this basic feasible solution (resulting from the artificial problem) as the initial basic feasible solution for the original LP problem (after deleting the components corresponding to artificial variables). Thus, using artificial variables, we can attack a general linear programming problem by applying the *two-phase simplex method*. In phase I, we introduce artificial variables and the artificial objective function, and find a basic feasible solution. In phase II, we use the basic feasible solution resulting from phase I to initialize the simplex algorithm to solve the original LP problem. The two-phase simplex method is illustrated in Figure 16.1.

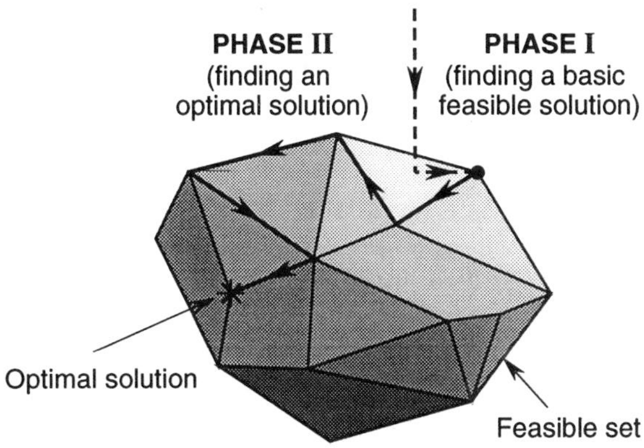


**Figure 16.1.** Graphical illustration of the two-phase simplex method

***Example 16.4.*** Consider the following linear programming problem:

$$\begin{array}{ll} \text{minimize} & 2x_1+3x_2 \\ \text{subject to} & 4x_1+2x_2 \geqslant 12, \\ & x_1+4x_2 \geqslant 6, \\ & x_1,x_2 \geqslant 0. \end{array}$$

First, we express the problem in standard form by introducing surplus variables:

$$\begin{array}{ll} \text{minimize} & 2x_1+3x_2 \\ \text{subject to} & 4x_1+2x_2-x_3=12, \\ & x_1+4x_2-x_4=6, \\ & x_1,\ldots,x_4 \geqslant 0. \end{array}$$

For the above LP problem, there is no obvious basic feasible solution that we can use to initialize the simplex method. Therefore, we will use the two-phase method.

*Phase I.* We introduce artificial variables $x_5, x_6 \geqslant 0$, and an artificial objective function $x_5+x_6$. We form the corresponding tableau for the problem:

| | $a_1$ | $a_2$ | $a_3$ | $a_4$ | $a_5$ | $a_6$ | $b$ |
|---|---|---|---|---|---|---|---|
| | 4 | 2 | −1 | 0 | 1 | 0 | 12 |
| | 1 | 4 | 0 | −1 | 0 | 1 | 6 |
| $c^T$ | 0 | 0 | 0 | 0 | 1 | 1 | 0 |

To initiate the simplex procedure, we must update the last row of the above tableau to transform it into canonical form. We obtain

| $a_1$ | $a_2$ | $a_3$ | $a_4$ | $a_5$ | $a_6$ | $b$ |
|---|---|---|---|---|---|---|
| 4 | 2 | −1 | 0 | 1 | 0 | 12 |
| 1 | 4 | 0 | −1 | 0 | 1 | 6 |
| −5 | −6 | 1 | 1 | 0 | 0 | −18 |

The basic feasible solution corresponding to the above tableau is not optimal. Therefore, we proceed with the simplex method to obtain the next tableau:

| | | | | | | |
|---|---|---|---|---|---|---|
| $\frac{7}{2}$ | 0 | −1 | $\frac{1}{2}$ | 1 | $-\frac{1}{2}$ | 9 |
| $\frac{1}{4}$ | 1 | 0 | $-\frac{1}{4}$ | 0 | $\frac{1}{4}$ | $\frac{3}{2}$ |
| $-\frac{7}{2}$ | 0 | 1 | $-\frac{1}{2}$ | 0 | $\frac{3}{2}$ | −9 |

We still have not yet reached an optimal basic feasible solution. Performing another iteration, we get

$$\begin{matrix} 1 & 0 & -\frac{2}{7} & \frac{1}{7} & \frac{2}{7} & -\frac{1}{7} & \frac{18}{7} \\ 0 & 1 & \frac{1}{14} & -\frac{2}{7} & -\frac{1}{14} & \frac{2}{7} & \frac{6}{7} \\ 0 & 0 & 0 & 0 & 1 & 1 & 0 \end{matrix}$$

Both of the artificial variables have been driven out of the basis, and the current basic feasible solution is optimal. We now proceed to phase II.

*Phase II.* We start by deleting the columns corresponding to the artificial variables in the last tableau in phase I, and revert back to the original objective function. We obtain

| | $a_1$ | $a_2$ | $a_3$ | $a_4$ | $b$ |
|---|---|---|---|---|---|
| | 1 | 0 | $-\frac{2}{7}$ | $\frac{1}{7}$ | $\frac{18}{7}$ |
| | 0 | 1 | $\frac{1}{14}$ | $-\frac{2}{7}$ | $\frac{6}{7}$ |
| $c^T$ | 2 | 3 | 0 | 0 | 0 |

We transform the last row so that the zeros appear in the basis columns; that is, transform the above tableau into canonical form:

$$\begin{matrix} 1 & 0 & -\frac{2}{7} & \frac{1}{7} & \frac{18}{7} \\ 0 & 1 & \frac{1}{14} & -\frac{2}{7} & \frac{6}{7} \\ 0 & 0 & \frac{5}{14} & \frac{4}{7} & -\frac{54}{7} \end{matrix}$$

All the reduced cost coefficients are nonnegative. Hence, the optimal solution is

$$x = [\tfrac{18}{7}, \tfrac{6}{7}, 0, 0]^T$$

and the optimal cost is 54/7. ■

## 16.7. THE REVISED SIMPLEX METHOD

Consider an LP problem in standard form with the matrix $A$ of size $m \times n$. Suppose we use the simplex method to solve the problem. Experience suggests that if $m$ is much smaller than $n$, then, in most instances, pivots will occur in only a small fraction of the columns of the matrix $A$. The operation of pivoting involves updating all the columns of the tableau. However, if a particular column of $A$ never enters any basis during the whole simplex procedure, then computations performed on this column are never used. Therefore, if $m$ is much smaller than $n$, the effort expended on performing operations on many of the columns $A$ may be wasted. The *revised simplex method* reduces the amount of computation leading to an optimal solution by eliminating operations on columns of $\boldsymbol{A}$ that do not enter the bases.

To be specific, suppose that we are at a particular iteration in the simplex algorithm. Let $\boldsymbol{B}$ be the matrix composed of the columns of $\boldsymbol{A}$ forming the current basis, and let $\boldsymbol{D}$ be the matrix composed of the remaining columns of $\boldsymbol{A}$. The sequence of elementary row operations on the tableau leading to this iteration (represented by matrices $\boldsymbol{E}_1, \ldots, \boldsymbol{E}_k$) corresponds to premultiplying $\boldsymbol{B}$, $\boldsymbol{D}$, and $\boldsymbol{b}$ by $\boldsymbol{B}^{-1} = \boldsymbol{E}_k, \ldots, \boldsymbol{E}_1$. In particular, the vector of current values of the basic variables is $\boldsymbol{B}^{-1}\boldsymbol{b}$. Observe that computation of the current basic feasible solution does not require computation of $\boldsymbol{B}^{-1}\boldsymbol{D}$; all we need is the matrix $\boldsymbol{B}^{-1}$. In the revised simplex method, we do not compute $\boldsymbol{B}^{-1}\boldsymbol{D}$. Instead, we only keep track of the basic variables and the revised tableau, which is the tableau $[\boldsymbol{B}^{-1}, \boldsymbol{B}^{-1}\boldsymbol{b}]$. Note that this tableau is only of size $m \times (m+1)$ [compared to the tableau in the original simplex method, which is $m \times (n+1)$]. To see how to update the revised tableau, suppose we choose the column $\boldsymbol{a}_q$ to enter the basis. Let $\boldsymbol{y}_q = \boldsymbol{B}^{-1}\boldsymbol{a}_q$, $\boldsymbol{y}_0 = [y_{01}, \ldots, y_{0m}]^T = \boldsymbol{B}^{-1}\boldsymbol{b}$, and $p = \arg\min_i\{y_{i0}/y_{iq} \colon y_{iq} > 0\}$ (as in the original simplex method). Then, to update the revised tableau, we form the augmented revised tableau $[\boldsymbol{B}^{-1}, \boldsymbol{y}_0, \boldsymbol{y}_q]$, and pivot about the $p$th element of the last column. We claim that the first $m+1$ columns of the resulting matrix comprise the updated revised tableau (i.e., we simply remove the last column of the updated augmented revised tableau to obtain the updated revised tableau). To see this, write $\boldsymbol{B}^{-1}$ as $\boldsymbol{B}^{-1} = \boldsymbol{E}_k \cdots \boldsymbol{E}_1$, and let the matrix $\boldsymbol{E}_{k+1}$ represent the pivoting operation above (i.e., $\boldsymbol{E}_{k+1}\boldsymbol{y}_q = \boldsymbol{e}_p$, the $p$th column of the $m \times m$ identity matrix). The matrix $\boldsymbol{E}_{k+1}$ is given by

$$\boldsymbol{E}_{k+1} = \begin{bmatrix} 1 & & -\dfrac{y_{1q}}{y_{iq}} & & 0 \\ & \ddots & \vdots & & \\ & & \dfrac{1}{y_{iq}} & & \\ & & \vdots & \ddots & \\ 0 & & -\dfrac{y_{mq}}{y_{iq}} & & 1 \end{bmatrix}$$

Then, the updated augmented tableau resulting from the above pivoting operation is $[\boldsymbol{E}_{k+1}\boldsymbol{B}^{-1}, \boldsymbol{E}_{k+1}\boldsymbol{y}_0, \boldsymbol{e}_p]$. Let $\boldsymbol{B}_{\text{new}}$ be the new basis. Then, we have $\boldsymbol{B}_{\text{new}}^{-1} = \boldsymbol{E}_{k+1} \cdots \boldsymbol{E}_1$. But notice that $\boldsymbol{B}_{\text{new}}^{-1} = \boldsymbol{E}_{k+1}\boldsymbol{B}^{-1}$, and the values of the basic variables corresponding to $\boldsymbol{B}_{\text{new}}$ are given by $\boldsymbol{y}_{0\text{new}} = \boldsymbol{E}_{k+1}\boldsymbol{y}_0$. Hence, the updated tableau is indeed $[\boldsymbol{B}_{\text{new}}^{-1}, \boldsymbol{y}_{\text{new}}] = [\boldsymbol{E}_{k+1}\boldsymbol{B}^{-1}, \boldsymbol{E}_{k+1}\boldsymbol{y}_0]$.

We summarize the above discussion in the following algorithm.

### The Revised Simplex Method

1. Form a revised tableau corresponding to an initial basic feasible solution $[\boldsymbol{B}^{-1}, \boldsymbol{y}_0]$.

2. Calculate the current relative cost coefficients vectors via

$$\boldsymbol{r}_D^T = \boldsymbol{c}_D^T - \boldsymbol{\lambda}^T \boldsymbol{D},$$

where

$$\boldsymbol{\lambda}^T = \boldsymbol{c}_B^T \boldsymbol{B}^{-1}.$$

3. If $r_j \geqslant 0$ for all $j$, stop—the current basic feasible solution is optimal.
4. Select a $q$ such that $r_q < 0$ (e.g., the $q$ corresponding to the most negative $r_q$), and compute

$$\boldsymbol{y}_q = \boldsymbol{B}^{-1}\boldsymbol{a}_q.$$

5. If no $y_{iq} > 0$, stop—the problem is unbounded; else, compute $p = \arg\min_i \{y_{i0}/y_{iq} : y_{iq} > 0\}$.
6. Form the augmented revised tableau $[\boldsymbol{B}^{-1}, \boldsymbol{y}_0, \boldsymbol{y}_q]$, and pivot about the $p$th element of the last column. Form the updated revised tableau by taking the first $m + 1$ columns of the resulting augmented revised tableau (i.e., remove the last column).
7. Go to step 2.

The reason for computing $\boldsymbol{r}_D$ in two steps as indicated in step 2 is as follows. We first note that $\boldsymbol{r}_D = \boldsymbol{c}_D^T - \boldsymbol{c}_B^T \boldsymbol{B}^{-1}\boldsymbol{D}$. To compute $\boldsymbol{c}_B^T \boldsymbol{B}^{-1}\boldsymbol{D}$, we can either do the multiplication in the order $(\boldsymbol{c}_B^T \boldsymbol{B}^{-1})\boldsymbol{D}$ or $\boldsymbol{c}_B^T(\boldsymbol{B}^{-1}\boldsymbol{D})$. The former involves two vector–matrix multiplications, whereas the latter involves a matrix–matrix multiplication followed by a vector–matrix multiplication. Clearly the former is more efficient.

As in the original simplex method, we can use the two-phase method to solve a given LP problem using the revised simplex method. In particular, we use the revised tableau from the final step of phase I as the initial revised tableau in phase II. We illustrate the method in the following example.

***Example 16.5.*** Consider solving the following LP problem using the revised simplex method:

$$\begin{aligned} \text{maximize} \quad & 3x_1 + 5x_2 \\ \text{subject to} \quad & x_1 + x_2 \leqslant 4 \\ & 5x_1 + 3x_2 \geqslant 8 \\ & x_1, x_2 \geqslant 0. \end{aligned}$$

First, we express the problem in standard form by introducing one slack and one surplus variable, to obtain

$$\begin{aligned} \text{minimize} \quad & -3x_1 - 5x_2 \\ \text{subject to} \quad & x_1 + x_2 + x_3 = 4 \\ & 5x_1 + 3x_2 - x_4 = 8 \\ & x_1, \ldots, x_4 \geqslant 0. \end{aligned}$$

There is no obvious basic feasible solution to the above LP problem. Therefore, we use the two-phase method.

*Phase I.* We introduce one artificial variable $x_5$ and an artificial objective function $x_5$. The tableau for the artificial problem is

| | $a_1$ | $a_2$ | $a_3$ | $a_4$ | $a_5$ | $b$ |
|---|---|---|---|---|---|---|
| | 1 | 1 | 1 | 0 | 0 | 4 |
| | 5 | 3 | 0 | −1 | 1 | 8 |
| $c^T$ | 0 | 0 | 0 | 0 | 1 | 0 |

We start with an initial basic feasible solution and corresponding $B^{-1}$ as shown in the following revised tableau

| Variable | $B^{-1}$ | | $y_0$ |
|---|---|---|---|
| $x_3$ | 1 | 0 | 4 |
| $x_5$ | 0 | 1 | 8 |

We compute

$$\lambda^T = c_B^T B^{-1} = [0, 1],$$
$$r_D^T = c_D^T - \lambda^T D = [0,0,0] - [5,3,-1] = [-5,-3,1] = [r_1, r_2, r_4].$$

Since $r_1$ is the most negative reduced cost coefficient, we bring $a_1$ into the basis. To do this, we first compute $y_1 = B^{-1}a_1$. In this case $y_1 = a_1$. We get the augmented revised tableau:

| Variable | $B^{-1}$ | | $y_0$ | $y_1$ |
|---|---|---|---|---|
| $x_3$ | 1 | 0 | 4 | 1 |
| $x_5$ | 0 | 1 | 8 | 5 |

We then compute $p = \arg\min_i \{y_{i0}/y_{iq} : y_{iq} > 0\} = 2$, and pivot about the second element of the last column to get the updated revised tableau:

| Variable | $B^{-1}$ | | $y_0$ |
|---|---|---|---|
| $x_3$ | 1 | $-\frac{1}{5}$ | $\frac{12}{5}$ |
| $x_1$ | 0 | $\frac{1}{5}$ | $\frac{8}{5}$ |

We next compute

$$\lambda^T = c_B^T B^{-1} = [0, 0]$$
$$r_D^T = c_D^T - \lambda^T D = [0,0,1] = [r_2, r_4, r_5] \geqslant 0^T.$$

The reduced cost coefficients are all nonnegative. Hence the solution to the artificial problem is $[8/5, 0, 12/5, 0, 0]^T$. The initial basic feasible solution for phase II is therefore $[8/5, 0, 12/5, 0]^T$.

*Phase II.* The tableau for the original problem (in standard form) is:

| | $a_1$ | $a_2$ | $a_3$ | $a_4$ | $b$ |
|---|---|---|---|---|---|
| | 1 | 1 | 1 | 0 | 4 |
| | 5 | 3 | 0 | $-1$ | 8 |
| $c^T$ | $-3$ | $-5$ | 0 | 0 | 0 |

As the initial revised tableau for phase II, we take the final revised tableau from phase I. We then compute

$$\lambda^T = c_B^T B^{-1} = [0, -3]\begin{bmatrix} 1 & -\frac{1}{5} \\ 0 & \frac{1}{5} \end{bmatrix} = \left[0, -\frac{3}{5}\right],$$

$$r_D^T = c_D^T - \lambda^T D = [-5, 0] - \left[0, -\frac{3}{5}\right]\begin{bmatrix} 1 & 0 \\ 3 & -1 \end{bmatrix} = \left[-\frac{16}{5}, -\frac{3}{5}\right] = [r_2, r_4].$$

We bring $a_2$ into the basis, and compute $y_2 = B^{-1}a_2$ to get:

| Variable | $B^{-1}$ | | $y_0$ | $y_2$ |
|---|---|---|---|---|
| $x_3$ | 1 | $-\frac{1}{5}$ | $\frac{12}{5}$ | $\frac{2}{5}$ |
| $x_1$ | 0 | $\frac{1}{5}$ | $\frac{8}{5}$ | $\frac{3}{5}$ |

In this case, we get $p = 2$. We update this tableau by pivoting about the second element of the last last column to get

| Variable | $B^{-1}$ | | $y_0$ |
|---|---|---|---|
| $x_3$ | 1 | $-\frac{1}{3}$ | $\frac{4}{3}$ |
| $x_2$ | 0 | $\frac{1}{3}$ | $\frac{8}{3}$ |

We compute

$$\lambda^T = c_B^T B^{-1} = [0, -5]\begin{bmatrix} 1 & -\frac{1}{3} \\ 0 & \frac{1}{3} \end{bmatrix} = \left[0, -\frac{5}{3}\right],$$

$$r_D^T = c_D^T - \lambda^T D = [-3, 0] - \left[0, -\frac{5}{3}\right]\begin{bmatrix} 1 & 0 \\ 5 & -1 \end{bmatrix} = \left[\frac{16}{3}, -\frac{5}{3}\right] = [r_1, r_4].$$

We now bring $\boldsymbol{a}_4$ into the basis:

| Variable | $\boldsymbol{B}^{-1}$ | | $\boldsymbol{y}_0$ | $\boldsymbol{y}_4$ |
|---|---|---|---|---|
| $x_3$ | 1 | $-\frac{1}{3}$ | $\frac{4}{3}$ | $\frac{1}{3}$ |
| $x_2$ | 0 | $\frac{1}{3}$ | $\frac{8}{3}$ | $-\frac{1}{3}$ |

We update the tableau to obtain

| Variable | $\boldsymbol{B}^{-1}$ | | $\boldsymbol{y}_0$ |
|---|---|---|---|
| $x_4$ | 3 | $-1$ | 4 |
| $x_2$ | 1 | 0 | 4 |

We compute

$$\boldsymbol{\lambda}^T = \boldsymbol{c}_B^T \boldsymbol{B}^{-1} = [0, -5]\begin{bmatrix} 3 & -1 \\ 1 & 0 \end{bmatrix} = [-5, 0],$$

$$\boldsymbol{r}_D^T = \boldsymbol{c}_D^T - \boldsymbol{\lambda}^T \boldsymbol{D} = [-3, 0] - [-5, 0]\begin{bmatrix} 1 & 1 \\ 5 & 0 \end{bmatrix} = [2, 5] = [r_1, r_3].$$

The reduced cost coefficients are all positive. Hence $[0, 4, 0, 4]^T$ is optimal. The optimal solution to the original problem is $[0, 4]^T$. ■

## EXERCISES

**16.1** Use the simplex method to solve the following linear program:

$$\begin{array}{ll} \text{maximize} & x_1 + x_2 + 3x_3 \\ \text{subject to} & x_1 + x_3 = 1 \\ & x_2 + x_3 = 2 \\ & x_1, x_2, x_3 \geqslant 0. \end{array}$$

**16.2** Consider the linear program:

$$\begin{array}{ll} \text{maximize} & 2x_1 + x_2 \\ \text{subject to} & 0 \leqslant x_1 \leqslant 5 \\ & 0 \leqslant x_2 \leqslant 7 \\ & x_1 + x_2 \leqslant 9. \end{array}$$

Convert the problem to standard form and solve it using the simplex method.

**16.3** Find the solution and the value of the optimal cost for the following problem using the revised simplex method:

$$\begin{array}{ll} \text{minimize} & x_1 + x_2 \\ \text{subject to} & x_1 + 2x_2 \geqslant 3 \\ & 2x_1 + x_2 \geqslant 3 \\ & x_1, x_2 \geqslant 0. \end{array}$$

*Hint*: Start with $x_1$ and $x_2$ as basic variables.

**16.4** Solve the following linear programs using the revised simplex method:

**a.** Maximize $-4x_1 - 3x_2$ subject to

$$\begin{aligned} 5x_1 + x_2 &\geqslant 11 \\ -2x_1 - x_2 &\leqslant -8 \\ x_1 + 2x_2 &\geqslant 7 \\ x_1, x_2 &\geqslant 0 \end{aligned}$$

**b.** Maximize $6x_1 + 4x_2 + 7x_3 + 5x_4$ subject to

$$\begin{aligned} x_1 + 2x_2 + x_3 + 2x_4 &\leqslant 20 \\ 6x_1 + 5x_2 + 3x_3 + 2x_4 &\leqslant 100 \\ 3x_1 + 4x_2 + 9x_3 + 12x_4 &\leqslant 75 \\ x_1, x_2, x_3, x_4 &\geqslant 0 \end{aligned}$$

**16.5** Consider a standard form linear programming problem, with

$$\boldsymbol{A} = \begin{bmatrix} 0 & 2 & 0 & 1 \\ 1 & 1 & 0 & 0 \\ 0 & 3 & 1 & 0 \end{bmatrix}, \quad \boldsymbol{b} = \begin{bmatrix} 7 \\ 8 \\ 9 \end{bmatrix}, \quad \boldsymbol{c} = \begin{bmatrix} 6 \\ c_2 \\ 4 \\ 5 \end{bmatrix}.$$

Suppose we are told that the relative cost coefficient vector corresponding to some basis is $\boldsymbol{r}^T = [0, 1, 0, 0]$.

**a.** Find an optimal feasible solution to the given problem.

**b.** Find $c_2$.

**16.6** Consider the linear programming problem:

$$\begin{array}{ll} \text{minimize} & c_1 x_1 + c_2 x_2 \\ \text{subject to} & 2x_1 + x_2 = 2 \\ & x_1, x_2 \geqslant 0, \end{array}$$

where $c_1, c_2 \in \mathbb{R}$. Suppose that the problem has an optimal feasible solution that is not basic.

**a.** Find all basic feasible solutions.

**b.** Find all possible values of $c_1$ and $c_2$.

**c.** At each basic feasible solution, compute the relative cost coefficients for all nonbasic variables.

**16.7** Consider the following linear programming problem (attributed to Beale [see Ref. 17, p. 43]):

$$\begin{aligned} \text{minimize} \quad & -\tfrac{3}{4}x_4 + 20x_5 - \tfrac{1}{2}x_6 + 6x_7 \\ \text{subject to} \quad & x_1 + \tfrac{1}{4}x_4 - 8x_5 - x_6 + 9x_7 = 0 \\ & x_2 + \tfrac{1}{2}x_4 - 12x_5 - \tfrac{1}{2}x_6 + 3x_7 = 0 \\ & x_3 + x_6 = 1 \\ & x_1, \ldots, x_7 = 0. \end{aligned}$$

**a.** Apply the simplex algorithm to the problem using the rule that $q$ is the index corresponding to the most negative $r_q$. (As usual, if more than one index $i$ minimizes $y_{i0}/y_{iq}$, let $p$ be the smallest such index.) Start with $x_1, x_2$, and $x_3$ as initial basic variables. Notice that cycling occurs.

**b.** Repeat part a using *Bland's rule* for choosing $q$ and $p$:

$$q = \min\{i : r_i < 0\}$$

$$p = \min\{j : y_{j0}/y_{jq} = \min_i \{y_{i0}/y_{iq} : y_{iq} > 0\}\}.$$

Note that Bland's rule for choosing $p$ corresponds to our usual rule that if more than one index $i$ minimizes $y_{i0}/y_{iq}$, we let $p$ be the smallest such index.

**16.8** Write a simple MATLAB function that implements the simplex algorithm. The inputs are $\boldsymbol{c}, \boldsymbol{A}, \boldsymbol{b}$ and $\boldsymbol{v}$, where $\boldsymbol{v}$ is the vector of indices of basic columns. Assume that the augmented matrix $[\boldsymbol{A}, \boldsymbol{b}]$ is already in canonical form; that is, the $v_i$th column of $\boldsymbol{A}$ is $[0, \ldots, 1, \ldots, 0]^T$, where 1 occurs in the $i$th position. The function should output the final solution and the vector of indices of basic columns. Test the MATLAB function on the problem in Example 16.2.

**16.9** Write a MATLAB routine that implements the two-phase simplex method. It may be useful to use the MATLAB function of Exercise 16.8. Test the routine on the problem in Example 16.5.

**16.10** Write a simple MATLAB function that implements the revised simplex algorithm. The inputs are $\boldsymbol{c}$, $\boldsymbol{A}$, $\boldsymbol{b}$, $\boldsymbol{v}$, and $\boldsymbol{B}^{-1}$, where $\boldsymbol{v}$ is the vector of indices of basic columns; that is, the $i$th column of $\boldsymbol{B}$ is the $v_i$th column of $\boldsymbol{A}$. The function should output the final solution, the vector of indices of basic columns, and the final $\boldsymbol{B}^{-1}$. Test the MATLAB function on the problem in Example 16.2.

**16.11** Write a MATLAB routine that implements the two-phase revised simplex method. It may be useful to use the MATLAB function of Exercise 16.10. Test the routine on the problem in Example 16.5.

# 17

# Duality

## 17.1. DUAL LINEAR PROGRAMS

Associated with every LP problem is a corresponding *dual* linear programming problem. The dual problem is constructed from the cost and constraints of the original, or *primal*, problem. Being an LP problem, the dual can be solved using the simplex method. However, as we shall see, the solution to the dual can also be obtained from the solution of the primal problem, and vice versa. Solving an LP problem via its dual may be simpler in certain cases, and also often provides further insight into the nature of the problem. In this chapter, we study basic properties of duality, and provide an interpretive example of duality. Duality can be used to improve the performance of the simplex algorithm (leading to the so called primal–dual algorithm), as well as to develop non-simplex algorithms for solving LP problems (such as Khachiyan's algorithm and Karmarkar's algorithm). We will not discuss this aspect of duality any further in this chapter. For an in depth discussion of the primal–dual method, as well as other aspects of duality, see, for example, Ref. 51. For a treatment of Khachiyan's algorithm and Karmarkar's algorithm, see Chapter 18.

Suppose we are given a linear programming problem of the form

$$\begin{aligned} \text{minimize} \quad & \boldsymbol{c}^T \boldsymbol{x} \\ \text{subject to} \quad & \boldsymbol{A}\boldsymbol{x} \geqslant \boldsymbol{b}, \\ & \boldsymbol{x} \geqslant \boldsymbol{0}. \end{aligned}$$

We will refer to the above as the primal problem. We define the corresponding dual problem as

$$\begin{aligned} \text{maximize} \quad & \boldsymbol{\lambda}^T \boldsymbol{b} \\ \text{subject to} \quad & \boldsymbol{\lambda}^T \boldsymbol{A} \leqslant \boldsymbol{c}^T, \\ & \boldsymbol{\lambda} \geqslant \boldsymbol{0}. \end{aligned}$$

We refer to the variable $\boldsymbol{\lambda} \in \mathbb{R}^m$ as the *dual vector*. Note that the cost vector $\boldsymbol{c}$ in the primal has moved to the constraints in the dual. The vector $\boldsymbol{b}$ on the right-hand side of $\boldsymbol{Ax} \geqslant \boldsymbol{b}$ becomes part of the cost in the dual. Thus, the roles of $\boldsymbol{b}$ and $\boldsymbol{c}$ are reversed. The form of duality defined above is called the *symmetric form of duality*.

Note that the dual of the dual problem is the primal problem. To see this, we first represent the dual problem in the form

$$\begin{aligned} &\text{minimize} \quad \boldsymbol{\lambda}^T(-\boldsymbol{b}) \\ &\text{subject to} \quad \boldsymbol{\lambda}^T(-\boldsymbol{A}) \geqslant -\boldsymbol{c}^T, \\ &\qquad\qquad\quad \boldsymbol{\lambda} \geqslant \boldsymbol{0}. \end{aligned}$$

Therefore, by the symmetric form of duality, the dual to the above is

$$\begin{aligned} &\text{maximize} \quad (-\boldsymbol{c}^T)\boldsymbol{x} \\ &\text{subject to} \quad (-\boldsymbol{A})\boldsymbol{x} \leqslant -\boldsymbol{b}, \\ &\qquad\qquad\quad \boldsymbol{x} \geqslant \boldsymbol{0}. \end{aligned}$$

Upon rewriting, we get the original primal problem.

Consider now an LP problem in standard form. This form has equality constraints, $\boldsymbol{Ax} = \boldsymbol{b}$. To formulate the corresponding dual problem, we first convert the equality constraints into equivalent inequality constraints. Specifically, observe that $\boldsymbol{Ax} = \boldsymbol{b}$ is equivalent to

$$\begin{aligned} \boldsymbol{Ax} &\geqslant \boldsymbol{b} \\ -\boldsymbol{Ax} &\geqslant -\boldsymbol{b}. \end{aligned}$$

Thus, the original problem with the equality constraints can be written in the form:

$$\begin{aligned} &\text{minimize} \quad \boldsymbol{c}^T\boldsymbol{x} \\ &\text{subject to} \quad \begin{bmatrix} \boldsymbol{A} \\ -\boldsymbol{A} \end{bmatrix}\boldsymbol{x} \geqslant \begin{bmatrix} \boldsymbol{b} \\ -\boldsymbol{b} \end{bmatrix}, \\ &\qquad\qquad\quad \boldsymbol{x} \geqslant \boldsymbol{0}. \end{aligned}$$

The above LP problem is in the form of the primal problem in the symmetric form of duality. The corresponding dual is therefore,

$$\begin{aligned} &\text{maximize} \quad [\boldsymbol{u}^T, \boldsymbol{v}^T]\begin{bmatrix} \boldsymbol{b} \\ -\boldsymbol{b} \end{bmatrix} \\ &\text{subject to} \quad [\boldsymbol{u}^T, \boldsymbol{v}^T]\begin{bmatrix} \boldsymbol{A} \\ -\boldsymbol{A} \end{bmatrix} \leqslant \boldsymbol{c}^T, \\ &\qquad\qquad\quad \boldsymbol{u}, \boldsymbol{v} \geqslant \boldsymbol{0}. \end{aligned}$$

After a simple manipulation, the above dual can be represented as

$$\begin{aligned} &\text{maximize} \quad (\boldsymbol{u}-\boldsymbol{v})^T\boldsymbol{b} \\ &\text{subject to} \quad (\boldsymbol{u}-\boldsymbol{v})^T\boldsymbol{A} \leqslant \boldsymbol{c}^T \\ &\qquad\qquad\qquad \boldsymbol{u},\boldsymbol{v} \geqslant \boldsymbol{0}. \end{aligned}$$

Let $\boldsymbol{\lambda}=\boldsymbol{u}-\boldsymbol{v}$. Then, the dual problem becomes

$$\begin{aligned} &\text{maximize} \quad \boldsymbol{\lambda}^T\boldsymbol{b} \\ &\text{subject to} \quad \boldsymbol{\lambda}^T\boldsymbol{A} \leqslant \boldsymbol{c}^T. \end{aligned}$$

Note that since $\boldsymbol{\lambda}=\boldsymbol{u}-\boldsymbol{v}$ and $\boldsymbol{u},\boldsymbol{v}\geqslant \boldsymbol{0}$, the dual vector $\boldsymbol{\lambda}$ is not restricted to be nonnegative. We have now derived the dual for a primal in standard form. The above form of duality is referred to as the *asymmetric form of duality*.

We summarize the above discussion in Tables 17.1 and 17.2.

Note that in the asymmetric form of duality, the dual of the dual is also the primal. We can show this by reversing the arguments we used to arrive at the asymmetric form of duality, and using the symmetric form of duality.

***Example 17.1.*** This example is adapted from Ref. 51. Recall the diet problem (Example 15.2). We have $n$ different types of food. Our goal is to create the most economical diet and at the same time meet or exceed nutritional requirements. Specifically, let $a_{ij}$ be the amount of the $i$th nutrient per unit of the $j$th food, $b_i$ the amount of the $i$th nutrient required, $1\leqslant i\leqslant m$, $c_j$ the cost per unit of the $j$th food, and $x_i$ the number of units of food $i$ in the diet. Then, the diet problem can be stated as follows:

$$\begin{aligned} &\text{minimize} \quad c_1x_1+c_2x_2+\cdots+c_nx_n \\ &\text{subject to} \quad a_{11}x_1+a_{12}x_2+\cdots+a_{1n}x_n\geqslant b_1 \\ &\qquad\qquad a_{21}x_1+a_{22}x_2+\cdots+a_{2n}x_n\geqslant b_2 \\ &\qquad\qquad\qquad \vdots \\ &\qquad\qquad a_{m1}x_1+a_{m2}x_2+\cdots+a_{mn}x_n\geqslant b_m \\ &\qquad\qquad\qquad x_1,\ldots,x_n\geqslant 0. \end{aligned}$$

Now, consider a health food store that sells nutrient pills (all $m$ types of nutrients are available). Let $\lambda_i$ be the price of a unit of the $i$th nutrient in the form of nutrient pills. Suppose we purchase nutrient pills from the health food store at the above price such that we exactly meet our nutritional requirements. Then, $\boldsymbol{\lambda}^T\boldsymbol{b}$ is the total revenue to the store. Note that since prices are nonnegative, we have $\boldsymbol{\lambda}\geqslant\boldsymbol{0}$. Consider now the task of substituting nutrient pills for natural food. The cost of

**Table 17.1 Symmetric Form of Duality**

| Primal | | Dual | |
|---|---|---|---|
| minimize | $c^T x$ | maximize | $\lambda^T b$ |
| subject to | $Ax \geqslant b$ | subject to | $\lambda^T A \leqslant c^T$ |
| | $x \geqslant 0$ | | $\lambda \geqslant 0$ |

**Table 17.2 Asymmetric Form of Duality**

| Primal | | Dual | |
|---|---|---|---|
| minimize | $c^T x$ | maximize | $\lambda^T b$ |
| subject to | $Ax = b$ | subject to | $\lambda^T A \leqslant c^T$ |
| | $x \geqslant 0$ | | |

buying pills to synthetically create the nutritional equivalent of the $i$th food is simply $\lambda_1 a_{1i} + \cdots + \lambda_m a_{mi}$. Since $c_i$ is the cost per unit of the $i$th food, then, if

$$\lambda_1 a_{1i} + \cdots + \lambda_m a_{mi} \leqslant c_i,$$

the cost of the unit of the $i$th food made synthetically from nutrient pills is less than or equal to the market price of a unit of the real food. Therefore, for the health food store to be competitive, the following must hold:

$$\begin{aligned} \lambda_1 a_{11} + \cdots + \lambda_m a_{m1} &\leqslant c_1 \\ &\vdots \\ \lambda_m a_{1n} + \cdots + \lambda_m a_{mn} &\leqslant c_n. \end{aligned}$$

The problem facing the health food store is to choose the prices $\lambda_1, \ldots, \lambda_m$ such that its revenue is maximized. This problem can be stated as

$$\begin{aligned} \text{maximize} \quad & \lambda^T b \\ \text{subject to} \quad & \lambda^T A \leqslant c^T \\ & \lambda \geqslant 0. \end{aligned}$$

Note that the above is simply the dual of the diet problem. ■

***Example 17.2.*** Consider the following linear programming problem:

$$\begin{aligned} \text{maximize} \quad & 2x_1 + 5x_2 + x_3 \\ \text{subject to} \quad & 2x_1 - x_2 + 7x_3 \leqslant 6 \\ & x_1 + 3x_2 + 4x_3 \leqslant 9 \\ & 3x_1 + 6x_2 + x_3 \leqslant 3 \\ & x_1, x_2, x_3 \geqslant 0. \end{aligned}$$

Find the corresponding dual problem and solve it.

We first write the primal problem in standard form by introducing slack variables $x_4$, $x_5$, and $x_6$. This primal problem in standard form is

$$\begin{aligned} \text{minimize} \quad & \boldsymbol{c}^T \boldsymbol{x} \\ \text{subject to} \quad & [\boldsymbol{A}, \boldsymbol{I}]\boldsymbol{x} = \boldsymbol{b} \\ & \boldsymbol{x} \geqslant \boldsymbol{0}, \end{aligned}$$

where $\boldsymbol{x} = [x_1, \ldots, x_6]^T$, and

$$\boldsymbol{A} = \begin{bmatrix} 2 & -1 & 7 \\ 1 & 3 & 4 \\ 3 & 6 & 1 \end{bmatrix}, \quad \boldsymbol{b} = \begin{bmatrix} 6 \\ 9 \\ 3 \end{bmatrix}, \quad \boldsymbol{c} = \begin{bmatrix} -2 \\ -5 \\ -1 \end{bmatrix}.$$

The corresponding dual problem (asymmetric form) is

$$\begin{aligned} \text{maximize} \quad & \boldsymbol{\lambda}^T \boldsymbol{b} \\ \text{subject to} \quad & \boldsymbol{\lambda}^T [\boldsymbol{A}, \boldsymbol{I}] \leqslant [\boldsymbol{c}^T, \boldsymbol{0}^T]. \end{aligned}$$

Note that the constraints in the dual can be written as

$$\begin{aligned} \boldsymbol{\lambda}^T \boldsymbol{A} &\leqslant \boldsymbol{c}^T \\ \boldsymbol{\lambda} &\leqslant \boldsymbol{0}. \end{aligned}$$

To solve the above dual problem, we use the simplex method. For this, we need to express the problem in standard form. We substitute $\boldsymbol{\lambda}$ by $-\boldsymbol{\lambda}$, and introduce surplus variables to get:

$$\begin{aligned} \text{minimize} \quad & 6\lambda_1 + 9\lambda_2 + 3\lambda_3 \\ \text{subject to} \quad & 2\lambda_1 + \lambda_2 + 3\lambda_3 - \lambda_4 = 2 \\ & -\lambda_1 + 3\lambda_2 + 6\lambda_3 - \lambda_5 = 5 \\ & 7\lambda_1 + 4\lambda_2 + \lambda_3 - \lambda_6 = 1 \\ & \lambda_1, \ldots, \lambda_6 \geqslant 0. \end{aligned}$$

There is no obvious basic feasible solution. Thus, we use the two-phase simplex method to solve the problem.

*Phase I.* We introduce artificial variables $\lambda_7, \lambda_8, \lambda_9$ and the artificial objective function $\lambda_7 + \lambda_8 + \lambda_9$. The tableau for the artificial problem is

| | $\lambda_1$ | $\lambda_2$ | $\lambda_3$ | $\lambda_4$ | $\lambda_5$ | $\lambda_6$ | $\lambda_7$ | $\lambda_8$ | $\lambda_9$ | $c$ |
|---|---|---|---|---|---|---|---|---|---|---|
| | 2 | 1 | 3 | −1 | 0 | 0 | 1 | 0 | 0 | 2 |
| | −1 | 3 | 6 | 0 | −1 | 0 | 0 | 1 | 0 | 5 |
| | 7 | 4 | 1 | 0 | 0 | −1 | 0 | 0 | 1 | 1 |
| Cost | 0 | 0 | 0 | 0 | 0 | 0 | 1 | 1 | 1 | 0 |

We start with an initial feasible solution and corresponding $B^{-1}$:

| Variable | $B^{-1}$ | | | $y_0$ |
|---|---|---|---|---|
| $\lambda_7$ | 1 | 0 | 0 | 2 |
| $\lambda_8$ | 0 | 1 | 0 | 5 |
| $\lambda_9$ | 0 | 0 | 1 | 1 |

We compute

$$r_D^T = [0,0,0,0,0,0] - [8,8,10,-1,-1,-1] = [-8,-8,-10,1,1,1]$$
$$= [r_1, r_2, r_3, r_4, r_5, r_6].$$

Since $r_3$ is the most negative reduced cost coefficient, we bring the third column into the basis. In this case, $y_3 = [3,6,1]^T$. We have

| Variable | $B^{-1}$ | | | $y_0$ | $y_3$ |
|---|---|---|---|---|---|
| $\lambda_7$ | 1 | 0 | 0 | 2 | 3 |
| $\lambda_8$ | 0 | 1 | 0 | 5 | 6 |
| $\lambda_9$ | 0 | 0 | 1 | 1 | 1 |

By inspection, $p = 1$, so we pivot about the first element of the last column. The updated tableau is:

| Variable | $B^{-1}$ | | | $y_0$ |
|---|---|---|---|---|
| $\lambda_3$ | $\frac{1}{3}$ | 0 | 0 | $\frac{2}{3}$ |
| $\lambda_8$ | $-2$ | 1 | 0 | 1 |
| $\lambda_9$ | $-\frac{1}{3}$ | 0 | 1 | $\frac{1}{3}$ |

We compute

$$r_D^T = \left[-\frac{4}{3}, -\frac{14}{3}, -\frac{7}{3}, 1, 1, \frac{10}{3}\right] = [r_1, r_2, r_4, r_5, r_6, r_7].$$

We bring the second column into the basis to get:

| Variable | $B^{-1}$ | | | $y_0$ | $y_2$ |
|---|---|---|---|---|---|
| $\lambda_3$ | $\frac{1}{3}$ | 0 | 0 | $\frac{2}{3}$ | $\frac{1}{3}$ |
| $\lambda_8$ | $-2$ | 1 | 0 | 1 | 1 |
| $\lambda_9$ | $-\frac{1}{3}$ | 0 | 1 | $\frac{1}{3}$ | $\frac{11}{3}$ |

We update the tableau to get

| Variable | $B^{-1}$ | | | $y_0$ |
|---|---|---|---|---|
| $\lambda_3$ | $\frac{4}{11}$ | 0 | $-\frac{1}{11}$ | $\frac{7}{11}$ |
| $\lambda_8$ | $-\frac{21}{11}$ | 1 | $-\frac{3}{11}$ | $\frac{10}{11}$ |
| $\lambda_2$ | $-\frac{1}{11}$ | 0 | $\frac{3}{11}$ | $\frac{1}{11}$ |

We compute

$$r_D^T = \left[\tfrac{74}{11}, -\tfrac{21}{11}, 1, -\tfrac{3}{11}, \tfrac{32}{11}, \tfrac{14}{11}\right] = [r_1, r_4, r_5, r_6, r_7, r_9].$$

We bring the fourth column into the basis:

| Variable | $B^{-1}$ | | | $y_0$ | $y_4$ |
|---|---|---|---|---|---|
| $\lambda_3$ | $\frac{4}{11}$ | 0 | $-\frac{1}{11}$ | $\frac{7}{11}$ | $-\frac{4}{11}$ |
| $\lambda_8$ | $-\frac{21}{11}$ | 1 | $-\frac{3}{11}$ | $\frac{10}{11}$ | $\frac{21}{11}$ |
| $\lambda_2$ | $-\frac{1}{11}$ | 0 | $\frac{3}{11}$ | $\frac{1}{11}$ | $\frac{1}{11}$ |

The updated tableau becomes

| Variable | $B^{-1}$ | | | $y_0$ |
|---|---|---|---|---|
| $\lambda_3$ | 0 | $\frac{4}{21}$ | $-\frac{3}{21}$ | $\frac{17}{21}$ |
| $\lambda_4$ | $-1$ | $\frac{11}{21}$ | $-\frac{3}{21}$ | $\frac{10}{21}$ |
| $\lambda_2$ | 0 | $-\frac{1}{21}$ | $\frac{6}{21}$ | $\frac{1}{21}$ |

We compute

$$r_D^T = [0, 0, 0, 1, 1, 1] = [r_1, r_5, r_6, r_7, r_8, r_9].$$

Since all the reduced cost coefficient are nonnegative, this terminates phase I.

*Phase II*. We use the last tableau in phase I as the initial tableau in phase II. Note that we now revert back to the original cost of the dual problem in standard form. We compute

$$r_D^T = \left[-\tfrac{62}{7}, \tfrac{1}{7}, \tfrac{15}{7}\right] = [r_1, r_5, r_6].$$

We bring the first column into the basis to obtain the augmented revised tableau

| Variable | $B^{-1}$ | | | $y_0$ | $y_1$ |
|---|---|---|---|---|---|
| $\lambda_3$ | 0 | $\frac{4}{21}$ | $-\frac{3}{21}$ | $\frac{17}{21}$ | $-\frac{25}{21}$ |
| $\lambda_4$ | $-1$ | $\frac{11}{21}$ | $-\frac{3}{21}$ | $\frac{10}{21}$ | $-\frac{74}{21}$ |
| $\lambda_2$ | 0 | $-\frac{1}{21}$ | $\frac{6}{21}$ | $\frac{1}{21}$ | $\frac{43}{21}$ |

We update the tableau to get

| Variable | $B^{-1}$ | | | $y_0$ |
|---|---|---|---|---|
| $\lambda_3$ | 0 | $\frac{7}{43}$ | $\frac{1}{43}$ | $\frac{36}{43}$ |
| $\lambda_4$ | $-1$ | $\frac{19}{43}$ | $\frac{15}{43}$ | $\frac{24}{43}$ |
| $\lambda_1$ | 0 | $-\frac{1}{43}$ | $\frac{6}{43}$ | $\frac{1}{43}$ |

We compute

$$r_D^T = [\tfrac{186}{43}, \tfrac{15}{43}, \tfrac{39}{43}] = [r_2, r_5, r_6].$$

Since all the reduced cost coefficients are nonnegative, the current basic feasible solution is optimal for the dual in standard form. Thus, an optimal solution to the original dual problem is

$$\boldsymbol{\lambda} = [-\tfrac{1}{43}, 0, -\tfrac{36}{43}]^T. \quad \blacksquare$$

## 17.2. PROPERTIES OF DUAL PROBLEMS

In this section, we present some basic results on dual linear programs. We begin with the weak duality lemma.

**Lemma 17.1. Weak Duality Lemma.** *Suppose $\boldsymbol{x}$ and $\boldsymbol{\lambda}$ are feasible solutions to primal and dual LP problems, respectively (either in the symmetric or asymmetric form). Then, $\boldsymbol{c}^T\boldsymbol{x} \geqslant \boldsymbol{\lambda}^T\boldsymbol{b}$.* □

*Proof.* We shall prove this lemma only for the asymmetric form of duality. The proof for the symmetric form involves only a slight modification—see Exercise 17.1.

Since $\boldsymbol{x}$ and $\boldsymbol{\lambda}$ are feasible, then $\boldsymbol{Ax} = \boldsymbol{b}$, $\boldsymbol{x} \geqslant \boldsymbol{0}$, and $\boldsymbol{\lambda}^T\boldsymbol{A} \leqslant \boldsymbol{c}^T$. Postmultiplying both sides of the inequality $\boldsymbol{\lambda}^T\boldsymbol{A} \leqslant \boldsymbol{c}^T$ by $\boldsymbol{x} \geqslant \boldsymbol{0}$ yields $\boldsymbol{\lambda}^T\boldsymbol{Ax} \leqslant \boldsymbol{c}^T\boldsymbol{x}$. But $\boldsymbol{Ax} = \boldsymbol{b}$, hence $\boldsymbol{\lambda}^T\boldsymbol{b} \leqslant \boldsymbol{c}^T\boldsymbol{x}$. ■

The weak duality lemma states that a feasible solution to either problem yields a bound on the optimal cost of the other problem. The cost in the dual is never above the cost in the primal. In particular, the optimal cost of the dual is less than or equal to the optimal cost of the primal (that is, maximum $\leqslant$ minimum). Hence, if the cost of one of the problems is unbounded, then the other problem has no feasible solution. In other words, if minimum $= -\infty$ or maximum $= +\infty$, then the feasible set in the other problem must be empty.

**Theorem 17.1.** *Suppose $\boldsymbol{x}_0$ and $\boldsymbol{\lambda}_0$ are feasible solutions to the primal and dual, respectively (either in symmetric or asymmetric form). If $\boldsymbol{c}^T\boldsymbol{x}_0 = \boldsymbol{\lambda}_0^T\boldsymbol{b}$, then $\boldsymbol{x}_0$ and $\boldsymbol{\lambda}_0$ are optimal solutions to their respective problems.* □

*Proof.* Let $\boldsymbol{x}$ be an arbitrary feasible solution to the primal problem. Since $\boldsymbol{\lambda}_0$ is a feasible solution to the dual, then, by the weak duality lemma, $\boldsymbol{c}^T\boldsymbol{x} \geqslant \boldsymbol{\lambda}_0^T\boldsymbol{b}$. So, if $\boldsymbol{c}^T\boldsymbol{x}_0 = \boldsymbol{\lambda}_0^T\boldsymbol{b}$, then $\boldsymbol{c}^T\boldsymbol{x}_0 = \boldsymbol{\lambda}_0^T\boldsymbol{b} \leqslant \boldsymbol{c}^T\boldsymbol{x}$. Hence, $\boldsymbol{x}_0$ is optimal for the primal.

On the other hand, let $\boldsymbol{\lambda}$ be an arbitrary feasible solution to the dual problem. Since $\boldsymbol{x}_0$ is a feasible solution to the primal, then, by the weak duality lemma,

$c^T x_0 \geqslant \lambda^T b$. Therefore, if $c^T x_0 = \lambda_0^T b$, then $\lambda^T b \leqslant c^T x_0 = \lambda_0^T b$. Hence, $\lambda_0$ is optimal for the dual. ■

We can interpret the above theorem as follows. The primal seeks to minimize its cost, and the dual seeks to maximize its cost. Since the weak duality lemma states that maximum $\leqslant$ minimum, then each problem seeks to reach the other. When their costs are equal for a pair of feasible solutions, both solutions are optimal, and we have maximum = minimum.

It turns out that the converse to the above theorem is also true; that is, maximum = minimum always holds. In fact, we can prove an even stronger result, known as the duality theorem.

**Theorem 17.2. Duality Theorem.** *If the primal problem (either in symmetric or asymmetric form) has an optimal solution, then so does the dual, and the optimal values of their respective objective functions are equal.* □

*Proof.* We first prove the result for the asymmetric form of duality. Assume that the primal has an optimal solution. Then, by the fundamental theorem of LP, there exists an optimal basic feasible solution. As is our usual notation, let $B$ be the matrix of the corresponding $m$ basic columns, $D$ the matrix of the $n-m$ nonbasic columns, $c_B$ the vector of elements of $c$ corresponding to basic variables, $c_D$ the vector of elements of $c$ corresponding to nonbasic variables, and $r_D$ the vector of reduced cost coefficients. Then, by Theorem 16.2,

$$r_D^T = c_D^T - c_B^T B^{-1} D \geqslant 0^T.$$

Hence,

$$c_B^T B^{-1} D \leqslant c_D^T.$$

Define

$$\lambda^T = c_B^T B^{-1}.$$

Then

$$c_B^T B^{-1} D = \lambda^T D \leqslant c_D^T.$$

We claim that $\lambda$ is a feasible solution to the dual. To see this, assume for convenience (and without loss of generality) that the basic columns are the first $m$ columns of $A$. Then,

$$\lambda^T A = \lambda^T [B, D] = [c_B^T, \lambda^T D] \leqslant [c_B^T, c_D^T] = c^T.$$

Hence, $\lambda^T A \leqslant c^T$ and thus $\lambda^T = c_B^T B^{-1}$ is feasible.

We claim that $\lambda$ is also an optimal feasible solution to the dual. To see this, note that

$$\lambda^T b = c_B^T B^{-1} b = c_B^T x_B.$$

Thus, by Theorem 17.1, $\lambda$ is optimal.

We now prove the symmetric case. First, we convert the primal problem for the symmetric form into the equivalent standard form by adding surplus variables:

$$\begin{aligned} \text{minimize} \quad & [\boldsymbol{c}^T, \boldsymbol{0}^T]\begin{bmatrix} \boldsymbol{x} \\ \boldsymbol{y} \end{bmatrix} \\ \text{subject to} \quad & [\boldsymbol{A}, -\boldsymbol{I}]\begin{bmatrix} \boldsymbol{x} \\ \boldsymbol{y} \end{bmatrix} = \boldsymbol{b}, \\ & \begin{bmatrix} \boldsymbol{x} \\ \boldsymbol{y} \end{bmatrix} \geqslant \boldsymbol{0}. \end{aligned}$$

Note that $\boldsymbol{x}$ is optimal for the original primal problem if, and only if, $[\boldsymbol{x}^T, (\boldsymbol{A}\boldsymbol{x} - \boldsymbol{b})^T]^T$ is optimal for the primal in standard form. The dual to the primal in standard form is equivalent to the dual to the original primal in symmetric form. Therefore, the above result for the asymmetric case applies to the symmetric case.

This completes the proof. ■

***Example 17.3.*** Recall Example 17.1, where we formulated the dual of the diet problem. From the duality theorem, the maximum revenue for the health food store is the same as the minimum cost of a diet that satisfies all of the nutritional requirements; that is, $\boldsymbol{c}^T\boldsymbol{x} = \boldsymbol{\lambda}^T\boldsymbol{b}$. ■

Consider a primal–dual pair in asymmetric form. Suppose we solve the primal problem using the simplex method. The proof of the duality theorem suggests a way of obtaining an optimal solution to the dual by using the last row of the final simplex tableau for the primal. First, we write the tableau for the primal problem:

$$\begin{bmatrix} \boldsymbol{A} & \boldsymbol{b} \\ \boldsymbol{c}^T & 0 \end{bmatrix} = \begin{bmatrix} \boldsymbol{B} & \boldsymbol{D} & \boldsymbol{b} \\ \boldsymbol{c}_B^T & \boldsymbol{c}_D^T & 0 \end{bmatrix}.$$

Suppose the matrix $\boldsymbol{B}$ is the basis for an optimal basic feasible solution. Then, the final simplex tableau is

$$\begin{bmatrix} \boldsymbol{I} & \boldsymbol{B}^{-1}\boldsymbol{D} & \boldsymbol{B}^{-1}\boldsymbol{b} \\ \boldsymbol{0}^T & \boldsymbol{r}_D^T & -\boldsymbol{c}_B^T\boldsymbol{B}^{-1}\boldsymbol{b} \end{bmatrix},$$

where $\boldsymbol{r}_D^T = \boldsymbol{c}_D^T - \boldsymbol{c}_B^T\boldsymbol{B}^{-1}\boldsymbol{D}$. In the proof of the duality theorem, we have shown that $\boldsymbol{\lambda}^T = \boldsymbol{c}_B^T\boldsymbol{B}^{-1}$ is an optimal solution to the dual. The vector $\boldsymbol{\lambda}$ can be obtained from the final tableau above. Specifically, if rank $\boldsymbol{D} = m$, then we can solve for $\boldsymbol{\lambda}$ using the vector $\boldsymbol{r}_D$, via the equation

$$\boldsymbol{\lambda}^T\boldsymbol{D} = \boldsymbol{c}_D^T - \boldsymbol{r}_D^T.$$

Of course, it may turn out that rank $\boldsymbol{D} < m$. In this case, as we shall now show, we have additional linear equations that allow us to solve for $\boldsymbol{\lambda}$. To this end, recall that $\lambda^T B = c_B^T$. Therefore, if we define $r^T = [0^T, r_D^T]$, then combining the equations $\lambda^T D = c_D^T - r_D^T$ and $\lambda^T B = c_B^T$ yields

$$\lambda^T A = c^T - r^T.$$

The vector $\lambda$ may be easy to obtain from the equation $\lambda^T D = c_D^T - r_D^T$ if $D$ takes certain special forms. In particular, this is the case if $D$ has an $m \times m$ identity matrix embedded in it, that is, by rearranging the positions of the columns of $D$, if necessary, $D$ has the form $D = [I_m, G]$, where $G$ is a $m \times (n - 2m)$ matrix. In this case, we can write the equation $\lambda^T D = c_D^T - r_D^T$ as

$$[\lambda^T, \lambda^T G] = [c_I^T, c_G^T] - [r_I^T, r_G^T].$$

Hence, $\lambda$ is given by

$$\lambda^T = c_I^T - r_I^T.$$

In other words, the solution to the dual is obtained by subtracting the reduced costs coefficients corresponding to the identity matrix in $D$ from the corresponding elements in the vector $c$ (that is, $c_I$).

For example, if we have a problem where we introduced slack variables, and the basic variables for the optimal basic feasible solution do not include any of the slack variables, then the matrix $D$ has an identity matrix embedded in it. In addition, in this case we have $c_I = 0$. Therefore, $\lambda = -r_I$ is an optimal solution to the dual.

***Example 17.4.*** In Example 17.2, the tableau for the primal in standard form is

| | $a_1$ | $a_2$ | $a_3$ | $a_4$ | $a_5$ | $a_6$ | $b$ |
|---|---|---|---|---|---|---|---|
| | 2 | $-1$ | 7 | 1 | 0 | 0 | 6 |
| | 1 | 3 | 4 | 0 | 1 | 0 | 9 |
| | 3 | 6 | 1 | 0 | 0 | 1 | 3 |
| $c^T$ | $-2$ | $-5$ | $-1$ | 0 | 0 | 0 | 0 |

If we now solve the problem using the simplex method, we get the following final simplex tableau:

| | | | | | | | |
|---|---|---|---|---|---|---|---|
| | $\frac{15}{43}$ | 0 | 1 | $\frac{6}{43}$ | 0 | $\frac{1}{43}$ | $\frac{39}{43}$ |
| | $-\frac{74}{43}$ | 0 | 0 | $-\frac{21}{43}$ | 1 | $-\frac{25}{43}$ | $\frac{186}{43}$ |
| | $\frac{19}{43}$ | 1 | 0 | $-\frac{1}{43}$ | 0 | $\frac{7}{43}$ | $\frac{15}{43}$ |
| $r^T$ | $\frac{24}{43}$ | 0 | 0 | $\frac{1}{43}$ | 0 | $\frac{36}{43}$ | $\frac{114}{43}$ |

We can now find the solution of the dual from the above simplex tableau using the equation $\boldsymbol{\lambda}^T \boldsymbol{D} = \boldsymbol{c}_D^T - \boldsymbol{r}_D^T$:

$$[\lambda_1, \lambda_2, \lambda_3] \begin{bmatrix} 2 & 1 & 0 \\ 1 & 0 & 0 \\ 3 & 0 & 1 \end{bmatrix} = [-2, 0, 0] - [\tfrac{24}{43}, \tfrac{1}{43}, \tfrac{36}{43}].$$

Solving the above, we get

$$\boldsymbol{\lambda}^T = [-\tfrac{1}{43}, 0, -\tfrac{36}{43}]$$

which agrees with our solution in Example 17.2. ■

We end this chapter by presenting the following theorem, which describes an alternative form of the relationship between the optimal solutions to the primal and dual problems.

**Theorem 17.3. Complementary Slackness Condition.** *The feasible solutions $\boldsymbol{x}$ and $\boldsymbol{\lambda}$ to a dual pair of problems (either in symmetric or asymmetric form) are optimal if, and only if,*

1. $(\boldsymbol{c}^T - \boldsymbol{\lambda}^T \boldsymbol{A})\boldsymbol{x} = 0$; *and*
2. $\boldsymbol{\lambda}^T(\boldsymbol{A}\boldsymbol{x} - \boldsymbol{b}) = 0$. □

*Proof.* We first prove the result for the asymmetric case. Note that condition 2 holds trivially for this case. Therefore, we only consider condition 1. $\Rightarrow$: If the two solutions are optimal, then by Theorem 17.2, $\boldsymbol{c}^T\boldsymbol{x} = \boldsymbol{\lambda}^T\boldsymbol{b}$. Since $\boldsymbol{A}\boldsymbol{x} = \boldsymbol{b}$, we also have $(\boldsymbol{c}^T - \boldsymbol{\lambda}^T\boldsymbol{A})\boldsymbol{x} = 0$. $\Leftarrow$: If $(\boldsymbol{c}^T - \boldsymbol{\lambda}^T\boldsymbol{A})\boldsymbol{x} = 0$, then $\boldsymbol{c}^T\boldsymbol{x} = \boldsymbol{\lambda}^T\boldsymbol{A}\boldsymbol{x} = \boldsymbol{\lambda}^T\boldsymbol{b}$. Therefore, by Theorem 17.1, $\boldsymbol{x}$ and $\boldsymbol{\lambda}$ are optimal.

We now prove the result for the symmetric case. $\Rightarrow$: We first show condition 1. If the two solutions are optimal, then by Theorem 17.2, $\boldsymbol{c}^T\boldsymbol{x} = \boldsymbol{\lambda}^T\boldsymbol{b}$. Since $\boldsymbol{A}\boldsymbol{x} \geqslant \boldsymbol{b}$ and $\boldsymbol{\lambda} \geqslant \boldsymbol{0}$, we have $(\boldsymbol{c}^T - \boldsymbol{\lambda}^T\boldsymbol{A})\boldsymbol{x} \leqslant 0$. On the other hand, since $\boldsymbol{\lambda}^T\boldsymbol{A} \leqslant \boldsymbol{c}^T$ and $\boldsymbol{x} \geqslant \boldsymbol{0}$, we have $(\boldsymbol{c}^T - \boldsymbol{\lambda}^T\boldsymbol{A})\boldsymbol{x} \geqslant 0$. Hence, $(\boldsymbol{c}^T - \boldsymbol{\lambda}^T\boldsymbol{A})\boldsymbol{x} = 0$. To show condition 2, note that since $\boldsymbol{A}\boldsymbol{x} \geqslant \boldsymbol{b}$ and $\boldsymbol{\lambda} \geqslant \boldsymbol{0}$, we have $\boldsymbol{\lambda}^T(\boldsymbol{A}\boldsymbol{x} - \boldsymbol{b}) \geqslant 0$. On the other hand, since $\boldsymbol{\lambda}^T\boldsymbol{A} \leqslant \boldsymbol{c}^T$ and $\boldsymbol{x} \geqslant \boldsymbol{0}$, we have $\boldsymbol{\lambda}^T(\boldsymbol{A}\boldsymbol{x} - \boldsymbol{b}) \leqslant 0$. $\Leftarrow$: Combining conditions 1 and 2, we get $\boldsymbol{c}^T\boldsymbol{x} = \boldsymbol{\lambda}^T\boldsymbol{A}\boldsymbol{x} = \boldsymbol{\lambda}^T\boldsymbol{b}$. Hence, by Theorem 17.1, $\boldsymbol{x}$ and $\boldsymbol{\lambda}$ are optimal. ■

Note that if $\boldsymbol{x}$ and $\boldsymbol{\lambda}$ are feasible solutions for the dual pair of problems, we can write condition 1, that is, $(\boldsymbol{c}^T - \boldsymbol{\lambda}^T\boldsymbol{A})\boldsymbol{x} = 0$, as $x_i > 0$ implies $\boldsymbol{\lambda}^T\boldsymbol{a}_i = c_i$, $i = 1, \ldots, n$; that is, for any component of $\boldsymbol{x}$ that is positive, the corresponding constraint for the dual must be an equality at $\boldsymbol{\lambda}$. Also, observe that the statement $x_i > 0$ implies $\boldsymbol{\lambda}^T\boldsymbol{a}_i = c_i$ is equivalent to $\boldsymbol{\lambda}^T\boldsymbol{a}_i < c_i$ implies $x_i = 0$. A similar representation can be written for condition 2.

Consider the asymmetric form of duality. Recall that for the case of an optimal basic feasible solution $\boldsymbol{x}$, $\boldsymbol{r}^T = \boldsymbol{c}^T - \boldsymbol{\lambda}^T\boldsymbol{A}$ is the vector of reduced cost coefficients.

Therefore, in this case, the complementary slackness condition can be written as $\boldsymbol{r}^T\boldsymbol{x} = 0$.

## EXERCISES

**17.1** Prove the weak duality lemma for the symmetric form of duality.

**17.2** Consider the following linear program:

$$\begin{aligned} \text{maximize} \quad & 2x_1 + 3x_2 \\ \text{subject to} \quad & x_1 + 2x_2 \leqslant 4 \\ & 2x_1 + x_2 \leqslant 5 \\ & x_1, x_2 \geqslant 0 \end{aligned}$$

**a.** Use the simplex method to solve the above problem.
**b.** Write down the dual of the above linear program, and solve the dual.

**17.3** Consider the linear program

$$\begin{aligned} \text{minimize} \quad & 4x_1 + 3x_2 \\ \text{subject to} \quad & 5x_1 + x_2 \geqslant 11 \\ & 2x_1 + x_2 \geqslant 8 \\ & x_1 + 2x_2 \geqslant 7 \\ & x_1, x_2 \geqslant 0 \end{aligned}$$

Write down the corresponding dual problem, and find the solution to the dual. (Compare the above problem with the one in Exercise 16.4a.)

**17.4** Consider the linear program

$$\begin{aligned} \text{minimize} \quad & x_1 + \cdots + x_n, \quad x_1, \ldots, x_n \in \mathbb{R} \\ \text{subject to} \quad & a_1 x_1 + \cdots + a_n x_n = 1 \\ & x_1, \ldots, x_n \geqslant 0, \end{aligned}$$

where $0 < a_1 < a_2 < \cdots < a_n$.

**a.** Write down the dual to the above problem, and find a solution to the dual in terms of $a_1, \ldots, a_n$.
**b.** State the duality theorem, and use it to find a solution to the primal problem above.

**c.** Suppose we apply the simplex algorithm to the primal problem. Show that if we start at a nonoptimal initial basic feasible solution, the algorithm terminates in one step if, and only if, we use the rule where the next nonbasic column to enter the basis is the one with the most negative relative cost coefficient.

**17.5** Consider an LP problem in standard form. Suppose $\boldsymbol{x}$ is a feasible solution to the problem. Show that if there exist $\boldsymbol{\lambda}$ and $\boldsymbol{\mu}$ such that

$$\begin{aligned} \boldsymbol{A}^T\boldsymbol{\lambda} + \boldsymbol{\mu} &= \boldsymbol{c} \\ \boldsymbol{\mu}^T\boldsymbol{x} &= 0 \\ \boldsymbol{\mu} &\geqslant \boldsymbol{0}, \end{aligned}$$

then $\boldsymbol{x}$ is an optimal feasible solution to the LP problem, and $\boldsymbol{\lambda}$ is an optimal feasible solution to the dual. The above are called the *Karush–Kuhn–Tucker optimality conditions for LP*, which are discussed in detail in Chapters 20 and 21.

**17.6** Consider the linear program:

$$\begin{aligned} &\text{maximize} \quad \boldsymbol{c}^T\boldsymbol{x}, \\ &\text{subject to} \quad \boldsymbol{A}\boldsymbol{x} \leqslant \boldsymbol{b}, \end{aligned}$$

where $\boldsymbol{c} \in \mathbb{R}^n$, $\boldsymbol{b} \in \mathbb{R}^m$, and $\boldsymbol{A} \in \mathbb{R}^{m \times n}$. Use the symmetric form of duality to derive the dual of this linear program, and show that the constraint in the dual involving $\boldsymbol{A}$ can be written as an equality constraint. *Hint*: Write $\boldsymbol{x} = \boldsymbol{u} - \boldsymbol{v}$, with $\boldsymbol{u}, \boldsymbol{v} \geqslant \boldsymbol{0}$.

**17.7** Consider the linear program:

$$\begin{aligned} &\text{minimize} \quad x_1 + x_2 \\ &\text{subject to} \quad x_1 + 2x_2 \geqslant 3 \\ &\qquad\qquad 2x_1 + x_2 \geqslant 3 \\ &\qquad\qquad x_1, x_2 \geqslant 0. \end{aligned}$$

The solution to the problem is $[1, 1]^T$ (see Exercise 16.3). Write down the dual to the above problem, solve the dual, and verify that the duality theorem holds.

**17.8** Consider the problem

$$\begin{aligned} &\text{minimize} \quad \boldsymbol{c}^T\boldsymbol{x}, \quad \boldsymbol{x} \in \mathbb{R}^n \\ &\text{subject to} \quad \boldsymbol{x} \geqslant \boldsymbol{0}. \end{aligned}$$

For this problem we have the following theorem.

**Theorem.** *A solution to the above problem exists if, and only if,* $\mathbf{c} \geqslant \mathbf{0}$. *Moreover, if a solution exists,* $\mathbf{0}$ *is a solution.* □

Use the duality theorem to prove the above theorem (see also Exercise 21.8).

**17.9** Let $\boldsymbol{A}$ be a given matrix and $\boldsymbol{b}$ a given vector. Show that there exists a vector $\boldsymbol{x}$ such that $\boldsymbol{Ax} \geqslant \boldsymbol{b}$ and $\boldsymbol{x} \geqslant \mathbf{0}$ if, and only if, for any given vector $\boldsymbol{y}$ satisfying $\boldsymbol{A}^T\boldsymbol{y} \leqslant \mathbf{0}$ and $\boldsymbol{y} \geqslant \mathbf{0}$, we have $\boldsymbol{b}^T\boldsymbol{y} \leqslant 0$.

**17.10** Let $\boldsymbol{A}$ be a given matrix and $\boldsymbol{b}$ a given vector. Show that there exists a vector $\boldsymbol{x}$ such that $\boldsymbol{Ax} = \boldsymbol{b}$ and $\boldsymbol{x} \geqslant \mathbf{0}$ if, and only if, for any given vector $\boldsymbol{y}$ satisfying $\boldsymbol{A}^T\boldsymbol{y} \leqslant \mathbf{0}$, we have $\boldsymbol{b}^T\boldsymbol{y} \leqslant 0$. This result is known as *Farkas's transposition theorem.*

**17.11** Let $\boldsymbol{A}$ be a given matrix and $\boldsymbol{b}$ a given vector. Show that there exists a vector $\boldsymbol{x}$ such that $\boldsymbol{Ax} \leqslant \boldsymbol{b}$ if, and only if, for any given vector $\boldsymbol{y}$ satisfying $\boldsymbol{A}^T\boldsymbol{y} = \mathbf{0}$ and $\boldsymbol{y} \geqslant \mathbf{0}$, we have $\boldsymbol{b}^T\boldsymbol{y} \geqslant 0$. This result is known as *Gale's transposition theorem.*

**17.12** Let $\boldsymbol{A}$ be a given matrix and $\boldsymbol{b}$ a given vector. Show that there exists a vector $\boldsymbol{x}$ such that $\boldsymbol{Ax} < \mathbf{0}$ if, and only if, for any given vector $\boldsymbol{y}$ satisfying $\boldsymbol{A}^T\boldsymbol{y} = \mathbf{0}$ and $\boldsymbol{y} \geqslant \mathbf{0}$, we have $\boldsymbol{y} = \mathbf{0}$ (i.e., $\boldsymbol{y} = \mathbf{0}$ is the only vector satisfying $\boldsymbol{A}^T\boldsymbol{y} = \mathbf{0}$ and $\boldsymbol{y} \geqslant \mathbf{0}$). This result is known as *Gordan's transposition theorem.*

# 18

# Non-Simplex Methods

## 18.1. INTRODUCTION

In the previous chapters, we studied the simplex method, and its variant, the revised simplex method, for solving linear programming (LP) problems. The method remains widely used in practice for solving LP problems. However, the amount of time required to compute the solution using the simplex method grows rapidly as the number of components $n$ of the variable $\boldsymbol{x} \in \mathbb{R}^n$ increases. Specifically, it turns out that the relationship between the required amount of time for the algorithm to find a solution and the size $n$ of $\boldsymbol{x}$ is exponential in the worst case. An example of an LP problem for which this relationship is evident was devised by Klee and Minty in 1972 [Ref. 45]. Below, we give a version of the Klee–Minty example [from Ref. 7]. Let $n$ be given. Let $\boldsymbol{c}^T = [10^{n-1}, 10^{n-2}, \ldots, 10^1, 1]$, $\boldsymbol{b} = [1, 10^2, 10^4, \ldots, 10^{2(n-1)}]^T$, and

$$\boldsymbol{A} = \begin{bmatrix} 1 & 0 & 0 & \cdots & 0 \\ 2 \times 10^1 & 1 & 0 & \cdots & 0 \\ 2 \times 10^2 & 2 \times 10^1 & 1 & \cdots & 0 \\ \vdots & \vdots & \ddots & \ddots & \vdots \\ 2 \times 10^{n-1} & 2 \times 10^{n-2} & \cdots & 2 \times 10^1 & 1 \end{bmatrix}.$$

Consider the following LP problem:

$$\begin{aligned} &\text{maximize} && \boldsymbol{c}^T\boldsymbol{x} \\ &\text{subject to} && \boldsymbol{A}\boldsymbol{x} \leqslant \boldsymbol{b} \\ &&& \boldsymbol{x} \geqslant \boldsymbol{0}. \end{aligned}$$

The simplex algorithm applied to the above LP problem requires $2^n - 1$ steps to find the solution. Clearly, in this example the relationship between the required amount of time for the simplex algorithm to find a solution and the size $n$ of the variable $\boldsymbol{x}$ is exponential. This relationship is also called the *complexity* of the algorithm. The simplex algorithm is therefore said to have *exponential complexity*. The complexity of the simplex algorithm is also often written as $\mathcal{O}(2^n - 1)$.

Naturally, we would expect that any algorithm that solves LP problems would have the property that the time required to arrive at a solution increases with the size $n$ of the variable $\boldsymbol{x}$. However, the issue at hand is the rate at which this increase occurs. As we have seen above, the simplex algorithm has the property that this rate of increase is exponential. For a number of years, computer scientists have distinguished between *exponential complexity* and *polynomial complexity*. If an algorithm for solving LP problems has polynomial complexity, then the time required to obtain the solution is bounded by a polynomial in $n$. Obviously, polynomial complexity is more desirable than exponential complexity. Therefore, the existence of an algorithm for solving LP problems with polynomial complexity is an important issue. This issue was partially resolved in 1979 by Khachiyan (also transliterated as Hačijan) [Ref. 44], who proposed an algorithm that has a complexity $\mathcal{O}(n^4 L)$, where, roughly speaking, $L$ represents the number of bits used in the computations. The reason that we consider Khachiyan's algorithm (also called the *ellipsoid algorithm*) as only a partial resolution of the above issue is that the complexity depends on $L$, which implies that the time required to solve a given LP problem increases with the required accuracy of the computations. The existence of a method for solving LP problems with a polynomial complexity bound based only on the size of the variable $n$ (and possibly the number of constraints) remains a difficult open problem [Ref. 30]. In any case, computational experience with Khachiyan's algorithm has shown that it is not a practical alternative to the simplex method [Ref. 10]. The theoretical complexity advantage of Khachiyan's method relative to the simplex method remains to be demonstrated in practice.

Another non-simplex algorithm for solving LP problems was proposed in 1984 by Karmarkar [Ref. 42]. Karmarkar's algorithm has a complexity of $\mathcal{O}(n^{3.5} L)$, which is lower than that of Khachiyan's algorithm. It is reported that well-implemented versions of this algorithm are very efficient, especially when the problem involves over a few thousand variables [Ref. 30]. The algorithm is superior to the simplex algorithm from a complexity viewpoint, and is also potentially more efficient for large practical problems.

This chapter is devoted to a discussion of non-simplex methods for solving LP problems. In the next section, we discuss some ideas underlying Khachiyan's algorithm. We then present Karmarkar's algorithm in the section to follow.

## 18.2. KHACHIYAN'S ALGORITHM

Our description of the Khachiyan's algorithm is based on Refs. 6 and 7. The method relies on the concept of duality (Chapter 17). Our exposition of Khachiyan's algorithm is geared toward a basic understanding of the method. For a detailed rigorous treatment of the method, we refer the reader to Ref. 57.

Consider the linear programming problem:

$$\begin{aligned} \text{minimize} \quad & \boldsymbol{c}^T\boldsymbol{x} \\ \text{subject to} \quad & \boldsymbol{A}\boldsymbol{x} \geqslant \boldsymbol{b} \\ & \boldsymbol{x} \geqslant \boldsymbol{0}. \end{aligned}$$

We write the corresponding dual problem as follows:

$$\begin{aligned} \text{maximize} \quad & \boldsymbol{\lambda}^T\boldsymbol{b} \\ \text{subject to} \quad & \boldsymbol{\lambda}^T\boldsymbol{A} \leqslant \boldsymbol{c}^T \\ & \boldsymbol{\lambda} \geqslant \boldsymbol{0}. \end{aligned}$$

Recall that the above two LP problems constitute the symmetric form of duality. From Theorem 17.1, if $\boldsymbol{x}$ and $\boldsymbol{\lambda}$ are feasible solutions to the primal and dual problems, respectively, and $\boldsymbol{c}^T\boldsymbol{x} = \boldsymbol{\lambda}^T\boldsymbol{b}$, then $\boldsymbol{x}$ and $\boldsymbol{\lambda}$ are optimal solutions to their respective problems. Using this result, we see that to solve the primal problem it is enough to find a vector $[\boldsymbol{x}^T, \boldsymbol{\lambda}^T]^T$ that satisfies the following set of relations:

$$\begin{aligned} \boldsymbol{c}^T\boldsymbol{x} &= \boldsymbol{b}^T\boldsymbol{\lambda} \\ \boldsymbol{A}\boldsymbol{x} &\geqslant \boldsymbol{b} \\ \boldsymbol{A}^T\boldsymbol{\lambda} &\leqslant \boldsymbol{c} \\ x &\geqslant \boldsymbol{0} \\ \boldsymbol{\lambda} &\geqslant \boldsymbol{0}. \end{aligned}$$

Note that the equality $\boldsymbol{c}^T\boldsymbol{x} = \boldsymbol{b}^T\boldsymbol{\lambda}$ is equivalent to the two inequalitites

$$\begin{aligned} \boldsymbol{c}^T\boldsymbol{x} - \boldsymbol{b}^T\boldsymbol{\lambda} &\leqslant 0 \\ -\boldsymbol{c}^T\boldsymbol{x} + \boldsymbol{b}^T\boldsymbol{\lambda} &\leqslant 0. \end{aligned}$$

Taking this into account, we can represent the previous set of relations as

$$\begin{bmatrix} \boldsymbol{c}^T & -\boldsymbol{b}^T \\ -\boldsymbol{c}^T & \boldsymbol{b}^T \\ -\boldsymbol{A} & \boldsymbol{0} \\ -\boldsymbol{I}_n & \boldsymbol{0} \\ \boldsymbol{0} & \boldsymbol{A}^T \\ \boldsymbol{0} & -\boldsymbol{I}_m \end{bmatrix} \begin{bmatrix} x \\ \lambda \end{bmatrix} \leqslant \begin{bmatrix} 0 \\ 0 \\ -\boldsymbol{b} \\ \boldsymbol{0} \\ c \\ \boldsymbol{0} \end{bmatrix}.$$

Therefore, we have reduced the problem of finding an optimal solution to the primal-dual pair into one of finding a vector $[\boldsymbol{x}^T, \boldsymbol{\lambda}^T]^T$ that satisfies the above system of inequalities. In other words, if we can find a vector that satisfies the above system of inequalities, then this vector gives an optimal solution to the primal-dual pair. On the other hand, if there does not exist a vector satisfying the above system of inequalities, then the primal-dual pair has no optimal feasible solution. In the subsequent discussion, we will simply represent the system of inequalities as

$$\boldsymbol{P}\boldsymbol{z} \leqslant \boldsymbol{q},$$

where

$$\boldsymbol{P} = \begin{bmatrix} \boldsymbol{c}^T & -\boldsymbol{b}^T \\ -\boldsymbol{c}^T & \boldsymbol{b}^T \\ -\boldsymbol{A} & \boldsymbol{0} \\ -\boldsymbol{I}_n & \boldsymbol{0} \\ \boldsymbol{0} & \boldsymbol{A}^T \\ \boldsymbol{0} & -\boldsymbol{I}_m \end{bmatrix}, \quad \boldsymbol{z} = \begin{bmatrix} \boldsymbol{x} \\ \boldsymbol{\lambda} \end{bmatrix}, \quad \boldsymbol{q} = \begin{bmatrix} 0 \\ 0 \\ -\boldsymbol{b} \\ \boldsymbol{0} \\ \boldsymbol{c} \\ \boldsymbol{0} \end{bmatrix}.$$

In our discussion of Khachiyan's algorithm, we will not be specifically using the above forms of $\boldsymbol{P}$, $\boldsymbol{q}$, and $\boldsymbol{z}$; we simply treat $\boldsymbol{P}\boldsymbol{z} \leqslant \boldsymbol{q}$ as a generic matrix inequality, with $\boldsymbol{P}$, $\boldsymbol{q}$, and $\boldsymbol{z}$ as generic entities. Let $r$ and $s$ be the sizes of $\boldsymbol{q}$ and $\boldsymbol{z}$, respectively; that is, $\boldsymbol{P} \in \mathbb{R}^{r \times s}$, $\boldsymbol{z} \in \mathbb{R}^s$, and $\boldsymbol{q} \in \mathbb{R}^r$.

Khachiyan's method solves the LP problem by first determining if there exists a vector $\boldsymbol{z}$ that satisfies the above inequality $\boldsymbol{P}\boldsymbol{z} \leqslant \boldsymbol{q}$ (that is, the algorithm decides if the above system of linear inequalities is *consistent*). If the system of inequalities is consistent, then the algorithm finds a vector $\boldsymbol{z}$ satisfying the system. In the following, we refer to any vector satisfying the above system of inequalities as a solution to the system. We assume that the entries in $\boldsymbol{P}$ and $\boldsymbol{q}$ are all rational numbers. This is not a restriction in practice, since any representation of our LP problem on a digital computer will involve only rational numbers. In fact, we further assume that the entries in $\boldsymbol{P}$ and $\boldsymbol{q}$ are all integers. We can do this without loss of generality since we can always multiply both sides of the inequality $\boldsymbol{P}\boldsymbol{z} \leqslant \boldsymbol{q}$ by a sufficiently large number to get only integer entries on both sides.

Before discussing Khachiyan's algorithm, we first introduce the idea of an *ellipsoid*. To this end, let $\boldsymbol{z} \in \mathbb{R}^s$ be a given vector, and let $\boldsymbol{Q}$ be an $s \times s$ nonsingular matrix. Then, the ellipsoid associated with $\boldsymbol{Q}$ centered at $\boldsymbol{z}$ is defined as the set

$$E_{\boldsymbol{Q}}(\boldsymbol{z}) = \{\boldsymbol{z} + \boldsymbol{Q}\boldsymbol{y} : \boldsymbol{y} \in \mathbb{R}^s, \|\boldsymbol{y}\| \leqslant 1\}.$$

The main idea underlying Khachiyan's algorithm is as follows. Khachiyan's algorithm is an iterative procedure, where we update at each iteration a vector $\boldsymbol{z}^{(k)}$ and a matrix $\boldsymbol{Q}_k$. Associated with $\boldsymbol{z}^{(k)}$ and $\boldsymbol{Q}_k$ is an ellipsoid $E_{\boldsymbol{Q}_k}(\boldsymbol{z}^{(k)})$. At each step of the algorithm, the associated ellipsoid contains a solution to the given system of linear inequalities. The algorithm updates $\boldsymbol{z}^{(k)}$ and $\boldsymbol{Q}_k$ in such a way that the ellipsoid at the next step is smaller than that of the current step, but at the same time is guaranteed to contain a solution to the given system of inequalities, if one exists. If we find that the current point $z^{(k)}$ satisfies $Pz^{(k)} \leqslant q$, then we terminate the algorithm and conclude that $z^{(k)}$ is a solution. Otherwise, we continue to iterate. The algorithm has a fixed prespecified maximum number of iterations $N$ to be performed, where $N$ is a number that depends on $L$ and $s$. Note that we are not free to choose $N$—it is computed using a formula that uses the values of $L$ and $s$. The constant $L$ is itself a quantity that we have to compute beforehand, using a formula that involves $P$ and $q$. When we have completed $N$ iterations without finding a solution in an earlier step, we terminate the algorithm. The associated ellipsoid will then have shrunk to the extent that it is smaller than the precision of computation. At this stage, we will either discover a solution inside the ellipsoid, if indeed a solution exists, or we will find no solution inside the ellipsoid, in which case we conclude that no solution exists.

As we can see from the above description, Khachiyan's approach is a radical departure from the classical simplex method for solving LP problems. The method has attracted a lot of attentiion, and many studies have been devoted to it. However, as we pointed out earlier, the algorithm is of little practical value for solving real-world LP problems. Therefore, we will not delve any further into the details of Khachiyan's algorithm. We refer the interested reader to Ref. 57.

Despite its practical drawbacks, Khachiyan's method has inspired other researchers to pursue the development of computationally efficient algorithms for solving LP problems with polynomial complexity. One such algorithm is due to Karmarkar, which we discus in the next section.

## 18.3. KARMARKAR'S ALGORITHM

### 18.3.1. Basic Ideas

Karmarkar's method for solving LP problems differs fundamentally from the classical simplex method in various respects. First, Karmarkar's method starts inside the feasible set and moves within it toward an optimal vertex. For this reason, we say that Karmarkar's algorithm is an *interior point method.* In contrast, the simplex method jumps from vertex to vertex of the feasible set seeking an optimal vertex. Another difference betwen the two methods is that the simplex method stops when it finds the absolute best solution. On the other hand, Karmarkar's method stops when it finds a solution that has an objective function value that is less than or equal to a prespecified fraction of the original guess. A third difference between the two methods is that the simplex method starts with LP problems in standard form, whereas Karmarkar's method starts with LP problems in a special canonical form, which we call *Karmarkar's canonical form.* We discuss this canonical form in Section 18.3.2. For more details on Karmarkar's algorithm, see Refs. 21, 30, 42, and 72. For an informal introduction to the algorithm, see Ref. 63.

### 18.3.2. Karmarkar's Canonical Form

To apply Karmarkar's algorithm to a given LP problem, we must first transform the given problem into a particular form, which we refer to as Karmarkar's canonical form. Karmarkar's canonical form is written as:

$$\begin{aligned} \text{minimize} \quad & \boldsymbol{c}^T \boldsymbol{x} \\ \text{subject to} \quad & \boldsymbol{A}\boldsymbol{x} = \boldsymbol{0} \\ & \sum_{i=1}^{n} x_i = 1 \\ & \boldsymbol{x} \geqslant \boldsymbol{0}, \end{aligned}$$

where $\boldsymbol{x} = [x_1, \ldots, x_n]^T$. As in our discussion of Khachiyan's method, we assume without loss of generality that the entries of $\boldsymbol{A}$ and $\boldsymbol{c}$ are integers.

We now introduce some notation that allows convenient manipulation of the canonical form. First, let $\boldsymbol{e} = [1, \ldots, 1]^T$ be the vector in $\mathbb{R}^n$ with each component equal to 1. Let $\Omega$ denote the nullspace of $\boldsymbol{A}$; that is, the subspace

$$\Omega = \{\boldsymbol{x} \in \mathbb{R}^n : \boldsymbol{A}\boldsymbol{x} = \boldsymbol{0}\}.$$

Define the simplex in $\mathbb{R}^n$ by

$$\Delta = \{\boldsymbol{x} \in \mathbb{R}^n : \boldsymbol{e}^T \boldsymbol{x} = 1, \boldsymbol{x} \geqslant \boldsymbol{0}\}.$$

We denote the center of the simplex $\Delta$ by

$$\boldsymbol{a}_0 = \boldsymbol{e}/n = [1/n, \ldots, 1/n]^T.$$

Clearly $\boldsymbol{a}_0 \in \Delta$. With the above notation, Karmarkar's canonical form can be rewritten as

$$\begin{aligned} \text{minimize} \quad & \boldsymbol{c}^T \boldsymbol{x} \\ \text{subject to} \quad & \boldsymbol{x} \in \Omega \cap \Delta. \end{aligned}$$

Note that the constraint set (or feasible set) $\Omega \cap \Delta$ can be represented as

$$\begin{aligned} \Omega \cap \Delta &= \{\boldsymbol{x} \in \mathbb{R}^n : \boldsymbol{A}\boldsymbol{x} = \boldsymbol{0}, \boldsymbol{e}^T \boldsymbol{x} = 1, \boldsymbol{x} \geqslant \boldsymbol{0}\} \\ &= \left\{ \boldsymbol{x} \in \mathbb{R}^n : \begin{bmatrix} \boldsymbol{A} \\ \boldsymbol{e}^T \end{bmatrix} \boldsymbol{x} = \begin{bmatrix} \boldsymbol{0} \\ 1 \end{bmatrix}, \boldsymbol{x} \geqslant \boldsymbol{0} \right\}. \end{aligned}$$

***Example 18.1.*** Consider the following LP problem, taken from Ref. 73:

$$\begin{aligned} \text{minimize} \quad & 5x_1 + 4x_2 + 8x_3 \\ \text{subject to} \quad & x_1 + x_2 + x_3 = 1 \\ & x_1, x_2, x_3 \geqslant 0. \end{aligned}$$

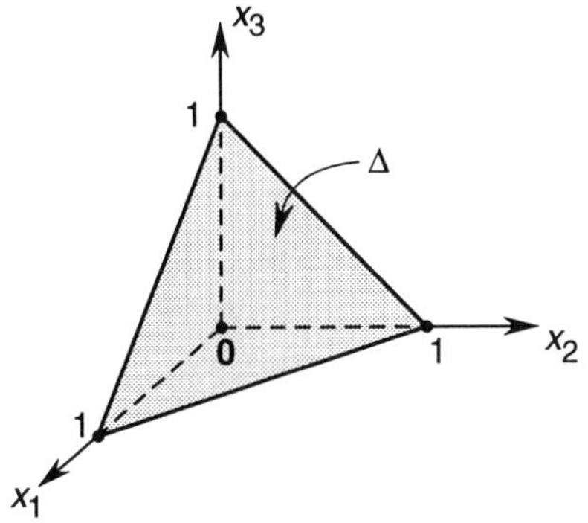


**Figure 18.1.** The feasible set for Example 18.1

Clearly the above problem is already in Karmarkar's canonical form, with $c^T = [5, 4, 8]$, and $A = O$. The feasible set for this example is illustrated in Figure 18.1. ■

***Example 18.2.*** Consider the following LP problem, taken from Ref. 63:

$$\begin{aligned} \text{minimize} \quad & 3x_1 + 3x_2 - x_3 \\ \text{subject to} \quad & 2x_1 - 3x_2 + x_3 = 0 \\ & x_1 + x_2 + x_3 = 1 \\ & x_1, x_2, x_3 \geqslant 0. \end{aligned}$$

The above problem is in Karmarkar's canonical form, with $c^T = [3, 3, -1]$, and $A = [2, -3, 1]$. The feasible set for this example is illustrated in Figure 18.2 [adapted from Ref. 63]. ■

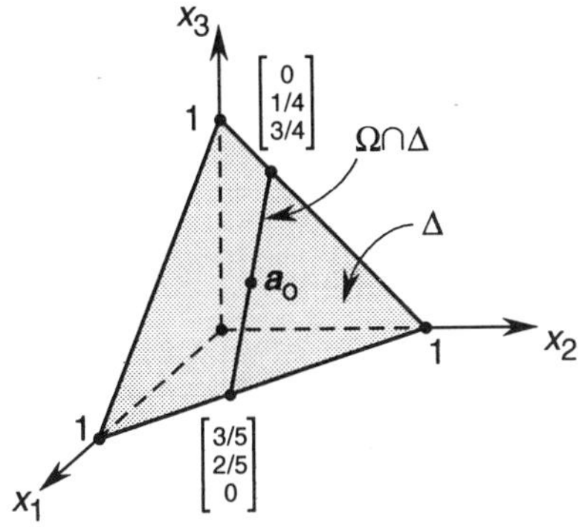


**Figure 18.2.** The feasible set for Example 18.2

We will show later that any LP problem can be converted into an equivalent problem in Karmarkar's canonical form.

### 18.3.3. Karmarkar's Restricted Problem

Karmarkar's algorithm solves LP problems in Karmarkar's canonical form, with the following assumptions:

**A.** The center $\boldsymbol{a}_0$ of the simplex $\Delta$ is a feasible point (that is, $\boldsymbol{a}_0 \in \Omega$).

**B.** The minimum value of the objective function over the feasible set is zero.

**C.** The $(m+1) \times n$ matrix

$$\begin{bmatrix} A \\ \boldsymbol{e}^T \end{bmatrix}$$

has rank $m+1$.

**D.** We are given a termination parameter $q > 0$, such that if we obtain a feasible point $\boldsymbol{x}$ satisfying

$$\frac{\boldsymbol{c}^T \boldsymbol{x}}{\boldsymbol{c}^T \boldsymbol{a}_0} \leqslant 2^{-q},$$

then we consider the problem solved.

Any LP problem that is in Karmarkar's canonical form and that also satisfies the above four assumptions is called a *Karmarkar's restricted problem.* In the following, we discuss the above assumptions and their interpretations.

We begin by looking at assumption A. We point out that this assumption is not restrictive, since any LP problem that has an optimal feasible solution can be converted into a problem in Karmarkar's canonical form that satisfies assumption A. We discuss this in the next subsection.

We next turn our attention to assumption B. Any LP problem in Karmarkar's canonical form can be converted into one that satisfies assumption B, provided we know beforehand the minimum value of its objective function over the feasible set. Specifically, suppose we are given an LP problem where the minimum value of the objective function is $M$. As in Ref. 63, consider the function $f(\boldsymbol{x}) = \boldsymbol{c}^T \boldsymbol{x} - M$. Then, using the property that $\boldsymbol{e}^T \boldsymbol{x} = 1$ on the feasible set, we have that, for any feasible $\boldsymbol{x}$,

$$f(\boldsymbol{x}) = \boldsymbol{c}^T \boldsymbol{x} - M = \boldsymbol{c}^T \boldsymbol{x} - M \boldsymbol{e}^T \boldsymbol{x} = (\boldsymbol{c}^T - M \boldsymbol{e}^T)\boldsymbol{x} = \tilde{\boldsymbol{c}}^T \boldsymbol{x},$$

where $\tilde{\boldsymbol{c}}^T = \boldsymbol{c}^T - M \boldsymbol{e}^T$. Notice that the above objective function has a minimum value of zero, and is a linear function of $\boldsymbol{x}$. We can replace the original objective function with the new objective function above, without altering the solution.

***Example 18.3.*** Recall the LP problem in Example 18.1:

$$\begin{aligned} \text{minimize} \quad & 5x_1 + 4x_2 + 8x_3 \\ \text{subject to} \quad & x_1 + x_2 + x_3 = 1 \\ & x_1, x_2, x_3 \geqslant 0. \end{aligned}$$

The problem satisfies assumption A (and assumption C), but not assumption B, because the minimum value of the objective function over the feasible set is 4. To convert the above into a problem that satisfies assumption B, we replace $\boldsymbol{c}^T = [5, 4, 8]$ by $\tilde{\boldsymbol{c}}^T = [1, 0, 4]$. ■

***Example 18.4.*** The reader can easily verify that the LP problem in Example 18.2 satisfies assumptions A, B, and C. ■

Assumption C is a technical assumption that is required in the implementation of the algorithm. Its significance will be clear when we discuss the update equation in Karmarkar's algorithm.

Assumption D is the basis for the stopping criterion of Karmarkar's algorithm. In particular, we stop when we have found a feasible point satisfying $\boldsymbol{c}^T\boldsymbol{x}/\boldsymbol{c}^T\boldsymbol{a}_0 \leqslant 2^{-q}$. Such a stopping criterion is inherent in any algorithm that uses finite precision arithmetic. Observe that the above stopping criterion depends on the value of $\boldsymbol{c}^T\boldsymbol{a}_0$. It will turn out that Karmarkar's algorithm uses $\boldsymbol{a}_0$ as the starting point. Therefore, we can see that the accuracy of the final solution in the algorithm is influenced by the starting point.

### 18.3.4. From General Form to Karmarkar's Canonical Form

We now show how any LP problem can be coverted into an equivalent problem in Karmarkar's canonical form. By *equivalent*, we mean that the solution to one can be used to determine the solution to the other, and vice versa. To this end, recall that any LP problem can be transformed into an equivalent problem in standard form. Therefore, it suffices to show that any LP problem in standard form can be transformed into an equivalent problem in Karmarkar's canonical form. In fact, the transformation given below [taken from Ref. 42] will also guarantee that assumption A of the previous subsection is satisfied.

To proceed, consider a given LP problem in standard form:

$$\begin{aligned} \text{minimize} \quad & \boldsymbol{c}^T\boldsymbol{x}, \quad \boldsymbol{x} \in \mathbb{R}^n \\ \text{subject to} \quad & \boldsymbol{A}\boldsymbol{x} = \boldsymbol{b} \\ & \boldsymbol{x} \geqslant \boldsymbol{0}. \end{aligned}$$

We assume that we are given a point $\boldsymbol{a} = [a_1, \ldots, a_n]^T$ that is a *strictly interior* feasible point; that is $\boldsymbol{A}\boldsymbol{a} = \boldsymbol{b}$ and $\boldsymbol{a} > \boldsymbol{0}$. We show later how this assumption can be enforced. Let $P_+$ denote the *positive orthant* of $\mathbb{R}^n$, given by $P_+ = \{\boldsymbol{x}\in\mathbb{R}^n : \boldsymbol{x} \geqslant \boldsymbol{0}\}$. Let $\Delta = \{\boldsymbol{x}\in\mathbb{R}^{n+1} : \boldsymbol{e}^T\boldsymbol{x} = 1,\ \boldsymbol{x} \geqslant \boldsymbol{0}\}$ be the simplex in $\mathbb{R}^{n+1}$. Define the map $\boldsymbol{T}: P_+ \to \Delta$ by

$$\boldsymbol{T}(\boldsymbol{x}) = [T_1(\boldsymbol{x}), \ldots, T_{n+1}(\boldsymbol{x})]^T$$

with

$$T_i(\boldsymbol{x}) = \frac{x_i/a_i}{x_1/a_1 + \cdots + x_n/a_n + 1}, \quad i = 1, \ldots, n$$

$$T_{n+1}(\boldsymbol{x}) = \frac{1}{x_1/a_1 + \cdots + x_n/a_n + 1}.$$

We call the map $\boldsymbol{T}$ a *projective transformation* of the positive orthant $P_+$ into the simplex $\Delta$ [for an introduction to projective transformations, see Ref. 39]. The transformation $\boldsymbol{T}$ has several interesting properties (see Exercises 18.1–18.3). In particular, we can find a vector $\boldsymbol{c}'\in\mathbb{R}^{n+1}$ and a matrix $A'\in\mathbb{R}^{m\times(n+1)}$ such that for each $\boldsymbol{x}\in\mathbb{R}^n$,

$$\boldsymbol{c}^T\boldsymbol{x} = 0 \quad \Leftrightarrow \quad \boldsymbol{c}'^T\boldsymbol{T}(\boldsymbol{x}) = 0,$$

and

$$\boldsymbol{A}\boldsymbol{x} = \boldsymbol{b} \quad \Leftrightarrow \quad \boldsymbol{A}'\boldsymbol{T}(\boldsymbol{x}) = \boldsymbol{0}$$

(see Exercises 18.2 and 18.3 for the forms of $\boldsymbol{A}'$ and $\boldsymbol{c}'$). Note that for each $\boldsymbol{x}\in\mathbb{R}^n$, $\boldsymbol{e}^T\boldsymbol{T}(\boldsymbol{x}) = 1$; that is, $\boldsymbol{T}(\boldsymbol{x})\in\Delta$. Furthermore, note that for each $\boldsymbol{x}\in\mathbb{R}^n$,

$$\boldsymbol{x} \geqslant \boldsymbol{0} \quad \Leftrightarrow \quad \boldsymbol{T}(\boldsymbol{x}) \geqslant \boldsymbol{0}.$$

Taking the above into account, consider the following LP problem:

$$\begin{aligned} \text{minimize} \quad & \boldsymbol{c}'^T\boldsymbol{y} \\ \text{subject to} \quad & \boldsymbol{A}'\boldsymbol{y} = \boldsymbol{0} \\ & \boldsymbol{e}^T\boldsymbol{y} = 1 \\ & \boldsymbol{y} \geqslant \boldsymbol{0}. \end{aligned}$$

Note that the above LP problem is in Karmarkar's canonical form. Furthermore, in light of the definitions of $\boldsymbol{c}'$ and $\boldsymbol{A}'$, the above LP problem is equivalent to the original LP problem in standard form. Hence, we have completed converting the LP problem in standard form into an equivalent problem in Karmarkar's canonical form. Note that since $\boldsymbol{a}$ is a strictly interior feasible point, and $\boldsymbol{a}_0 = \boldsymbol{T}(\boldsymbol{a})$ is the center of the simplex $\Delta$ (see Exercise 18.1) then $\boldsymbol{a}_0$ is a feasible point of the transformed problem. Hence, assumption A of the previous subsection is satisfied for the above problem.

In the above, we started with the assumption that we are given a point $\boldsymbol{a}$ that is a strictly interior feasible point of the original LP problem in standard form. To see how this assumption can be made to hold, we now show that we can transform any given LP problem into an equivalent problem in standard form where such a point $\boldsymbol{a}$ is explicitly given. To this end, consider a given LP problem of the form:

$$\begin{aligned} \text{minimize} \quad & \boldsymbol{c}^T\boldsymbol{x} \\ \text{subject to} \quad & \boldsymbol{A}\boldsymbol{x} \geqslant \boldsymbol{b} \\ & \boldsymbol{x} \geqslant \boldsymbol{0}. \end{aligned}$$

Note that any LP problem can be converted into an equivalent problem of the above form. To see this, recall that any LP problem can be transformed into an equivalent problem in standard form. But, any problem in standard form can be represented as above, since the constraint $\boldsymbol{A}\boldsymbol{x}=\boldsymbol{b}$ can be written as $\boldsymbol{A}\boldsymbol{x} \geqslant \boldsymbol{b}$, $-\boldsymbol{A}\boldsymbol{x} \geqslant -\boldsymbol{b}$. We next write the dual to the above problem:

$$\begin{aligned} \text{maximize} \quad & \boldsymbol{\lambda}^T\boldsymbol{b} \\ \text{subject to} \quad & \boldsymbol{\lambda}^T\boldsymbol{A} \leqslant \boldsymbol{c}^T \\ & \boldsymbol{\lambda} \geqslant \boldsymbol{0}. \end{aligned}$$

As we did in our discussion of Khachiyan's algorithm, we now combine the primal and dual problems to get

$$\begin{aligned} \boldsymbol{c}^T\boldsymbol{x} - \boldsymbol{b}^T\boldsymbol{\lambda} &= 0 \\ \boldsymbol{A}\boldsymbol{x} &\geqslant \boldsymbol{b} \\ \boldsymbol{A}^T\boldsymbol{\lambda} &\leqslant \boldsymbol{c} \\ \boldsymbol{x} &\geqslant \boldsymbol{0} \\ \boldsymbol{\lambda} &\geqslant \boldsymbol{0}. \end{aligned}$$

As we pointed out in the previous section on Khachiyan's algorithm, the original LP problem is solved if, and only if, we can find a pair $(\boldsymbol{x}, \boldsymbol{\lambda})$ that satisfies the above set of relations. This follows from the Theorem 17.1. We now introduce slack and surplus variables $\boldsymbol{u}$ and $\boldsymbol{v}$ to get the following equivalent set of relations:

$$\begin{aligned} \boldsymbol{c}^T\boldsymbol{x} - \boldsymbol{b}^T\boldsymbol{\lambda} &= 0 \\ \boldsymbol{A}\boldsymbol{x} - \boldsymbol{v} &= \boldsymbol{b} \\ \boldsymbol{A}^T\boldsymbol{\lambda} + \boldsymbol{u} &= \boldsymbol{c} \\ \boldsymbol{x}, \boldsymbol{\lambda}, \boldsymbol{u}, \boldsymbol{v} &\geqslant \boldsymbol{0}. \end{aligned}$$

Let $\boldsymbol{x}_0 \in \mathbb{R}^n$, $\boldsymbol{\lambda}_0 \in \mathbb{R}^m$, $\boldsymbol{u}_0 \in \mathbb{R}^n$, and $\boldsymbol{v}_0 \in \mathbb{R}^m$ be points that satisfy $\boldsymbol{x}_0 > \boldsymbol{0}$, $\boldsymbol{\lambda}_0 > \boldsymbol{0}$, $\boldsymbol{u}_0 > \boldsymbol{0}$, and $\boldsymbol{v}_0 > \boldsymbol{0}$. For example, we could choose $\boldsymbol{x}_0 = [1, \ldots, 1]^T$, and likewise with $\boldsymbol{\lambda}_0$, $\boldsymbol{u}_0$, and $\boldsymbol{v}_0$. Consider the LP problem

$$\begin{aligned} \text{minimize} \quad & z \\ \text{subject to} \quad & \boldsymbol{c}^T\boldsymbol{x} - \boldsymbol{b}^T\boldsymbol{\lambda} + (-\boldsymbol{c}^T\boldsymbol{x}_0 + \boldsymbol{b}^T\boldsymbol{\lambda}_0)z = 0 \\ & \boldsymbol{A}\boldsymbol{x} - \boldsymbol{v} + (\boldsymbol{b} - \boldsymbol{A}\boldsymbol{x}_0 + \boldsymbol{v}_0)z = \boldsymbol{b} \\ & \boldsymbol{A}^T\boldsymbol{\lambda} + \boldsymbol{u} + (\boldsymbol{c} - \boldsymbol{A}^T\boldsymbol{\lambda}_0)z = \boldsymbol{c} \\ & \boldsymbol{x}, \boldsymbol{\lambda}, \boldsymbol{u}, \boldsymbol{v}, z \geqslant \boldsymbol{0}. \end{aligned}$$

We refer to the above as *Karmarkar's artificial problem.* Observe that the following point is a strictly interior feasible point for the above problem:

$$\begin{bmatrix} \boldsymbol{x} \\ \boldsymbol{\lambda} \\ \boldsymbol{u} \\ \boldsymbol{v} \\ z \end{bmatrix} = \begin{bmatrix} \boldsymbol{x}_0 \\ \boldsymbol{\lambda}_0 \\ \boldsymbol{u}_0 \\ \boldsymbol{v}_0 \\ 1 \end{bmatrix}.$$

Furthermore, the minimum value of the objective function for Karmarkar's artificial problem is zero if, and only if, the previous set of relations has a solution; that is there exist $\boldsymbol{x}$, $\boldsymbol{\lambda}$, $\boldsymbol{u}$, and $\boldsymbol{v}$ satisfying

$$\begin{aligned} \boldsymbol{c}^T\boldsymbol{x} - \boldsymbol{b}^T\boldsymbol{\lambda} &= 0 \\ \boldsymbol{A}\boldsymbol{x} - \boldsymbol{v} &= \boldsymbol{b} \\ \boldsymbol{A}^T\boldsymbol{\lambda} + \boldsymbol{u} &= \boldsymbol{c} \\ \boldsymbol{x}, \boldsymbol{\lambda}, \boldsymbol{u}, \boldsymbol{v} &\geqslant \boldsymbol{0}. \end{aligned}$$

Therefore, Karmarkar's artificial LP problem is equivalent to the original LP problem:

$$\begin{aligned} \text{minimize} \quad & \boldsymbol{c}^T\boldsymbol{x} \\ \text{subject to} \quad & \boldsymbol{A}\boldsymbol{x} \geqslant \boldsymbol{b} \\ & \boldsymbol{x} \geqslant \boldsymbol{0}. \end{aligned}$$

Note that the main difference between the original LP problem above and Karmarkar's artificial problem is that we have an explicit strictly interior feasible point for Karmarkar's artificial problem, and hence we have satisfied the assumption that we imposed at the beginning of this subsection.

### 18.3.5. The Algorithm

We are now ready to describe Karmarkar's algorithm. Keep in mind that the LP problem we are solving is a Karmarkar's restricted problem (that is, a problem in Karmarkar's canonical form) and satisfies assumptions A, B, C, and D. For convenience, we restate the problem:

$$\begin{aligned} &\text{minimize} \quad \boldsymbol{c}^T\boldsymbol{x}, \quad \boldsymbol{x}\in\mathbb{R}^n \\ &\text{subject to} \quad \boldsymbol{x}\in\Omega\cap\Delta, \end{aligned}$$

where $\Omega = \{\boldsymbol{x}\in\mathbb{R}^n: \boldsymbol{A}\boldsymbol{x} = \boldsymbol{0}\}$, and $\Delta = \{\boldsymbol{x}\in\mathbb{R}^n: \boldsymbol{e}^T\boldsymbol{x} = 1, \boldsymbol{x} \geqslant \boldsymbol{0}\}$. Karmarkar's algorithm is an iterative algorithm that, given an initial point $\boldsymbol{x}^{(0)}$ and parameter $q$, generates a sequence $\boldsymbol{x}^{(1)}, \boldsymbol{x}^{(2)}, \ldots, \boldsymbol{x}^{(N)}$. Karmarkar's algorithm is described by the following steps:

1. *Initialize.* Set $k := 0$; $\boldsymbol{x}^{(0)} = \boldsymbol{a}_0 = \boldsymbol{e}/n$.
2. *Update.* Set $\boldsymbol{x}^{(k+1)} = \boldsymbol{\Psi}(\boldsymbol{x}^{(k)})$, where $\boldsymbol{\Psi}$ is an update map described below.
3. *Check stopping criterion.* If the condition $\boldsymbol{c}^T\boldsymbol{x}^{(k)}/\boldsymbol{c}^T\boldsymbol{x}^{(0)} \leqslant 2^{-q}$ is satisfied, then stop.
4. *Iterate.* Set $k := k + 1$, go to 2.

We describe the update map $\boldsymbol{\Psi}$ as follows. First, consider the first step in the algorithm: $\boldsymbol{x}^{(0)} = \boldsymbol{a}_0$. To compute $\boldsymbol{x}^{(1)}$, we use the familiar update equation

$$\boldsymbol{x}^{(1)} = \boldsymbol{x}^{(0)} + \alpha \boldsymbol{d}^{(0)}$$

where $\alpha$ is a step size and $\boldsymbol{d}^{(0)}$ is an update direction. The step size $\alpha$ is chosen to be a value in $(0, 1)$. Karmarkar recommends a value of $1/4$ in his original paper [Ref. 42]. The update direction $\boldsymbol{d}^{(0)}$ is chosen as follows. First, note that the gradient of the objective function is $\boldsymbol{c}$. Therefore, the direction of maximum rate of decrease of the objective function is $-\boldsymbol{c}$. However, in general, we cannot simply update along this direction, since $\boldsymbol{x}^{(1)}$ is required to lie in the constraint set

$$\begin{aligned} \Omega\cap\Delta &= \{\boldsymbol{x}\in\mathbb{R}^n: \boldsymbol{A}\boldsymbol{x} = \boldsymbol{0}, \boldsymbol{e}^T\boldsymbol{x} = 1, \boldsymbol{x} \geqslant \boldsymbol{0}\} \\ &= \left\{\boldsymbol{x}\in\mathbb{R}^n: \begin{bmatrix} \boldsymbol{A} \\ \boldsymbol{e}^T \end{bmatrix}\boldsymbol{x} = \begin{bmatrix} \boldsymbol{0} \\ 1 \end{bmatrix}, \boldsymbol{x} \geqslant \boldsymbol{0}\right\} \\ &= \left\{\boldsymbol{x}\in\mathbb{R}^n: \boldsymbol{B}_0\boldsymbol{x} = \begin{bmatrix} \boldsymbol{0} \\ 1 \end{bmatrix}, \boldsymbol{x} \geqslant \boldsymbol{0}\right\}, \end{aligned}$$

where $\boldsymbol{B}_0 \in \mathbb{R}^{(m+1)\times n}$ is given by

$$\boldsymbol{B}_0 = \begin{bmatrix} \boldsymbol{A} \\ \boldsymbol{e}^T \end{bmatrix}.$$

Note that since $x^{(0)} \in \Omega \cap \Delta$, then for $x^{(1)} = x^{(0)} + \alpha d^{(0)}$ to also lie in $\Omega \cap \Delta$, the vector $d^{(0)}$ must be an element of the nullspace of $B_0$. Hence, we choose $d^{(0)}$ to be in the direction of the orthogonal projection of $-c$ onto the nullspace of $B_0$. This projection is accomplished by the matrix $P_0$ given by

$$P_0 = I_{m+1} - B_0^T(B_0 B_0^T)^{-1} B_0.$$

Note that $B_0 B_0^T$ is nonsingular by assumption C. Specifically, we choose $d^{(0)}$ to be the vector $d^{(0)} = -r\hat{c}^{(0)}$, where

$$\hat{c}^{(0)} = \frac{P_0 c}{\|P_0 c\|},$$

and $r = 1/\sqrt{n(n-1)}$. The scalar $r$ is incorporated into the update vector $d^{(0)}$ for the following reason. First, observe that $r$ is the radius of the largest sphere inscribed in the simplex $\Delta$ (see Exercise 18.4). Therefore, the vector $d^{(0)} = r\hat{c}^{(0)}$ points in the direction of the projection $\hat{c}^{(0)}$ of $c$ onto the nullspace of $B_0$, and $x^{(1)} = x^{(0)} + \alpha d^{(0)}$ is guaranteed to lie in the constraint set $\Omega \cap \Delta$. In fact, $x^{(0)}$ lies in the set $\Omega \cap \Delta \cap \{x : \|x - a_0\| \leq r\}$. Finally, we note that $x^{(1)}$ is a strictly interior point of $\Delta$.

The general update step $x^{(k+1)} = \Psi(x^{(k)})$ is performed as follows. We first give a brief description of the basic idea, which is similar to the update from $x^{(0)}$ to $x^{(1)}$ described above. However, note that $x^{(k)}$ is in general not at the center of the simplex. Therefore, let us first transform this point to the center. To do this, let $D_k$ be a diagonal matrix whose diagonal entries are the components of the vector $x^{(k)}$; that is,

$$D_k = \begin{bmatrix} x_1^{(k)} & \cdots & 0 \\ \vdots & \ddots & \vdots \\ 0 & \cdots & x_n^{(k)} \end{bmatrix}.$$

It turns out that, since $x^{(0)}$ is a strictly interior point of $\Delta$, then, for all $k$, $x^{(k)}$ is a strictly interior point of $\Delta$ (see Exercise 18.7). Therefore, $D_k$ is nonsingular and

$$D_k^{-1} = \begin{bmatrix} 1/x_1^{(k)} & \cdots & 0 \\ \vdots & \ddots & \vdots \\ 0 & \cdots & 1/x_n^{(k)} \end{bmatrix}.$$

Consider the mapping $U_k : \Delta \to \Delta$ given by $U_k(x) = D_k^{-1}x / e^T D_k^{-1} x$. Note that $U_k(x^{(k)}) = e/n = a_0$. We use $U_k$ to change the variable from $x$ to $\bar{x} = U_k(x)$. We do this so that $x^{(k)}$ is mapped into the center of the simplex, as indicated above. Note that $U_k$ is an invertible mapping, with $x = U_k^{-1}(\bar{x}) = D_k \bar{x} / e^T D_k \bar{x}$. Letting $\bar{x}^{(k)} = U_k(x^{(k)}) = a_0$, we can now apply the procedure that we described before for getting $x^{(1)}$ from $x^{(0)} = a_0$. Specifically, we update $\bar{x}^{(k)}$ to obtain $\bar{x}^{(k+1)}$ using the update formula $\bar{\boldsymbol{x}}^{(k+1)} = \bar{\boldsymbol{x}}^{(k)} + \alpha \boldsymbol{d}^{(k)}$. To compute $\boldsymbol{d}^{(k)}$, we need to state the original LP problem in the new variable $\bar{\boldsymbol{x}}$:

$$\begin{aligned} \text{minimize} \quad & \boldsymbol{c}^T \boldsymbol{D}_k \bar{\boldsymbol{x}} \\ \text{subject to} \quad & \boldsymbol{A}\boldsymbol{D}_k \bar{\boldsymbol{x}} = \boldsymbol{0} \\ & \bar{\boldsymbol{x}} \in \Delta. \end{aligned}$$

The reader can easily verify that the above LP problem in the new variable $\bar{\boldsymbol{x}}$ is equivalent to the original LP problem in the sense that $\boldsymbol{x}^*$ is an optimal solution to the original problem if, and only if, $\boldsymbol{U}_k(\boldsymbol{x}^*)$ is an optimal solution to the transformed problem. To see this, simply note that $\bar{\boldsymbol{x}} = \boldsymbol{U}_k(\boldsymbol{x}) = \boldsymbol{D}_k^{-1}\boldsymbol{x}/\,\boldsymbol{e}^T\boldsymbol{D}_k^{-1}\boldsymbol{x}$, and rewrite the objective function and constraints accordingly (see Exercise 18.5). As before, let

$$\boldsymbol{B}_k = \begin{bmatrix} \boldsymbol{A}\boldsymbol{D}_k \\ \boldsymbol{e}^T \end{bmatrix}.$$

We choose $\boldsymbol{d}^{(k)} = -r\hat{\boldsymbol{c}}^{(k)}$, where $\hat{\boldsymbol{c}}^{(k)}$ is the normalized projection of $-(\boldsymbol{c}^T\boldsymbol{D}_k)^T = -\boldsymbol{D}_k\boldsymbol{c}$ onto the nullspace of $\boldsymbol{B}_k$, and $r = 1/\sqrt{n(n-1)}$ as before. To determine $\hat{\boldsymbol{c}}^{(k)}$, we define the projector matrix $\boldsymbol{P}_k$ by

$$\boldsymbol{P}_k = \boldsymbol{I}_{m+1} - \boldsymbol{B}_k^T(\boldsymbol{B}_k\boldsymbol{B}_k^T)^{-1}\boldsymbol{B}_k.$$

Note that $\boldsymbol{B}_k\boldsymbol{B}_k^T$ is nonsingular (see Exercise 18.6). The vector $\hat{\boldsymbol{c}}^{(k)}$ is, therefore, given by

$$\hat{\boldsymbol{c}}^{(k)} = \frac{\boldsymbol{P}_k\boldsymbol{D}_k\boldsymbol{c}}{\|\boldsymbol{P}_k\boldsymbol{D}_k\boldsymbol{c}\|}.$$

The direction vector $\boldsymbol{d}^{(k)}$ is then

$$\boldsymbol{d}^{(k)} = -r\hat{\boldsymbol{c}}^{(k)} = -r\frac{\boldsymbol{P}_k\boldsymbol{D}_k\boldsymbol{c}}{\|\boldsymbol{P}_k\boldsymbol{D}_k\boldsymbol{c}\|}.$$

The updated vector $\bar{\boldsymbol{x}}^{(k+1)} = \bar{\boldsymbol{x}}^{(k)} + \alpha\boldsymbol{d}^{(k)}$ is guaranteed to lie in the transformed feasible set $\{\bar{\boldsymbol{x}}: \boldsymbol{A}\boldsymbol{D}_k\bar{\boldsymbol{x}} = \boldsymbol{0}\} \cap \Delta$. The final step is to apply the inverse transformation $\boldsymbol{U}_k^{-1}$ to obtain $\boldsymbol{x}^{(k+1)}$:

$$\boldsymbol{x}^{(k+1)} = \boldsymbol{U}_k^{-1}(\bar{\boldsymbol{x}}^{(k+1)}) = \frac{\boldsymbol{D}_k\bar{\boldsymbol{x}}^{(k+1)}}{\boldsymbol{e}^T\boldsymbol{D}_k\bar{\boldsymbol{x}}^{(k+1)}}.$$

Note that $\boldsymbol{x}^{(k+1)}$ lies in the set $\Omega \cap \Delta$. Indeed, we have already seen that $\boldsymbol{U}_k$ and $\boldsymbol{U}_k^{-1}$ map $\Delta$ into $\Delta$. To see that $\boldsymbol{A}\boldsymbol{x}^{(k+1)} = \boldsymbol{0}$, we simply premultiply the above expression by $\boldsymbol{A}$ and use the fact that $\boldsymbol{A}\boldsymbol{D}_k\bar{\boldsymbol{x}}^{(k+1)} = \boldsymbol{0}$.

We now summarize the update $x^{(k+1)} = \Psi(x^{(k)})$ as:

**1.** Compute the matrices:

$$D_k = \begin{bmatrix} x_1^{(k)} & \cdots & 0 \\ \vdots & \ddots & \vdots \\ 0 & \cdots & x_n^{(k)} \end{bmatrix}$$

$$B_k = \begin{bmatrix} AD_k \\ e^T \end{bmatrix}.$$

**2.** Compute the orthogonal projector onto the nullspace of $B_k$:

$$P_k = I_{m+1} - B_k^T (B_k B_k^T)^{-1} B_k.$$

**3.** Compute the normalized orthogonal projection of $c$ onto the nullspace of $B_k$:

$$\hat{c}^{(k)} = \frac{P_k D_k c}{\| P_k D_k c \|}.$$

**4.** Compute the direction vector:

$$d^{(k)} = -r\hat{c}^{(k)},$$

where $r = 1/\sqrt{n(n-1)}$.

**5.** Compute $\bar{x}^{(k+1)}$ using

$$\bar{x}^{(k+1)} = a_0 + \alpha d^{(k)},$$

where $\alpha$ is the prespecified step size, $\alpha \in (0, 1)$.

**6.** Compute $x^{(k+1)}$ by applying the inverse transformation $U_k^{-1}$:

$$x^{(k+1)} = U_k^{-1}(\bar{x}^{(k+1)}) = \frac{D_k \bar{x}^{(k+1)}}{e^T D_k \bar{x}^{(k+1)}}.$$

The matrix $P_k$ in step 2 is needed solely for computing $P_k D_k c$ in step 3. In fact, the two steps can be combined in an efficient way without having to explicitly compute $P_k$, as follows. We first solve a set of linear equations $B_k B_k^T y = B_k D_k c$ (for the variable $y$), and then compute $P_k D_k c$ using the expression $P_k D_k c = D_k c - B_k^T y$.

For further reading on non-simplex methods in linear programming, see Refs. 21, 30, and 69.

## EXERCISES

**18.1** Let $\boldsymbol{a}\in\mathbb{R}^n$, $\boldsymbol{a}>\mathbf{0}$. Let $\boldsymbol{T}=[T_1,\ldots,T_{n+1}]$ be the projective transformation of the positive orthant $P_+$ of $\mathbb{R}^n$ into the simplex $\Delta$ in $\mathbb{R}^{n+1}$, given by

$$T_i(x)=\begin{cases}\dfrac{x_i/a_i}{x_1/a_1+\cdots+x_n/a_n+1}, & \text{if } 1\leqslant i\leqslant n\\[2ex] \dfrac{1}{x_1/a_1+\cdots+x_n/a_n+1}, & \text{if } i=n+1.\end{cases}$$

Prove the following properties of $\boldsymbol{T}$ [see Ref. 42]:

**a.** $\boldsymbol{T}$ is a one-to-one mapping; that is, $\boldsymbol{T}(\boldsymbol{x})=\boldsymbol{T}(\boldsymbol{y})$ implies that $\boldsymbol{x}=\boldsymbol{y}$.

**b.** $\boldsymbol{T}$ maps $P_+$ onto $\Delta\backslash\{\boldsymbol{x}:x_{n+1}=0\}=\{\boldsymbol{x}\in\Delta:x_{n+1}>0\}$; that is, for each $\boldsymbol{y}\in\{\boldsymbol{x}\in\Delta:x_{n+1}>0\}$, there exists $\boldsymbol{x}\in P_+$ such that $\boldsymbol{y}=\boldsymbol{T}(\boldsymbol{x})$.

**c.** The inverse transformation of $\boldsymbol{T}$ exists on $\{\boldsymbol{x}\in\Delta:x_{n+1}>0\}$, and is given by $\boldsymbol{T}^{-1}=[T_1^{-1},\ldots,T_n^{-1}]^T$, with $T_i^{-1}(\boldsymbol{y})=a_iy_i/y_{n+1}$.

**d.** $\boldsymbol{T}$ maps $\boldsymbol{a}$ to the center of the simplex $\Delta$; that is, $\boldsymbol{T}(\boldsymbol{a})=\boldsymbol{e}/(n+1)=[1/(n+1),\ldots,1/(n+1)]\in\mathbb{R}^{n+1}$.

**e.** Suppose $\boldsymbol{x}$ satisfies $\boldsymbol{Ax}=\boldsymbol{b}$, and $\boldsymbol{y}=\boldsymbol{T}(\boldsymbol{x})$. Let $\boldsymbol{x}'=[y_1a_1,\ldots,y_na_n]^T$. Then, $\boldsymbol{Ax}'=\boldsymbol{b}y_{n+1}$.

**18.2** Let $\boldsymbol{T}$ be the projective transformation in Exercise 18.1, and $\boldsymbol{A}\in\mathbb{R}^{m\times n}$ a given matrix. Prove that there exists a matrix $\boldsymbol{A}'\in\mathbb{R}^{m\times(n+1)}$ such that $\boldsymbol{Ax}=\boldsymbol{b}$ if, and only if, $\boldsymbol{A}'\boldsymbol{T}(\boldsymbol{x})=\mathbf{0}$. *Hint*: Let the $i$th column of $\boldsymbol{A}'$ be given by $a_i$ times the $i$th column of $\boldsymbol{A}$, $i=1,\ldots,n$, and the $(n+1)$st column of $\boldsymbol{A}'$ be given by $-\boldsymbol{b}$.

**18.3** Let $\boldsymbol{T}$ be the projective transformation in Exercise 18.1, and $\boldsymbol{c}\in\mathbb{R}^n$ a given vector. Prove that there exists a vector $\boldsymbol{c}'\in\mathbb{R}^{n+1}$ such that $\boldsymbol{c}^T\boldsymbol{x}=0$ if, and only if, $\boldsymbol{c}'^T\boldsymbol{T}(\boldsymbol{x})=0$. *Hint:* Use property c in Exercise 18.1, with the $\boldsymbol{c}'=[c_1',\ldots,c_{n+1}']^T$ given by $c_i'=a_ic_i$, $i=1,\ldots,n$, and $c_{n+1}'=0$.

**18.4** Let $\Delta=\{\boldsymbol{x}\in\mathbb{R}^n:\boldsymbol{e}^T\boldsymbol{x}=1,\boldsymbol{x}\geqslant\mathbf{0}\}$ be the simplex in $\mathbb{R}^n$, $n>1$, and let $\boldsymbol{a}_0=\boldsymbol{e}/n$ be its center. A sphere of radius $r$ centered at $\boldsymbol{a}_0$ is the set $\{\boldsymbol{x}\in\mathbb{R}^n:\|\boldsymbol{x}-\boldsymbol{a}_0\|\leqslant r\}$. The sphere is said to be inscribed in $\Delta$ if $\{\boldsymbol{x}\in\mathbb{R}^n:\|\boldsymbol{x}-\boldsymbol{a}_0\|=r,\boldsymbol{e}^T\boldsymbol{x}=1\}\subset\Delta$. Show that the largest such sphere has radius $r=1/\sqrt{n(n-1)}$.

**18.5** Consider the following Karmarkar's restricted problem:

$$\begin{aligned}\text{minimize}\quad & \boldsymbol{c}^T\boldsymbol{x}\\ \text{subject to}\quad & \boldsymbol{Ax}=\mathbf{0}\\ & \boldsymbol{x}\in\Delta.\end{aligned}$$

Let $\boldsymbol{x}_0 \in \Delta$ be a strictly interior point of $\Delta$ and $\boldsymbol{D}$ be a diagonal matrix whose diagonal entries are the components of $\boldsymbol{x}_0$. Define the map $\boldsymbol{U}:\Delta \to \Delta$ by $\boldsymbol{U}(\boldsymbol{x}) = \boldsymbol{D}^{-1}\boldsymbol{x}/\boldsymbol{e}^T\boldsymbol{D}^{-1}\boldsymbol{x}$. Let $\bar{\boldsymbol{x}} = \boldsymbol{U}(\boldsymbol{x})$ represent a change of variable. Show that the following transformed LP problem in the variable $\bar{\boldsymbol{x}}$

$$\begin{aligned} \text{minimize} \quad & \boldsymbol{c}^T\boldsymbol{D}\bar{\boldsymbol{x}} \\ \text{subject to} \quad & \boldsymbol{A}\boldsymbol{D}\bar{\boldsymbol{x}} = \boldsymbol{0} \\ & \bar{\boldsymbol{x}} \in \Delta \end{aligned}$$

is equivalent to the original LP problem above, in the sense that $\boldsymbol{x}^*$ is an optimal solution to the original problem if, and only if, $\bar{\boldsymbol{x}}^* = \boldsymbol{U}(\boldsymbol{x}^*)$ is an optimal solution to the transformed problem.

**18.6** Let $\boldsymbol{A} \in \mathbb{R}^{m \times n}$, $m < n$, and $\Omega = \{\boldsymbol{x}:\boldsymbol{A}\boldsymbol{x} = \boldsymbol{0}\}$. Suppose $\boldsymbol{A}$ satisfies

$$\operatorname{rank}\begin{bmatrix} \boldsymbol{A} \\ \boldsymbol{e}^T \end{bmatrix} = m + 1.$$

Let $\boldsymbol{x}_0 \in \Delta \cap \Omega$ be a strictly interior point of $\Delta \subset \mathbb{R}^n$, and $\boldsymbol{D}$ be a diagonal matrix whose diagonal entries are the components of $\boldsymbol{x}_0$. Consider the matrix $\boldsymbol{B}$ defined by

$$\boldsymbol{B} = \begin{bmatrix} \boldsymbol{A}\boldsymbol{D} \\ \boldsymbol{e}^T \end{bmatrix}$$

Show that rank $\boldsymbol{B} = m + 1$, and hence $\boldsymbol{B}\boldsymbol{B}^T$ is nonsingular.

**18.7** Show that in Karmarkar's algorithm, $\boldsymbol{x}^{(k)}$ is a strictly interior point of $\Delta$.

# Part IV

# NONLINEAR CONSTRAINED OPTIMIZATION

# 19

# Problems with Equality Constraints

## 19.1. INTRODUCTION

In this part, we discuss methods for solving a class of nonlinear constrained optimization problems that can be formulated as

$$\begin{aligned} &\text{minimize} && f(\boldsymbol{x}) \\ &\text{subject to} && h_i(\boldsymbol{x}) = 0, \quad i = 1, \ldots, m \\ & && g_j(\boldsymbol{x}) \leqslant 0, \quad j = 1, \ldots, p, \end{aligned}$$

where $\boldsymbol{x} \in \mathbb{R}^n$, $f: \mathbb{R}^n \to \mathbb{R}$, $h_i: \mathbb{R}^n \to \mathbb{R}$, $g_j: \mathbb{R}^n \to \mathbb{R}$, and $m \leqslant n$. In vector notation, the problem above can be represented in the following standard form:

$$\begin{aligned} &\text{minimize} && f(\boldsymbol{x}) \\ &\text{subject to} && \boldsymbol{h}(\boldsymbol{x}) = \boldsymbol{0} \\ & && \boldsymbol{g}(\boldsymbol{x}) \leqslant \boldsymbol{0}, \end{aligned}$$

where $\boldsymbol{h}: \mathbb{R}^n \to \mathbb{R}^m$ and $\boldsymbol{g}: \mathbb{R}^n \to \mathbb{R}^p$. As usual, we adopt the following terminology.

***Definition 19.1.*** Any point satisfying the constraints is called a *feasible point*. The set of all feasible points

$$\{\boldsymbol{x} \in \mathbb{R}^n : \boldsymbol{h}(\boldsymbol{x}) = \boldsymbol{0}, \boldsymbol{g}(\boldsymbol{x}) \leqslant \boldsymbol{0}\}$$

is called the *feasible set*. ■

Optimization problems of the above form are not new to us. Indeed, linear programming problems of the form

$$\begin{aligned} &\text{minimize} && \boldsymbol{c}^T \boldsymbol{x} \\ &\text{subject to} && \boldsymbol{A}\boldsymbol{x} = \boldsymbol{b} \\ & && \boldsymbol{x} \geqslant \boldsymbol{0} \end{aligned}$$

that we studied in Part III are of this type.

As we remarked in Part II, there is no loss of generality by considering only minimization problems. For, if we are confronted with a maximization problem, it can be easily transformed into the minimization problem by observing that

$$\text{maximize } f(\boldsymbol{x}) = \text{minimize } -f(\boldsymbol{x}).$$

We illustrate the problems we will study in this part by considering the following simple numerical example.

***Example 19.1.***

$$\begin{aligned} \text{minimize} \quad & (x_1 - 1)^2 + x_2 - 2 \\ \text{subject to} \quad & x_2 - x_1 = 1, \\ & x_1 + x_2 \leqslant 2. \end{aligned}$$

This problem is already in the standard form given earlier, with $f(x_1, x_2) = (x_1 - 1)^2 + x_2 - 2$, $h(x_1, x_2) = x_2 - x_1 - 1$, and $g(x_1, x_2) = x_1 + x_2 - 2$. This problem turns out to be simple enough to be solved graphically (see Figure 19.1). In the figure, the set of points that satisfy the constraints (the feasible set) is marked by the heavy solid line. The inverted parabolas represent level sets of the objective function $f$—the lower the level set, the smaller the objective function value. Therefore, the solution can be obtained by finding the lowest level set that intersects the feasible set. In this case, the minimizer lies on the level set with $f = -\frac{1}{4}$. The minimizer of the objective function is $\boldsymbol{x}^* = [\frac{1}{2}, \frac{3}{2}]^T$. ■

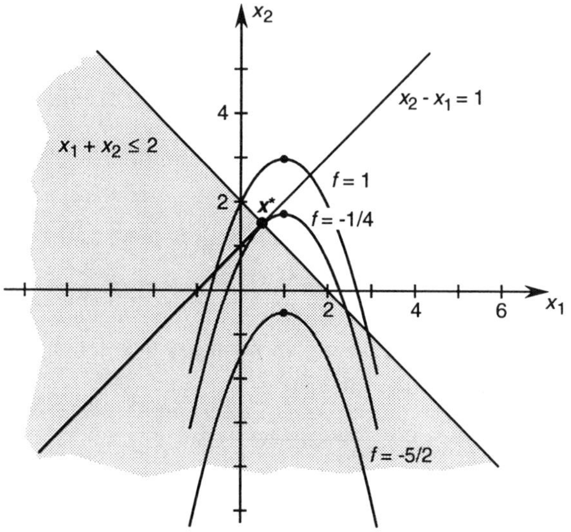


**Figure 19.1.** A graphic solution to the problem in Example 19.1

In the remainder of this chapter, we discuss constrained optimization problems with only equality constraints. The general constrained optimization problem is discussed in the chapters to follow.

## 19.2. PROBLEM FORMULATION

The class of optimization problems we analyze in this chapter is

$$\begin{aligned} &\text{minimize} \quad f(\boldsymbol{x}) \\ &\text{subject to} \quad \boldsymbol{h}(\boldsymbol{x}) = \boldsymbol{0}, \end{aligned}$$

where $\boldsymbol{x} \in \mathbb{R}^n$, $f: \mathbb{R}^n \to \mathbb{R}$, $\boldsymbol{h}: \mathbb{R}^n \to \mathbb{R}^m$, $\boldsymbol{h} = [h_1, \ldots, h_m]^T$, and $m \leqslant n$. We assume that the function $\boldsymbol{h}$ is continuously differentiable (that is, $\boldsymbol{h} \in \mathscr{C}^1$).

We introduce the following definition.

***Definition 19.2.*** A point $\boldsymbol{x}^*$ satisfying the constraints $h_1(\boldsymbol{x}^*) = 0, \ldots, h_m(\boldsymbol{x}^*) = 0$ is said to be a *regular point* of the constraints if the gradient vectors $\nabla h_1(\boldsymbol{x}^*), \ldots, \nabla h_m(\boldsymbol{x}^*)$ are linearly independent. ■

Let $D\boldsymbol{h}(\boldsymbol{x}^*)$ be the Jacobian matrix of $\boldsymbol{h} = [h_1, \ldots, h_m]^T$ at $\boldsymbol{x}^*$, given by

$$D\boldsymbol{h}(\boldsymbol{x}^*) = \begin{bmatrix} Dh_1(\boldsymbol{x}^*) \\ \vdots \\ Dh_m(\boldsymbol{x}^*) \end{bmatrix} = \begin{bmatrix} \nabla h_1(\boldsymbol{x}^*)^T \\ \vdots \\ \nabla h_m(\boldsymbol{x}^*)^T \end{bmatrix}.$$

Then, $\boldsymbol{x}^*$ is regular if, and only if, rank $D\boldsymbol{h}(\boldsymbol{x}^*) = m$; that is, the Jacobian matrix is of full rank.

The set of equality constraints $h_1(\boldsymbol{x}) = 0, \ldots, h_m(\boldsymbol{x}) = 0$, $h_i: \mathbb{R}^n \to \mathbb{R}$, describes a surface

$$S = \{\boldsymbol{x} \in \mathbb{R}^n : h_1(\boldsymbol{x}) = 0, \ldots, h_m(\boldsymbol{x}) = 0\}.$$

Assuming the points in $S$ are regular, the dimension of the surface $S$ is $n - m$.

***Example 19.2.*** Let $n = 3$ and $m = 1$ (i.e., we are operating in $\mathbb{R}^3$). Assuming that all points in $S$ are regular, the set $S$ is a two-dimensional surface. For example, let

$$h_1(\boldsymbol{x}) = x_2 - x_3^2 = 0.$$

Note that $\nabla h_1(\boldsymbol{x}) = [0, 1, -2x_3]^T$, and hence, for any $\boldsymbol{x} \in \mathbb{R}^3$, $\nabla h_1(\boldsymbol{x}) \neq \boldsymbol{0}$. In this case,

$$\dim S = \dim\{\boldsymbol{x} : h_1(\boldsymbol{x}) = 0\} = n - m = 2.$$

See Figure 19.2 for a graphic illustration. ■

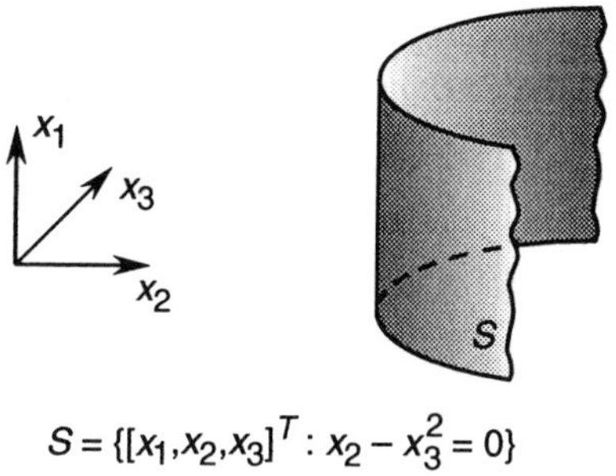


**Figure 19.2.** A two-dimensonal surface in $\mathbb{R}^3$

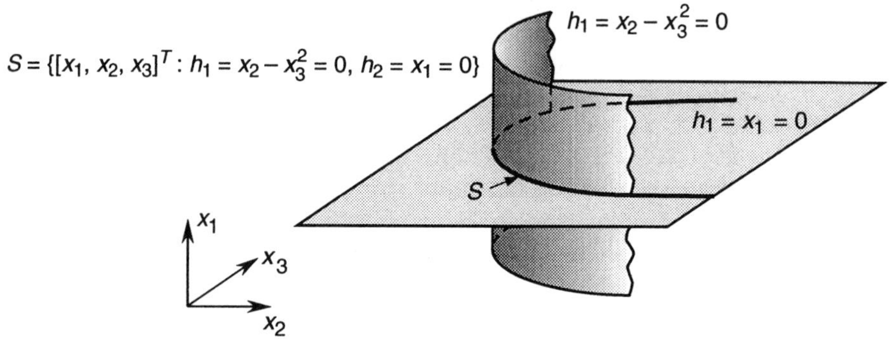


**Figure 19.3.** A one-dimensional surface in $\mathbb{R}^3$

***Example 19.3.*** Let $n = 3$ and $m = 2$. Assuming regularity, the feasible set $S$ is a one-dimensional object (that is, a curve in $\mathbb{R}^3$). For example, let

$$h_1(\boldsymbol{x}) = x_1,$$
$$h_2(\boldsymbol{x}) = x_2 - x_3^2.$$

In this case, $\nabla h_1(\boldsymbol{x}) = [1, 0, 0]^T$, and $\nabla h_2(\boldsymbol{x}) = [0, 1, -2x_3]^T$. Hence, the vectors $\nabla h_1(\boldsymbol{x})$ and $\nabla h_2(\boldsymbol{x})$ are linearly independent in $\mathbb{R}^3$. Thus

$$\dim S = \dim\{\boldsymbol{x} : h_1(\boldsymbol{x}) = 0,\ h_2(\boldsymbol{x}) = 0\} = n - m = 1.$$

See Figure 19.3 for a graphic illustration. ■

## 19.3. TANGENT AND NORMAL SPACES

In this section, we discuss the notion of a tangent space and normal space at a point on a surface. We begin by defining a curve on a surface $S$.

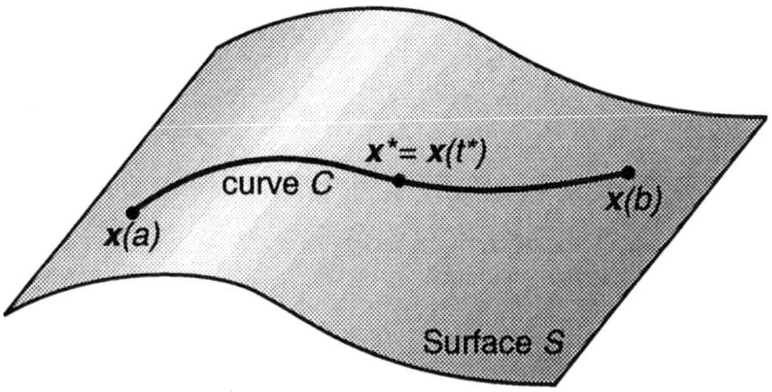


**Figure 19.4.** A curve on a surface

***Definition 19.3.*** A *curve* $C$ on a surface $S$ is a set of points $\{x(t) \in S : t \in (a, b)\}$, continuously parameterized by $t \in (a, b)$. ■

A graphic illustration of the definition of a curve is given in Figure 19.4. The definition of a curve implies that all the points on the curve satisfy the equation describing the surface. The curve $C$ passes through a point $x^*$ if there exists $t^* \in (a, b)$ such that $x(t^*) = x^*$.

Intuitively, we can think of a curve $C = \{x(t) : t \in (a, b)\}$ as the path traversed by a point $x$ traveling on the surface $S$. The position of the point at time $t$ is given by $x(t)$.

***Definition 19.4.*** The curve $C = \{x(t) : t \in (a, b)\}$ is *differentiable* if

$$\dot{x}(t) = \frac{dx}{dt}(t) = \begin{bmatrix} \dot{x}_1(t) \\ \vdots \\ \dot{x}_n(t) \end{bmatrix}$$

exists for all $t \in (a, b)$. The curve $C = \{x(t) : t \in (a, b)\}$ is *twice differentiable* if

$$\ddot{x}(t) = \frac{d^2x}{dt^2}(t) = \begin{bmatrix} \ddot{x}_1(t) \\ \vdots \\ \ddot{x}_n(t) \end{bmatrix}$$

exists for all $t \in (a, b)$. ■

Note that both $\dot{x}(t)$ and $\ddot{x}(t)$ are $n$-dimensional vectors. We can think of $\dot{x}(t)$ and $\ddot{x}(t)$ as the velocity and acceleration, respectively, of a point traversing the curve $C$ with position $x(t)$ at time $t$. The vector $\dot{x}(t)$ points in the direction of the instantaneous motion of $x(t)$. Therefore, the vector $\dot{x}(t^*)$ is *tangent* to the curve $C$ at $x^*$ (see Figure 19.5).

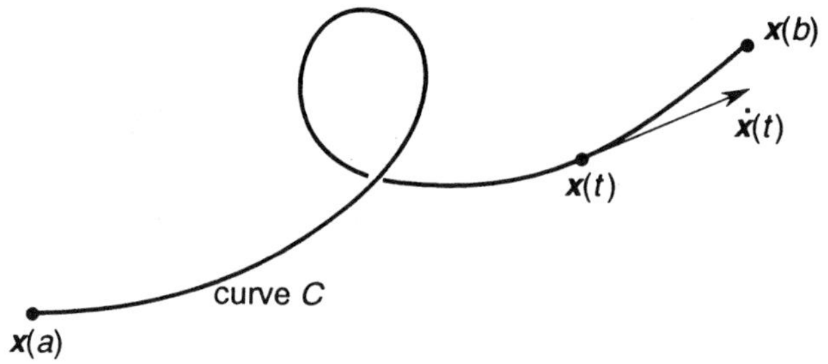


**Figure 19.5.** Geometric interpretation of the differentiability of a curve

We are now ready to introduce the notions of a tangent space. For this, recall the set

$$S = \{\boldsymbol{x} \in \mathbb{R}^n : \boldsymbol{h}(\boldsymbol{x}) = \boldsymbol{0}\},$$

where $\boldsymbol{h} \in \mathscr{C}^1$. We think of $S$ as a surface in $\mathbb{R}^n$.

***Definition 19.5.*** The *tangent space* at a point $\boldsymbol{x}^*$ on the surface $S = \{\boldsymbol{x} \in \mathbb{R}^n : \boldsymbol{h}(\boldsymbol{x}^*) = \boldsymbol{0}\}$ is the set

$$T(\boldsymbol{x}^*) = \{\boldsymbol{y} : D\boldsymbol{h}(\boldsymbol{x}^*)\boldsymbol{y} = \boldsymbol{0}\}.$$

■

Note that the tangent space $T(\boldsymbol{x}^*)$ is the nullspace of the matrix $D\boldsymbol{h}(\boldsymbol{x}^*)$, that is,

$$T(\boldsymbol{x}^*) = \mathscr{N}(D\boldsymbol{h}(\boldsymbol{x}^*)).$$

The tangent space is therefore a subspace of $\mathbb{R}^n$.

Assuming $\boldsymbol{x}^*$ is regular, the dimension of the tangent space is $n - m$, where $m$ is the number of equality constraints $h_i(\boldsymbol{x}^*) = 0$. Note that the tangent space passes through the origin. However, it is often convenient to picture the tangent space as a plane that passes through the point $\boldsymbol{x}^*$. For this, we define the *tangent plane* at $\boldsymbol{x}^*$ to be the set

$$TP(\boldsymbol{x}^*) = T(\boldsymbol{x}^*) + \boldsymbol{x}^* = \{\boldsymbol{x} + \boldsymbol{x}^* : \boldsymbol{x} \in T(\boldsymbol{x}^*)\}.$$

Figure 19.6 illustrates the notion of a tangent plane. Figure 19.7 illustrates the relationship between the tangent plane and the tangent space.

***Example 19.4.*** Let

$$S = \{\boldsymbol{x} \in \mathbb{R}^3 : h_1(\boldsymbol{x}) = x_1 = 0,\ h_2(\boldsymbol{x}) = x_1 - x_2 = 0\}.$$

Then, $S$ is the $x_3$ axis in $\mathbb{R}^3$ (see Figure 19.8).

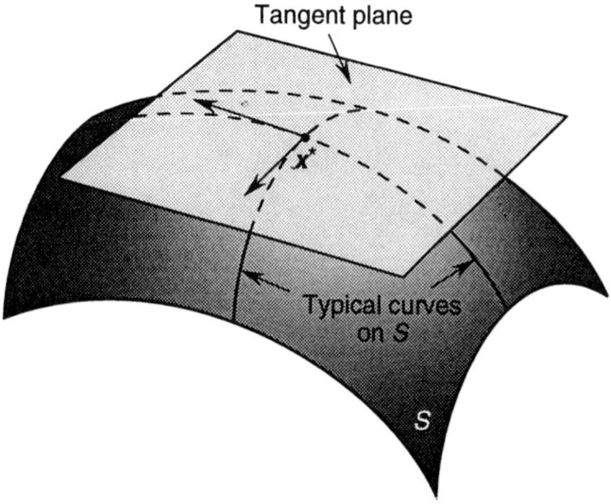


**Figure 19.6.** The tangent plane to the surface $S$ at the point $\boldsymbol{x}^*$

We have

$$D\boldsymbol{h}(\boldsymbol{x}) = \begin{bmatrix} \nabla h_1(\boldsymbol{x})^T \\ \nabla h_2(\boldsymbol{x})^T \end{bmatrix} = \begin{bmatrix} 1 & 0 & 0 \\ 1 & -1 & 0 \end{bmatrix}.$$

Since $\nabla h_1$ and $\nabla h_2$ are linearly independent when evaluated at any $\boldsymbol{x} \in S$, all the points of $S$ are regular. The tangent space at an arbitrary point of $S$ is

$$\begin{aligned} T(\boldsymbol{x}) &= \{\boldsymbol{y} : \nabla h_1(\boldsymbol{x})^T \boldsymbol{y} = 0,\ \nabla h_2(\boldsymbol{x})^T \boldsymbol{y} = 0\} \\ &= \left\{ \boldsymbol{y} : \begin{bmatrix} 1 & 0 & 0 \\ 1 & -1 & 0 \end{bmatrix} \begin{bmatrix} y_1 \\ y_2 \\ y_3 \end{bmatrix} = \boldsymbol{0} \right\} \\ &= \{[0, 0, \alpha]^T : \alpha \in \mathbb{R}\} \\ &= \text{the } x_3 \text{ axis in } \mathbb{R}^3. \end{aligned}$$

In this example, the tangent space $T(\boldsymbol{x})$ at any point $\boldsymbol{x} \in S$ is a one-dimensional subspace of $\mathbb{R}^3$. ■

Intuitively, we would expect the definition of the tangent space at a point on a surface to be the collection of all "tangent vectors" to the surface at that point. We have seen that the derivative of a curve on a surface at a point is a tangent vector to the curve and hence to the surface. The above intuition agrees with our definition whenever $\boldsymbol{x}^*$ is regular, as stated in the theorem below.

**Theorem 19.1.** *Suppose $\boldsymbol{x}^* \in S$ is a regular point, and $T(\boldsymbol{x}^*)$ is the tangent space at $\boldsymbol{x}^*$. Then, $\boldsymbol{y} \in T(\boldsymbol{x}^*)$ if, and only if, there exists a differentiable curve in $S$ passing through $\boldsymbol{x}^*$ with derivative $\boldsymbol{y}$ at $\boldsymbol{x}^*$.* □

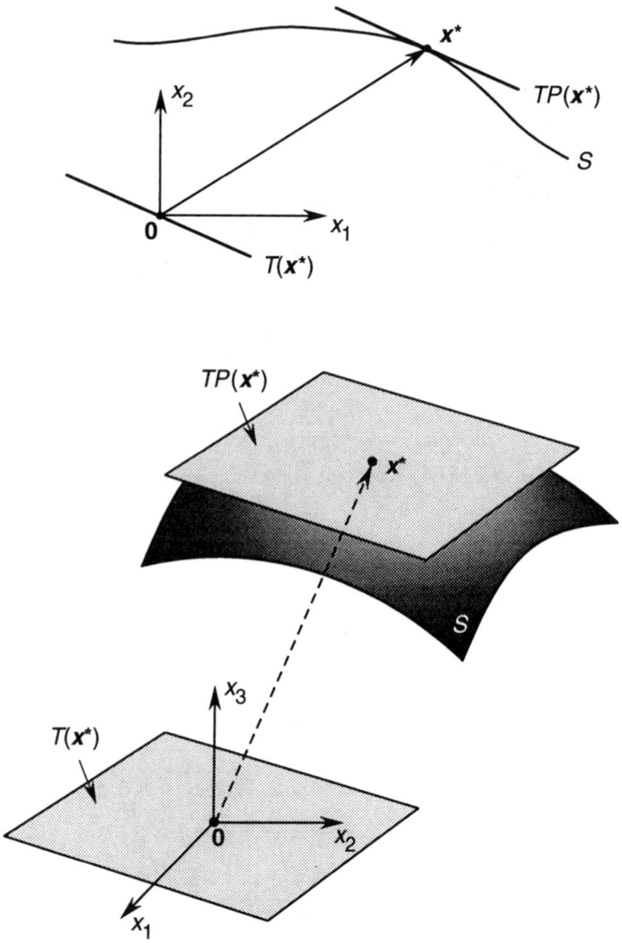


**Figure 19.7.** Tangent spaces and planes in $\mathbb{R}^2$ and $\mathbb{R}^3$

*Proof.* $\Leftarrow$: Suppose there exists a curve $\{\boldsymbol{x}(t): t \in (a, b)\}$ in $S$ such that $\boldsymbol{x}(t^*) = \boldsymbol{x}^*$ and $\dot{\boldsymbol{x}}(t^*) = \boldsymbol{y}$ for some $t^* \in (a, b)$. Then,

$$\boldsymbol{h}(\boldsymbol{x}(t)) = \boldsymbol{0}$$

for all $t \in (a, b)$. If we differentiate the function $\boldsymbol{h}(\boldsymbol{x}(t))$ with respect to $t$ using the chain rule, we obtain

$$\frac{d}{dt}\boldsymbol{h}(\boldsymbol{x}(t)) = D\boldsymbol{h}(\boldsymbol{x}(t))\dot{\boldsymbol{x}}(t) = \boldsymbol{0}$$

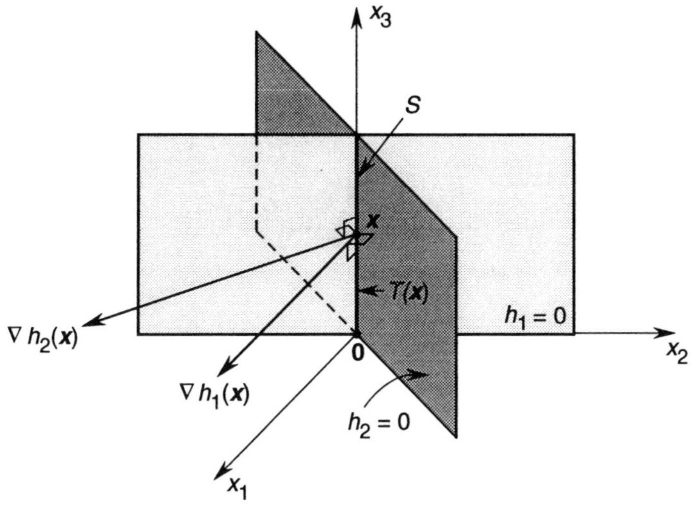


**Figure 19.8.** The surface $S = \{x \in \mathbb{R}^3 : x_1 = 0, x_1 - x_2 = 0\}$

for all $t \in (a, b)$. Therefore, at $t^*$, we get

$$Dh(x^*)y = 0,$$

and hence $y \in T(x^*)$.

$\Rightarrow$: To prove this, we need to use the implicit function theorem. [We refer the reader to Ref. 51, p. 298.] ■

We now introduce the notion of a normal space.

***Definition 19.6.*** The *normal space* $N(x^*)$ at a point $x^*$ on the surface $S = \{x \in \mathbb{R}^n : h(x^*) = 0\}$ is the set

$$N(x^*) = \{x \in \mathbb{R}^n : x = Dh(x^*)^T z,\ z \in \mathbb{R}^m\}.$$

■

We can express the normal space $N(x^*)$ as

$$N(x^*) = \mathscr{R}(Dh(x^*)^T);$$

that is, the range of the matrix $Dh(x^*)^T$. Note that the normal space $N(x^*)$ is the subspace of $\mathbb{R}^n$ spanned by the vectors $\nabla h_1(x^*), \ldots, \nabla h_m(x^*)$, that is,

$$\begin{aligned} N(x^*) &= \operatorname{span}[\nabla h_1(x^*), \ldots, \nabla h_m(x^*)] \\ &= \{x \in \mathbb{R}^n : x = z_1 \nabla h_1(x^*) + \cdots + z_m \nabla h_m(x^*),\ z_1, \ldots, z_m \in \mathbb{R}\}. \end{aligned}$$

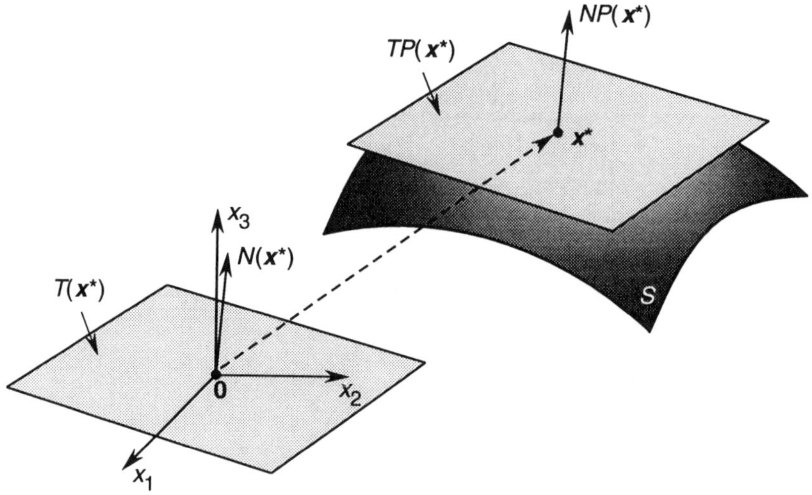


**Figure 19.9.** Normal space in $\mathbb{R}^3$

Note that the normal space contains the zero vector. Assuming that $\boldsymbol{x}^*$ is regular, the dimension of the normal space $N(\boldsymbol{x}^*)$ is $m$. As in the case of the tangent space, it is often convenient to picture the normal space $N(\boldsymbol{x}^*)$ as passing through the point $\boldsymbol{x}^*$ (rather than through the origin of $\mathbb{R}^n$). For this, we define the *normal plane* at $\boldsymbol{x}^*$ as the set

$$NP(\boldsymbol{x}^*) = N(\boldsymbol{x}^*) + \boldsymbol{x}^* = \{\boldsymbol{x} + \boldsymbol{x}^* \in \mathbb{R}^n : \boldsymbol{x} \in N(\boldsymbol{x}^*)\}.$$

Figure 19.9 illustrates the normal space and plane in $\mathbb{R}^3$ (that is, $n = 3$ and $m = 1$).

We now show that the tangent space and normal space are orthogonal complements of each other (see Section 3.3).

**Lemma 19.1.** *We have* $T(\boldsymbol{x}^*) = N(\boldsymbol{x}^*)^\perp$ *and* $T(\boldsymbol{x}^*)^\perp = N(\boldsymbol{x}^*)$. □

*Proof.* By definition of $T(\boldsymbol{x}^*)$, we may write

$$T(\boldsymbol{x}^*) = \{\boldsymbol{y} \in \mathbb{R}^n : \boldsymbol{x}^T \boldsymbol{y} = 0, \boldsymbol{x} \in N(\boldsymbol{x}^*)\}.$$

Hence, by definition of $N(\boldsymbol{x}^*)$, we have $T(\boldsymbol{x}^*) = N(\boldsymbol{x}^*)^\perp$. By Exercise 3.6, we also have $T(\boldsymbol{x}^*)^\perp = N(\boldsymbol{x}^*)$. ■

By the above lemma, we can write (see Section 3.3)

$$\mathbb{R}^n = N(\boldsymbol{x}^*) \oplus T(\boldsymbol{x}^*),$$

that is, given any vector $\boldsymbol{v} \in \mathbb{R}^n$, there are unique vectors $\boldsymbol{w} \in N(\boldsymbol{x}^*)$ and $\boldsymbol{y} \in T(\boldsymbol{x}^*)$ such that

$$\boldsymbol{v} = \boldsymbol{w} + \boldsymbol{y}.$$

## 19.4. LAGRANGE CONDITIONS

In this section, we present a first-order necessary condition for extremum problems with equality constraints. The result is the well-known Lagrange multiplier theorem. To better understand the idea underlying this theorem, we first consider functions of two variables and only one equality constraint. Let $h: \mathbb{R}^2 \to \mathbb{R}$ be the constraint function. Recall that at each point $\boldsymbol{x}$ of the domain, the gradient vector $\nabla h(\boldsymbol{x})$ is orthogonal to the level set that passes through that point. Indeed, let us choose a point $\boldsymbol{x}^* = [x_1^*, x_2^*]^T$ such that $h(\boldsymbol{x}^*) = 0$, and assume $\nabla h(\boldsymbol{x}^*) \neq \boldsymbol{0}$. The level set through the point $\boldsymbol{x}^*$ is the set $\{\boldsymbol{x}: h(\boldsymbol{x}) = 0\}$. We then parameterize this level set in a neighborhood of $\boldsymbol{x}^*$ by a curve $\{\boldsymbol{x}(t)\}$; that is, a continuously differentiable vector function $\boldsymbol{x}: \mathbb{R} \to \mathbb{R}^2$ such that

$$\boldsymbol{x}(t) = \begin{bmatrix} x_1(t) \\ x_2(t) \end{bmatrix}, \quad t \in (a, b), \quad \boldsymbol{x}^* = \boldsymbol{x}(t^*), \quad \dot{\boldsymbol{x}}(t^*) \neq \boldsymbol{0}, \quad t^* \in (a, b).$$

We can now show that $\nabla h(\boldsymbol{x}^*)$ is orthogonal to $\dot{\boldsymbol{x}}(t^*)$. Indeed, since $h$ is zero on the curve $\{\boldsymbol{x}(t): t \in (a, b)\}$, we have that for all $t \in (a, b)$,

$$h(\boldsymbol{x}(t)) = 0.$$

Hence, for all $t \in (a, b)$,

$$\frac{d}{dt} h(\boldsymbol{x}(t)) = 0.$$

Applying the chain rule, we get

$$\frac{d}{dt} h(\boldsymbol{x}(t)) = \nabla h(\boldsymbol{x}(t))^T \dot{\boldsymbol{x}}(t) = 0.$$

Therefore, $\nabla h(\boldsymbol{x}^*)$ is orthogonal to $\dot{\boldsymbol{x}}(t^*)$.

Now suppose that $\boldsymbol{x}^*$ is a minimizer of $f: \mathbb{R}^2 \to \mathbb{R}$ on the set $\{\boldsymbol{x}: h(\boldsymbol{x}) = 0\}$. We claim that $\nabla f(\boldsymbol{x}^*)$ is orthogonal to $\dot{\boldsymbol{x}}(t^*)$. To see this, it is enough to observe that the composite function of $t$ given by

$$\phi(t) = f(\boldsymbol{x}(t))$$

achieves a minimum at $t^*$. Consequently, the first-order necessary condition for the unconstrained extremum problem implies

$$\frac{d\phi}{dt}(t^*) = 0.$$

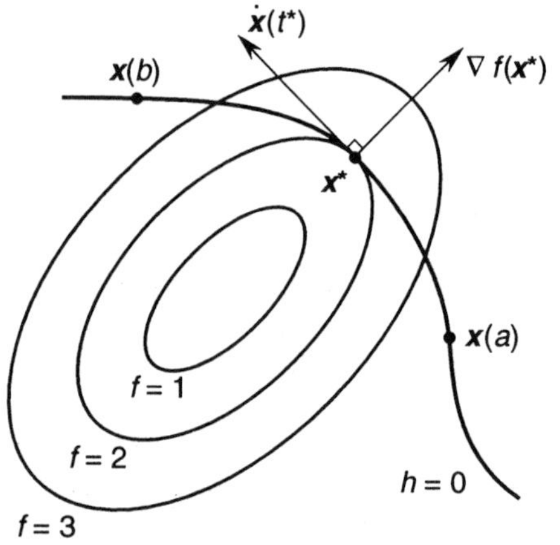


**Figure 19.10.** The gradient $\nabla f(x^*)$ is orthogonal to the curve $\{x(t)\}$ at the point $x^*$ that is a minimizer of $f$ on the curve

Applying the chain rule yields

$$0 = \frac{d}{dt}\phi(t^*) = \nabla f(x(t^*))^T \dot{x}(t^*) = \nabla f(x^*)^T \dot{x}(t^*).$$

Thus, $\nabla f(x^*)$ is orthogonal to $\dot{x}(t^*)$. The fact that $\dot{x}(t^*)$ is tangent to the curve $\{x(t)\}$ at $x^*$, means that $\nabla f(x^*)$ is orthogonal to the curve at $x^*$ (see Figure 19.10).

Recall that $\nabla h(x^*)$ is also orthogonal to $\dot{x}(t^*)$. Therefore, the vectors $\nabla h(x^*)$ and $\nabla f(x^*)$ are parallel, that is, $\nabla f(x^*)$ is a scalar multiple of $\nabla h(x^*)$. The above observation allow us now to formulate the Lagrange theorem for functions of two variables with one constraint.

**Theorem 19.2. Lagrange's Theorem for** $n = 2$, $m = 1$. *Let the point* $x^*$ *be a minimizer of* $f: \mathbb{R}^2 \to \mathbb{R}$ *subject to the constraint* $h(x) = 0$, $h: \mathbb{R}^2 \to \mathbb{R}$. *Then,* $\nabla f(x^*)$ *and* $\nabla h(x^*)$ *are parallel. That is, if* $\nabla h(x^*) \neq \mathbf{0}$, *then there exists a scalar* $\lambda^*$ *such that*

$$\nabla f(x^*) + \lambda^* \nabla h(x^*) = \mathbf{0}.$$

□

In the above theorem, we refer to $\lambda^*$ as the Lagrange multiplier. Note that the theorem also holds for maximizers. Figure 19.11 gives an illustration of Lagrange's theorem for the case where $x^*$ is a maximizer of $f$ over the set $\{x: h(x) = 0\}$. Also, note that the Lagrange condition is only necessary but not sufficient. In Figure 19.12, we illustrate a variety of points where the Lagrange condition is satisfied, including a case where the point is not an extremizer (neither a maximizer nor a minimizer).

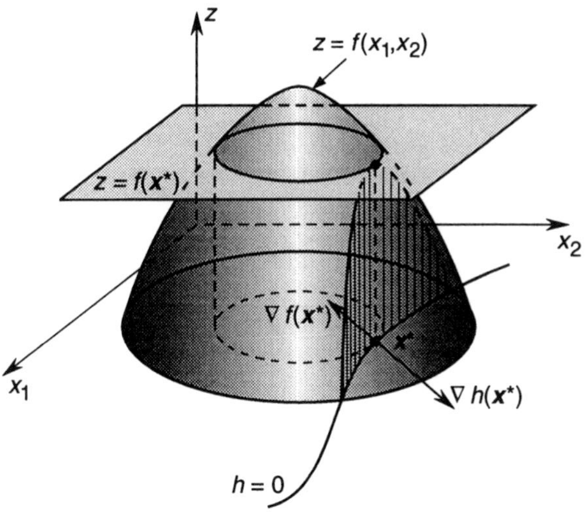


**Figure 19.11.** Illustration of Lagrange's theorem for $n = 2$, $m = 1$

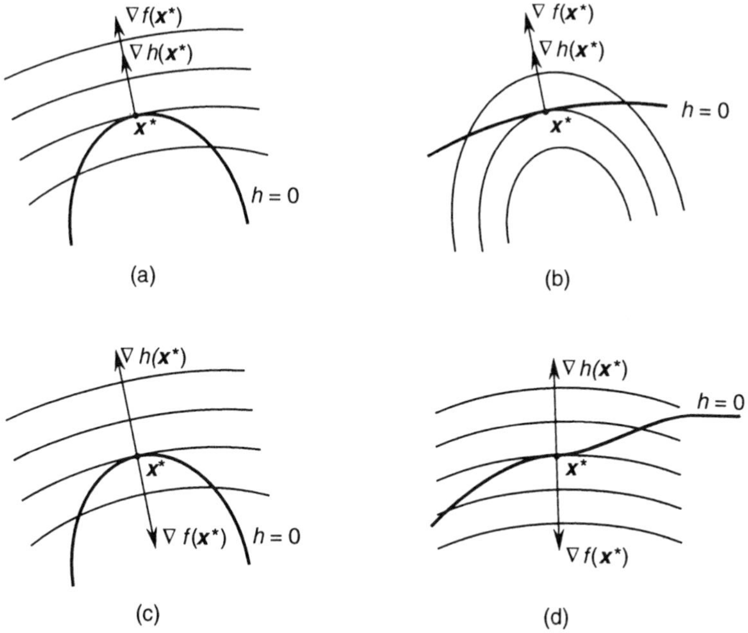


**Figure 19.12.** Four examples where the Lagrange condition is satisfied: (a) maximizer, (b) minimizer, (c) minimizer, (d) not an extremizer. Adapted from Ref. 70

We now generalize the Lagrange theorem for the case when $f:\mathbb{R}^n \to \mathbb{R}$ and $\boldsymbol{h}:\mathbb{R}^n \to \mathbb{R}^m$, $m \leqslant n$.

**Theorem 19.3. Lagrange Multiplier Theorem.** *Let $\boldsymbol{x}^*$ be a local minimizer (or maximizer) of $f:\mathbb{R}^n \to \mathbb{R}$, subject to $\boldsymbol{h}(\boldsymbol{x}) = \boldsymbol{0}$, $\boldsymbol{h}:\mathbb{R}^n \to \mathbb{R}^m$, $m \leqslant n$. Assume that $\boldsymbol{x}^*$ is a regular point. Then, there exists $\boldsymbol{\lambda}^* \in \mathbb{R}^m$ such that*

$$Df(\boldsymbol{x}^*) + \boldsymbol{\lambda}^{*T} D\boldsymbol{h}(\boldsymbol{x}^*) = \boldsymbol{0}^T.$$

□

*Proof.* We need to prove that

$$\nabla f(\boldsymbol{x}^*) = -D\boldsymbol{h}(\boldsymbol{x}^*)^T \boldsymbol{\lambda}^*$$

for some $\boldsymbol{\lambda}^* \in \mathbb{R}^m$; that is, $\nabla f(\boldsymbol{x}^*) \in \mathscr{R}(D\boldsymbol{h}(\boldsymbol{x}^*)^T) = N(\boldsymbol{x}^*)$. But, by Lemma 19.1, $N(\boldsymbol{x}^*) = T(\boldsymbol{x}^*)^\perp$. Therefore, it remains to show that $\nabla f(\boldsymbol{x}^*) \in T(\boldsymbol{x}^*)^\perp$.

We proceed as follows. Suppose that

$$\boldsymbol{y} \in T(\boldsymbol{x}^*).$$

Then, by Theorem 19.1, there exists a differentiable curve $\{\boldsymbol{x}(t): t \in (a, b)\}$ such that, for all $t \in (a, b)$,

$$\boldsymbol{h}(\boldsymbol{x}(t)) = \boldsymbol{0},$$

and there exists $t^* \in (a, b)$ satisfying

$$\boldsymbol{x}(t^*) = \boldsymbol{x}^*, \quad \dot{\boldsymbol{x}}(t^*) = \boldsymbol{y}.$$

Consider now the composite function $\phi(t) = f(\boldsymbol{x}(t))$. Note that $t^*$ is a local minimizer of this function. By the first-order necessary condition for unconstrained local minimizers (see Theorem 6.1),

$$\frac{d\phi}{dt}(t^*) = 0.$$

Applying the chain rule yields

$$\frac{d\phi}{dt}(t^*) = Df(\boldsymbol{x}^*)\dot{\boldsymbol{x}}(t^*) = Df(\boldsymbol{x}^*)\boldsymbol{y} = \nabla f(\boldsymbol{x}^*)^T \boldsymbol{y} = 0.$$

So all $\boldsymbol{y} \in T(\boldsymbol{x}^*)$ satisfy

$$\nabla f(\boldsymbol{x}^*)^T \boldsymbol{y} = 0,$$

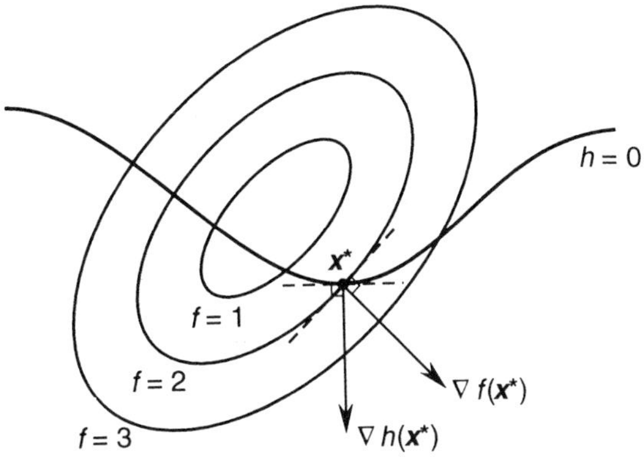


**Figure 19.13.** An example where the Lagrange condition does not hold

that is

$$\nabla f(\boldsymbol{x}^*) \in T(\boldsymbol{x}^*)^{\perp}.$$

This completes the proof. ■

The Lagrange theorem states that, if $\boldsymbol{x}^*$ is an extremizer, then the gradient of the objective function $f$ can be expressed as a linear combination of the gradients of the constraints. We refer to the vector $\boldsymbol{\lambda}^*$ in the above theorem as the *Lagrange multiplier vector*.

Observe that $\boldsymbol{x}^*$ cannot be an extremizer if

$$\nabla f(\boldsymbol{x}^*) \notin N(\boldsymbol{x}^*).$$

This situation is illustrated in Figure 19.13.

It is convenient to introduce the so-called Lagrangian function $l: \mathbb{R}^n \times \mathbb{R}^m \to \mathbb{R}$, given by

$$l(\boldsymbol{x}, \boldsymbol{\lambda}) \triangleq f(\boldsymbol{x}) + \boldsymbol{\lambda}^T \boldsymbol{h}(\boldsymbol{x}).$$

The Lagrange multiplier theorem for a local minimizer $\boldsymbol{x}^*$ can be represented using the Lagrangian function as

$$Dl(\boldsymbol{x}^*, \boldsymbol{\lambda}^*) = \boldsymbol{0}^T$$

for some $\boldsymbol{\lambda}^*$, where the derivative operation $D$ is with respect to the entire argument $[\boldsymbol{x}^T, \boldsymbol{\lambda}^T]^T$. In other words, the necessary condition in the Lagrange multiplier theorem is equivalent to the first-order necessary condition for unconstrained optimization applied to the Lagrangian function.

To see the above, denote the derivative of $l$ with respect to $\boldsymbol{x}$ as $D_x l$, and the derivative of $l$ with respect to $\boldsymbol{\lambda}$ as $D_\lambda l$. Then,

$$Dl(\boldsymbol{x}, \boldsymbol{\lambda}) = [D_x l(\boldsymbol{x}, \boldsymbol{\lambda}), D_\lambda l(\boldsymbol{x}, \boldsymbol{\lambda})].$$

Note that $D_x l(\boldsymbol{x}, \boldsymbol{\lambda}) = Df(\boldsymbol{x}) + \boldsymbol{\lambda}^T D\boldsymbol{h}(\boldsymbol{x})$ and $D_\lambda l(\boldsymbol{x}, \boldsymbol{\lambda}) = \boldsymbol{h}(\boldsymbol{x})^T$. Therefore, the Lagrange theorem for a local minimizer $\boldsymbol{x}^*$ can be stated as

$$D_x l(\boldsymbol{x}^*, \boldsymbol{\lambda}^*) = \boldsymbol{0}^T$$
$$D_\lambda l(\boldsymbol{x}^*, \boldsymbol{\lambda}^*) = \boldsymbol{0}^T$$

for some $\boldsymbol{\lambda}^*$, which is equivalent to

$$Dl(\boldsymbol{x}^*, \boldsymbol{\lambda}^*) = \boldsymbol{0}^T.$$

We refer to the condition $Dl(\boldsymbol{x}^*, \boldsymbol{\lambda}^*) = \boldsymbol{0}^T$ as the *Lagrange condition.*

The Lagrange multiplier theorem is used to find possible extremizers. This entails solving the equations:

$$D_x l(\boldsymbol{x}, \boldsymbol{\lambda}) = \boldsymbol{0}^T$$
$$D_\lambda l(\boldsymbol{x}, \boldsymbol{\lambda}) = \boldsymbol{0}^T.$$

The above represents $n + m$ equations in $n + m$ unknowns. Keep in mind that the Lagrange condition is only necessary, but not sufficient; that is, a point $\boldsymbol{x}^*$ satisfying the above equations need not be an extremizer.

***Example 19.5.*** Consider the problem of extremizing the objective function

$$f(\boldsymbol{x}) = x_1^2 + x_2^2$$

on the ellipse

$$\{[x_1, x_2]^T : h(\boldsymbol{x}) = x_1^2 + 2x_2^2 - 1 = 0\}.$$

We have

$$\nabla f(\boldsymbol{x}) = [2x_1, 2x_2]^T,$$
$$\nabla h(\boldsymbol{x}) = [2x_1, 4x_2]^T.$$

Thus

$$D_x l(\boldsymbol{x}, \lambda) = D_x[f(\boldsymbol{x}) + \lambda h(\boldsymbol{x})] = [2x_1 + 2\lambda x_1, 2x_2 + 4\lambda x_2],$$

and

$$D_\lambda l(\boldsymbol{x}, \lambda) = h(\boldsymbol{x}) = x_1^2 + 2x_2^2 - 1.$$

Setting $D_x l(\boldsymbol{x}, \lambda) = \boldsymbol{0}^T$ and $D_\lambda l(\boldsymbol{x}, \lambda) = 0$ we obtain three equations in three unknowns

$$\begin{aligned} 2x_1 + 2\lambda x_1 &= 0 \\ 2x_2 + 4\lambda x_2 &= 0 \\ x_1^2 + 2x_2^2 &= 1. \end{aligned}$$

From the first of the above equations, we get either $x_1 = 0$ or $\lambda = -1$. For the case where $x_1 = 0$, the second and third equations imply that $\lambda = -1/2$ and $x_2 = \pm 1/\sqrt{2}$. For the case where $\lambda = -1$, the second and third equations imply that $x_1 = \pm 1$ and $x_2 = 0$. Thus, the points that satisfy the Lagrange condition for extrema are

$$\boldsymbol{x}^{(1)} = \begin{bmatrix} 0 \\ 1/\sqrt{2} \end{bmatrix}, \quad \boldsymbol{x}^{(2)} = \begin{bmatrix} 0 \\ -1/\sqrt{2} \end{bmatrix}, \quad \boldsymbol{x}^{(3)} = \begin{bmatrix} 1 \\ 0 \end{bmatrix}, \quad \boldsymbol{x}^{(4)} = \begin{bmatrix} -1 \\ 0 \end{bmatrix}.$$

Since

$$f(\boldsymbol{x}^{(1)}) = f(\boldsymbol{x}^{(2)}) = \tfrac{1}{2}$$

and

$$f(\boldsymbol{x}^{(3)}) = f(\boldsymbol{x}^{(4)}) = 1$$

we conclude that if there are minimizers, they are located at $\boldsymbol{x}^{(1)}$ and $\boldsymbol{x}^{(2)}$, and if there are maximizers, they are located at $\boldsymbol{x}^{(3)}$ and $\boldsymbol{x}^{(4)}$. It turns out that, indeed, $\boldsymbol{x}^{(1)}$ and $\boldsymbol{x}^{(2)}$ are minimizers and $\boldsymbol{x}^{(3)}$ and $\boldsymbol{x}^{(4)}$ are maximizers. This problem can be solved graphically, as illustrated in Figure 19.14. ■

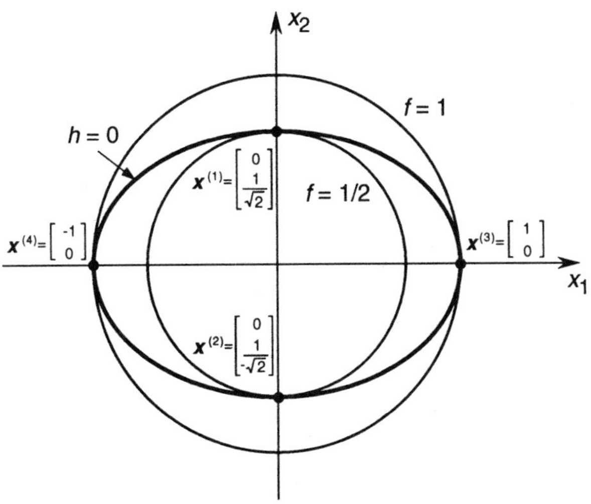


**Figure 19.14.** Graphical solution of the problem in Example 19.5

In the above example both the objective function $f$ and the constraint function $h$ are quadratic functions. In the next example, we take a closer look at a class of problems where both the objective function $f$ and the constraint $h$ are quadratic functions of $n$ variables.

***Example 19.6.*** Consider the following problem:

$$\text{maximize} \quad \frac{\boldsymbol{x}^T \boldsymbol{Q} \boldsymbol{x}}{\boldsymbol{x}^T \boldsymbol{P} \boldsymbol{x}},$$

where $\boldsymbol{Q} = \boldsymbol{Q}^T \geqslant 0$ and $\boldsymbol{P} = \boldsymbol{P}^T > 0$. Note that if a point $\boldsymbol{x} = [x_1, \ldots, x_n]^T$ is a solution to the problem, then so is any nonzero scalar multiple of it,

$$t\boldsymbol{x} = [tx_1, \ldots, tx_n]^T, \quad t \neq 0.$$

Indeed

$$\frac{(t\boldsymbol{x})^T \boldsymbol{Q}(t\boldsymbol{x})}{(t\boldsymbol{x})^T \boldsymbol{P}(t\boldsymbol{x})} = \frac{t^2 \boldsymbol{x}^T \boldsymbol{Q} \boldsymbol{x}}{t^2 \boldsymbol{x}^T \boldsymbol{P} \boldsymbol{x}} = \frac{\boldsymbol{x}^T \boldsymbol{Q} \boldsymbol{x}}{\boldsymbol{x}^T \boldsymbol{P} \boldsymbol{x}}.$$

Therefore, to avoid the multiplicity of solutions, we further impose the constraint

$$\boldsymbol{x}^T \boldsymbol{P} \boldsymbol{x} = 1.$$

The optimization problem becomes

$$\begin{aligned} &\text{maximize} \quad \boldsymbol{x}^T \boldsymbol{Q} \boldsymbol{x} \\ &\text{subject to} \quad \boldsymbol{x}^T \boldsymbol{P} \boldsymbol{x} = 1. \end{aligned}$$

Let us write

$$\begin{aligned} f(\boldsymbol{x}) &= \boldsymbol{x}^T \boldsymbol{Q} \boldsymbol{x} \\ h(\boldsymbol{x}) &= 1 - \boldsymbol{x}^T \boldsymbol{P} \boldsymbol{x}. \end{aligned}$$

Any feasible point for this problem is regular (see Exercise 19.4). We now apply Lagrange's method. We first form the Lagrangian function

$$l(\boldsymbol{x}, \lambda) = \boldsymbol{x}^T \boldsymbol{Q} \boldsymbol{x} + \lambda(1 - \boldsymbol{x}^T \boldsymbol{P} \boldsymbol{x}).$$

Applying the Lagrange condition yields

$$\begin{aligned} D_x l(\boldsymbol{x}, \lambda) &= 2\boldsymbol{x}^T \boldsymbol{Q} - 2\lambda \boldsymbol{x}^T \boldsymbol{P} = \boldsymbol{0}^T, \\ D_\lambda l(\boldsymbol{x}, \lambda) &= 1 - \boldsymbol{x}^T \boldsymbol{P} \boldsymbol{x} = 0. \end{aligned}$$

The first of the above conditions can be represented as

$$\boldsymbol{Q}\boldsymbol{x} - \lambda \boldsymbol{P} \boldsymbol{x} = \boldsymbol{0}$$

or

$$(\lambda \boldsymbol{P} - \boldsymbol{Q})\boldsymbol{x} = \boldsymbol{0}.$$

This representation is possible because $\boldsymbol{P}=\boldsymbol{P}^T$ and $\boldsymbol{Q}=\boldsymbol{Q}^T$. By assumption, $\boldsymbol{P}>0$, hence $\boldsymbol{P}^{-1}$ exists. Premultiplying $(\lambda\boldsymbol{P}-\boldsymbol{Q})\boldsymbol{x}=\boldsymbol{0}$ by $\boldsymbol{P}^{-1}$, we obtain

$$(\lambda\boldsymbol{I}_n-\boldsymbol{P}^{-1}\boldsymbol{Q})\boldsymbol{x}=\boldsymbol{0}.$$

or, equivalently,

$$\boldsymbol{P}^{-1}\boldsymbol{Q}\boldsymbol{x}=\lambda\boldsymbol{x}.$$

Therefore, the solution, if it exists, is an eigenvector of $\boldsymbol{P}^{-1}\boldsymbol{Q}$, and the Lagrange multiplier is the corresponding eigenvalue. As usual, let $\boldsymbol{x}^*$ and $\lambda^*$ be the optimal solution. Since $\boldsymbol{x}^{*T}\boldsymbol{P}\boldsymbol{x}^*=1$, and $\boldsymbol{P}^{-1}\boldsymbol{Q}\boldsymbol{x}^*=\lambda^*\boldsymbol{x}^*$, we have

$$\lambda^*=\boldsymbol{x}^{*T}\boldsymbol{Q}\boldsymbol{x}^*.$$

Hence, $\lambda^*$ is the maximum of the objective function, and therefore is, in fact, the maximal eigenvalue of $\boldsymbol{P}^{-1}\boldsymbol{Q}$. ■

In the above problems, we are able to find points that are candidates for extremizers of the given objective function subject to equality constraints. These critical points are the only candidates because they are the only points that satisfy the Lagrange conditions. To classify such critical points as minimizers, maximizers, or neither, we need a stronger condition—possibly a necessary and sufficient condition. In the next section, we discuss a second-order necessary condition and a second-order sufficient condition for minimizers.

## 19.5. SECOND-ORDER CONDITIONS

We assume that $f:\mathbb{R}^n\to\mathbb{R}$ and $\boldsymbol{h}:\mathbb{R}^n\to\mathbb{R}^m$ are twice continuously differentiable (i.e., $f,\boldsymbol{h}\in\mathscr{C}^2$). Let

$$l(\boldsymbol{x},\boldsymbol{\lambda})=f(\boldsymbol{x})+\boldsymbol{\lambda}^T\boldsymbol{h}(\boldsymbol{x})=f(\boldsymbol{x})+\lambda_1h_1(\boldsymbol{x})+\cdots+\lambda_mh_m(\boldsymbol{x})$$

be the Lagrangian function. Let $\boldsymbol{L}(\boldsymbol{x},\boldsymbol{\lambda})$ be the Hessian matrix of $l(\boldsymbol{x},\boldsymbol{\lambda})$ with respect to $\boldsymbol{x}$; that is,

$$\boldsymbol{L}(\boldsymbol{x},\boldsymbol{\lambda})=\boldsymbol{F}(\boldsymbol{x})+\lambda_1\boldsymbol{H}_1(\boldsymbol{x})+\cdots+\lambda_m\boldsymbol{H}_m(\boldsymbol{x}),$$

where $\boldsymbol{F}(\boldsymbol{x})$ is the Hessian matrix of $f$ at $\boldsymbol{x}$, and $\boldsymbol{H}_k(\boldsymbol{x})$ is the Hessian matrix of $h_k$ at $\boldsymbol{x}$, $k=1,\ldots,m$, given by

$$\boldsymbol{H}_k(\boldsymbol{x})=\begin{bmatrix}\dfrac{\partial^2h_k}{\partial x_1^2}(\boldsymbol{x}) & \cdots & \dfrac{\partial^2h_k}{\partial x_n\partial x_1}(\boldsymbol{x})\\ \vdots & & \vdots\\ \dfrac{\partial^2h_k}{\partial x_1\partial x_n}(\boldsymbol{x}) & \cdots & \dfrac{\partial^2h_k}{\partial^2 x_n}(\boldsymbol{x})\end{bmatrix}.$$

We introduce the notation $[\boldsymbol{\lambda H}(\boldsymbol{x})]$:

$$[\boldsymbol{\lambda H}(\boldsymbol{x})] = \lambda_1 \boldsymbol{H}_1(\boldsymbol{x}) + \cdots + \lambda_m \boldsymbol{H}_m(\boldsymbol{x}).$$

Using the above notation, we can write

$$\boldsymbol{L}(\boldsymbol{x}, \boldsymbol{\lambda}) = \boldsymbol{F}(\boldsymbol{x}) + [\boldsymbol{\lambda H}(\boldsymbol{x})].$$

**Theorem 19.4. Second-Order Necessary Conditions.** *Let $\boldsymbol{x}^*$ be a local minimizer of $f:\mathbb{R}^n \to \mathbb{R}$ subject to $\boldsymbol{h}(\boldsymbol{x}) = \boldsymbol{0}$, $\boldsymbol{h}:\mathbb{R}^n \to \mathbb{R}^m$, $m \leqslant n$, and $f, \boldsymbol{h} \in \mathscr{C}^2$. Suppose $\boldsymbol{x}^*$ is regular. Then, there exists $\boldsymbol{\lambda}^* \in \mathbb{R}^m$ such that*

1. $Df(\boldsymbol{x}^*) + \boldsymbol{\lambda}^{*T} D\boldsymbol{h}(\boldsymbol{x}^*) = \boldsymbol{0}^T$ and
2. *for all $\boldsymbol{y} \in \boldsymbol{T}(\boldsymbol{x}^*)$, we have $\boldsymbol{y}^T \boldsymbol{L}(\boldsymbol{x}^*, \boldsymbol{\lambda}^*)\boldsymbol{y} \geqslant 0$.*

□

*Proof.* The existence of $\boldsymbol{\lambda}^* \in \mathbb{R}^m$ such that $Df(\boldsymbol{x}^*) + \boldsymbol{\lambda}^{*T} D\boldsymbol{h}(\boldsymbol{x}^*) = \boldsymbol{0}^T$ follows from the Lagrange multiplier theorem. It remains to prove the second part of the result. Suppose $\boldsymbol{y} \in T(\boldsymbol{x}^*)$; that is, $\boldsymbol{y}$ belongs to the tangent space to $S = \{\boldsymbol{x} \in \mathbb{R}^n : \boldsymbol{h}(\boldsymbol{x}) = \boldsymbol{0}\}$ at $\boldsymbol{x}^*$. Since $\boldsymbol{h} \in \mathscr{C}^2$, following the argument of Theorem 19.1, there exists a twice differentiable curve $\{\boldsymbol{x}(t) : t \in (a, b)\}$ on $S$ such that

$$\boldsymbol{x}(t^*) = \boldsymbol{x}^*, \quad \dot{\boldsymbol{x}}(t^*) = \boldsymbol{y}$$

for some $t^* \in (a, \mathrm{b})$. Observe that by assumption, $t^*$ is a local minimizer of the function $\phi(t) = f(\boldsymbol{x}(t))$. From the second-order necessary condition for unconstrained minimization (see Theorem 6.2), we obtain

$$\frac{d^2\phi}{dt^2}(t^*) \geqslant 0.$$

Using the following formula

$$\frac{d}{dt}(\boldsymbol{y}^T(t)\boldsymbol{z}(t)) = \boldsymbol{z}(t)^T \frac{d\boldsymbol{y}}{dt}(t) + \boldsymbol{y}(t)^T \frac{d\boldsymbol{z}}{dt}(t)$$

and applying the chain rule yields

$$\begin{aligned} \frac{d^2\phi}{dt^2}(t^*) &= \frac{d}{dt}[Df(\boldsymbol{x}(t^*))\dot{\boldsymbol{x}}(t^*)] \\ &= \dot{\boldsymbol{x}}(t^*)^T \boldsymbol{F}(\boldsymbol{x}^*)\dot{\boldsymbol{x}}(t^*) + Df(\boldsymbol{x}^*)\ddot{\boldsymbol{x}}(t^*) \\ &= \boldsymbol{y}^T \boldsymbol{F}(\boldsymbol{x}^*)\boldsymbol{y} + Df(\boldsymbol{x}^*)\ddot{\boldsymbol{x}}(t^*) \geqslant 0. \end{aligned}$$

Since, for all $t \in (a, b)$, we have

$$\boldsymbol{h}(\boldsymbol{x}(t)) = \boldsymbol{0},$$

then

$$\frac{d^2}{dt^2} \boldsymbol{\lambda}^{*T} \boldsymbol{h}(\boldsymbol{x}(t)) = 0.$$

Thus, for all $t \in (a, b)$,

$$\begin{aligned}
\frac{d^2}{dt^2} \boldsymbol{\lambda}^{*T} \boldsymbol{h}(\boldsymbol{x}(t)) &= \frac{d}{dt}\left[\boldsymbol{\lambda}^{*T} \frac{d}{dt} \boldsymbol{h}(\boldsymbol{x}(t))\right] \\
&= \frac{d}{dt}\left[\sum_{k=1}^{m} \lambda_k^* \frac{d}{dt} h_k(\boldsymbol{x}(t))\right] \\
&= \frac{d}{dt}\left[\sum_{k=1}^{m} \lambda_k^* Dh_k(\boldsymbol{x}(t))\dot{\boldsymbol{x}}(t)\right] \\
&= \sum_{k=1}^{m} \lambda_k^* \frac{d}{dt}(Dh_k(\boldsymbol{x}(t))\dot{\boldsymbol{x}}(t)) \\
&= \sum_{k=1}^{m} \lambda_k^* [\dot{\boldsymbol{x}}(t)^T \boldsymbol{H}_k(\boldsymbol{x}(t))\dot{\boldsymbol{x}}(t) + Dh_k(\boldsymbol{x}(t))\ddot{\boldsymbol{x}}(t)] \\
&= \dot{\boldsymbol{x}}^T(t)[\boldsymbol{\lambda}^* \boldsymbol{H}(\boldsymbol{x}(t))]\dot{\boldsymbol{x}}(t) + \boldsymbol{\lambda}^{*T} D\boldsymbol{h}(\boldsymbol{x}(t))\ddot{\boldsymbol{x}}(t) \\
&= 0.
\end{aligned}$$

In particular, the above is true for $t = t^*$; that is,

$$\boldsymbol{y}^T[\boldsymbol{\lambda}^* \boldsymbol{H}(\boldsymbol{x}^*)]\boldsymbol{y} + \boldsymbol{\lambda}^{*T} D\boldsymbol{h}(\boldsymbol{x}^*)\ddot{\boldsymbol{x}}(t^*) = 0.$$

Adding the above equation to the inequality

$$\boldsymbol{y}^T \boldsymbol{F}(\boldsymbol{x}^*)\boldsymbol{y} + Df(\boldsymbol{x}^*)\ddot{\boldsymbol{x}}(t^*) \geqslant 0$$

yields

$$\boldsymbol{y}^T(\boldsymbol{F}(\boldsymbol{x}^*) + [\boldsymbol{\lambda}^* \boldsymbol{H}(\boldsymbol{x}^*)])\boldsymbol{y} + (Df(\boldsymbol{x}^*) + \boldsymbol{\lambda}^{*T} D\boldsymbol{h}(\boldsymbol{x}^*))\ddot{\boldsymbol{x}}(t^*) \geqslant 0.$$

But, by the Lagrange multiplier theorem, $Df(\boldsymbol{x}^*) + \boldsymbol{\lambda}^{*T} D\boldsymbol{h}(\boldsymbol{x}^*) = \boldsymbol{0}^T$. Therefore,

$$\boldsymbol{y}^T(\boldsymbol{F}(\boldsymbol{x}^*) + [\boldsymbol{\lambda}^* \boldsymbol{H}(\boldsymbol{x}^*)])\boldsymbol{y} = \boldsymbol{y}^T \boldsymbol{L}(\boldsymbol{x}^*, \boldsymbol{\lambda}^*)\boldsymbol{y} \geqslant 0,$$

which proves the result. ■

Observe that $\boldsymbol{L}(\boldsymbol{x}, \boldsymbol{\lambda})$ plays a similar role as the Hessian matrix $\boldsymbol{F}(\boldsymbol{x})$ of the objective function $f$ did in the unconstrained minimization case. However, we now require that $\boldsymbol{L}(\boldsymbol{x}^*, \boldsymbol{\lambda}^*) \geqslant 0$ only on $T(\boldsymbol{x}^*)$ rather than on $\mathbb{R}^n$.

The above conditions are necessary, but not sufficient, for a point to be a local minimizer. We now present, without a proof, sufficient conditions for a point to be a strict local minimizer.

**Theorem 19.5. Second-Order Sufficient Conditions.** *Suppose $f, \boldsymbol{h} \in \mathscr{C}^2$ and there exists a point $\boldsymbol{x}^* \in \mathbb{R}^n$ and $\boldsymbol{\lambda}^* \in \mathbb{R}^m$ such that*

1. $Df(\boldsymbol{x}^*) + \boldsymbol{\lambda}^{*T} D\boldsymbol{h}(\boldsymbol{x}^*) = \boldsymbol{0}^T$ and
2. *for all* $\boldsymbol{y} \in T(\boldsymbol{x}^*)$, $\boldsymbol{y} \neq \boldsymbol{0}$, *we have* $\boldsymbol{y}^T \boldsymbol{L}(\boldsymbol{x}^*, \boldsymbol{\lambda}^*)\boldsymbol{y} > 0$.

*Then, $\boldsymbol{x}^*$ is a strict local minimizer of $f$ subject to $\boldsymbol{h}(\boldsymbol{x}) = \boldsymbol{0}$.* □

*Proof.* The interested reader can consult Ref. 51 (p. 307) for a proof of this result. ■

The above theorem states that if a point $\boldsymbol{x}^*$ satisfies the Lagrange condition, and $\boldsymbol{L}(\boldsymbol{x}^*, \boldsymbol{\lambda}^*)$ is positive definite on $T(\boldsymbol{x}^*)$, then $\boldsymbol{x}^*$ is a strict local minimizer. A similar result to Theorem 19.5 holds for a strict local maximizer, the only difference being that $\boldsymbol{L}(\boldsymbol{x}^*, \boldsymbol{\lambda}^*)$ be negative definite on $T(\boldsymbol{x}^*)$. We illustrate this condition in the following example.

***Example 19.7.*** Consider the following problem:

$$\text{maximize} \quad \frac{\boldsymbol{x}^T \boldsymbol{Q} \boldsymbol{x}}{\boldsymbol{x}^T \boldsymbol{P} \boldsymbol{x}},$$

where

$$\boldsymbol{Q} = \begin{bmatrix} 4 & 0 \\ 0 & 1 \end{bmatrix}, \quad \boldsymbol{P} = \begin{bmatrix} 2 & 0 \\ 0 & 1 \end{bmatrix}.$$

As pointed out earlier, we can represent the above problem in the equivalent form

$$\begin{aligned} &\text{maximize} \quad \boldsymbol{x}^T \boldsymbol{Q} \boldsymbol{x} \\ &\text{subject to} \quad \boldsymbol{x}^T \boldsymbol{P} \boldsymbol{x} = 1. \end{aligned}$$

The Lagrangian function for the transformed problem is given by

$$l(\boldsymbol{x}, \lambda) = \boldsymbol{x}^T \boldsymbol{Q} \boldsymbol{x} + \lambda(1 - \boldsymbol{x}^T \boldsymbol{P} \boldsymbol{x}).$$

The Lagrange condition yields

$$(\lambda \boldsymbol{I} - \boldsymbol{P}^{-1} \boldsymbol{Q}) \boldsymbol{x} = \boldsymbol{0},$$

where

$$\boldsymbol{P}^{-1} \boldsymbol{Q} = \begin{bmatrix} 2 & 0 \\ 0 & 1 \end{bmatrix}.$$

There are only two values of $\lambda$ that satisfy $(\lambda \boldsymbol{I} - \boldsymbol{P}^{-1}\boldsymbol{Q})\boldsymbol{x} = \boldsymbol{0}$, namely, the eigenvalues of $\boldsymbol{P}^{-1}\boldsymbol{Q}$: $\lambda_1 = 2$, $\lambda_2 = 1$. We recall from our previous discussion of this problem that the Lagrange multiplier corresponding to the solution is the maximum eigenvalue of $\boldsymbol{P}^{-1}\boldsymbol{Q}$, namely, $\lambda^* = \lambda_1 = 2$. The corresponding eigenvector is the maximizer, that is, the solution to the problem. The eigenvector corresponding to the eigenvalue $\lambda^* = 2$ satisfying the constraint $\boldsymbol{x}^T\boldsymbol{P}\boldsymbol{x} = 1$ is $\pm\boldsymbol{x}^*$, where

$$\boldsymbol{x}^* = \left[\frac{1}{\sqrt{2}}, 0\right]^T.$$

At this point, all we have established is that the pairs $(\pm\boldsymbol{x}^*, \lambda^*)$ satisfy the Lagrange condition. We now show that the points $\pm\boldsymbol{x}^*$ are in fact strict local maximizers. We do this for the point $\boldsymbol{x}^*$. A similar procedure applies to $-\boldsymbol{x}^*$. We first compute the Hessian matrix of the Lagrangian function. We have

$$\boldsymbol{L}(\boldsymbol{x}^*, \lambda^*) = 2\boldsymbol{Q} - 2\lambda\boldsymbol{P} = \begin{bmatrix} 0 & 0 \\ 0 & -2 \end{bmatrix}.$$

The tangent space $T(\boldsymbol{x}^*)$ to $\{\boldsymbol{x}: 1 - \boldsymbol{x}^T\boldsymbol{P}\boldsymbol{x}\}$ is

$$\begin{aligned} T(\boldsymbol{x}^*) &= \{\boldsymbol{y} \in \mathbb{R}^2 : \boldsymbol{x}^{*T}\boldsymbol{P}\boldsymbol{y} = 0\} \\ &= \{\boldsymbol{y} : [\sqrt{2}, 0]\boldsymbol{y} = 0\} \\ &= \{\boldsymbol{y} : \boldsymbol{y} = [0, a]^T,\ a \in \mathbb{R}\}. \end{aligned}$$

Note that for each $\boldsymbol{y} \in T(\boldsymbol{x}^*)$, $\boldsymbol{y} \neq \boldsymbol{0}$,

$$\boldsymbol{y}^T\boldsymbol{L}(\boldsymbol{x}^*, \lambda^*)\boldsymbol{y} = [0, a]\begin{bmatrix} 0 & 0 \\ 0 & -2 \end{bmatrix}\begin{bmatrix} 0 \\ a \end{bmatrix} = -2a^2 < 0.$$

Hence, $\boldsymbol{L}(\boldsymbol{x}^*, \lambda^*) < 0$ on $T(\boldsymbol{x}^*)$, and thus $\boldsymbol{x}^* = [1/\sqrt{2}, 0]^T$ is a strict local maximizer. The same is true for the point $-\boldsymbol{x}^*$. Note that

$$\frac{\boldsymbol{x}^{*T}\boldsymbol{Q}\boldsymbol{x}^*}{\boldsymbol{x}^{*T}\boldsymbol{P}\boldsymbol{x}^*} = 2,$$

which, as expected, is the value of the maximal eigenvalue of $\boldsymbol{P}^{-1}\boldsymbol{Q}$. Finally, we point out that any scalar multiple $t\boldsymbol{x}^*$ of $\boldsymbol{x}^*$, $t \neq 0$, is a solution to the original problem of maximizing $\boldsymbol{x}^T\boldsymbol{Q}\boldsymbol{x}/\boldsymbol{x}^T\boldsymbol{P}\boldsymbol{x}$. ■

## 19.6. MINIMIZING $\|x\|$ SUBJECT TO $Ax = b$: A LAGRANGE PERSPECTIVE

Consider a system of linear equations

$$\boldsymbol{A}\boldsymbol{x} = \boldsymbol{b},$$

where $\boldsymbol{A} \in \mathbb{R}^{m \times n}$, $\boldsymbol{b} \in \mathbb{R}^m$, $m < n$, and rank $\boldsymbol{A} = m$. This system of equations has an infinite number of solutions (see Section 2.3). We now show, using the Lagrange multiplier theorem, that there is only one solution that is closest to the origin of $\mathbb{R}^n$, that is, the solution to $\boldsymbol{Ax} = \boldsymbol{b}$ whose Euclidean norm $\|\boldsymbol{x}\|$ is minimal. This problem can be stated as

$$\begin{aligned} &\text{minimize} \quad \|\boldsymbol{x}\| \\ &\text{subject to} \quad \boldsymbol{Ax} = \boldsymbol{b}. \end{aligned}$$

The objective function $f(\boldsymbol{x}) = \|\boldsymbol{x}\|$ above is not differentiable at $\boldsymbol{x} = \boldsymbol{0}$. This precludes the use of the results from the previous sections—those results require differentiability of the objective function. We can overcome this difficulty by considering an equivalent optimization problem:

$$\begin{aligned} &\text{minimize} \quad \tfrac{1}{2}\|\boldsymbol{x}\|^2 \\ &\text{subject to} \quad \boldsymbol{Ax} = \boldsymbol{b}. \end{aligned}$$

The objective function $\|\boldsymbol{x}\|^2/2$ has the same minimizer as the previous objective function $\|\boldsymbol{x}\|$. Indeed, if $\boldsymbol{x}^*$ is such that for all $\boldsymbol{x} \in \mathbb{R}^n$ satisfying $\boldsymbol{Ax} = \boldsymbol{b}$, $\|\boldsymbol{x}^*\| \leqslant \|\boldsymbol{x}\|$, then $\|\boldsymbol{x}^*\|^2/2 \leqslant \|\boldsymbol{x}\|^2/2$. The same is true for the converse.

To solve the above problem, with objective function $\|\boldsymbol{x}\|^2/2$ and constraint equation $\boldsymbol{h}(\boldsymbol{x}) = \boldsymbol{b} - \boldsymbol{Ax} = \boldsymbol{0}$, we form the Lagrangian function

$$l(\boldsymbol{x}, \boldsymbol{\lambda}) = \tfrac{1}{2}\|\boldsymbol{x}\|^2 + \boldsymbol{\lambda}^T(\boldsymbol{b} - \boldsymbol{Ax}).$$

The Lagrange condition yields

$$D_x l(\boldsymbol{x}^*, \boldsymbol{\lambda}^*) = \boldsymbol{x}^{*T} - \boldsymbol{\lambda}^{*T}\boldsymbol{A} = \boldsymbol{0}^T.$$

Rewriting, we get

$$\boldsymbol{x}^* = \boldsymbol{A}^T\boldsymbol{\lambda}^*.$$

Premultiplying both sides of the above by $\boldsymbol{A}$ gives

$$\boldsymbol{Ax}^* = \boldsymbol{AA}^T\boldsymbol{\lambda}^*.$$

Using the fact that $\boldsymbol{Ax}^* = \boldsymbol{b}$, and noting that $\boldsymbol{AA}^T$ is invertible because rank $\boldsymbol{A} = m$, we can solve for $\boldsymbol{\lambda}^*$ to obtain

$$\boldsymbol{\lambda}^* = (\boldsymbol{AA}^T)^{-1}\boldsymbol{b}.$$

Therefore, $\boldsymbol{x}^* = \boldsymbol{A}^T(\boldsymbol{AA}^T)^{-1}\boldsymbol{b}$. The point $\boldsymbol{x}^*$ is the only candidate for a minimizer. To establish that $\boldsymbol{x}^*$ is indeed a minimizer, we verify that $\boldsymbol{x}^*$ satisfies the second-order sufficient conditions. For this, we first find the Hessian matrix of the Lagrangian function at $(\boldsymbol{x}^*, \boldsymbol{\lambda}^*)$. We have

$$\boldsymbol{L}(\boldsymbol{x}^*, \boldsymbol{\lambda}^*) = \boldsymbol{I}_n,$$

which is positive definite. Thus, the point $\boldsymbol{x}^*$ is a strict local minimizer of $\|\boldsymbol{x}\|^2/2$, or equivalently $\|\boldsymbol{x}\|$, subject to $\boldsymbol{A}\boldsymbol{x} = \boldsymbol{b}$.

## EXERCISES

**19.1** Find local extremizers for the following optimization problems:

**a.**

$$\begin{aligned} &\text{minimize} && x_1^2 + 2x_1x_2 + 3x_2^2 + 4x_1 + 5x_2 + 6x_3 \\ &\text{subject to} && x_1 + 2x_2 = 3 \\ &&& 4x_1 + 5x_3 = 6 \end{aligned}$$

**b.**

$$\begin{aligned} &\text{maximize} && 4x_1 + x_2^2 \\ &\text{subject to} && x_1^2 + x_2^2 = 9 \end{aligned}$$

**c.**

$$\begin{aligned} &\text{maximize} && x_1x_2 \\ &\text{subject to} && x_1^2 + 4x_2^2 = 1 \end{aligned}$$

**19.2** Find local extremizers of

**a.** $f(x_1, x_2, x_3) = x_1^2 + 3x_2^2 + x_3$ subject to $x_1^2 + x_2^2 + x_3^2 = 16$
**b.** $f(x_1, x_2) = x_1^2 + x_2^2$ subject to $3x_1^2 + 4x_1x_2 + 6x_2^2 = 140$

**19.3** Consider the problem

$$\begin{aligned} &\text{minimize} && 2x_1 + 3x_2 - 4, \quad x_1, x_2 \in \mathbb{R} \\ &\text{subject to} && x_1x_2 = 6 \end{aligned}$$

**a.** Use the Lagrange multiplier theorem to find all possible local minimizers and maximizers.
**b.** Use the second-order sufficient conditions to specify which points are strict local minimizers and which are strict local maximizers.
**c.** Are the points in b global minimizers or maximizers? Explain.

**19.4** Let $\boldsymbol{P} = \boldsymbol{P}^T$ be a positive definite matrix. Show that any point $\boldsymbol{x}$ satisfying $1 - \boldsymbol{x}^T\boldsymbol{P}\boldsymbol{x} = 0$ is a regular point.

**19.5** Consider the problem:

$$\begin{array}{ll}\text{maximize} & ax_1 + bx_2, \quad x_1, x_2 \in \mathbb{R} \\ \text{subject to} & x_1^2 + x_2^2 = 2\end{array}$$

where $a, b \in \mathbb{R}$. Show that if $[1, 1]^T$ is a solution to the problem, then $a = b$.

**19.6** Consider the problem:

$$\begin{array}{ll}\text{minimize} & x_1x_2 - 2x_1, \quad x_1, x_2 \in \mathbb{R} \\ \text{subject to} & x_1^2 - x_2^2 = 0.\end{array}$$

**a.** Apply the Lagrange multiplier theorem directly to the problem to show that, if a solution exists, it must be either $[1, 1]^T$ or $[-1, 1]^T$.

**b.** Use the second-order necessary conditions to show that $[-1, 1]^T$ cannot possibly be the solution.

**c.** Use the second-order sufficient conditions to show that $[1, 1]^T$ is a strict local minimizer.

**19.7** Let $A \in \mathbb{R}^{m \times n}, m \leqslant n$, rank $A = m$, and $\boldsymbol{x}_0 \in \mathbb{R}^n$. Let $\boldsymbol{x}^*$ be the point on the nullspace of $A$ that is closest to $\boldsymbol{x}_0$ (in the sense of Euclidean norm).

**a.** Show that $\boldsymbol{x}^*$ is orthogonal to $\boldsymbol{x}^* - \boldsymbol{x}_0$.

**b.** Find a formula for $\boldsymbol{x}^*$ in terms of $\boldsymbol{A}$ and $\boldsymbol{x}_0$.

**19.8** Let $\boldsymbol{L}$ be an $n \times n$ real symmetric matrix, and let $\mathcal{M}$ be a subspace of $\mathbb{R}^n$ with dimension $m < n$. Let $\{\boldsymbol{b}_1, \ldots, \boldsymbol{b}_m\} \subset \mathbb{R}^n$ be a basis for $\mathcal{M}$, and let $\boldsymbol{B}$ be the $n \times m$ matrix with $\boldsymbol{b}_i$ as the $i$th column. Let $\boldsymbol{L}_{\mathcal{M}}$ be the $m \times m$ matrix defined by $\boldsymbol{L}_{\mathcal{M}} = \boldsymbol{B}^T\boldsymbol{L}\boldsymbol{B}$. Show that $\boldsymbol{L}$ is positive semidefinite (definite) on $\mathcal{M}$ if, and only if, $\boldsymbol{L}_{\mathcal{M}}$ is positive semidefinite (definite). *Note*: This result is useful for checking that the Hessian of the Lagrangian at a point is positive definite on the tangent space at that point.

**19.9** Consider the sequence $\{x_k\}$, $x_k \in \mathbb{R}$, generated by the recursion

$$x_{k+1} = ax_k + bu_k, \quad k \geqslant 0 \quad (a, b \in \mathbb{R},\ a, b \neq 0)$$

where $u_0, u_1, u_2, \ldots$ is a sequence of control inputs, and the initial condition $x_0 \neq 0$ is given. We wish to find values of control inputs $u_0$ and $u_1$ such that $x_2 = 0$, and the average input energy $(u_0^2 + u_1^2)/2$ is minimized. Denote the optimal inputs by $u_0^*$ and $u_1^*$.

**a.** Find expressions for $u_0^*$ and $u_1^*$ in terms of $a$, $b$, and $x_0$.

**b.** Use the second-order sufficient conditions to show that the point $\boldsymbol{u}^* = [u_0^*, u_1^*]^T$ in part a is a strict local minimizer.

# 20

# Problems with Inequality Constraints

## 20.1. KARUSH–KUHN–TUCKER CONDITIONS

In the previous chapter, we analyzed constrained optimization problems involving only equality constraints. In this chapter, we discuss extremum problems that also involve inequality constraints. The treatment in this chapter parallels that of the previous chapter. In particular, as we shall see, problems with inequality constraints can also be treated using Lagrange multipliers.

We consider the following problem:

$$\begin{aligned} \text{minimize} \quad & f(\boldsymbol{x}) \\ \text{subject to} \quad & \boldsymbol{h}(\boldsymbol{x}) = \boldsymbol{0}, \\ & \boldsymbol{g}(\boldsymbol{x}) \leqslant \boldsymbol{0}, \end{aligned}$$

where $f:\mathbb{R}^n \to \mathbb{R}$, $\boldsymbol{h}:\mathbb{R}^n \to \mathbb{R}^m$, $m \leqslant n$, and $\boldsymbol{g}:\mathbb{R}^n \to \mathbb{R}^p$. For the above general problem, we adopt the following definitions.

***Definition 20.1.*** An inequality constraint $g_j \leqslant 0$ is said to be *active* at $\boldsymbol{x}$ if $g_j(\boldsymbol{x}) = 0$. It is *inactive* at $\boldsymbol{x}$ if $g_j(\boldsymbol{x}) < 0$. ■

By convention, we consider an equality constraint $h_i = 0$ to be always active.

***Definition 20.2.*** Let $\boldsymbol{x}^*$ satisfy $\boldsymbol{h}(\boldsymbol{x}^*) = \boldsymbol{0}$, $\boldsymbol{g}(\boldsymbol{x}^*) \leqslant \boldsymbol{0}$, and let $J(\boldsymbol{x}^*)$ be the index set of active inequality constraints, that is,

$$J(\boldsymbol{x}^*) \triangleq \{j : g_j(\boldsymbol{x}^*) = 0\}.$$

Then, we say that $\boldsymbol{x}^*$ is a *regular point* if the vectors

$$\nabla h_i(\boldsymbol{x}^*), \quad \nabla g_j(\boldsymbol{x}^*), \quad 1 \leqslant i \leqslant m, \quad j \in J(\boldsymbol{x}^*)$$

are linearly independent. ■

We now prove a first-order necessary condition for a point to be a local minimizer. We call this condition the *Karush-Kuhn-Tucker (KKT) condition.* In the literature, this condition is sometimes also called the Kuhn–Tucker condition.

**Theorem 20.1. Karush–Kuhn–Tucker Theorem.** *Let $f, \boldsymbol{h}, \boldsymbol{g} \in \mathscr{C}^1$. Let $\boldsymbol{x}^*$ be a regular point and a local minimizer for the problem of minimizing $f$ subject to $\boldsymbol{h}(\boldsymbol{x}) = \boldsymbol{0}, \boldsymbol{g}(\boldsymbol{x}) \leqslant \boldsymbol{0}$. Then, there exist $\boldsymbol{\lambda}^* \in \mathbb{R}^m$ and $\boldsymbol{\mu}^* \in \mathbb{R}^p$ such that*

1. $\boldsymbol{\mu}^* \geqslant \boldsymbol{0}$
2. $Df(\boldsymbol{x}^*) + \boldsymbol{\lambda}^{*T} D\boldsymbol{h}(\boldsymbol{x}^*) + \boldsymbol{\mu}^{*T} D\boldsymbol{g}(\boldsymbol{x}^*) = \boldsymbol{0}^T$ *and*
3. $\boldsymbol{\mu}^{*T} \boldsymbol{g}(\boldsymbol{x}^*) = 0.$ □

In the above theorem, we refer to $\boldsymbol{\lambda}^*$ as the Lagrange multiplier vector, and $\boldsymbol{\mu}^*$ as the Karush–Kuhn–Tucker (KKT) multiplier vector. We refer to their components as Lagrange multipliers and Karush–Kuhn–Tucker (KKT) multipliers, respectively.

Before proving this theorem, let us first discuss its meaning. Observe that $\mu_j^* \geqslant 0$ (by part 1) and $g_j(\boldsymbol{x}^*) \leqslant 0$. Therefore, the condition

$$\boldsymbol{\mu}^{*T} \boldsymbol{g}(\boldsymbol{x}^*) = \mu_1^* g_1(\boldsymbol{x}^*) + \cdots + \mu_p^*(\boldsymbol{x}^*) g_p(\boldsymbol{x}^*) = 0$$

implies that if $g_j(\boldsymbol{x}^*) < 0$, then $\mu_j^* = 0$; that is, for all $j \notin J(\boldsymbol{x}^*)$, we have $\mu_j^* = 0$. In other words, the KKT multipliers $\mu_j^*$ corresponding to inactive constraints are zero. The other KKT multipliers, $\mu_i^*$, $i \in J(\boldsymbol{x}^*)$, are nonnegative; they may or may not be equal to zero.

***Example 20.1.*** A graphic illustration of the Karush–Kuhn–Tucker (KKT) theorem is given in Figure 20.1. In this two-dimensional example, we have only inequality constraints $g_j(x) \leqslant 0, j = 1, 2, 3$. Note that the point $\boldsymbol{x}^*$ in the figure is indeed a minimizer. The constraint $g_3(\boldsymbol{x}) \leqslant 0$ is inactive; that is, $g_3(\boldsymbol{x}^*) < 0$, and, hence $\mu_3^* = 0$. By the KKT theorem, we have

$$\nabla f(\boldsymbol{x}^*) + \mu_1^* \nabla g_1(\boldsymbol{x}^*) + \mu_2^* \nabla g_2(\boldsymbol{x}^*) = \boldsymbol{0}$$

or, equivalently,

$$\nabla f(\boldsymbol{x}^*) = -\mu_1^* \nabla g_1(\boldsymbol{x}^*) - \mu_2^* \nabla g_2(\boldsymbol{x}^*),$$

where $\mu_1^* > 0$ and $\mu_2^* > 0$. It is easy to graphically interpret the KKT condition above for this example. Specifically, we can see from the Figure 20.1 that $\nabla f(\boldsymbol{x}^*)$ must be a linear combination of the vectors $-\nabla g_1(\boldsymbol{x}^*)$ and $-\nabla g_2(\boldsymbol{x}^*)$ with positive coefficients. This is exactly reflected in the above equation, where the coefficients $\mu_1^*$ and $\mu_2^*$ are the KKT multipliers. ■

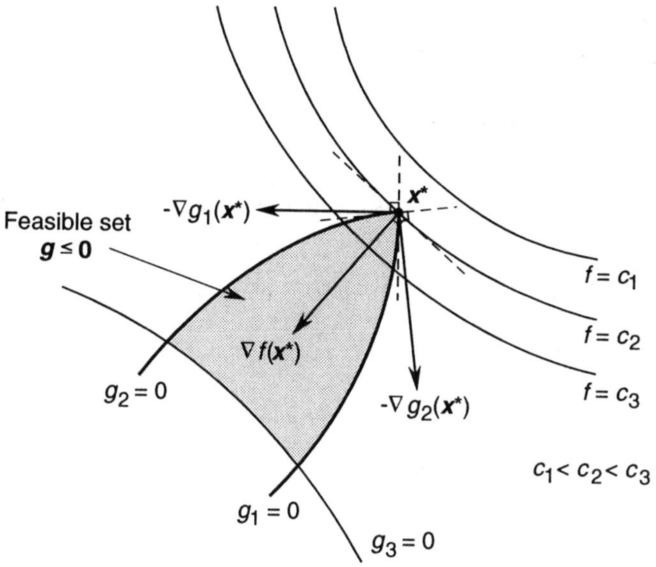


**Figure 20.1.** Graphic illustration of the Karush–Kuhn–Tucker Theorem

We now prove the KKT theorem.

*Proof of Karush–Kuhn–Tucker Theorem.* Let $\boldsymbol{x}^*$ be a regular local minimizer of $f$ on the set $\{\boldsymbol{x}:\boldsymbol{h}(\boldsymbol{x})=\mathbf{0},\boldsymbol{g}(\boldsymbol{x})\leqslant\mathbf{0}\}$. Then, $\boldsymbol{x}^*$ is also a regular local minimizer of $f$ on the set $\{x:\boldsymbol{h}(\boldsymbol{x})=\mathbf{0},g_j(\boldsymbol{x})=0,j\in J(\boldsymbol{x}^*)\}$ (see Exercise 20.6). Note that the latter constraint set involves only equality constraints. Therefore, from the Lagrange theorem, it follows that there exist vectors $\boldsymbol{\lambda}^*\in\mathbb{R}^m$ and $\boldsymbol{\mu}^*\in\mathbb{R}^p$ such that

$$Df(\boldsymbol{x}^*)+\boldsymbol{\lambda}^{*T}D\boldsymbol{h}(\boldsymbol{x}^*)+\boldsymbol{\mu}^{*T}D\boldsymbol{g}(\boldsymbol{x}^*)=\mathbf{0}^T,$$

where for all $j\notin J(\boldsymbol{x}^*)$, we have $\mu_j^*=0$. To complete the proof, it remains to show that for all $j\in J(\boldsymbol{x}^*)$, we have $\mu_j^*\geqslant 0$ (and hence for all $j=1,\ldots,p$, we have $\mu_j^*\geqslant 0$, i.e., $\boldsymbol{\mu}^*\geqslant\mathbf{0}$). We use a proof by contradiction. So suppose that there exists $j\in J(\boldsymbol{x}^*)$ such that $\mu_j^*<0$. Let $\hat{S}$ and $\hat{T}(\boldsymbol{x}^*)$ be the surface and tangent space defined by all other active constraints at $\boldsymbol{x}^*$. Specifically,

$$\hat{S}=\{\boldsymbol{x}:\boldsymbol{h}(\boldsymbol{x})=\mathbf{0},g_i(\boldsymbol{x})=0,i\in J(\boldsymbol{x}^*),i\neq j\},$$

and

$$\hat{T}(\boldsymbol{x}^*)=\{\boldsymbol{y}:D\boldsymbol{h}(\boldsymbol{x}^*)\boldsymbol{y}=\mathbf{0},\ Dg_i(\boldsymbol{x}^*)\boldsymbol{y}=0,i\in J(\boldsymbol{x}^*),i\neq j\}.$$

We claim that, by the regularity of $\boldsymbol{x}^*$, there exists $\boldsymbol{y}\in\hat{T}(\boldsymbol{x}^*)$ such that

$$Dg_j(\boldsymbol{x}^*)\boldsymbol{y}\neq 0.$$

To see this, suppose that for all $y \in \hat{T}(x^*)$, $\nabla g_j(x^*)^T y = Dg_j(x^*) y = 0$. This implies that $\nabla g_j(x^*) \in \hat{T}(x^*)^{\perp}$. By Lemma 19.1, this in turn implies that

$$\nabla g_j(x^*) \in \operatorname{span}[\nabla h_k(x^*), k = 1, \ldots, m, \nabla g_i(x^*), i \in J(x^*), i \neq j]$$

But this contradicts the fact that $x^*$ is a regular point, which proves our claim. Without loss of generality, we assume that we have $y$ such that $Dg_j(x^*) y < 0$.

Consider the Lagrange condition, rewritten as

$$Df(x^*) + \lambda^{*T} Dh(x^*) + \mu_j^* Dg_j(x^*) + \sum_{i \neq j} \mu_i^* Dg_i(x^*) = \mathbf{0}^T.$$

If we postmultiply the above by $y$ and use the fact that $y \in \hat{T}(x^*)$, we get

$$Df(x^*) y = -\mu_j^* Dg_j(x^*) y.$$

Since $Dg_j(x^*) y < 0$, and we have assumed that $\mu_j^* < 0$, we have

$$Df(x^*) y < 0.$$

Since $y \in \hat{T}(x^*)$, by Theorem 19.1, we can find a differentiable curve $\{x(t): t \in (a, b)\}$ on $\hat{S}$ such that there exists $t^* \in (a, b)$ with $x(t^*) = x^*$ and $\dot{x}(t^*) = y$. Now,

$$\frac{d}{dt} f(x(t^*)) = Df(x^*) y < 0.$$

The above means that there is a $\delta > 0$ such that for all $t \in (t^*, t^* + \delta]$, we have

$$f(x(t)) < f(x(t^*)) = f(x^*).$$

On the other hand,

$$\frac{d}{dt} g_j(x(t^*)) = Dg_j(x^*) y < 0,$$

and for some $\varepsilon > 0$ and all $t \in [t^*, t^* + \varepsilon]$, we have that $g_j(x(t)) \leqslant 0$. Therefore, for all $t \in (t^*, t^* + \min(\delta, \varepsilon)]$, we have that $g_j(x(t)) \leqslant 0$ and $f(x(t)) < f(x^*)$. Since the points $x(t)$, $t \in (t^*, t^* + \min(\delta, \varepsilon)]$, are in $\hat{S}$, they are feasible points with lower objective function values than $x^*$. This contradicts the assumption that $x^*$ is a local minimizer, and hence the proof is completed. ■

In the case when the objective function is to be maximized, that is, when the optimization problem has the form

$$\begin{aligned} &\text{maximize} && f(x) \\ &\text{subject to} && h(x) = \mathbf{0} \\ & && g(x) \leqslant \mathbf{0}, \end{aligned}$$

the KKT conditions are

1. $\boldsymbol{\mu}^* \geqslant \mathbf{0}$
2. $-Df(\boldsymbol{x}^*) + \boldsymbol{\lambda}^{*T} D\boldsymbol{h}(\boldsymbol{x}^*) + \boldsymbol{\mu}^{*T} D\boldsymbol{g}(\boldsymbol{x}^*) = \mathbf{0}^T$ and
3. $\boldsymbol{\mu}^{*T} \boldsymbol{g}(\boldsymbol{x}^*) = 0.$

These conditions are easily derived by converting the maximization problem above into a minimization problem, by multiplying the objective function by $-1$. The above conditions can be further rewritten as

1. $\boldsymbol{\mu}^* \leqslant \mathbf{0}$
2. $Df(\boldsymbol{x}^*) + \boldsymbol{\lambda}^{*T} D\boldsymbol{h}(\boldsymbol{x}^*) + \boldsymbol{\mu}^{*T} D\boldsymbol{g}(\boldsymbol{x}^*) = \mathbf{0}^T$ and
3. $\boldsymbol{\mu}^{*T} \boldsymbol{g}(\boldsymbol{x}^*) = 0.$

The above conditions are obtained from the previous ones by changing the signs of $\boldsymbol{\mu}^*$ and $\boldsymbol{\lambda}^*$ and multiplying part 2 by $-1$.

We can similarly derive KKT conditions for the case when the inequality constraint is of the form $\boldsymbol{g}(\boldsymbol{x}) \geqslant \mathbf{0}$. Specifically, consider the problem

$$\begin{aligned} \text{minimize} \quad & f(\boldsymbol{x}) \\ \text{subject to} \quad & \boldsymbol{h}(\boldsymbol{x}) = \mathbf{0} \\ & \boldsymbol{g}(\boldsymbol{x}) \geqslant \mathbf{0}. \end{aligned}$$

We multiply the inequality constraint function by $-1$, to obtain $-\boldsymbol{g}(\boldsymbol{x}) \leqslant \mathbf{0}$. Thus, the KKT conditions for this case are

1. $\boldsymbol{\mu}^* \geqslant \mathbf{0}$;
2. $Df(\boldsymbol{x}^*) + \boldsymbol{\lambda}^{*T} D\boldsymbol{h}(\boldsymbol{x}^*) - \boldsymbol{\mu}^{*T} D\boldsymbol{g}(\boldsymbol{x}^*) = \mathbf{0}^T$; and
3. $\boldsymbol{\mu}^{*T} \boldsymbol{g}(\boldsymbol{x}^*) = 0.$

Changing the sign of $\boldsymbol{\mu}^*$, as before, we obtain

1. $\boldsymbol{\mu}^* \leqslant \mathbf{0}$;
2. $Df(\boldsymbol{x}^*) + \boldsymbol{\lambda}^{*T} D\boldsymbol{h}(\boldsymbol{x}^*) + \boldsymbol{\mu}^{*T} D\boldsymbol{g}(\boldsymbol{x}^*) = \mathbf{0}^T$; and
3. $\boldsymbol{\mu}^{*T} \boldsymbol{g}(\boldsymbol{x}^*) = 0.$

For the problem

$$\begin{aligned} \text{maximize} \quad & f(\boldsymbol{x}) \\ \text{subject to} \quad & \boldsymbol{h}(\boldsymbol{x}) = \mathbf{0} \\ & \boldsymbol{g}(\boldsymbol{x}) \geqslant \mathbf{0}, \end{aligned}$$

the KKT conditions are exactly the same as in Theorem 20.1.

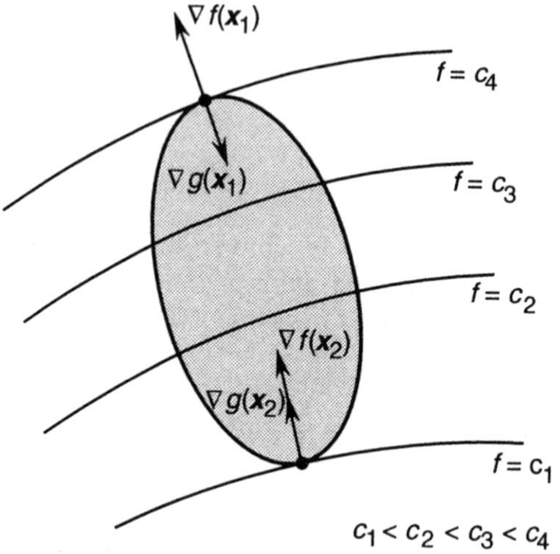


**Figure 20.2.** Points satisfying the Karush–Kuhn–Tucker conditions ($\boldsymbol{x}_1$ is a maximizer and $\boldsymbol{x}_2$ is a minimizer)

***Example 20.2.*** In Figure 20.2, the two points $\boldsymbol{x}_1$ and $\boldsymbol{x}_2$ are feasible points; that is, $g(\boldsymbol{x}_1) \geqslant 0$ and $g(\boldsymbol{x}_2) \geqslant 0$, and they satisfy the KKT conditions.

The point $\boldsymbol{x}_1$ is a maximizer. The KKT conditions for this point (with KKT multiplier $\mu_1$) are

1. $\mu_1 \geqslant 0$
2. $\nabla f(\boldsymbol{x}_1) + \mu_1 \nabla g(\boldsymbol{x}_1) = \mathbf{0}$
3. $\mu_1 g(\boldsymbol{x}_1) = 0$

The point $\boldsymbol{x}_2$ is a minimizer of $f$. The KKT conditions for this point (with KKT multiplier $\mu_2$) are

1. $\mu_2 \leqslant 0$
2. $\nabla f(\boldsymbol{x}_2) + \mu_2 \nabla g(\boldsymbol{x}_2) = \mathbf{0}$
3. $\mu_2 g(\boldsymbol{x}_2) = 0$

■

***Example 20.3.*** Consider the problem

$$\begin{aligned} &\text{minimize} && f(x_1, x_2) \\ &\text{subject to} && x_1, x_2 \geqslant 0, \end{aligned}$$

where

$$f(x_1, x_2) = x_1^2 + x_2^2 + x_1 x_2 - 3x_1.$$

The KKT conditions for this problem are

1. $\boldsymbol{\mu} = [\mu_1, \mu_2]^T \leqslant \mathbf{0}$
2. $Df(\boldsymbol{x}) + \boldsymbol{\mu}^T = \mathbf{0}^T$
3. $\boldsymbol{\mu}^T \boldsymbol{x} = 0$

We have

$$Df(\boldsymbol{x}) = [2x_1 + x_2 - 3, x_1 + 2x_2].$$

This gives

$$2x_1 + x_2 + \mu_1 = 3$$
$$x_1 + 2x_2 + \mu_2 = 0$$
$$\mu_1 x_1 + \mu_2 x_2 = 0.$$

We now have four variables, three equations, and the inequality constraints on each variable. To find a solution $(\boldsymbol{x}^*, \boldsymbol{\mu}^*)$, we first try

$$\mu_1^* = 0, \quad x_2^* = 0,$$

which gives

$$x_1^* = \tfrac{3}{2}, \quad \mu_2^* = -\tfrac{3}{2}.$$

The above satisfies all the KKT and feasibility conditions. In a similar fashion, we can try

$$\mu_2^* = 0, \quad x_1^* = 0,$$

which gives

$$x_2^* = 0, \quad \mu_1^* = 3.$$

This point clearly violates the nonpositivity constraint on $\mu_1^*$.

The feasible point above satisfying the KKT condition is only a candidate for a minimizer. However, there is no guarantee that the point is indeed a minimizer, since the KKT conditions are, in general, only necessary. A sufficient condition for a point to be a minimizer is given in the next section. ■

The above example is a special case of a more general problem of the form

$$\begin{aligned} &\text{minimize} \quad f(\boldsymbol{x}) \\ &\text{subject to} \quad \boldsymbol{x} \geqslant \mathbf{0}. \end{aligned}$$

The KKT conditions for this problem have the form

$$\begin{aligned} \boldsymbol{\mu} &\leqslant \mathbf{0} \\ \nabla f(\boldsymbol{x}) + \boldsymbol{\mu} &= \mathbf{0} \\ \boldsymbol{\mu}^T \boldsymbol{x} &= 0. \end{aligned}$$

From the above, we deduce

$$\begin{aligned} \nabla f(\boldsymbol{x}) &\geqslant \mathbf{0} \\ \boldsymbol{x}^T \nabla f(\boldsymbol{x}) &= 0 \\ \boldsymbol{x} &\geqslant \mathbf{0}. \end{aligned}$$

Some possible points in $\mathbb{R}^2$ that satisfy the above conditions are depicted in Figure 20.3.

For further results related to the KKT conditions, we refer the reader to Ref. 54 (Chapter 7).

## 20.2. SECOND-ORDER CONDITIONS

As in the case of extremum problems with equality constraints, we can also give second-order necessary and sufficient conditions for extremum problems involving inequality constraints. For this, we need to define the following matrix:

$$\boldsymbol{L}(\boldsymbol{x}, \boldsymbol{\lambda}, \boldsymbol{\mu}) = \boldsymbol{F}(\boldsymbol{x}) + [\boldsymbol{\lambda} \boldsymbol{H}(\boldsymbol{x})] + [\boldsymbol{\mu} \boldsymbol{G}(\boldsymbol{x})]$$

where $\boldsymbol{F}(\boldsymbol{x})$ is the Hessian matrix of $f$ at $\boldsymbol{x}$, and the notation $[\boldsymbol{\lambda} \boldsymbol{H}(\boldsymbol{x})]$ represents

$$[\boldsymbol{\lambda} \boldsymbol{H}(\boldsymbol{x})] = \lambda_1 \boldsymbol{H}_1(\boldsymbol{x}) + \cdots + \lambda_m \boldsymbol{H}_m(\boldsymbol{x}),$$

as before. Similarly, the notation $[\boldsymbol{\mu} \boldsymbol{G}(\boldsymbol{x})]$ represents

$$[\boldsymbol{\mu} \boldsymbol{G}(\boldsymbol{x})] = \mu_1 \boldsymbol{G}_1(\boldsymbol{x}) + \cdots + \mu_p \boldsymbol{G}_p(\boldsymbol{x}),$$

where $\boldsymbol{G}_k(\boldsymbol{x})$ is the Hessian of $g_k$ at $\boldsymbol{x}$, given by

$$\boldsymbol{G}_k(\boldsymbol{x}) = \begin{bmatrix} \dfrac{\partial^2 g_k}{\partial x_1^2}(\boldsymbol{x}) & \cdots & \dfrac{\partial^2 g_k}{\partial x_n \partial x_1}(\boldsymbol{x}) \\ \vdots & & \vdots \\ \dfrac{\partial^2 g_k}{\partial x_1 \partial x_n}(\boldsymbol{x}) & \cdots & \dfrac{\partial^2 g_k}{\partial^2 x_n}(\boldsymbol{x}) \end{bmatrix}.$$

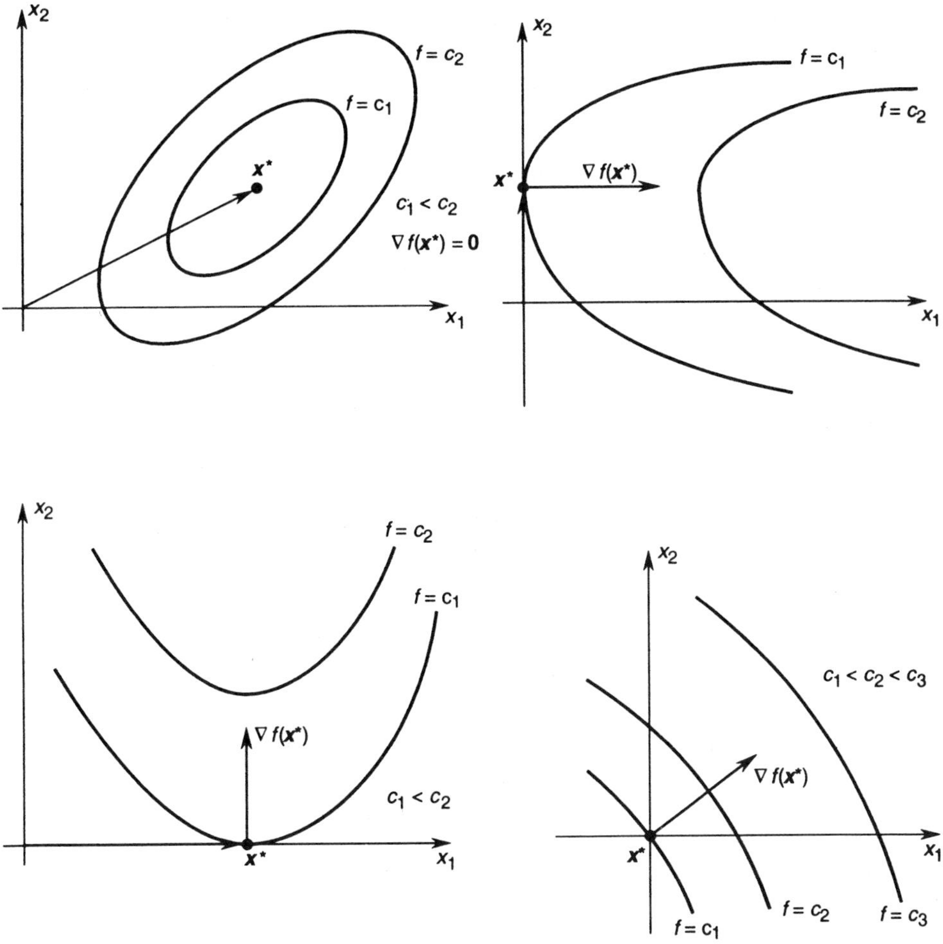


**Figure 20.3.** Some possible points satisfying the Karush–Kuhn–Tucker conditions for problems with positive constraints. Adapted from Ref. 9

In the following theorem, we use

$$T(\boldsymbol{x}^*) = \{\boldsymbol{y} \in \mathbb{R}^n : D\boldsymbol{h}(\boldsymbol{x}^*)\boldsymbol{y} = \boldsymbol{0}, \quad Dg_j(\boldsymbol{x}^*)\boldsymbol{y} = 0, j \in J(\boldsymbol{x}^*)\},$$

that is, the tangent space to the surface defined by active constraints.

**Theorem 20.2. Second-Order Necessary Conditions.** *Let $\boldsymbol{x}^*$ be a local minimizer of $f: \mathbb{R}^n \to \mathbb{R}$ subject to $\boldsymbol{h}(\boldsymbol{x}) = \boldsymbol{0}$, $\boldsymbol{g}(\boldsymbol{x}) \leqslant \boldsymbol{0}$, $\boldsymbol{h}: \mathbb{R}^n \to \mathbb{R}^m$, $m \leqslant n$, $\boldsymbol{g}: \mathbb{R}^n \to \mathbb{R}^p$, and $f, \boldsymbol{h}, \boldsymbol{g} \in \mathscr{C}^2$. Suppose $\boldsymbol{x}^*$ is regular. Then, there exist $\boldsymbol{\lambda}^* \in \mathbb{R}^m$ and $\boldsymbol{\mu}^* \in \mathbb{R}^p$ such that*

1. $\boldsymbol{\mu}^* \geqslant \boldsymbol{0}$, $Df(\boldsymbol{x}^*) + \boldsymbol{\lambda}^{*T} D\boldsymbol{h}(\boldsymbol{x}^*) + \boldsymbol{\mu}^{*T} D\boldsymbol{g}(\boldsymbol{x}^*) = \boldsymbol{0}^T$, $\boldsymbol{\mu}^{*T}\boldsymbol{g}(\boldsymbol{x}^*) = 0$; *and*
2. *for all* $\boldsymbol{y} \in T(\boldsymbol{x}^*)$, *we have* $\boldsymbol{y}^T \boldsymbol{L}(\boldsymbol{x}^*, \boldsymbol{\lambda}^*, \boldsymbol{\mu}^*)\boldsymbol{y} \geqslant 0$. □

*Proof.* Part 1 is simply a result of the Karush–Kuhn–Tucker theorem. To prove part 2, we note that since the point $\boldsymbol{x}^*$ is a local minimizer over $\{\boldsymbol{x}:\boldsymbol{h}(\boldsymbol{x}^*)=\mathbf{0}, \boldsymbol{g}(\boldsymbol{x}^*)\leqslant\mathbf{0}\}$, then it is also a local minimizer over $\{\boldsymbol{x}:\boldsymbol{h}(\boldsymbol{x}^*)=\mathbf{0}, g_j(\boldsymbol{x}^*)=\mathbf{0}, j\in J(\boldsymbol{x}^*)\}$; that is, the point $\boldsymbol{x}^*$ is a local minimizer with active constraints taken as equality constraints. Hence, the second-order necessary conditions for equality constraints (Theorem 19.4) are applicable here, which completes the proof. ■

We now state the second-order sufficient conditions for extremum problems involving inequality constraints. In the formulation of the result, we use the following set:

$$\tilde{T}(\boldsymbol{x}^*,\boldsymbol{\mu}^*)=\{\boldsymbol{y}:D\boldsymbol{h}(\boldsymbol{x}^*)\boldsymbol{y}=\mathbf{0},\quad Dg_i(\boldsymbol{x}^*)\boldsymbol{y}=0,\quad i\in\tilde{J}(\boldsymbol{x},\boldsymbol{\mu}^*)\}$$

where $\tilde{J}(\boldsymbol{x}^*,\boldsymbol{\mu}^*)=\{i:g_i(\boldsymbol{x}^*)=0,\ \mu_i^*>0\}$. Note that $\tilde{J}(\boldsymbol{x}^*,\boldsymbol{\mu}^*)$ is a subset of $J(\boldsymbol{x}^*)$, that is, $\tilde{J}(\boldsymbol{x}^*,\boldsymbol{\mu}^*)\subset J(\boldsymbol{x}^*)$. This, in turn, implies that $T(\boldsymbol{x}^*)$ is a subset of $\tilde{T}(\boldsymbol{x}^*,\boldsymbol{\mu}^*)$, that is, $T(\boldsymbol{x}^*)\subset\tilde{T}(\boldsymbol{x}^*,\boldsymbol{\mu}^*)$.

**Theorem 20.3. Second-Order Sufficient Conditions.** *Suppose $f,\boldsymbol{g},\boldsymbol{h}\in\mathscr{C}^2$ and there exists a feasible point $\boldsymbol{x}^*\in\mathbb{R}^n$ and vectors $\boldsymbol{\lambda}^*\in\mathbb{R}^m$ and $\boldsymbol{\mu}^*\in\mathbb{R}^p$ such that*

1. $\boldsymbol{\mu}^*\geqslant\mathbf{0}$, $Df(\boldsymbol{x}^*)+\boldsymbol{\lambda}^{*T}D\boldsymbol{h}(\boldsymbol{x}^*)+\boldsymbol{\mu}^{*T}D\boldsymbol{g}(\boldsymbol{x}^*)=\mathbf{0}^T$, $\boldsymbol{\mu}^{*T}\boldsymbol{g}(\boldsymbol{x}^*)=0$; *and*
2. *for all* $\boldsymbol{y}\in\tilde{T}(\boldsymbol{x}^*,\boldsymbol{\mu}^*)$, $\boldsymbol{y}\neq\mathbf{0}$, *we have* $\boldsymbol{y}^T\boldsymbol{L}(\boldsymbol{x}^*,\boldsymbol{\lambda}^*,\boldsymbol{\mu}^*)\boldsymbol{y}>0$.

*Then, $\boldsymbol{x}^*$ is a strict local minimizer of $f$ subject to $\boldsymbol{h}(\boldsymbol{x})=\mathbf{0}$, $\boldsymbol{g}(\boldsymbol{x})\leqslant\mathbf{0}$.* □

*Proof.* For a proof of this theorem, we refer the reader to Ref. 51 (p. 317). ■

A similar result to Theorem 20.3 holds for a strict local maximizer, the only difference being that we need $\boldsymbol{\mu}^*\leqslant\mathbf{0}$ and that $\boldsymbol{L}(\boldsymbol{x}^*,\boldsymbol{\lambda}^*)$ be negative definite on $\tilde{T}(\boldsymbol{x}^*,\boldsymbol{\mu}^*)$.

With the above result, we can now analytically solve the problem in Example 19.1, which we previously solved graphically.

***Example 20.4.*** We wish to minimize $f(\boldsymbol{x})=(x_1-1)^2+x_2-2$ subject to

$$h(\boldsymbol{x})=x_2-x_1-1=0,$$
$$g(\boldsymbol{x})=x_1+x_2-2\leqslant 0.$$

For all $\boldsymbol{x}\in\mathbb{R}^2$, we have

$$Dh(\boldsymbol{x})=[-1,1],\quad Dg(\boldsymbol{x})=[1,1].$$

Thus, $\nabla h(\boldsymbol{x})$ and $\nabla g(\boldsymbol{x})$ are linearly independent and hence all feasible points are regular. We first write the KKT conditions. Since $Df(\boldsymbol{x})=[2x_1-2,1]$, we have

$$Df(\boldsymbol{x})+\lambda Dh(\boldsymbol{x})+\mu Dg(\boldsymbol{x})=[2x_1-2-\lambda+\mu,1+\lambda+\mu]=\mathbf{0}^T$$
$$\mu(x_1+x_2-2)=0,$$
$$\mu\geqslant 0.$$

To find points that satisfy the above conditions, we first try $\mu > 0$, which implies $x_1 + x_2 - 2 = 0$. Thus, we are faced with a system of four linear equations

$$\begin{aligned} 2x_1 - 2 - \lambda + \mu &= 0 \\ 1 + \lambda + \mu &= 0 \\ x_2 - x_1 - 1 &= 0 \\ x_1 + x_2 - 2 &= 0. \end{aligned}$$

Solving the above system of equations we obtain

$$x_1 = \tfrac{1}{2}, \quad x_2 = \tfrac{3}{2}, \quad \lambda = -1, \quad \mu = 0.$$

However, the above is not a legitimate solution to the KKT conditions, since we obtained $\mu = 0$, which contradicts the assumption that $\mu > 0$.

In the second try, we assume $\mu = 0$. Thus, we have to solve the following system of equations

$$\begin{aligned} 2x_1 - 2 - \lambda &= 0 \\ 1 + \lambda &= 0 \\ x_2 - x_1 - 1 &= 0, \end{aligned}$$

and the solutions must satisfy

$$g(x_1, x_2) = x_1 + x_2 - 2 \leqslant 0.$$

Solving the above equations, we obtain

$$x_1 = \tfrac{1}{2}, \quad x_2 = \tfrac{3}{2}, \quad \lambda = -1.$$

Note that $\boldsymbol{x}^* = [\frac{1}{2}, \frac{3}{2}]^T$ satisfies the constraint $g(\boldsymbol{x}^*) \leqslant 0$. The point $\boldsymbol{x}^*$ satisfying the KKT necessary conditions is therefore the candidate for being a minimizer.

We now verify that $\boldsymbol{x}^* = [\frac{1}{2}, \frac{3}{2}]^T$, $\lambda^* = -1$, and $\mu^* = 0$ satisfy the second-order sufficient conditions. For this, we form the matrix

$$\begin{aligned} \boldsymbol{L}(\boldsymbol{x}^*, \lambda^*, \mu^*) &= \boldsymbol{F}(\boldsymbol{x}^*) + \lambda^* \boldsymbol{H}(\boldsymbol{x}^*) + \mu^* \boldsymbol{G}(\boldsymbol{x}^*) \\ &= \begin{bmatrix} 2 & 0 \\ 0 & 0 \end{bmatrix} + (-1) \begin{bmatrix} 0 & 0 \\ 0 & 0 \end{bmatrix} + (0) \begin{bmatrix} 0 & 0 \\ 0 & 0 \end{bmatrix} \\ &= \begin{bmatrix} 2 & 0 \\ 0 & 0 \end{bmatrix}. \end{aligned}$$

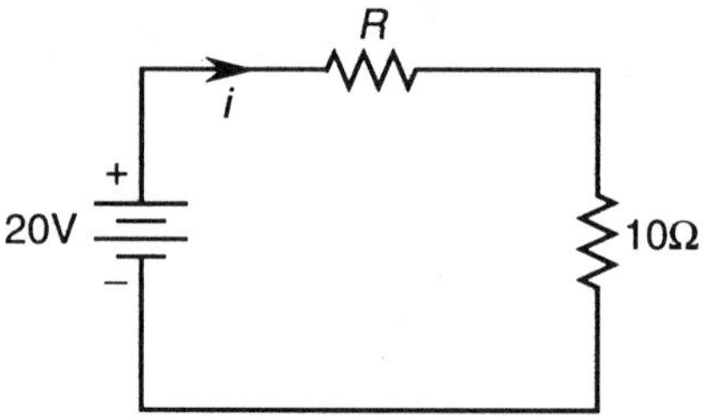


**Figure 20.4.** Exercise 20.3.

We then find the subspace

$$\tilde{T}(\boldsymbol{x}^*, \mu^*) = \{\boldsymbol{y} : Dh(\boldsymbol{x}^*)\boldsymbol{y} = 0\}.$$

Note that, since $\mu^* = 0$, the active constraint $g(\boldsymbol{x}^*) = 0$ does not enter the computation of $\tilde{T}(\boldsymbol{x}^*, \mu^*)$. Note also that in this case, $T(\boldsymbol{x}^*) = \{\boldsymbol{0}\}$. We have

$$\tilde{T}(\boldsymbol{x}^*, \mu^*) = \{\boldsymbol{y} : [-1, 1]\boldsymbol{y} = 0\} = \{[a, a]^T : a \in \mathbb{R}\}.$$

We then check for positive definiteness of $\boldsymbol{L}(\boldsymbol{x}^*, \lambda^*, \mu^*)$ on $\tilde{T}(\boldsymbol{x}^*, \mu^*)$. We have

$$\boldsymbol{y}^T \boldsymbol{L}(\boldsymbol{x}^*, \lambda^*, \mu^*)\boldsymbol{y} = [a, a]\begin{bmatrix} 2 & 0 \\ 0 & 0 \end{bmatrix}\begin{bmatrix} a \\ a \end{bmatrix} = 2a^2.$$

Thus, $\boldsymbol{L}(\boldsymbol{x}^*, \lambda^*, \mu^*)$ is positive definite on $\tilde{T}(\boldsymbol{x}^*, \mu^*)$. Observe that $\boldsymbol{L}(\boldsymbol{x}^*, \lambda^*, \mu^*)$ is in fact only positive semidefinite on $\mathbb{R}^2$.

By the second-order sufficient conditions, we conclude that $\boldsymbol{x}^* = [\frac{1}{2}, \frac{3}{2}]^T$ is a strict local minimizer. ■

## EXERCISES

**20.1** Find local extremizers for

**a.** $x_1^2 + x_2^2 - 2x_1 - 10x_2 + 26$ subject to $\frac{1}{5}x_2 - x_1^2 \leqslant 0$, $5x_1 + \frac{1}{2}x_2 \leqslant 5$.
**b.** $x_1^2 + x_2^2$ subject to $x_1 \geqslant 0$, $x_2 \geqslant 0$, $x_1 + x_2 \geqslant 5$.
**c.** $x_1^2 + 6x_1x_2 - 4x_1 - 2x_2$ subject to $x_1^2 + 2x_2 \leqslant 1$, $2x_1 - 2x_2 \leqslant 1$.

**20.2** Find local minimizers for $x_1^2 + x_2^2$ subject to $x_1^2 + 2x_1x_2 + x_2^2 = 1$, $x_1^2 - x_2 \leqslant 0$.

**20.3** Consider the circuit in Figure 20.4.

Formulate and solve the Karush–Kuhn–Tucker conditions for the following problems:

**a.** Find the value of the resistor $R \geqslant 0$ such that the power absorbed by this resistor is maximized.

**b.** Find the value of the resistor $R \geqslant 0$ such that the power delivered to the $10\Omega$ resistor is maximized.

**20.4** Consider a square room, with corners located at $[0,0]^T$, $[0,2]^T$, $[2,0]^T$, and $[2,2]^T$ (in $\mathbb{R}^2$). We wish to find the point in the room that is closest to the point $[3,4]^T$.

**a.** Guess which point in the box is the closest point in the room to the point $[3,4]^T$.

**b.** Prove that the point you have guessed is a strict local minimizer by using the second-order sufficient conditions.

*Hint*: Minimizing the distance is the same as minimizing the square distance.

**20.5** Consider a linear programming problem in standard form (see Chapter 15).

**a.** Write down the Karush–Kuhn–Tucker conditions for the problem.

**b.** Use part a to show that if there exists an optimal feasible solution to the linear program, then there exists a feasible solution to the corresponding dual problem that achieves an objective function value that is the same as the optimal value of the primal (compare this with Theorem 17.1).

**c.** Use parts a and b to prove that if $\boldsymbol{x}^*$ is an optimal feasible solution of the primal, then there exists a feasible solution $\boldsymbol{\lambda}^*$ to the dual such that $(\boldsymbol{c}^T - \boldsymbol{\lambda}^{*T}\boldsymbol{A})\boldsymbol{x}^* = 0$ (compare this with Theorem 17.3).

**20.6** Consider the constraint set $S = \{\boldsymbol{x}:\boldsymbol{h}(\boldsymbol{x}) = \boldsymbol{0}, \boldsymbol{g}(\boldsymbol{x}) \leqslant \boldsymbol{0}\}$. Let $\boldsymbol{x}^* \in S$ be a regular local minimizer of $f$ over $S$, and $J(\boldsymbol{x}^*)$ the index set of active inequality constraints. Show that $\boldsymbol{x}^*$ is also a regular local minimizer of $f$ over the set $S' = \{\boldsymbol{x}:\boldsymbol{h}(\boldsymbol{x}) = \boldsymbol{0},\ g_j(\boldsymbol{x}) = 0,\ j \in J(\boldsymbol{x}^*)\}$.

**20.7** Solve the following optimization problem using the second-order sufficient conditions:

$$\begin{aligned} \text{minimize} \quad & x_1^2 + x_2^2 \\ \text{subject to} \quad & x_1^2 - x_2 - 4 \leqslant 0 \\ & x_2 - x_1 - 2 \leqslant 0. \end{aligned}$$

See Figure 21.1 for a graphic illustration of the problem.

**20.8** Solve the following optimization problem using the second-order sufficient conditions:

$$\begin{aligned} &\text{minimize} \quad x_1^2 + x_2^2 \\ &\text{subject to} \quad x_1 - x_2^2 - 4 \geqslant 0 \\ &\qquad\qquad\quad x_1 - 10 \leqslant 0. \end{aligned}$$

See Figure 21.2 for a graphic illustration of the problem.

**20.9** Consider the problem:

$$\begin{aligned} &\text{minimize} \quad x_1^2 + x_2^2 \\ &\text{subject to} \quad 4 - x_1 - x_2^2 \leqslant 0 \\ &\qquad\qquad\quad 3x_2 - x_1 \leqslant 0 \\ &\qquad\qquad\quad -3x_2 - x_1 \leqslant 0. \end{aligned}$$

Figure 21.3 gives a graphic illustration of the problem. Deduce from the figure that the problem has two strict local minimizers, and use the second-order sufficient conditions to verify the graphic solutions.

**20.10** Consider the problem:

$$\begin{aligned} &\text{minimize} \quad (x_1 - a)^2 + (x_2 - b)^2, \quad x_1, x_2 \in \mathbb{R} \\ &\text{subject to} \quad x_1^2 + x_2^2 \leqslant 1 \end{aligned}$$

where $a, b \in \mathbb{R}$ are given constants satisfying $a^2 + b^2 \geqslant 1$.

**a.** Let $\boldsymbol{x}^* = [x_1^*, x_2^*]^T$ be a solution to the above problem. Use the first-order necessary conditions for unconstrained optimization to show that $(x_1^*)^2 + (x_2^*)^2 = 1$.

**b.** Use the Karush–Kuhn–Tucker theorem to show that the solution $\boldsymbol{x}^* = [x_1^*, x_2^*]^T$ is unique, and has the form $x_1^* = \alpha a$, $x_2^* = \alpha b$, where $\alpha \in \mathbb{R}$ is a positive constant.

**c.** Find an expression for $\alpha$ (from part b) in terms of $a$ and $b$.

**20.11** Consider the problem:

$$\begin{aligned} &\text{minimize} \quad x_1^2 + (x_2 + 1)^2, \quad x_1, x_2 \in \mathbb{R} \\ &\text{subject to} \quad x_2 \geqslant \exp(x_1) \end{aligned}$$

[$\exp(x) = e^x$ is the exponential of $x$]. Let $\boldsymbol{x}^* = [x_1^*, x_2^*]^T$ be the solution to the problem.

**a.** Write down the Karush–Kuhn–Tucker conditions that must be satisfied by $\boldsymbol{x}^*$.

**b.** Prove that $x_2^* = \exp(x_1^*)$.

**c.** Prove that $-2 < x_1^* < 0$.

**20.12** Consider the problem:

$$\begin{aligned} &\text{minimize} \quad \boldsymbol{c}^T\boldsymbol{x} + 8 \\ &\text{subject to} \quad \tfrac{1}{2}\|\boldsymbol{x}\|^2 \leqslant 1, \end{aligned}$$

where $\boldsymbol{c} \in \mathbb{R}^n$, $\boldsymbol{c} \neq \boldsymbol{0}$. Suppose $\boldsymbol{x}^* = \alpha\boldsymbol{e}$ is a solution to the problem, where $\alpha \in \mathbb{R}$ and $\boldsymbol{e} = [1, \ldots, 1]^T$, and the corresponding objective value is 4.

**a.** Show that $\|\boldsymbol{x}^*\|^2 = 2$.

**b.** Find $\alpha$ and $\boldsymbol{c}$ (they may depend on $n$).

**20.13** Consider the problem

$$\begin{aligned} &\text{minimize} \quad f(\boldsymbol{x}) \\ &\text{subject to} \quad \boldsymbol{h}(\boldsymbol{x}) = \boldsymbol{0}. \end{aligned}$$

We can convert the above into the equivalent optimization problem

$$\begin{aligned} &\text{minimize} \quad f(\boldsymbol{x}) \\ &\text{subject to} \quad \tfrac{1}{2}\|\boldsymbol{h}(\boldsymbol{x})\|^2 \leqslant 0. \end{aligned}$$

Write down the Karush–Kuhn–Tucker conditions for the equivalent problem (with inequality constraint), and explain why the Karush–Kuhn–Tucker theorem cannot be applied in this case.

# 21

# Convex Optimization Problems

## 21.1. INTRODUCTION

The optimization problems posed at the beginning of this part are in general very difficult to solve. The source of these difficulties may be in the objective function or the constraints. Even if the objective function is simple and well-behaved, the nature of the constraints may make the problem difficult to solve. We illustrate some of these difficulties in the following examples.

***Example 21.1.*** Consider the optimization problem

$$\begin{aligned} \text{minimize} \quad & x_1^2 + x_2^2 \\ \text{subject to} \quad & x_2 - x_1 - 2 \leqslant 0 \\ & x_1^2 - x_2 - 4 \leqslant 0. \end{aligned}$$

The problem is depicted in Figure 21.1. As we can see in Figure 21.1, the constrained minimizer is the same as the unconstrained minimizer. At the minimizer, all the constraints are inactive. If we had only known about this fact, we could have approached this problem as an unconstrained optimization problem using techniques from Part II. ■

***Example 21.2.*** Consider the optimization problem

$$\begin{aligned} \text{minimize} \quad & x_1^2 + x_2^2 \\ \text{subject to} \quad & x_1 - 10 \leqslant 0 \\ & x_1 - x_2^2 - 4 \geqslant 0. \end{aligned}$$

The problem is depicted in Figure 21.2. At the solution, only one constraint is active. If we had only known about this, we could have handled this problem as a constrained optimization problem using the Lagrange multiplier method.

■

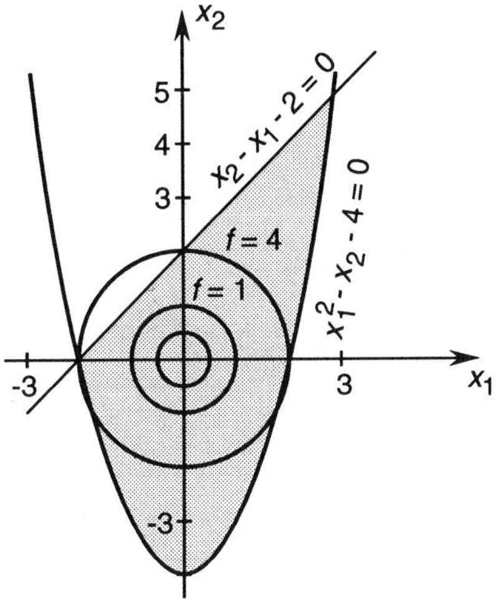


**Figure 21.1.** Situation where the constrained and the unconstrained minimizers are the same

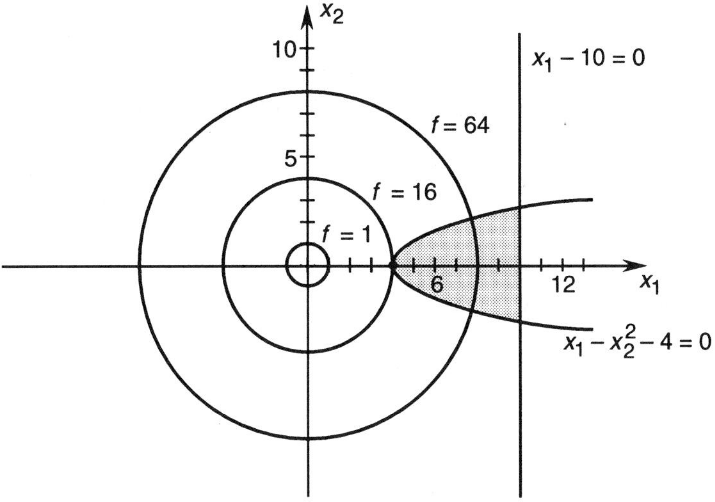


**Figure 21.2.** Situation where only one constraint is active

***Example 21.3.*** Consider the optimization problem

$$\begin{aligned} \text{minimize} \quad & x_1^2 + x_2^2 \\ \text{subject to} \quad & 4 - x_1 - x_2^2 \leqslant 0 \\ & 3x_2 - x_1 \leqslant 0 \\ & -3x_2 - x_1 \leqslant 0. \end{aligned}$$

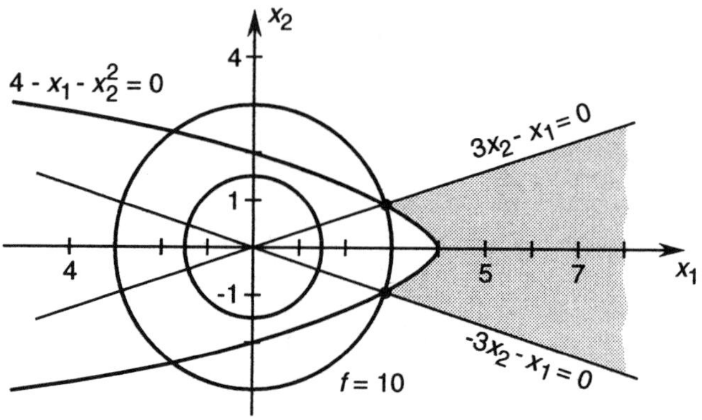


**Figure 21.3.** Situation where the constraints introduce local minimizers

The problem is depicted in Figure 21.3. This example illustrates the situation where the constraints introduce local minimizers, even though the objective function itself has only one unconstrained global minimizer. ■

Some of the difficulties illustrated in the above examples can be eliminated if we restrict our problems to convex feasible regions. Admittedly, some important real-life problems do not fit into this framework. On the other hand, it is possible to give results of a global nature for this class of optimization problems. In the next section, we introduce the notion of a convex function, which plays an important role in our subsequent treatment of such problems.

## 21.2. CONVEX FUNCTIONS

We begin with a definition of the graph of a real-valued function.

***Definition 21.1.*** The *graph* of $f:\Omega \to \mathbb{R}, \Omega \to \mathbb{R}^n$, is the set of points in $\Omega \times \mathbb{R}$ given by

$$\{(\boldsymbol{x}, f(\boldsymbol{x})) : \boldsymbol{x} \in \Omega\}.$$ ■

We can visualize the graph of $f$ as the set of points on a plot of $f(\boldsymbol{x})$ versus $\boldsymbol{x}$ (see Figure 21.4). We next define the epigraph of a real-valued function.

***Definition 21.2.*** The *epigraph* of a function $f:\Omega \to \mathbb{R}, \Omega \subset \mathbb{R}^n$, denoted $\text{epi}(f)$, is the set of points in $\Omega \times \mathbb{R}$ given by

$$\text{epi}(f) = \{(\boldsymbol{x}, \beta) : \boldsymbol{x} \in \Omega, \quad \beta \in \mathbb{R}, \quad \beta \geq f(\boldsymbol{x})\}.$$ ■

The epigraph $\text{epi}(f)$ of a function $f$ is the set of points in $\Omega \times \mathbb{R}$ on or above the graph of $f$ (see Figure 21.4). We can also think of $\text{epi}(f)$ as a subset of $\mathbb{R}^{n+1}$.

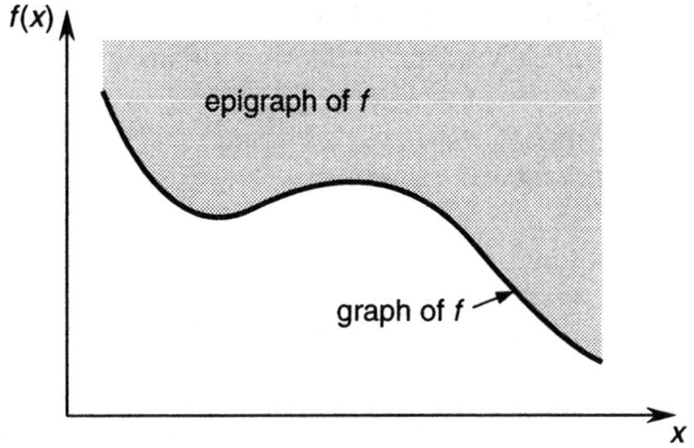


**Figure 21.4.** The graph and epigraph of a function $f : \mathbb{R} \to \mathbb{R}$

Recall that a set $\Omega \subset \mathbb{R}^n$ is convex if for every $\boldsymbol{x}_1, \boldsymbol{x}_2 \in \Omega$ and $\alpha \in (0, 1)$, $\alpha \boldsymbol{x}_1 + (1-\alpha)\boldsymbol{x}_2 \in \Omega$ (see Section 4.3). We now introduce the notion of a *convex function.*

***Definition 21.3.*** A function $f:\Omega \to \mathbb{R}$, $\Omega \subset \mathbb{R}^n$, is *convex* on $\Omega$ if its epigraph is a convex set. ■

**Theorem 21.1.** *If a function $f:\Omega \to \mathbb{R}$, $\Omega \subset \mathbb{R}^n$, is convex on $\Omega$, then $\Omega$ is a convex set.* □

*Proof.* We prove this theorem by contraposition. Suppose that $\Omega$ is not a convex set. Then, there exist two points $\boldsymbol{y}_1$ and $\boldsymbol{y}_2$ such that for some $\alpha \in (0, 1)$,

$$\boldsymbol{z} = \alpha \boldsymbol{y}_1 + (1-\alpha)\boldsymbol{y}_2 \notin \Omega.$$

Let

$$\beta_1 = f(\boldsymbol{y}_1), \quad \beta_2 = f(\boldsymbol{y}_2).$$

Then, the pairs

$$(\boldsymbol{y}_1, \beta_1), \quad (\boldsymbol{y}_2, \beta_2)$$

belong to the graph of $f$, and hence also the epigraph of $f$. Let

$$\boldsymbol{w} = \alpha(\boldsymbol{y}_1, \beta_1) + (1-\alpha)(\boldsymbol{y}_2, \beta_2).$$

We have

$$\boldsymbol{w} = (\boldsymbol{z}, \alpha\beta_1 + (1-\alpha)\beta_2).$$

But note that $\boldsymbol{w} \neq \operatorname{epi}(f)$, since $\boldsymbol{z} \notin \Omega$. Therefore, $\operatorname{epi}(f)$ is not convex, and hence $f$ is not a convex function. ■

The next theorem gives a very useful characterization of convex functions. This characterization is often used as a definition for a convex function.

**Theorem 21.2.** *A function $f:\Omega \to \mathbb{R}$ defined on a convex set $\Omega \subset \mathbb{R}^n$ is convex if, and only if, for all $\boldsymbol{x}, \boldsymbol{y} \in \Omega$ and all $\alpha \in (0, 1)$, we have*

$$f(\alpha\boldsymbol{x} + (1-\alpha)\boldsymbol{y}) \leqslant \alpha f(\boldsymbol{x}) + (1-\alpha) f(\boldsymbol{y}). \qquad \square$$

*Proof.* $\Leftarrow$: Assume that for all $\boldsymbol{x}, \boldsymbol{y} \in \Omega$ and $\alpha \in (0, 1)$,

$$f(\alpha\boldsymbol{x} + (1-\alpha)\boldsymbol{y}) \leqslant \alpha f(\boldsymbol{x}) + (1-\alpha) f(\boldsymbol{y}).$$

Let $(\boldsymbol{x}, a)$ and $(\boldsymbol{y}, b)$ be two points in $\text{epi}(f)$, where $a, b \in \mathbb{R}$. From the definition of $\text{epi}(f)$ it follows that

$$f(\boldsymbol{x}) \leqslant a, \quad f(\boldsymbol{y}) \leqslant b.$$

Therefore, using the first inequality above, we have

$$f(\alpha\boldsymbol{x} + (1-\alpha)\boldsymbol{y}) \leqslant \alpha a + (1-\alpha) b.$$

Since $\Omega$ is convex, $\alpha\boldsymbol{x} + (1-\alpha)\boldsymbol{y} \in \Omega$. Hence

$$(\alpha\boldsymbol{x} + (1-\alpha)\boldsymbol{y}, \alpha a + (1-\alpha)b) \in \text{epi}(f)$$

which implies that $\text{epi}(f)$ is a convex set, and hence $f$ is a convex function.

$\Rightarrow$: Assume that $f:\Omega \to \mathbb{R}$ is a convex function. Let $\boldsymbol{x}, \boldsymbol{y} \in \Omega$ and

$$f(\boldsymbol{x}) = a, \quad f(\boldsymbol{y}) = b.$$

Thus

$$(\boldsymbol{x}, a), (\boldsymbol{y}, b) \in \text{epi}(f).$$

Since $f$ is a convex function, its epigraph is a convex subset of $\mathbb{R}^{n+1}$. Therefore, for all $\alpha \in (0, 1)$, we have

$$\alpha(\boldsymbol{x}, a) + (1-\alpha)(\boldsymbol{y}, b) = (\alpha\boldsymbol{x} + (1-\alpha)\boldsymbol{y}, \alpha a + (1-\alpha)b) \in \text{epi}(f),$$

The above implies that for all $\alpha \in (0, 1)$,

$$f(\alpha\boldsymbol{x} + (1-\alpha)\boldsymbol{y}) \leqslant \alpha a + (1-\alpha)b = \alpha f(\boldsymbol{x}) + (1-\alpha) f(\boldsymbol{y}).$$

Thus, the proof is completed. ■

A geometric interpretation of the above theorem is given in Figure 21.5. The theorem states that if $f:\Omega \to \mathbb{R}$ is a convex function over a convex set $\Omega$, then for all

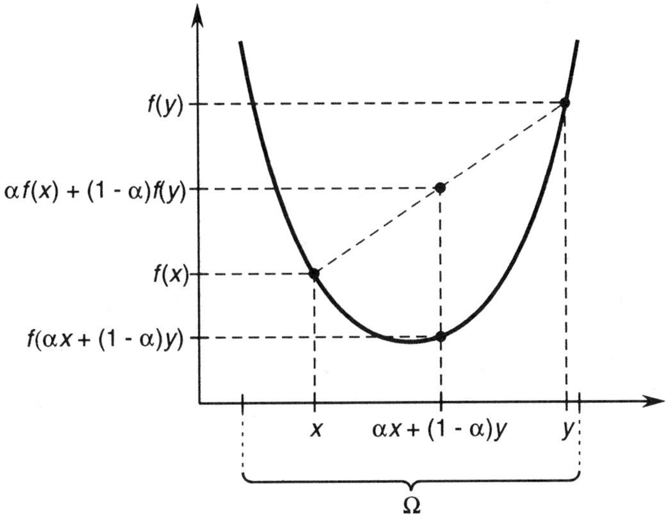


**Figure 21.5.** A geometric interpretation of Theorem 21.2

$\boldsymbol{x},\boldsymbol{y}\in\Omega$, the points on the line segment in $\mathbb{R}^{n+1}$ connecting $(\boldsymbol{x}, f(\boldsymbol{x}))$ and $(\boldsymbol{y}, f(\boldsymbol{y}))$ must lie on or above the graph of $f$.

***Definition 21.4.*** A function $f:\Omega\to\mathbb{R}$ on a convex set $\Omega\subset\mathbb{R}^n$ is *strictly convex* if for all $\boldsymbol{x},\boldsymbol{y}\in\Omega$, $\boldsymbol{x}\neq\boldsymbol{y}$, and $\alpha\in(0,1)$, we have

$$f(\alpha\boldsymbol{x}+(1-\alpha)\boldsymbol{y})<\alpha f(\boldsymbol{x})+(1-\alpha)f(\boldsymbol{y}).$$ ■

From the above definition, we see that for a strictly convex function, all points on the open line segment connecting the points $(\boldsymbol{x}, f(\boldsymbol{x}))$ and $(\boldsymbol{y}, f(\boldsymbol{y}))$ lie (strictly) above the graph of $f$.

***Definition 21.5.*** A function $f:\Omega\to\mathbb{R}$ on a convex set $\Omega\subset\mathbb{R}^n$ is (strictly) *concave* if $-f$ is (strictly) convex. ■

Note that the graph of a strictly concave function always lies above the line segment connecting any two points on its graph.

***Example 21.4.*** Let $f(\boldsymbol{x})=x_1x_2$. Is $f$ convex over $\Omega=\{\boldsymbol{x}:x_1\geqslant 0, x_2\geqslant 0\}$?

The answer is no. Take, for example, $\boldsymbol{x}=[1,2]^T\in\Omega$ and $\boldsymbol{y}=[2,1]^T\in\Omega$. Then

$$\alpha\boldsymbol{x}+(1-\alpha)\boldsymbol{y}=\begin{bmatrix}2-\alpha\\1+\alpha\end{bmatrix}.$$

Hence

$$f(\alpha\boldsymbol{x}+(1-\alpha)\boldsymbol{y})=(2-\alpha)(1+\alpha)=2+\alpha-\alpha^2,$$

and

$$\alpha f(\boldsymbol{x}) + (1-\alpha)f(\boldsymbol{y}) = 2.$$

If, for example, $\alpha = \frac{1}{2} \in (0, 1)$, then

$$f(\tfrac{1}{2}\boldsymbol{x} + \tfrac{1}{2}\boldsymbol{y}) = \tfrac{9}{4} > \tfrac{1}{2}f(\boldsymbol{x}) + \tfrac{1}{2}f(\boldsymbol{y})$$

which shows that $f$ is not convex over $\Omega$. ■

The above numerical example is an illustration of the following general result.

**Proposition 21.1.** *A quadratic form $f:\Omega \to \mathbb{R}$, $\Omega \subset \mathbb{R}^n$, given by $f(\boldsymbol{x}) = \boldsymbol{x}^T\boldsymbol{Q}\boldsymbol{x}$, $\boldsymbol{Q} \in \mathbb{R}^{n\times n}$, $\boldsymbol{Q} = \boldsymbol{Q}^T$, is convex on $\Omega$ if, and only if, for all $\boldsymbol{x}, \boldsymbol{y} \in \Omega$, $(\boldsymbol{x}-\boldsymbol{y})^T\boldsymbol{Q}(\boldsymbol{x}-\boldsymbol{y}) \geq 0$.* □

*Proof.* The result follows from Theorem 21.2. Indeed, the function $f(\boldsymbol{x}) = \boldsymbol{x}^T\boldsymbol{Q}\boldsymbol{x}$ is convex if, and only if, for every $\alpha \in (0, 1)$ and every $\boldsymbol{x}, \boldsymbol{y} \in \mathbb{R}^n$, we have

$$f(\alpha\boldsymbol{x} + (1-\alpha)\boldsymbol{y}) \leqslant \alpha f(\boldsymbol{x}) + (1-\alpha)f(\boldsymbol{y}),$$

or equivalently

$$\alpha f(\boldsymbol{x}) + (1-\alpha)f(\boldsymbol{y}) - f(\alpha\boldsymbol{x} + (1-\alpha)\boldsymbol{y}) \geqslant 0.$$

Substituting for $f$ into the left-hand side of the above equation yields

$$\begin{aligned}
&\alpha\boldsymbol{x}^T\boldsymbol{Q}\boldsymbol{x} + (1-\alpha)\boldsymbol{y}^T\boldsymbol{Q}\boldsymbol{y} - (\alpha\boldsymbol{x} + (1-\alpha)\boldsymbol{y})^T\boldsymbol{Q}(\alpha\boldsymbol{x} + (1-\alpha)\boldsymbol{y})\\
&\quad = \alpha\boldsymbol{x}^T\boldsymbol{Q}\boldsymbol{x} + \boldsymbol{y}^T\boldsymbol{Q}\boldsymbol{y} - \alpha\boldsymbol{y}^T\boldsymbol{Q}\boldsymbol{y} - \alpha^2\boldsymbol{x}^T\boldsymbol{Q}\boldsymbol{x}\\
&\qquad - (2\alpha - 2\alpha^2)\boldsymbol{x}^T\boldsymbol{Q}\boldsymbol{y} - (1 - 2\alpha + \alpha^2)\boldsymbol{y}^T\boldsymbol{Q}\boldsymbol{y}\\
&\quad = \alpha(1-\alpha)\boldsymbol{x}^T\boldsymbol{Q}\boldsymbol{x} - 2\alpha(1-\alpha)\boldsymbol{x}^T\boldsymbol{Q}\boldsymbol{y} + \alpha(1-\alpha)\boldsymbol{y}^T\boldsymbol{Q}\boldsymbol{y}\\
&\quad = \alpha(1-\alpha)(\boldsymbol{x}-\boldsymbol{y})^T\boldsymbol{Q}(\boldsymbol{x}-\boldsymbol{y}).
\end{aligned}$$

Therefore, $f$ is convex if, and only if,

$$\alpha(1-\alpha)(\boldsymbol{x}-\boldsymbol{y})^T\boldsymbol{Q}(\boldsymbol{x}-\boldsymbol{y}) \geqslant 0,$$

which proves the result. ■

***Example 21.5.*** In the previous example, $f(\boldsymbol{x}) = x_1x_2$, which can be written as $f(\boldsymbol{x}) = \boldsymbol{x}^T\boldsymbol{Q}\boldsymbol{x}$, where

$$\boldsymbol{Q} = \frac{1}{2}\begin{bmatrix} 0 & 1 \\ 1 & 0 \end{bmatrix}.$$

Let $\Omega=\{x:x\geqslant 0\}$, and $x=[2,2]^T\in\Omega, y=[1,3]^T\in\Omega$. We have

$$y-x=\begin{bmatrix}-1\\1\end{bmatrix}$$

and

$$(y-x)^T Q(y-x)=\frac{1}{2}[-1,1]\begin{bmatrix}0&1\\1&0\end{bmatrix}\begin{bmatrix}-1\\1\end{bmatrix}=-1<0.$$

Hence, by the above theorem, $f$ is not convex on $\Omega$. ■

Differentiable convex functions can be characterized using the following theorem.

**Theorem 21.3.** *Let $f:\Omega\to\mathbb{R}, f\in\mathscr{C}^1$, be defined on an open convex set $\Omega\subset\mathbb{R}^n$. Then, $f$ is convex on $\Omega$ if, and only if, for all $x,y\in\Omega$,*

$$f(y)\geqslant f(x)+Df(x)(y-x).$$ □

*Proof.* $\Rightarrow$: Suppose $f:\Omega\to\mathbb{R}$ is differentiable and convex. Then, by Theorem 21.2, for any $y,x\in\Omega$ and $\alpha\in(0,1)$ we have

$$f(\alpha y+(1-\alpha)x)\leqslant\alpha f(y)+(1-\alpha)f(x).$$

Rearranging terms yields

$$f(x+\alpha(y-x))-f(x)\leqslant\alpha(f(y)-f(x)).$$

Upon dividing both sides of the above inequality by $\alpha$ we get

$$\frac{f(x+\alpha(y-x))-f(x)}{\alpha}\leqslant f(y)-f(x).$$

If we now take the limit as $\alpha\to 0$ and apply the definition of the directional derivative of $f$ at $x$ in the direction $y-x$ (see Section 6.2), we get

$$Df(x)(y-x)\leqslant f(y)-f(x)$$

or

$$f(y)\geqslant f(x)+Df(x)(y-x).$$

$\Leftarrow$: Assume that $\Omega$ is convex, $f:\Omega\to\mathbb{R}$ is differentiable, and, for all $x,y\in\Omega$,

$$f(y)\geqslant f(x)+Df(x)(y-x).$$

Let $u,v\in\Omega$ and $\alpha\in(0,1)$. Since $\Omega$ is convex,

$$w=\alpha u+(1-\alpha)v\in\Omega.$$

We also have

$$f(\boldsymbol{u}) \geqslant f(\boldsymbol{w}) + Df(\boldsymbol{w})(\boldsymbol{u} - \boldsymbol{w})$$

and

$$f(\boldsymbol{v}) \geqslant f(\boldsymbol{w}) + Df(\boldsymbol{w})(\boldsymbol{v} - \boldsymbol{w}).$$

Multiplying the first of the above inequalities by $\alpha$ and the second by $(1 - \alpha)$, and then adding them together yields

$$\alpha f(\boldsymbol{u}) + (1 - \alpha) f(\boldsymbol{v}) \geqslant f(\boldsymbol{w}) + Df(\boldsymbol{w})(\alpha \boldsymbol{u} + (1 - \alpha)\boldsymbol{v} - \boldsymbol{w}).$$

But

$$\boldsymbol{w} = \alpha \boldsymbol{u} + (1 - \alpha)\boldsymbol{v}.$$

Hence

$$\alpha f(\boldsymbol{u}) + (1 - \alpha) f(\boldsymbol{v}) \geqslant f(\alpha \boldsymbol{u} + (1 - \alpha)\boldsymbol{v}).$$

Hence, by Theorem 21.2, $f$ is a convex function. ■

A geometric interpretation of the above theorem is given in Figure 21.6. To explain the interpretation, let $\boldsymbol{x}_0 \in \Omega$. The function $g(\boldsymbol{x}) = f(\boldsymbol{x}_0) + Df(\boldsymbol{x}_0)(\boldsymbol{x} - \boldsymbol{x}_0)$ is the linear approximation to $f$ at $\boldsymbol{x}_0$. The theorem says that the graph of $f$ always lies above its linear approximation at any point. In other words, the linear approximation to a convex function $f$ at any point of its domain lies below epi$(f)$.

For functions that are twice continuously differentiable the following theorem gives another possible characterization of convexity.

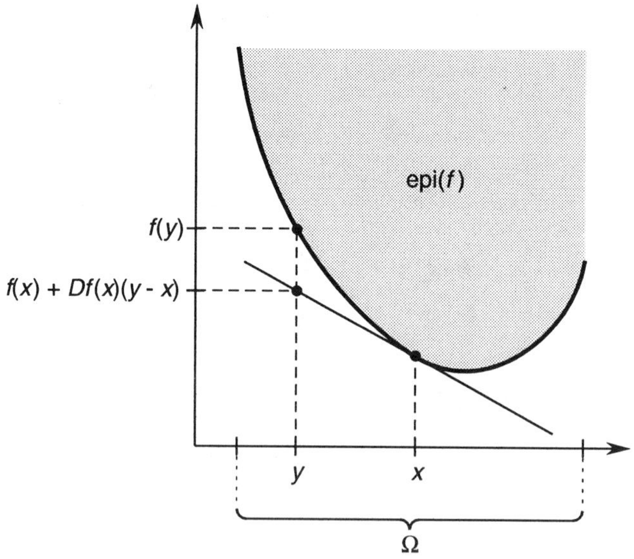


**Figure 21.6.** Geometric interpretation of Theorem 21.3

**Theorem 21.4.** *Let $f:\Omega \to \mathbb{R}$, $f \in \mathscr{C}^2$, be defined on an open convex set $\Omega \subset \mathbb{R}^n$. Then, $f$ is convex (strictly convex) on $\Omega$ if, and only if, for each $\boldsymbol{x} \in \Omega$, the Hessian $\boldsymbol{F}(\boldsymbol{x})$ of $f$ at $\boldsymbol{x}$ is a positive semidefinite (positive definite) matrix.* □

*Proof.* $\Leftarrow$: Since $f \in \mathscr{C}^2$, by Taylor's theorem there exists $\alpha \in (0, 1)$ such that

$$f(\boldsymbol{y}) = f(\boldsymbol{x}) + Df(\boldsymbol{x})(\boldsymbol{y} - \boldsymbol{x}) + \tfrac{1}{2}(\boldsymbol{y} - \boldsymbol{x})^T \boldsymbol{F}(\boldsymbol{x} + \alpha(\boldsymbol{y} - \boldsymbol{x}))(\boldsymbol{y} - \boldsymbol{x}).$$

Since $\boldsymbol{F}(\boldsymbol{x} + \alpha(\boldsymbol{y} - \boldsymbol{x}))$ is positive semidefinite,

$$(\boldsymbol{y} - \boldsymbol{x})^T \boldsymbol{F}(\alpha \boldsymbol{y} + (1 - \alpha)\boldsymbol{x})(\boldsymbol{y} - \boldsymbol{x}) \geqslant 0.$$

Therefore, we have

$$f(\boldsymbol{y}) \geqslant f(\boldsymbol{x}) + Df(\boldsymbol{x})(\boldsymbol{y} - \boldsymbol{x}),$$

which implies that $f$ is convex, by Theorem 21.3.

$\Rightarrow$: We use contraposition. Assume that there exists $\boldsymbol{x} \in \Omega$ such that $\boldsymbol{F}(\boldsymbol{x})$ is not positive semidefinite. Therefore, there exists $\boldsymbol{d} \in \mathbb{R}^n$ such that $\boldsymbol{d}^T \boldsymbol{F}(\boldsymbol{x})\boldsymbol{d} < 0$. By assumption, $\Omega$ is open; thus, the point $\boldsymbol{x}$ is an interior point. By the continuity of the Hessian matrix, there exists a nonzero $s \in \mathbb{R}$ such that $\boldsymbol{x} + s\boldsymbol{d} \in \Omega$, and, if we write $\boldsymbol{y} = \boldsymbol{x} + s\boldsymbol{d}$, then for all points $\boldsymbol{z}$ on the line segment joining $\boldsymbol{x}$ and $\boldsymbol{y}$, we have $\boldsymbol{d}^T \boldsymbol{F}(\boldsymbol{z})\boldsymbol{d} < 0$. By Taylor's theorem, there exists $\alpha \in (0, 1)$ such that

$$\begin{aligned} f(\boldsymbol{y}) &= f(\boldsymbol{x}) + Df(\boldsymbol{x})(\boldsymbol{y} - \boldsymbol{x}) + \tfrac{1}{2}(\boldsymbol{y} - \boldsymbol{x})^T \boldsymbol{F}(\boldsymbol{x} + \alpha(\boldsymbol{y} - \boldsymbol{x}))(\boldsymbol{y} - \boldsymbol{x}) \\ &= f(\boldsymbol{x}) + Df(\boldsymbol{x})(\boldsymbol{y} - \boldsymbol{x}) + \tfrac{1}{2}s^2 \boldsymbol{d}^T \boldsymbol{F}(\boldsymbol{x} + \alpha s\boldsymbol{d})\boldsymbol{d}. \end{aligned}$$

Because $\alpha \in (0, 1)$, the point $\boldsymbol{x} + \alpha s\boldsymbol{d}$ is on the segment joining $\boldsymbol{x}$ and $\boldsymbol{y}$, and, therefore,

$$\boldsymbol{d}^T \boldsymbol{F}(\boldsymbol{x} + \alpha s\boldsymbol{d})\boldsymbol{d} < 0.$$

Since $s \neq 0$, we have $s^2 > 0$, and hence

$$f(\boldsymbol{y}) < f(\boldsymbol{x}) + Df(\boldsymbol{x})(\boldsymbol{y} - \boldsymbol{x}).$$

Therefore, by Theorem 21.3, $f$ is not a convex function.

For the case of strict convexity, the same line of argument holds, with inequalities replaced by strict inequalities. ■

Note that by definition of concavity, a function $f:\Omega \to \mathbb{R}$, $f \in \mathscr{C}^2$, is concave over the convex set $\Omega \subset \mathbb{R}^n$ if, and only if, for all $\boldsymbol{x} \in \Omega$, the Hessian $\boldsymbol{F}(\boldsymbol{x})$ of $f$ is negative semidefinite.

***Example 21.6.*** Determine whether the following functions are convex, concave, or neither:

1. $f:\mathbb{R}\to\mathbb{R}, f(x) = -8x^2$
2. $f:\mathbb{R}^3\to\mathbb{R}, f(\boldsymbol{x}) = 4x_1^2 + 3x_2^2 + 5x_3^2 + 6x_1x_2 + x_1x_3 - 3x_1 - 2x_2 + 15$
3. $f:\mathbb{R}^2\to\mathbb{R}, f(\boldsymbol{x}) = 2x_1x_2 - x_1^2 - x_2^2$

Answers:

1. We use Theorem 21.4. We first compute the Hessian, which in this case is just the second derivative: $(d^2f/dx^2)(x) = -16 < 0$ for all $x\in\mathbb{R}$. Hence, $f$ is concave over $\mathbb{R}$.
2. The Hessian matrix of $f$ is

$$\boldsymbol{F}(\boldsymbol{x}) = \begin{bmatrix} 8 & 6 & 1 \\ 6 & 6 & 0 \\ 1 & 0 & 10 \end{bmatrix}.$$

The leading principal minors of $\boldsymbol{F}(\boldsymbol{x})$ are

$$\Delta_1 = 8 > 0,$$

$$\Delta_2 = \det\begin{bmatrix} 8 & 6 \\ 6 & 6 \end{bmatrix} = 12 > 0,$$

$$\Delta_3 = \det \boldsymbol{F}(\boldsymbol{x}) = 114 > 0.$$

Hence, $\boldsymbol{F}(\boldsymbol{x})$ is positive definite for all $\boldsymbol{x}\in\mathbb{R}^3$. Therefore, $f$ is a convex function over $\mathbb{R}^3$.

3. The Hessian of $f$ is

$$\boldsymbol{F}(\boldsymbol{x}) = \begin{bmatrix} -2 & 2 \\ 2 & -2 \end{bmatrix},$$

which is negative semidefinite for all $\boldsymbol{x}\in\mathbb{R}^2$. Hence, $f$ is concave on $\mathbb{R}^2$.

■

## 21.3. OPTIMIZATION OF CONVEX FUNCTIONS

In this section, we consider optimization problems where the objective function is a convex function, and the constraint set is a convex set. We refer to such problems as *convex programming problems*. Optimization problems that can be classified as convex programming problems include linear programs and optimization problems with quadratic objective function and linear constraints. Convex programming problems are interesting for several reasons. Specifically, as we shall see, local minimizers are global for such problems. Furthermore, first-order necessary conditions become sufficient conditions for minimization.

Our first theorem below states that in convex programming problems, local minimizers are also global.

**Theorem 21.5.** *Let $f:\Omega \to \mathbb{R}$ be a convex function defined on a convex set $\Omega \subset \mathbb{R}^n$. Then, a point is a global minimizer of $f$ over $\Omega$ if, and only if, it is local minimizer of $f$.* □

*Proof.* $\Rightarrow$: This is obvious.

$\Leftarrow$: We prove this by contraposition. Suppose that $\boldsymbol{x}^*$ is not a global minimizer of $f$ over $\Omega$. Then, for some $\boldsymbol{y} \in \Omega$, we have $f(\boldsymbol{y}) < f(\boldsymbol{x}^*)$. By assumption, the function $f$ is convex, and hence for all $\alpha \in (0, 1)$,

$$f(\alpha \boldsymbol{y} + (1-\alpha)\boldsymbol{x}^*) \leqslant \alpha f(\boldsymbol{y}) + (1-\alpha) f(\boldsymbol{x}^*).$$

Since $f(\boldsymbol{y}) < f(\boldsymbol{x}^*)$, we have

$$\alpha f(\boldsymbol{y}) + (1-\alpha) f(\boldsymbol{x}^*) = \alpha(f(\boldsymbol{y}) - f(\boldsymbol{x}^*)) + f(\boldsymbol{x}^*) < f(\boldsymbol{x}^*).$$

Thus, for all $\alpha \in (0, 1)$,

$$f(\alpha \boldsymbol{y} + (1-\alpha)\boldsymbol{x}^*) < f(\boldsymbol{x}^*).$$

Hence, there exist points that are arbitrarily close to $\boldsymbol{x}^*$ and have lower objective function value. For example, the sequence $\{\boldsymbol{y}_n\}$ of points given by

$$\boldsymbol{y}_n = \frac{1}{n}\boldsymbol{y} + \left(1 - \frac{1}{n}\right)\boldsymbol{x}^*$$

converges to $\boldsymbol{x}^*$, and $f(\boldsymbol{y}_n) < f(\boldsymbol{x}^*)$. Hence, $\boldsymbol{x}^*$ is not a local minimizer, which completes the proof. ■

We now show that the set of global optimizers is convex. For this, we need the following lemma.

**Lemma 21.1.** *Let $g:\Omega \to \mathbb{R}$ be a convex function defined on a convex set $\Omega \subset \mathbb{R}^n$. Then, for each $c \in \mathbb{R}$, the set*

$$\Gamma_c = \{\boldsymbol{x} \in \Omega : g(\boldsymbol{x}) \leqslant c\}$$

*is a convex set.* □

*Proof.* Let $\boldsymbol{x}, \boldsymbol{y} \in \Gamma_c$. Then, $g(\boldsymbol{x}), g(\boldsymbol{y}) \leqslant c$. Since $g$ is convex, for all $\alpha \in (0, 1)$,

$$g(\alpha\boldsymbol{x} + (1-\alpha)\boldsymbol{y}) \leqslant \alpha g(\boldsymbol{x}) + (1-\alpha)g(\boldsymbol{y}) \leqslant c.$$

Hence, $\alpha\boldsymbol{x} + (1-\alpha)\boldsymbol{y} \in \Gamma_c$, which implies that $\Gamma_c$ is convex. ■

**Corollary 21.1.** *Let $f: \Omega \to \mathbb{R}$ be a convex function defined on a convex set $\Omega \subset \mathbb{R}^n$. Then, the set of all global minimizers of $f$ over $\Omega$ is a convex set.* □

*Proof.* This is immediate from the previous lemma by setting

$$c = \min_{\boldsymbol{x} \in \Omega} f(\boldsymbol{x}).$$ ■

We now show that if the objective function is continuously differentiable and convex, then the first-order necessary condition (see Theorem 6.1) for a point to be a minimizer is also sufficient. We use the following lemma.

**Lemma 21.2.** *Let $f: \Omega \to \mathbb{R}$, $f \in \mathscr{C}^1$, be a convex function defined on the convex set $\Omega \subset \mathbb{R}^n$. Suppose the point $\boldsymbol{x}^* \in \Omega$ is such that for all $\boldsymbol{x} \in \Omega$, $\boldsymbol{x} \neq \boldsymbol{x}^*$, we have*

$$Df(\boldsymbol{x}^*)(\boldsymbol{x} - \boldsymbol{x}^*) \geqslant 0.$$

*Then, $\boldsymbol{x}^*$ is a global minimizer of $f$ over $\Omega$.* □

*Proof.* Since the function $f$ is convex, by Theorem 21.3, for all $\boldsymbol{x} \in \Omega$, we have

$$f(\boldsymbol{x}) \geqslant f(\boldsymbol{x}^*) + Df(\boldsymbol{x}^*)(\boldsymbol{x} - \boldsymbol{x}^*).$$

Hence, the condition

$$Df(\boldsymbol{x}^*)(\boldsymbol{x} - \boldsymbol{x}^*) \geqslant 0,$$

implies

$$f(\boldsymbol{x}) \geqslant f(\boldsymbol{x}^*),$$

and the proof is completed. ■

Observe that for any $\boldsymbol{x} \in \Omega$, the vector $\boldsymbol{x} - \boldsymbol{x}^*$ can be interpreted as a feasible direction at $\boldsymbol{x}^*$ (see Definition 6.2). Using the above lemma, we have the following theorem (cf., Theorem 6.1).

**Theorem 21.6.** *Let $f: \Omega \to \mathbb{R}$, $f \in \mathscr{C}^1$, be a convex function defined on the convex set $\Omega \subset \mathbb{R}^n$. Suppose the point $\boldsymbol{x}^* \in \Omega$ is such that for any feasible direction $\boldsymbol{d}$ at $\boldsymbol{x}^*$, we have*

$$\boldsymbol{d}^T \nabla f(\boldsymbol{x}^*) \geqslant 0.$$

*Then, $\boldsymbol{x}^*$ is a global minimizer of $f$ over $\Omega$.* □

*Proof.* Let $\boldsymbol{x} \in \Omega$, $\boldsymbol{x} \neq \boldsymbol{x}^*$. By convexity of $\Omega$,

$$\boldsymbol{x}^* + \alpha(\boldsymbol{x} - \boldsymbol{x}^*) = \alpha\boldsymbol{x} + (1 - \alpha)\boldsymbol{x}^* \in \Omega$$

for all $\alpha \in (0, 1)$. Hence, the vector $\boldsymbol{d} = \boldsymbol{x} - \boldsymbol{x}^*$ is a feasible direction at $\boldsymbol{x}^*$ (see Definition 6.2). By assumption,

$$Df(\boldsymbol{x}^*)(\boldsymbol{x} - \boldsymbol{x}^*) = \boldsymbol{d}^T \nabla f(\boldsymbol{x}^*) \geqslant 0.$$

Hence, by Lemma 21.2, $\boldsymbol{x}^*$ is a global minimizer of $f$ over $\Omega$. ■

From the above theorem, we easily deduce the following corollary (compare this with Corollary 6.1).

**Corollary 21.2.** *Let $f: \Omega \to \mathbb{R}$, $f \in \mathscr{C}^1$, be a convex function defined on the convex set $\Omega \subset \mathbb{R}^n$. Suppose the point $\boldsymbol{x}^* \in \Omega$ is such that*

$$\nabla f(\boldsymbol{x}^*) = \boldsymbol{0}.$$

*Then, $\boldsymbol{x}^*$ is a global minimizer of $f$ over $\Omega$.* □

We now consider the constrained optimization problem

$$\begin{aligned} &\text{minimize} \quad && f(\boldsymbol{x}) \\ &\text{subject to} \quad && \boldsymbol{h}(\boldsymbol{x}) = \boldsymbol{0}. \end{aligned}$$

We assume that the feasible set is convex, for example, when

$$\boldsymbol{h}(\boldsymbol{x}) = \boldsymbol{A}\boldsymbol{x} - \boldsymbol{b}.$$

The following theorem states that, provided the feasible set is convex, the Lagrange condition is sufficient for a point to be a minimizer.

**Theorem 21.7.** *Let $f: \mathbb{R}^n \to \mathbb{R}$, $f \in \mathscr{C}^1$, be a convex function on the set of feasible points*

$$\Omega = \{\boldsymbol{x} \in \mathbb{R}^n : \boldsymbol{h}(\boldsymbol{x}) = \boldsymbol{0}\},$$

*where $\boldsymbol{h}: \mathbb{R}^n \to \mathbb{R}^m$, $\boldsymbol{h} \in \mathscr{C}^1$, and $\Omega$ is convex. Suppose there exists $\boldsymbol{x}^* \in \Omega$ and $\boldsymbol{\lambda}^* \in \mathbb{R}^m$ such that*

$$Df(\boldsymbol{x}^*) + \boldsymbol{\lambda}^{*T} D\boldsymbol{h}(\boldsymbol{x}^*) = \boldsymbol{0}^T.$$

*Then, $\boldsymbol{x}^*$ is a global minimizer of $f$ over $\Omega$.* □

*Proof.* By Theorem 21.3, for all $\boldsymbol{x} \in \Omega$, we have

$$f(\boldsymbol{x}) \geqslant f(\boldsymbol{x}^*) + Df(\boldsymbol{x}^*)(\boldsymbol{x} - \boldsymbol{x}^*).$$

Substituting $Df(\boldsymbol{x}^*) = -\boldsymbol{\lambda}^{*T} D\boldsymbol{h}(\boldsymbol{x}^*)$ into the above inequality yields

$$f(\boldsymbol{x}) \geqslant f(\boldsymbol{x}^*) - \boldsymbol{\lambda}^T D\boldsymbol{h}(\boldsymbol{x}^*)(\boldsymbol{x} - \boldsymbol{x}^*).$$

Since $\Omega$ is convex, $(1-\alpha)\boldsymbol{x}^* + \alpha\boldsymbol{x} \in \Omega$ for all $\alpha \in (0, 1)$. Thus,

$$\boldsymbol{h}(\boldsymbol{x}^* + \alpha(\boldsymbol{x} - \boldsymbol{x}^*)) = \boldsymbol{h}((1-\alpha)\boldsymbol{x}^* + \alpha\boldsymbol{x}) = \boldsymbol{0}$$

for all $\alpha \in (0, 1)$. Premultiplying by $\boldsymbol{\lambda}^{*T}$, subtracting $\boldsymbol{\lambda}^{*T}\boldsymbol{h}(\boldsymbol{x}^*) = 0$, and dividing by $\alpha$, we get

$$\frac{\boldsymbol{\lambda}^{*T}\boldsymbol{h}(\boldsymbol{x}^* + \alpha(\boldsymbol{x} - \boldsymbol{x}^*)) - \boldsymbol{\lambda}^{*T}\boldsymbol{h}(\boldsymbol{x}^*)}{\alpha} = 0$$

for all $\alpha \in (0, 1)$. If we now take the limit as $\alpha \to 0$ and apply the definition of the directional derivative of $\boldsymbol{\lambda}^{*T}\boldsymbol{h}$ at $\boldsymbol{x}^*$ in the direction $\boldsymbol{x} - \boldsymbol{x}^*$ (see Section 6.2), we get

$$\boldsymbol{\lambda}^{*T} D\boldsymbol{h}(\boldsymbol{x}^*)(\boldsymbol{x} - \boldsymbol{x}^*) = 0.$$

Hence,

$$f(\boldsymbol{x}) \geqslant f(\boldsymbol{x}^*),$$

which implies that $\boldsymbol{x}^*$ is a global minimizer of $f$ over $\Omega$. ∎

Consider the general constrained optimization problem

$$\begin{aligned} &\text{minimize} && f(\boldsymbol{x}) \\ &\text{subject to} && \boldsymbol{h}(\boldsymbol{x}) = \boldsymbol{0} \\ & && \boldsymbol{g}(\boldsymbol{x}) \leqslant \boldsymbol{0}. \end{aligned}$$

As before, we assume that the feasible set is convex. This is the case if, for example, the two sets $\{\boldsymbol{x}: \boldsymbol{h}(\boldsymbol{x}) = \boldsymbol{0}\}$ and $\{\boldsymbol{x}: \boldsymbol{g}(\boldsymbol{x}) \leqslant \boldsymbol{0}\}$ are convex, because the feasible set is the intersection of these two sets (see also Theorem 4.1). We have already seen an example where the set $\{\boldsymbol{x}: \boldsymbol{h}(\boldsymbol{x}) = \boldsymbol{0}\}$ is convex. On the other hand, an example where the set $\{\boldsymbol{x}: \boldsymbol{g}(\boldsymbol{x}) \leqslant \boldsymbol{0}\}$ is convex is when the components of $\boldsymbol{g} = [g_1, \ldots, g_p]^T$ are all convex functions. Indeed, the set $\{\boldsymbol{x}: \boldsymbol{g}(\boldsymbol{x}) \leqslant \boldsymbol{0}\}$ is the intersection of the sets $\{\boldsymbol{x}: g_i(\boldsymbol{x}) \leqslant 0\}$, $i = 1, \ldots, p$. Because each of these sets is convex (see Lemma 21.1), their intersection is also convex.

We now prove that the Karush–Kuhn–Tucker conditions are sufficient for a point to be a minimizer to the above problem.

**Theorem 21.8.** *Let $f:\mathbb{R}^n \to \mathbb{R}$, $f \in \mathscr{C}^1$, be a convex function on the set of feasible points*

$$\Omega = \{\boldsymbol{x} \in \mathbb{R}^n : \boldsymbol{h}(\boldsymbol{x}) = \boldsymbol{0}, \boldsymbol{g}(\boldsymbol{x}) \leqslant \boldsymbol{0}\},$$

*where $\boldsymbol{h}:\mathbb{R}^n \to \mathbb{R}^m$, $\boldsymbol{g}:\mathbb{R}^n \to \mathbb{R}^p$, $\boldsymbol{h}, \boldsymbol{g} \in \mathscr{C}^1$, and $\Omega$ is convex. Suppose there exist $\boldsymbol{x}^* \in \Omega$, $\boldsymbol{\lambda}^* \in \mathbb{R}^m$, and $\boldsymbol{\mu}^* \in \mathbb{R}^p$, such that*

1. $\boldsymbol{\mu}^* \geqslant \boldsymbol{0}$;
2. $Df(\boldsymbol{x}^*) + \boldsymbol{\lambda}^{*T} D\boldsymbol{h}(\boldsymbol{x}^*) + \boldsymbol{\mu}^{*T} D\boldsymbol{g}(\boldsymbol{x}^*) = \boldsymbol{0}^T$; and
3. $\boldsymbol{\mu}^{*T} \boldsymbol{g}(\boldsymbol{x}^*) = 0$.

*Then, $\boldsymbol{x}^*$ is a global minimizer of $f$ over $\Omega$.* □

*Proof.* Suppose $\boldsymbol{x} \in \Omega$. By convexity of $f$ and Theorem 21.3,

$$f(\boldsymbol{x}) \geqslant f(\boldsymbol{x}^*) + Df(\boldsymbol{x}^*)(\boldsymbol{x} - \boldsymbol{x}^*).$$

Using condition 2, we get

$$f(\boldsymbol{x}) \geqslant f(\boldsymbol{x}^*) - \boldsymbol{\lambda}^{*T} D\boldsymbol{h}(\boldsymbol{x}^*)(\boldsymbol{x} - \boldsymbol{x}^*) - \boldsymbol{\mu}^{*T} D\boldsymbol{g}(\boldsymbol{x}^*)(\boldsymbol{x} - \boldsymbol{x}^*).$$

As in the proof of Theorem 21.7, we can show that $\boldsymbol{\lambda}^{*T} D\boldsymbol{h}(\boldsymbol{x}^*)(\boldsymbol{x} - \boldsymbol{x}^*) = 0$. We now claim that $\boldsymbol{\mu}^{*T} D\boldsymbol{g}(\boldsymbol{x}^*)(\boldsymbol{x} - \boldsymbol{x}^*) \leqslant 0$. To see this, note that since $\Omega$ is convex, $(1 - \alpha)\boldsymbol{x}^* + \alpha \boldsymbol{x} \in \Omega$ for all $\alpha \in (0, 1)$. Thus,

$$\boldsymbol{g}(\boldsymbol{x}^* + \alpha(\boldsymbol{x} - \boldsymbol{x}^*)) = \boldsymbol{g}((1 - \alpha)\boldsymbol{x}^* + \alpha \boldsymbol{x}) \leqslant \boldsymbol{0}$$

for all $\alpha \in (0, 1)$. Premultiplying by $\boldsymbol{\mu}^{*T} \geqslant \boldsymbol{0}$ (by condition 1), subtracting $\boldsymbol{\mu}^{*T} \boldsymbol{g}(\boldsymbol{x}^*) = 0$ (by condition 3), and dividing by $\alpha$, we get

$$\frac{\boldsymbol{\mu}^{*T} \boldsymbol{g}(\boldsymbol{x}^* + \alpha(\boldsymbol{x} - \boldsymbol{x}^*)) - \boldsymbol{\mu}^{*T} \boldsymbol{g}(\boldsymbol{x}^*)}{\alpha} \leqslant 0.$$

We now take the limit as $\alpha \to 0$ to obtain $\boldsymbol{\mu}^{*T} D\boldsymbol{g}(\boldsymbol{x}^*)(\boldsymbol{x} - \boldsymbol{x}^*) \leqslant 0$.

From the above, we have

$$\begin{aligned} f(\boldsymbol{x}) &\geqslant f(\boldsymbol{x}^*) - \boldsymbol{\lambda}^{*T} D\boldsymbol{h}(\boldsymbol{x}^*)(\boldsymbol{x} - \boldsymbol{x}^*) - \boldsymbol{\mu}^{*T} D\boldsymbol{g}(\boldsymbol{x}^*)(\boldsymbol{x} - \boldsymbol{x}^*) \\ &\geqslant f(\boldsymbol{x}^*) \end{aligned}$$

for all $\boldsymbol{x} \in \Omega$, and the proof is completed. ■

For extensions of the theory of convex optimization, we refer the reader to Ref. 81 (Chapter 10). The study of convex programming problems also serves as a prerequisite to nondifferentiable optimization [see, e.g., Ref. 18].

## EXERCISES

**21.1** Consider the function

$$f(\boldsymbol{x}) = \tfrac{1}{2}\boldsymbol{x}^T\boldsymbol{Q}\boldsymbol{x} - \boldsymbol{x}^T\boldsymbol{b}$$

where $\boldsymbol{Q} = \boldsymbol{Q}^T > 0$, and $\boldsymbol{x}, \boldsymbol{b} \in \mathbb{R}^n$. Define the function $\phi: \mathbb{R} \to \mathbb{R}$ by $\phi(\alpha) = f(\boldsymbol{x} + \alpha\boldsymbol{d})$, where $\boldsymbol{x}, \boldsymbol{d} \in \mathbb{R}^n$ are fixed vectors, and $\boldsymbol{d} \neq \boldsymbol{0}$. Show that $\phi(\alpha)$ is a strictly convex quadratic function of $\alpha$.

**21.2** Show that $f(\boldsymbol{x}) = x_1 x_2$ is a convex function on $\Omega = \{[a, ma]^T : a \in \mathbb{R}\}$, where $m$ is any given nonnegative constant.

**21.3** Suppose the set $\Omega = \{\boldsymbol{x} : h(\boldsymbol{x}) = c\}$ is convex, where $h: \mathbb{R}^n \to \mathbb{R}$ and $c \in \mathbb{R}$. Show that $h$ is convex and concave over $\Omega$.

**21.4** Let $\Omega \subset \mathbb{R}^n$ be an open convex set. Show that a symmetric matrix $\boldsymbol{Q} \in \mathbb{R}^n$ is positive semidefinite if, and only if, for each $\boldsymbol{x}, \boldsymbol{y} \in \Omega$, $(\boldsymbol{x} - \boldsymbol{y})^T\boldsymbol{Q}(\boldsymbol{x} - \boldsymbol{y}) \geqslant 0$. Show that a similar result for positive definiteness holds if we replace the "$\geqslant$" by "$>$" in the above inequality.

**21.5** Let $f: \mathbb{R}^n \to \mathbb{R}$, $f \in \mathscr{C}^1$, be a convex function on the set of feasible points

$$\Omega = \{\boldsymbol{x} \in \mathbb{R}^n : \boldsymbol{a}_i^T\boldsymbol{x} + b_i \geqslant 0, \quad i = 1, \ldots, p\},$$

where $\boldsymbol{a}_1, \ldots, \boldsymbol{a}_p \in \mathbb{R}^n$, and $b_1, \ldots, b_p \in \mathbb{R}$. Suppose there exist $\boldsymbol{x}^* \in S$, and $\boldsymbol{\mu}^* \in \mathbb{R}^p$, $\boldsymbol{\mu}^* \leqslant \boldsymbol{0}$, such that

$$Df(\boldsymbol{x}^*) + \sum_{j \in J(\boldsymbol{x}^*)} \mu_j^* \boldsymbol{a}_j^T = \boldsymbol{0}^T,$$

where $J(\boldsymbol{x}^*) = \{i : \boldsymbol{a}_i^T\boldsymbol{x}^* + b_i = 0\}$. Show that $\boldsymbol{x}^*$ is a global minimizer of $f$ over $\Omega$.

**21.6** A bank account starts out with 0 dollars. At the beginning of each month, we deposit some money into the bank account. Denote by $x_k$ the amount deposited in the $k$th month, $k = 1, 2, \ldots$. Suppose the monthly interest rate is $r > 0$, and the interest is paid into the account at the end of each month (and compounded). We wish to maximize the total amount of money accumulated at the end of $n$ months, such that the total money deposited during the $n$ months does not exceed $D$ dollars (where $D > 0$).

**a.** Show that the problem can be posed as a linear program, and represent the linear program in standard form.

**b.** Prove that the optimal strategy is to deposit $D$ dollars in the first month.

**21.7** Consider the problem: minimize $\|\boldsymbol{x}\|^2$ $(\boldsymbol{x} \in \mathbb{R}^n)$ subject to $\boldsymbol{a}^T\boldsymbol{x} \geqslant b$, where $\boldsymbol{a} \in \mathbb{R}^n$ is a nonzero vector, and $b \in \mathbb{R}$, $b > 0$. Suppose $\boldsymbol{x}^*$ is a solution to the problem.

**a.** Show that the constraint set is convex.

**b.** Use the Karush–Kuhn–Tucker theorem to show that $\boldsymbol{a}^T\boldsymbol{x}^* = b$.

**c.** Show that $\boldsymbol{x}^*$ is unique, and find an expression for $\boldsymbol{x}^*$ in terms of $\boldsymbol{a}$ and $b$.

**21.8** Consider the problem

$$\begin{array}{ll} \text{minimize} & \boldsymbol{c}^T\boldsymbol{x}, \quad \boldsymbol{x} \in \mathbb{R}^n \\ \text{subject to} & \boldsymbol{x} \geqslant \boldsymbol{0}. \end{array}$$

For this problem we have the following theorem (see also Exercise 17.8).

**Theorem.** *A solution to the above problem exists if, and only if,* $\boldsymbol{c} \geqslant \boldsymbol{0}$. *Moreover, if a solution exists,* $\boldsymbol{0}$ *is a solution.* □

**a.** Show that the above problem is a convex programming problem.

**b.** Use the first-order necessary condition (for set constraints) to prove the above theorem.

**c.** Use the Karush–Kuhn–Tucker condition to prove the above theorem.

**21.9** Consider a linear programming problem in standard form.

**a.** Derive the Karush–Kuhn–Tucker conditions for the problem.

**b.** Explain precisely why the Karush–Kuhn–Tucker conditions are sufficient for optimality in this case.

**c.** Write down the dual to the standard form primal problem (see Chapter 17).

**d.** Suppose $\boldsymbol{x}^*$ and $\boldsymbol{\lambda}^*$ are feasible solutions to the primal and dual, respectively. Use the Karush–Kuhn–Tucker conditions to prove that if the complementary slackness condition $(\boldsymbol{c}^T - \boldsymbol{\lambda}^{*T}\boldsymbol{A})\boldsymbol{x}^* = 0$ holds, then $\boldsymbol{x}^*$ is an optimal solution to the primal problem. Compare the above with Exercise 20.5.

# 22

# Algorithms for Constrained Optimization

## 22.1. PROJECTED GRADIENT METHODS

This chapter is devoted to a treatment of some algorithms for solving constrained optimization problems. We begin our presentation with a discussion of projected gradient methods for problems with linear equality constraints. We then consider penalty methods. This chapter is intended as an introduction to methods for solving constrained optimization problems. This area of optimization is still an active research area, and much more remains to be explored.

In the remainder of this section, we consider optimization problems of the form

$$\begin{aligned} &\text{minimize} \quad f(\boldsymbol{x}) \\ &\text{subject to} \quad \boldsymbol{A}\boldsymbol{x} = \boldsymbol{b}, \end{aligned}$$

where $f: \mathbb{R}^n \to \mathbb{R}$, $\boldsymbol{A} \in \mathbb{R}^{m \times n}$, $m < n$, rank $\boldsymbol{A} = m$, $\boldsymbol{b} \in \mathbb{R}^m$. We assume throughout that $f \in \mathscr{C}^1$. We shall discuss iterative methods for the optimization of $f$, based on the gradient $\nabla f$. The algorithms we consider here are similar to gradient algorithms considered in Chapter 8. In our subsequent discussion, we use the orthogonal projector $\boldsymbol{P}$ given by

$$\boldsymbol{P} = \boldsymbol{I}_n - \boldsymbol{A}^T(\boldsymbol{A}\boldsymbol{A}^T)^{-1}\boldsymbol{A}.$$

Orthogonal projectors were discussed in Section 3.3. Two important properties of the orthogonal projector $\boldsymbol{P}$ that we use in this section are (see Theorem 3.5)

1. $\boldsymbol{P} = \boldsymbol{P}^T$
2. $\boldsymbol{P}^2 = \boldsymbol{P}$

The orthogonal projector is needed because of the constraint set $\{\boldsymbol{x}: \boldsymbol{A}\boldsymbol{x} = \boldsymbol{b}\}$. Its specific use will be clear in the subsequent discussion.

Another property of the orthogonal projector that we need in our discussion is given in the following lemma.

**Lemma 22.1.** *Let $v \in \mathbb{R}^n$. Then, $Pv = 0$ if, and only if, $v \in \mathscr{R}(A^T)$. In other words, $\mathscr{N}(P) = \mathscr{R}(A^T)$. Moreover, $Av = 0$ if, and only if, $v \in \mathscr{R}(P)$, that is $\mathscr{N}(A) = \mathscr{R}(P)$.* □

*Proof.* ⇒:We have

$$\begin{aligned} Pv &= (I_n - A^T(AA^T)^{-1}A)v \\ &= v - A^T(AA^T)^{-1}Av. \end{aligned}$$

If $Pv = 0$, then

$$v = A^T(AA^T)^{-1}Av$$

and hence $v \in \mathscr{R}(A^T)$.

⇐: Suppose there exists $u \in \mathbb{R}^m$ such that $v = A^T u$. Then,

$$\begin{aligned} Pv &= (I_n - A^T(AA^T)^{-1}A)A^T u \\ &= A^T u - A^T(AA^T)^{-1}AA^T u \\ &= 0. \end{aligned}$$

Hence, we have proved that $\mathscr{N}(P) = \mathscr{R}(A^T)$.

Using a similar argument as above, we can show that $\mathscr{N}(A) = \mathscr{R}(P)$. ■

Recall that in unconstrained optimization, the first-order necessary condition for a point $x^*$ to be a local minimizer is $\nabla f(x^*) = 0$ (see Section 6.2). In optimization problems with equality constraints, the Lagrange condition plays the role of the first-order necessary condition (see Section 19.4). When the constraint set takes the form $\{x : Ax = b\}$, the Lagrange condition can be written as $P\nabla f(x^*) = 0$, as stated in the following proposition.

**Proposition 22.1.** *Let $x^* \in \mathbb{R}^n$ be a feasible point. Then, $P\nabla f(x^*) = 0$ if, and only if, $x^*$ satisfies the Lagrange condition.* □

*Proof.* By Lemma 22.1, $P\nabla f(x^*) = 0$ if, and only if, we have $\nabla f(x^*) \in \mathscr{R}(A^T)$. This is equivalent to the condition that there exists $\lambda \in \mathbb{R}^m$ such that $\nabla f(x^*) + A^T\lambda^* = 0$, which constitutes the Lagrange condition. ■

As in the case of unconstrained optimization, we consider algorithms that use gradients for the above constrained optimization problem. Recall that the vector $-\nabla f(x)$ points in the direction of maximum decrease of $f$ at $x$. This was the basis for gradient methods for unconstrained optimization, which have the form $x^{(k+1)} = x^{(k)} - \alpha_k \nabla f(x^{(k)})$, where $\alpha_k$ is the step size (see Chapter 8). The choice of the step size $\alpha_k$ depends on the particular gradient algorithm. For example, recall that in the steepest descent algorithm, $\alpha_k = \arg\min_{\alpha \geqslant 0} f(x^{(k)} - \alpha\nabla f(x^{(k)}))$.

In our constrained optimization problem, the vector $-\nabla f(\boldsymbol{x})$ is not necessarily a feasible direction. In other words, if $\boldsymbol{x}^{(k)}$ is a feasible point and we apply the algorithm $\boldsymbol{x}^{(k+1)} = \boldsymbol{x}^{(k)} - \alpha_k \nabla f(\boldsymbol{x}^{(k)})$, then $\boldsymbol{x}^{(k+1)}$ need not be feasible. This problem can be overcome by replacing $\nabla f(\boldsymbol{x}^{(k)})$ by a vector that points in a feasible direction. Note that the set of feasible directions is simply the nullspace $\mathcal{N}(\boldsymbol{A})$ of the matrix $\boldsymbol{A}$. Specifically, we update $\boldsymbol{x}^{(k)}$ according to the algorithm

$$\boldsymbol{x}^{(k+1)} = \boldsymbol{x}^{(k)} - \alpha_k \boldsymbol{P} \nabla f(\boldsymbol{x}^{(k)}).$$

We refer to the above as a *projected gradient algorithm.* The algorithm has the following property.

**Proposition 22.2.** *In a projected gradient algorithm, if $\boldsymbol{x}^{(0)}$ is feasible, then each $\boldsymbol{x}^{(k)}$ is feasible; that is, for each $k \geqslant 0$, $\boldsymbol{A}\boldsymbol{x}^{(k)} = \boldsymbol{b}$.* □

*Proof.* We proceed by induction. The result holds for $k = 0$ by assumption. Suppose now that $\boldsymbol{A}\boldsymbol{x}^{(k)} = \boldsymbol{b}$. We now show that $\boldsymbol{A}\boldsymbol{x}^{(k+1)} = \boldsymbol{b}$. To show this, first observe that $\boldsymbol{P}\nabla f(\boldsymbol{x}^{(k)}) \in \mathcal{N}(\boldsymbol{A})$. Therefore,

$$\begin{aligned} \boldsymbol{A}\boldsymbol{x}^{(k+1)} &= \boldsymbol{A}(\boldsymbol{x}^{(k)} - \alpha_k \boldsymbol{P}\nabla f(\boldsymbol{x}^{(k)})) \\ &= \boldsymbol{A}\boldsymbol{x}^{(k)} - \alpha_k \boldsymbol{A}\boldsymbol{P}\nabla f(\boldsymbol{x}^{(k)}) \\ &= \boldsymbol{b}, \end{aligned}$$

which completes the proof. ■

Projected gradient algorithms update $\boldsymbol{x}^{(k)}$ in the direction of $-\boldsymbol{P}\nabla f(\boldsymbol{x}^{(k)})$. This vector points in the direction of maximum rate of decrease of $f$ at $\boldsymbol{x}^{(k)}$ along the surface defined by $\boldsymbol{A}\boldsymbol{x} = \boldsymbol{b}$, as described in the following argument. Let $\boldsymbol{x}$ be any feasible point and $\boldsymbol{d}$ a feasible direction such that $\|\boldsymbol{d}\| = 1$. The rate of increase of $f$ at $\boldsymbol{x}$ in the direction $\boldsymbol{d}$ is $\langle \nabla f(\boldsymbol{x}), \boldsymbol{d} \rangle$. Next, we note that since $\boldsymbol{d}$ is a feasible direction, it lies in $\mathcal{N}(\boldsymbol{A})$ and hence by Lemma 22.1, we have $\boldsymbol{d} \in \mathcal{R}(\boldsymbol{P}) = \mathcal{R}(\boldsymbol{P}^T)$. So, there exists $\boldsymbol{v}$ such that $\boldsymbol{d} = \boldsymbol{P}\boldsymbol{v}$. Hence,

$$\langle \nabla f(\boldsymbol{x}), \boldsymbol{d} \rangle = \langle \nabla f(\boldsymbol{x}), \boldsymbol{P}^T \boldsymbol{v} \rangle = \langle \boldsymbol{P}\nabla f(\boldsymbol{x}), \boldsymbol{v} \rangle.$$

By the Cauchy–Schwarz Inequality,

$$\langle \boldsymbol{P}\nabla f(\boldsymbol{x}), \boldsymbol{v} \rangle \leqslant \|\boldsymbol{P}\nabla f(\boldsymbol{x})\| \, \|\boldsymbol{v}\|$$

with equality if, and only if, the direction of $\boldsymbol{v}$ is parallel with the direction of $\boldsymbol{P}\nabla f(\boldsymbol{x})$. Therefore, the vector $-\boldsymbol{P}\nabla f(\boldsymbol{x})$ points in the direction of maximum rate of decrease of $f$ at $\boldsymbol{x}$ among all feasible directions.

Following the discussion in Chapter 8 for gradient methods in unconstrained optimization, we suggest the following gradient method for our constrained problem. Suppose we have a starting point $\boldsymbol{x}^{(0)}$, which we assume is feasible, that is, $\boldsymbol{A}\boldsymbol{x}^{(0)} = \boldsymbol{b}$. Consider the point $\boldsymbol{x} = \boldsymbol{x}^{(0)} - \alpha \boldsymbol{P}\nabla f(\boldsymbol{x}^{(0)})$, where $\alpha \in \mathbb{R}$. As usual, the scalar $\alpha$ is called the step size. By the above discussion, $\boldsymbol{x}$ is also a feasible point. Using a Taylor series expansion of $f$ about $\boldsymbol{x}^{(0)}$ and the fact that $\boldsymbol{P} = \boldsymbol{P}^2 = \boldsymbol{P}^T\boldsymbol{P}$, we get

$$\begin{aligned} f(\boldsymbol{x}^{(0)} - \alpha \boldsymbol{P}\nabla f(\boldsymbol{x}^{(0)})) &= f(\boldsymbol{x}^{(0)}) - \alpha \nabla f(\boldsymbol{x}^{(0)})^T \boldsymbol{P}\nabla f(\boldsymbol{x}^{(0)}) + o(\alpha) \\ &= f(\boldsymbol{x}^{(0)}) - \alpha \| \boldsymbol{P}\nabla f(\boldsymbol{x}^{(0)}) \|^2 + o(\alpha). \end{aligned}$$

Thus, if $\boldsymbol{P}\nabla f(\boldsymbol{x}^{(0)}) \neq \boldsymbol{0}$ [that is, $\boldsymbol{x}^{(0)}$ does not satisfy the Lagrange condition], then we can choose an $\alpha$ sufficiently small such that $f(\boldsymbol{x}) < f(\boldsymbol{x}^{(0)})$, which means that $\boldsymbol{x} = \boldsymbol{x}^{(0)} - \alpha \boldsymbol{P}\nabla f(\boldsymbol{x}^{(0)})$ is an improvement over $\boldsymbol{x}^{(0)}$. This is the basis for the projected gradient algorithm,

$$\boldsymbol{x}^{(k+1)} = \boldsymbol{x}^{(k)} - \alpha_k \boldsymbol{P}\nabla f(\boldsymbol{x}^{(k)})$$

where the initial point $\boldsymbol{x}^{(0)}$ satisfies $\boldsymbol{A}\boldsymbol{x}^{(0)} = \boldsymbol{b}$, and $\alpha_k$ is some step size. As for unconstrained gradient methods, the choice of $\alpha_k$ determines the behavior of the algorithm. For small step sizes, the algorithm progresses slowly, while large step sizes may result in a zigzagging path. A well-known variant of the projected gradient algorithm is the projected steepest descent algorithm, where $\alpha_k$ is given by

$$\alpha_k = \arg\min_{\alpha \geqslant 0} f(\boldsymbol{x}^{(k)} - \alpha \boldsymbol{P}\nabla f(\boldsymbol{x}^{(k)})).$$

The following theorem states that the projected steepest descent algorithm is a descent algorithm, in the sense that at each step the value of the objective function decreases.

**Theorem 22.1.** *If $\{\boldsymbol{x}^{(k)}\}_{k=0}^{\infty}$ is the sequence of points generated by the projected steepest descent algorithm and if $\boldsymbol{P}\nabla f(\boldsymbol{x}^{(k)}) \neq \boldsymbol{0}$, then $f(\boldsymbol{x}^{(k+1)}) < f(\boldsymbol{x}^{(k)})$.* □

*Proof.* First, recall that

$$\boldsymbol{x}^{(k+1)} = \boldsymbol{x}^{(k)} - \alpha_k \boldsymbol{P}\nabla f(\boldsymbol{x}^{(k)}),$$

where $\alpha_k \geqslant 0$ is the minimizer of

$$\phi_k(\alpha) = f(\boldsymbol{x}^{(k)} - \alpha \boldsymbol{P}\nabla f(\boldsymbol{x}^{(k)}))$$

over all $\alpha \geqslant 0$. Thus, for $\alpha \geqslant 0$, we have

$$\phi_k(\alpha_k) \leqslant \phi_k(\alpha).$$

By the chain rule,

$$\begin{aligned}\phi_k'(0) &= \frac{d\phi_k}{d\alpha}(0) \\ &= -\nabla f(\boldsymbol{x}^{(k)} - 0\boldsymbol{P}\nabla f(\boldsymbol{x}^{(k)}))^T \boldsymbol{P}\nabla f(\boldsymbol{x}^{(k)}) \\ &= -\nabla f(\boldsymbol{x}^{(k)})^T \boldsymbol{P}\nabla f(\boldsymbol{x}^{(k)}).\end{aligned}$$

Using the fact that $\boldsymbol{P} = \boldsymbol{P}^2 = \boldsymbol{P}^T\boldsymbol{P}$, we get

$$\phi_k'(0) = -\nabla f(\boldsymbol{x}^{(k)})^T \boldsymbol{P}^T \boldsymbol{P}\nabla f(\boldsymbol{x}^{(k)}) = -\|\boldsymbol{P}\nabla f(\boldsymbol{x}^{(k)})\|^2 < 0$$

since $\boldsymbol{P}\nabla f(\boldsymbol{x}^{(k)}) \neq \boldsymbol{0}$ by assumption. Thus, there exists $\bar{\alpha} > 0$ such that $\phi_k(0) > \phi_k(\alpha)$ for all $\alpha \in (0, \bar{\alpha}]$. Hence,

$$f(\boldsymbol{x}^{(k+1)}) = \phi_k(\alpha_k) \leqslant \phi_k(\bar{\alpha}) < \phi_k(0) = f(\boldsymbol{x}^{(k)})$$

and the proof of the theorem is completed. ■

In the above theorem we needed the assumption that $\boldsymbol{P}\nabla f(\boldsymbol{x}^{(k)}) \neq \boldsymbol{0}$ to prove that the algorithm possesses a descent property. If for some $k$, we have $\boldsymbol{P}\nabla f(\boldsymbol{x}^{(k)}) = \boldsymbol{0}$, then, by Proposition 22.1, the point $\boldsymbol{x}^{(k)}$ satisfies the Lagrange condition. This condition can be used as a stopping condition for the algorithm. Note that in this case, $\boldsymbol{x}^{(k+1)} = \boldsymbol{x}^{(k)}$. For the case when $f$ is a convex function, the condition $\boldsymbol{P}\nabla f(\boldsymbol{x}^{(k)}) = \boldsymbol{0}$ is in fact equivalent to $\boldsymbol{x}^{(k)}$ being a global minimizer of $f$ over the constraint set $\{\boldsymbol{x}: \boldsymbol{A}\boldsymbol{x} = \boldsymbol{b}\}$. We show this in the following proposition.

**Proposition 22.3.** *The point $\boldsymbol{x}^* \in \mathbb{R}^n$ is a global minimizer of a convex function $f$ over $\{\boldsymbol{x}: \boldsymbol{A}\boldsymbol{x} = \boldsymbol{b}\}$ if, and only if, $\boldsymbol{P}\nabla f(\boldsymbol{x}^*) = \boldsymbol{0}$.* □

*Proof.* We first write $\boldsymbol{h}(\boldsymbol{x}) = \boldsymbol{A}\boldsymbol{x} - \boldsymbol{b}$. Then, the constraints can be written as $\boldsymbol{h}(\boldsymbol{x}) = \boldsymbol{0}$, and the problem is of the form considered in previous chapters. Note that $D\boldsymbol{h}(\boldsymbol{x}) = \boldsymbol{A}$. Hence, $\boldsymbol{x}^* \in \mathbb{R}^n$ is a global minimizer of $f$ if, and only if, the Lagrange condition holds (see Theorem 21.7). By Proposition 22.1, this is true if, and only if, $\boldsymbol{P}\nabla f(\boldsymbol{x}^*) = \boldsymbol{0}$, and the proof is completed. ■

For an application of the projected steepest descent algorithm to minimum fuel and minimum amplitude control problems in linear discrete systems, see, for example, Ref. 46.

## 22.2. PENALTY METHODS

In this section, we consider constrained optimization problems of the form

$$\begin{aligned} \text{minimize} \quad & f(\boldsymbol{x}) \\ \text{subject to} \quad & g_1(\boldsymbol{x}) \leqslant 0 \\ & g_2(\boldsymbol{x}) \leqslant 0 \\ & \vdots \\ & g_p(\boldsymbol{x}) \leqslant 0 \end{aligned}$$

where $f: \mathbb{R}^n \to \mathbb{R}$, $g_i: \mathbb{R}^n \to \mathbb{R}$, $i = 1, \ldots, p$. Considering only inequality constraints is not restrictive, because an equality constraint of the form $\boldsymbol{h}(\boldsymbol{x}) = \boldsymbol{0}$ is equivalent to the inequality constraint $\|\boldsymbol{h}(\boldsymbol{x})\|^2 \leqslant 0$ (however, see Exercise 20.13 for a caveat). We now discuss a method for solving the above constrained optimization problem using techniques from unconstrained optimization. Specifically, we approximate the constrained optimization problem above by an unconstrained optimization problem

$$\text{minimize} \quad f(\boldsymbol{x}) + \gamma P(\boldsymbol{x}),$$

where $\gamma \in \mathbb{R}$ is a positive constant, and $P: \mathbb{R}^n \to \mathbb{R}$ is a given function. We then solve the associated unconstrained optimization problem and use the solution as an approximation to the minimizer of the original problem. The constant $\gamma$ is called the *penalty parameter*, and the function $P$ is called the *penalty function*. Formally, we define a penalty function as follows.

***Definition 22.1.*** A function $P: \mathbb{R}^n \to \mathbb{R}$ is called a *penalty function* for the above constrained optimization problem if it satisfies the following three conditions:

1. $P$ is continuous
2. $P(\boldsymbol{x}) \geqslant 0$ for all $\boldsymbol{x} \in \mathbb{R}^n$
3. $P(\boldsymbol{x}) = 0$ if, and only if, $\boldsymbol{x}$ is feasible, that is, $g_1(\boldsymbol{x}) \leqslant 0, \ldots, g_p(\boldsymbol{x}) \leqslant 0$. ■

Clearly, for the above unconstrained problem to be a good approximation to the original problem, the penalty function $P$ must be appropriately chosen. The role of the penalty function is to penalize points that are outside the feasible set. Therefore, it is natural that the penalty function be defined in terms of the constraint functions $g_1, \ldots, g_p$. A possible choice for $P$ is

$$P(\boldsymbol{x}) = \sum_{i=1}^{p} g_i^+(\boldsymbol{x}),$$

where

$$g_i^+(\boldsymbol{x}) = \max(0, g_i(\boldsymbol{x})) = \begin{cases} 0, & \text{if } g_i(\boldsymbol{x}) \leqslant 0, \\ g_i(\boldsymbol{x}), & \text{if } g_i(\boldsymbol{x}) > 0. \end{cases}$$

We refer to the above penalty function as the *absolute value penalty function*, because it is equal to $\sum |g_i(\boldsymbol{x})|$, where the summation is taken over all constraints that are violated at $\boldsymbol{x}$. We illustrate this penalty function in the following example.

***Example 22.1.*** Let $g_1, g_2: \mathbb{R} \to \mathbb{R}$ be defined by $g_1(x) = x - 2$, $g_2(x) = -(x+1)^3$. The feasible set defined by $\{x \in \mathbb{R}: g_1(x) \leqslant 0, g_2(x) \leqslant 0\}$ is simply the interval $[-1, 2]$. In this example, we have

$$g_1^+(x) = \max(0, g_1(x)) = \begin{cases} 0, & \text{if } x \leqslant 2, \\ x - 2, & \text{otherwise;} \end{cases}$$

$$g_2^+(x) = \max(0, g_2(x)) = \begin{cases} 0, & \text{if } x \geqslant -1, \\ -(x+1)^3, & \text{otherwise;} \end{cases}$$

and

$$P(x) = g_1^+(x) + g_2^+(x) = \begin{cases} x - 2, & \text{if } x > 2, \\ 0, & \text{if } -1 \leqslant x \leqslant 2, \\ -(x+1)^3, & \text{if } x < -1. \end{cases}$$

Figure 22.1 provides a graphic illustration of $g^+$ for this example. ■

The absolute value penalty function may not be differentiable at points $\boldsymbol{x}$ where $g_i(\boldsymbol{x}) = 0$, as is the case at the point $x = 2$ in Example 22.1 (notice in Example 22.1, though, that $P$ is differentiable at $x = -1$). Therefore, in such cases, we cannot use techniques for optimization that involve derivatives. A form of the penalty function that is guaranteed to be differentiable is the so called *Courant–Beltrami penalty function*, given by

$$P(\boldsymbol{x}) = \sum_{i=1}^{p} (g_i^+(\boldsymbol{x}))^2.$$

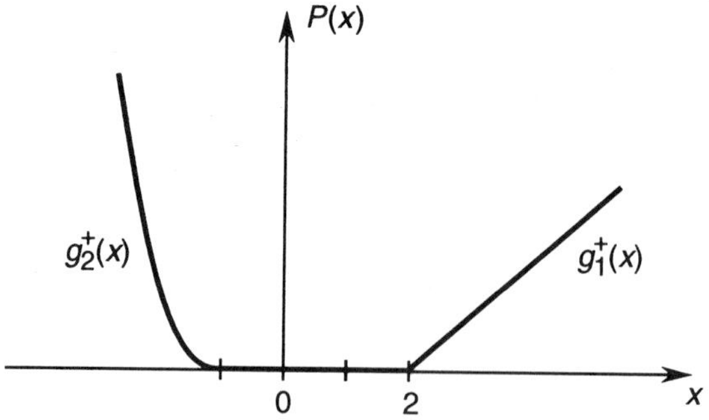


**Figure 22.1.** $g^+$ for Example 22.1

In the following discussion, we do not assume any particular form of the penalty function $P$. We only assume that $P$ satisfies conditions 1 to 3 given in Definition 22.1.

The penalty function method for solving constrained optimization problems involves constructing and solving an associated unconstrained optimization problem and using the solution to the unconstrained problem as the solution to the original constrained problem. Of course, the solution to the unconstrained problem (the approximated solution) may not be exactly equal to the solution to the constrained problem (the true solution). Whether or not the solution to the unconstrained problem is a good approximation to the true solution depends on the penalty parameter $\gamma$ and the penalty function $P$. We would expect that the larger the value of the penalty parameter $\gamma$, the closer the approximated solution will be to the true solution, since points that violate the constraints are penalized more heavily. Ideally, in the limit as $\gamma \to \infty$, the penalty method should yield the true solution to the constrained problem. In the remainder of this section, we analyze this property of the penalty function method.

In our analysis of the penalty method, we adopt the following setting. Recall that the original constrained optimization problem is

$$\begin{array}{ll} \text{minimize} & f(\boldsymbol{x}) \\ \text{subject to} & g_1(\boldsymbol{x}) \leqslant 0 \\ & g_2(\boldsymbol{x}) \leqslant 0 \\ & \vdots \\ & g_p(\boldsymbol{x}) \leqslant 0. \end{array}$$

Denote by $\boldsymbol{x}^*$ a solution (global minimizer) to the above problem. Let $P$ be a penalty function for the problem. For each $k = 1, 2, \ldots$, let $\gamma_k \in \mathbb{R}$ be a given positive constant. Define an associated function $q(\gamma_k, \cdot): \mathbb{R}^n \to \mathbb{R}$ by

$$q(\gamma_k, \boldsymbol{x}) = f(\boldsymbol{x}) + \gamma_k P(\boldsymbol{x}).$$

For each $k$, we can write the following associated unconstrained optimization problem:

$$\text{minimize} \quad q(\gamma_k, \boldsymbol{x}).$$

Denote by $\boldsymbol{x}_k$ a minimizer of $q(\gamma_k, \boldsymbol{x})$. The following technical lemma describes certain useful relationships between the constrained problem and the associated unconstrained problems.

**Lemma 22.2.** *Suppose $\{\gamma_k\}$ is a monotonically increasing sequence (that is, for each $k$, we have $\gamma_k \leqslant \gamma_{k+1}$). Then, for each $k$ we have*

1. $q(\gamma_{k+1}, \boldsymbol{x}_{k+1}) \geqslant q(\gamma_k, \boldsymbol{x}_k)$
2. $P(\boldsymbol{x}_{k+1}) \leqslant P(\boldsymbol{x}_k)$
3. $f(\boldsymbol{x}_{k+1}) \geqslant f(\boldsymbol{x}_k)$
4. $f(\boldsymbol{x}^*) \geqslant q(\gamma_k, \boldsymbol{x}_k) \geqslant f(\boldsymbol{x}_k)$. □

*Proof.* We first prove part 1. From the definition of $q$ and the fact that $\{\gamma_k\}$ is an increasing sequence, we have

$$q(\gamma_{k+1}, \boldsymbol{x}_{k+1}) = f(\boldsymbol{x}_{k+1}) + \gamma_{k+1} P(\boldsymbol{x}_{k+1}) \geqslant f(\boldsymbol{x}_{k+1}) + \gamma_k P(\boldsymbol{x}_{k+1}).$$

Now, since $\boldsymbol{x}_k$ is a minimizer of $q(\gamma_k, \boldsymbol{x})$,

$$q(\gamma_k, \boldsymbol{x}_k) = f(\boldsymbol{x}_k) + \gamma_k P(\boldsymbol{x}_k) \leqslant f(\boldsymbol{x}_{k+1}) + \gamma_k P(\boldsymbol{x}_{k+1}).$$

Combining the above, we get part 1.

We next prove part 2. Since $\boldsymbol{x}_k$ and $\boldsymbol{x}_{k+1}$ minimize $q(\gamma_k, \boldsymbol{x})$ and $q(\gamma_{k+1}, \boldsymbol{x})$, respectively, we can write

$$q(\gamma_k, \boldsymbol{x}_k) = f(\boldsymbol{x}_k) + \gamma_k P(\boldsymbol{x}_k) \leqslant f(\boldsymbol{x}_{k+1}) + \gamma_k P(\boldsymbol{x}_{k+1}),$$
$$q(\gamma_{k+1}, \boldsymbol{x}_{k+1}) = f(\boldsymbol{x}_{k+1}) + \gamma_{k+1} P(\boldsymbol{x}_{k+1}) \leqslant f(\boldsymbol{x}_k) + \gamma_{k+1} P(\boldsymbol{x}_k).$$

Adding the above inequalities yields

$$\gamma_k P(\boldsymbol{x}_k) + \gamma_{k+1} P(\boldsymbol{x}_{k+1}) \leqslant \gamma_{k+1} P(\boldsymbol{x}_k) + \gamma_k P(\boldsymbol{x}_{k+1}).$$

Rearranging, we get

$$(\gamma_{k+1} - \gamma_k) P(\boldsymbol{x}_{k+1}) \leqslant (\gamma_{k+1} - \gamma_k) P(\boldsymbol{x}_k).$$

We know by assumption that $\gamma_{k+1} \geqslant \gamma_k$. If $\gamma_{k+1} > \gamma_k$, then we get $P(\boldsymbol{x}_{k+1}) \leqslant P(\boldsymbol{x}_k)$. If, on the other hand $\gamma_{k+1} = \gamma_k$, then clearly $\boldsymbol{x}_{k+1} = \boldsymbol{x}_k$ and so $P(\boldsymbol{x}_{k+1}) = P(\boldsymbol{x}_k)$. Therefore, in either case, we arrive at part 2.

We now prove part 3. Since $\boldsymbol{x}_k$ is a minimizer of $q(\gamma_k, \boldsymbol{x})$, we obtain

$$q(\gamma_k, \boldsymbol{x}_k) = f(\boldsymbol{x}_k) + \gamma_k P(\boldsymbol{x}_k) \leqslant f(\boldsymbol{x}_{k+1}) + \gamma_k P(\boldsymbol{x}_{k+1}).$$

Therefore,

$$f(\boldsymbol{x}_{k+1}) \geqslant f(\boldsymbol{x}_k) + \gamma_k (P(\boldsymbol{x}_k) - P(\boldsymbol{x}_{k+1})).$$

From part 2, we have $P(\boldsymbol{x}_k) - P(\boldsymbol{x}_{k+1}) \geqslant 0$, and $\gamma_k > 0$ by assumption; therefore, we get

$$f(\boldsymbol{x}_{k+1}) \geqslant f(\boldsymbol{x}_k).$$

Finally, we now prove part 4. Since $\boldsymbol{x}_k$ is a minimizer of $q(\gamma_k, \boldsymbol{x})$, we get

$$f(\boldsymbol{x}^*) + \gamma_k P(\boldsymbol{x}^*) \geqslant q(\gamma_k, \boldsymbol{x}_k) = f(\boldsymbol{x}_k) + \gamma_k P(\boldsymbol{x}_k).$$

Since $\boldsymbol{x}^*$ is a minimizer for the constrained optimization problem, we have $P(\boldsymbol{x}^*) = 0$. Therefore,

$$f(\boldsymbol{x}^*) \geqslant f(\boldsymbol{x}_k) + \gamma_k P(\boldsymbol{x}_k).$$

Since $P(\boldsymbol{x}_k), \gamma_k \geqslant 0$,

$$f(\boldsymbol{x}^*) \geqslant q(\gamma_k, \boldsymbol{x}_k) \geqslant f(\boldsymbol{x}_k),$$

which completes the proof. ■

With the above lemma, we are now ready to prove the following theorem.

**Theorem 22.2.** *Suppose the objective function $f$ is continuous, and $\gamma_k \to \infty$ as $k \to \infty$. Then, the limit of any convergent subsequence of the sequence $\{\boldsymbol{x}_k\}$ is a solution to the constrained optimization problem.* □

*Proof.* Suppose $\{\boldsymbol{x}_{k_n}\}$ is a convergent subsequence of the sequence $\{\boldsymbol{x}_k\}$. Let $\hat{\boldsymbol{x}}$ be the limit of $\{\boldsymbol{x}_{k_n}\}$. By Lemma 22.2, the sequence $\{q(\gamma_n, \boldsymbol{x}_n)\}$ is nondecreasing and bounded above by $f(\boldsymbol{x}^*)$. Therefore, the sequence $\{q(\gamma_n, \boldsymbol{x}_n)\}$ has a limit $q^* = \lim_{n\to\infty} q(\gamma_n, \boldsymbol{x}_n)$ such that $q^* \leqslant f(\boldsymbol{x}^*)$ [see Ref. 3, p. 104]. Since the function $f$ is continuous and, by Lemma 22.2, we have $f(\boldsymbol{x}_{k_n}) \leqslant f(\boldsymbol{x}^*)$,

$$\lim_{n\to\infty} f(\boldsymbol{x}_{k_n}) = f\left(\lim_{n\to\infty} \boldsymbol{x}_{k_n}\right) = f(\hat{\boldsymbol{x}}) \leqslant f(\boldsymbol{x}^*).$$

Since the sequences $\{f(\boldsymbol{x}_{k_n})\}$ and $\{q(\gamma_k, \boldsymbol{x}_{k_n})\}$ both converge, the sequence $\{\gamma_{k_n} P(\boldsymbol{x}_{k_n})\} = \{q(\gamma_{k_n}, \boldsymbol{x}_{k_n}) - f(\boldsymbol{x}_{k_n})\}$ also converges, with

$$\lim_{n\to\infty} \gamma_{k_n} P(\boldsymbol{x}_{k_n}) = q^* - f(\hat{\boldsymbol{x}}).$$

By Lemma 22.2, the sequence $\{P(\boldsymbol{x}_n)\}$ is nonincreasing and bounded from below by 0. Therefore, $\{P(\boldsymbol{x}_n)\}$ converges [again see Ref. 3, p. 105], and hence so does $\{P(\boldsymbol{x}_{k_n})\}$. Since $\gamma_{k_n} \to \infty$, we conclude that

$$\lim_{n\to\infty} P(\boldsymbol{x}_{k_n}) = 0.$$

By continuity of $P$, we have

$$0 = \lim_{n\to\infty} P(\boldsymbol{x}_{k_n}) = P\left(\lim_{n\to\infty} \boldsymbol{x}_{k_n}\right) = P(\hat{\boldsymbol{x}}),$$

and hence $\hat{x}$ is a feasible point. Since $f(x^*) \geqslant f(\hat{x})$ from above, we conclude that $\hat{x}$ must be a solution to the constrained optimization problem. ∎

If we perform an infinite number of minimization runs with the penalty parameter $\gamma_k \to \infty$, then the above theorem ensures that the limit of any convergent subsequence is a minimizer $x^*$ to the original constrained optimization problem. There is clearly a practical limitation in applying this theorem. It is certainly desirable to find a minimizer to the original constrained optimization problem using a single minimization run for the unconstrained problem that approximates the original problem using a penalty function. It turns out that indeed this can be accomplished. However, it is necessary in this case that the penalty function be nondifferentiable [as shown in Ref. 8]. Optimization of nondifferentiable functions goes beyond the scope of methods introduced in this book. For further reading on the subject, see, for example, Ref. [18].

## EXERCISES

**22.1** Let $f:\mathbb{R}^n \to \mathbb{R}$ be given by $f(x) = \frac{1}{2}x^T Q x - x^T c$, where $Q = Q^T > 0$. We wish to minimize $f$ over $\{x: Ax = b\}$, where $A \in \mathbb{R}^{m \times n}$, $m < n$, and rank $A = m$. Show that the projected steepest descent algorithm for this case takes the form

$$x^{(k+1)} = x^{(k)} - \left( \frac{g^{(k)T} P g^{(k)}}{g^{(k)T} P Q P g^{(k)}} \right) P g^{(k)},$$

where

$$g^{(k)} = \nabla f(x^{(k)}) = Q x^{(k)} - c,$$

and $P = I_n - A^T(AA^T)^{-1}A$.

**22.2** Consider the problem

$$\begin{aligned} &\text{minimize} && \tfrac{1}{2}\|x\|^2 \\ &\text{subject to} && Ax = b, \end{aligned}$$

where $A \in \mathbb{R}^{m \times n}$, $m < n$, and rank $A = m$. Show that if $x^{(0)} \in \{x: Ax = b\}$, then the projected steepest descent algorithm converges to the solution in one step.

**22.3** Show that in the projected steepest descent algorithm, we have that for each $k$,

**a.** $g^{(k+1)T} P g^{(k)} = 0$

**b.** The vector $x^{(k+1)} - x^{(k)}$ is orthogonal to the vector $x^{(k+2)} - x^{(k+1)}$.

**22.4** Consider the problem

$$\begin{array}{ll} \text{minimize} & \frac{1}{2}\|\boldsymbol{x}\|^2 \\ \text{subject to} & \boldsymbol{A}\boldsymbol{x}=\boldsymbol{b}, \end{array}$$

where $\boldsymbol{A}\in\mathbb{R}^{m\times n}$, $\boldsymbol{b}\in\mathbb{R}^m$, $m\leqslant n$, and $\operatorname{rank}\boldsymbol{A}=m$. Let $\boldsymbol{x}^*$ be the solution. Suppose we solve the problem using the penalty method, with the penalty function

$$P(\boldsymbol{x})=\|\boldsymbol{A}\boldsymbol{x}-\boldsymbol{b}\|^2.$$

Let $\boldsymbol{x}_\gamma^*$ be the solution to the associated unconstrained problem with the penalty parameter $\gamma>0$; that is, $\boldsymbol{x}_\gamma^*$ is the solution to

$$\text{minimize} \quad \frac{1}{2}\|\boldsymbol{x}\|^2+\gamma\|\boldsymbol{A}\boldsymbol{x}-\boldsymbol{b}\|^2.$$

**a.** Suppose

$$\boldsymbol{A}=[1 \quad 1], \quad \boldsymbol{b}=[1].$$

Verify that $\boldsymbol{x}_\gamma^*$ converges to the solution $\boldsymbol{x}^*$ of the original constrained problem as $\gamma\to\infty$.

**b.** Prove that $\boldsymbol{x}_\gamma^*\to\boldsymbol{x}^*$ as $\gamma\to\infty$ holds in general. *Hint*: Use the following result: There exist orthogonal matrices $\boldsymbol{U}\in\mathbb{R}^{m\times m}$ and $\boldsymbol{V}^T\in\mathbb{R}^{n\times n}$ such that

$$\boldsymbol{A}=\boldsymbol{U}[\boldsymbol{S},\boldsymbol{O}]\boldsymbol{V}^T,$$

where

$$\boldsymbol{S}=\operatorname{diag}\left(\sqrt{\lambda_1(\boldsymbol{A}\boldsymbol{A}^T)},\ldots,\sqrt{\lambda_m(\boldsymbol{A}\boldsymbol{A}^T)}\right)$$

is a diagonal matrix with diagonal elements that are the eigenvalues of $\boldsymbol{A}\boldsymbol{A}^T$. The above result is called the *singular value decomposition* [see, e.g., Ref. 35, p. 411].

# Bibliography


1. The Math Works, Inc., Natick, MA, *MATLAB: High-Performance Numeric Computation and Visualization Software, User's Guide for UNIX Workstations*, Aug. 1992.
2. J. S. Arora, *Introduction to Optimum Design*. New York: McGraw-Hill, 1989.
3. R. G. Bartle, *The Elements of Real Analysis*, 2nd ed. New York, Wiley, 1976.
4. M. S. Bazaraa, H. D. Sherali, and C. M. Shetty, *Nonlinear Programming: Theory and Algorithms*, 2nd ed. New York: Wiley, 1993.
5. A. Ben-Israel and T. N. E. Greville, *Generalized Inverses: Theory and Applications*. New York: Wiley-Interscience, 1974.
6. C. C. Berresford, A. M. Rockett, and J. C. Stevenson, "Khachiyan's algorithm, Part 1: A new solution to linear programming problems," *Byte*, 5(8): 198–208, 1980.
7. C. C. Berresford, A. M. Rockett, and J. C. Stevenson, "Khachiyan's algorithm, Part 2: Problems with the algorithm," *Byte*, 5(9): 242–255, 1980.
8. D. P. Bertsekas, "Necessary and sufficient conditions for a penalty method to be exact," *Mathematical Programming*, 9: 87–99, 1975.
9. K. G. Binmore, *Calculus*. Cambridge: Cambridge University Press, 1986.
10. R. G. Bland, D. Goldfarb, and M. J. Todd, "The ellipsoid method: A survey," *Operations Research*, 29: 1039–1091, 1981.
11. L. Brickman, *Mathematical Introduction to Linear Programming and Game Theory*. New York: Springer-Verlag, 1989.
12. C. G. Broyden, "Quasi-Newton methods," in *Optimization Methods in Electronics and Communications* (K. W. Cattermole and J. J. O'Reilly, Eds.), Vol. 1: *Mathematical Topics in Telecommunications*. New York: Wiley, 1984, pp. 105–110.
13. B. D. Bunday, *Basic Optimization Methods*. London: Edward Arnold, 1984.
14. S. L. Campbell and C. D. Meyer, Jr., *Generalized Inverses of Linear Transformations*. New York: Dover, 1991.
15. S. D. Conte and C. de Boor, *Elementary Numerical Analysis: An Algorithmic Approach*, 3rd ed. New York: McGraw-Hill, 1980.
16. G. B. Dantzig, *Linear Programming and Extensions*. Princeton, NJ: Princeton University Press, 1963.
17. L. Davis, Ed., *Genetic Algorithms and Simulated Annealing, Research Notes in Artificial Intelligence*. London: Pitman, 1987.
18. V. F. Dem'yanov and L. V. Vasil'ev, *Nondifferentiable Optimization*. New York: Optimization Software, 1985.
19. J. E. Dennis, Jr., and R. B. Schnabel, *Numerical Methods for Unconstrained Optimization and Nonlinear Equations*. Englewood Cliffs, NJ: Prentice-Hall, 1983.
20. V. N. Faddeeva, *Computational Methods of Linear Algebra*. New York: Dover, 1959.
21. S.-C. Fang and S. Puthenpura, *Linear Optimization and Extensions: Theory and Algorithms*. Englewood Cliffs, NJ: Prentice-Hall, 1993.
22. R. Fletcher, *Practical Methods of Optimization*, 2nd ed. Chichester: Wiley, 1987.
23. F. R. Gantmacher, *The Theory of Matrices*, revised 2nd ed. Moscow: Nauka, 1966.
24. S. I. Gass, *An Illustrated Guide to Linear Programming*. New York: McGraw-Hill, 1970.

25. I. M. Gel'fand, *Lectures on Linear Algebra.* New York: Interscience, 1961.

26. P. E. Gill and W. Murray, "Safeguarded steplength algorithms for optimization using descent methods," NPL Tech. Rep. No. NAC 37, National Physical Laboratory, Division of Numerical Analysis and Computing, Teddington, Middlesex, Aug. 1974.

27. P. E. Gill, W. Murray, M. A. Saunders, and M. H. Wright, "Two step-length algorithms for numerical optimization," Tech. Rep. SOL 79-25, Systems Optimization Laboratory, Department of Operations Research, Stanford University, Stanford, CA 94305, Dec. 1979.

28. P. E. Gill, W. Murray, and M. H. Wright, *Practical Optimization.* London: Academic Press, 1981.

29. G. H. Golub and C. F. Van Loan, *Matrix Computations.* Baltimore: The Johns Hopkins University Press, 1983.

30. C. C. Gonzaga, "Path-following methods for linear programming," *SIAM Review*, 34(2): 167–224, 1992.

31. R. L. Harvey, *Neural Network Principles.* Englewood Cliffs, NJ: Prentice-Hall, 1994.

32. S. Haykin, *Neural Networks: A Comprehensive Foundation.* New York: Macmillan College, 1994.

33. J. Hertz, A. Krogh, and R. G. Palmer, *Introduction to the Theory of Neural Computation*, Vol. 1: *Santa Fe Institute Studies in the Sciences of Complexity.* Redwood City, CA: Addison-Wesley, 1991.

34. J. H. Holland, *Adaptation in Natural and Artificial Systems: An Introductory Analysis with Applications to Biology, Control, and Artificial Intelligence.* Cambridge, MA: The MIT Press, 1992.

35. R. A. Horn and C. R. Johnson, *Matrix Analysis.* Cambridge: Cambridge University Press, 1985.

36. A. S. Householder, *The Theory of Matrices in Numerical Analysis.* New York: Dover, 1975.

37. D. R. Hush and B. G. Horne, "Progress in supervised neural networks: What's new since Lippmann," *IEEE Signal Processing Magazine*, 10(1): 8–39, 1993.

38. S. Isaak and M. N. Manougian, *Basic Concepts of Linear Algebra.* New York: Norton, 1976.

39. W. E. Jenner, *Rudiments of Algebraic Geometry.* New York: Oxford University Press, 1963.

40. E. M. Johansson, F. U. Dowla, and D. M. Goodman, "Backpropagation learning for multi-layer feed-forward neural networks using the conjugate gradient method," Tech. Rep. UCRL-JC-104850, Lawrence Livermore National Laboratory, Berkeley, CA, Sept. 26 1990.

41. S. Kaczmarz, "Approximate solution of systems of linear equations," *International Journal of Control*, 57(6): 1269–1271, 1993. Reprint of Kaczmarz, S., "Angenäherte Auflösung von Systemen linearer Gleichunger," *Bulletin International de l'Academie Polonaise des Sciences. Lett A*, 355–357, 1937.

42. N. Karmarkar, "A new polynomial-time algorithm for linear programming," *Combinatorica*, 4(4): 373–395, 1984.

43. M. F. Kelly, P. A. Perker, and R. N. Scott, "The application of neural networks to myoelectric signal analysis: A preliminary study," *IEEE Transactions on Biomedical Engineering*, 37(3): 221–230, 1990.

44. L. G. Khachiyan (Hačijan), "A polynomial algorithm in linear programming," *Soviet Mathematics Doklady*, Providence, RI: American Mathematical Society, 20(1): 191–194, 1979.

45. V. Klee and G. J. Minty, "How good is the simplex algorithm?," in *Inequalities-III* (O. Shisha, Ed.). New York: Academic Press, 1972, pp. 159–175.

46. L. Kolev, "Iterative algorithm for the minimum fuel and minimum amplitude problems for linear discrete systems," *International Journal of Control*, 21(5): 779–784, 1975.

47. J. R. Koza, *Genetic Programming: On the Programming of Computers by Means of Nautral Selection.* Cambridge, MA: The MIT Press, 1992.

48. T. Kozek, T. Roska, and L. O. Chua, "Genetic algorithm for CNN template learning" *IEEE Transactions on Circuits and Systems—I: Fund. Theory and Applications*, 40(6): 392–402, 1993.

49. K. Kuratowski, *Introduction to Calculus*, Vol. 17: *International Series of Monographs in Pure and Applied Mathematics*, 2nd ed. Warsaw: Pergamon Press, 1969.
50. S. Lang, *Calculus of Several Variables*, 3rd ed. New York: Springer-Verlag, 1987.
51. D. G. Luenberger, *Linear and Nonlinear Programming*, 2nd ed. Reading, MA: Addison-Wesley, 1984.
52. D. G. Luenberger, *Optimization by Vector Space Methods*. New York: Wiley, 1969.
53. I. J. Maddox, *Elements of Functional Analysis*, 2nd ed. Cambridge: Cambridge University Press, 1988.
54. O. L. Mangasarian, *Nonlinear Programming*. New York: McGraw-Hill, 1969.
55. A. Mostowski and M. Stark, *Elements of Higher Algebra*. Warsaw: PWN—Polish Scientific Publishers, 1958.
56. T. M. Ozan, *Applied Mathematical Programming for Production and Engineering Management*. Englewood Cliffs, NJ: Prentice-Hall, 1986.
57. C. H. Papadimitriou and K. Steiglitz, *Combinatorial Optimization: Algorithms and Complexity*. Englewood Cliffs, NJ: Prentice-Hall, 1982.
58. P. C. Parks, "S. Kaczmarz (1895–1939)," *International Journal Control*, 57(6): 1263–1267, 1993.
59. A. L. Peressini, F. E. Sullivan, and J. J. Uhr, Jr., *The Mathematics of Nonlinear Programming*. New York: Springer-Verlag, 1988.
60. M. J. D. Powell, "Convergence properties of algorithms for nonlinear optimization," *SIAM Review*, 28(4): 487–500, 1986.
61. S. S. Rangwala and D. A. Dornfeld, "Learning and optimization of machining operations using computing abilities of neural networks," *IEEE Transactions on Systems, Man, and Cybernetics*, 19(2): 299–314, 1989.
62. G. V. Reklaitis, A. Ravindran, and K. M. Ragsdell, *Engineering Optimization: Methods and Applications*. New York: Wiley-Interscience, 1983.
63. A. M. Rockett and J. C. Stevenson, "Karmarkar's algorithm: A method for solving large linear programming problems," *Byte*, 12(10): 146–160, 1987.
64. H. L. Royden, *Real Analysis*, 2nd ed. New York: Macmillan, 1971.
65. W. Rudin, *Principles of Mathematical Analysis*, 3rd ed. New York: McGraw-Hill, 1976.
66. D. E. Rumelhart, J. L. McClelland, and P. R. Group, *Parallel Distributed Processing, Explorations in the Microstructure of Cognition, Volume 1: Foundations*. Cambridge, MA: The MIT Press, 1986.
67. D. Russell, *Optimization Theory*. New York: Benjamin, 1970.
68. S. L. Salas and E. Hille, *Calculus: One and Several Variables*, 4th ed. New York: Wiley, 1982.
69. A. Schrijver, *Theory of Linear and Integer Programming*. New York: Wiley, 1986.
70. R. T. Seeley, *Calculus of Several Variables; An Introduction*. Glenview, IL: Scott, Foresman, 1970.
71. W. A. Spivey, *Linear Programming: An Introduction*. New York: Macmillan, 1963.
72. R. E. Stone and C. A. Tovey, "The simplex and projective scaling algorithms as iteratively reweighted least squares methods," *SIAM Review*, 33(2): 220–237, June 1991.
73. G. Strang, *Introduction to Applied Mathematics*. Wellesley, MA: Wellesley-Cambridge Press, 1986.
74. G. Strang, *Linear Algebra and Its Applications*. New York: Academic Press, 1980.
75. T. W. Then and E. K. P. Chong, "Genetic algorithms in noisy environments," *Proceedings of the 9th IEEE Symposium on Intelligent Control*. Aug. 1994, pp. 225–230.
76. P. P. Varaiya, *Notes on Optimization*. New York: Van Nostrand Reinhold, 1972.
77. B. Widrow and M. A. Lehr, "30 years of adaptive neural networks: Perceptron, Madaline, and backpropagation," *Proceedings of the IEEE*, 78(9): 1415–1442, 1990.
78. D. J. Wilde, *Optimum Seeking Methods*. Englewood Cliffs, NJ: Prentice-Hall, 1964.

79. R. E. Williamson and H. F. Trotter, *Multivariable Mathematics*, 2nd ed. Englewood Cliffs, NJ: Prentice-Hall, 1979.

80. W. I. Zangwill, *Nonlinear Programming: A Unified Approach.* Englewood Cliffs, NJ: Prentice-Hall, 1969.

81. G. Zoutendijk, *Mathematical Programming Methods.* Amsterdam: North-Holland, 1976.

82. J. M. Zurada, *Introduction to Artificial Neural Systems.* St. Paul: West Publishing, 1992.

# Index